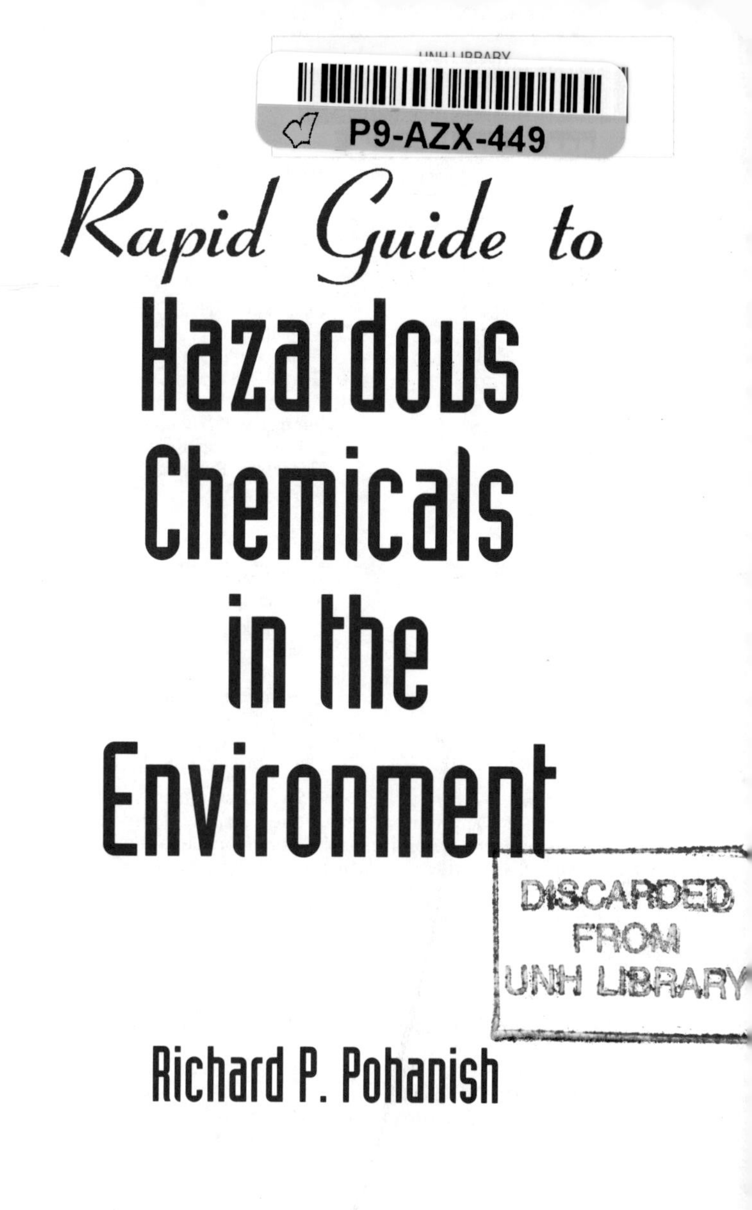

Rapid Guide to Hazardous Chemicals in the Environment

Richard P. Pohanish

VAN NOSTRAND REINHOLD

I(T)P® an International Thomson Company

New York • Albany • Bonn • Boston • Detroit • London • Madrid • Melbourne
Mexico City • Paris • San Francisco • Singapore • Tokyo • Toronto

I(T)P® an International Thomson Publishing Company
The ITP logo is a registered trademark used herein under license

Printed in the United States of America

For more information, contact:

Van Nostrand Reinhold
115 Fifth Avenue
New York, NY 10003

Chapman & Hall GmbH
Pappellee 3
69469 Weinheim
Germany

Chapman & Hall
2-6 Boundary Row
London
SE1 8HN
United Kingdom

International Thomson Publishing Asia
221 Henderson Road #05-10
Henderson Building
Singapore 0315

Thomas Nelson Australia
102 Dodds Street
South Melbourne, 3205
Victoria, Australia

International Thomson Publishing Japan
Hirakawacho Kyowa Building, 3F
2-2-1 Hirakawacho
Chiyoda-ku, 102 Tokyo
Japan

Nelson Canada
1120 Birchmount Road
Scarborough, Ontario
Canada MIK 5G4

International Thomson Editores
Seneca 53
Col. Polanco
11560 Mexico D.F. Mexico

1 2 3 4 5 6 7 8 9 10 COUWF 01 00 99 98 97

Library of Congress Catalog-in-Publication Data

Pohanish, Richard P.
Rapid Guide to hazardous chemicals in the environment / Richard P. Pohanish.
p. cm.
Includes index.
ISBN 0-442-02527-0 (pbk.)
1. Environmental toxicology—Handbooks, manuals, etc. I. Title.
RA1226.P64 1997
615.9'02—dc21 97-14013
CIP

http://www.vnr.com
product discounts • free email newsletters
software demos • online resources

email: info@vnr.com

A service of I(T)P®

NOTICE

This reference is intended to provide initial data about hazardous chemicals that are regulated under major federal and state laws and statutes. It cannot be assumed that all necessary information is contained in this work and that other additional information or assessments may not be required. Extreme care has been taken in the preparation of this work, and to the best knowledge of the publisher and author, the information presented is accurate. Information may not be available for some chemicals; consequently, an absence of data does not necessarily mean that a substance is not regulated under a specific law or statute. Neither the publisher nor the authors assume any liability or responsibility for completeness or accuracy of the information presented or any alleged damages resulting in connection with, or arising from, the use of any information appearing in this book. The publisher and authors strongly encourage all readers to use this book as a guide only. If data are required for legal purposes, the original source documents and appropriate agencies, which are referenced, should be consulted.

To Nora P. Calpin and Wanda P. Dembosky

Table of Contents

INTRODUCTION

This book was prepared to fill the need for a compact and convenient single-reference source of regulatory lists, standards, and relevant information related to chemicals of environmental concern. The primary data for this book appear in diverse sources, including the Federal Register, Code of Federal Regulations, EPA documents, and State lists. *Rapid Guide to Hazardous Chemicals in the Environment* is a chemical-by-chemical distillation of these diverse lists.

The information in this book can be accessed directly by using a regulatory name used by the EPA, a Chemical Abstract Service (CAS) number, or the Synonym index. For the sake of brevity, chemical synonyms have been confined to those that most often appear on regulatory lists.

In many source documents, more than one chemical name may be listed for a single Chemical Abstract Service (CAS) number. This is because the same chemical may appear on different lists under different names. When multiple chemical names have been encountered, all have been included along with CAS numbers and a reference to "see" the primary entry. For example, CAS number (8001-35-2) appears in U.S. EPA regulatory lists as toxaphene (from the Section 313 and Clean Air Act lists), camphechlor (from the Section 302 list), camphene, octochloro- (from the CERCLA list). All of these names are listed in this book with a reference: [see TOXAPHENE]. In addition, other regulatory names such as chlorinated camphene and campheclor appear as synonyms in the chemical record. All names, regardless of source, appear in the name index with reference to the main record.

Some regulatory lists, particularly the CERCLA and EPCRA section 313 lists, include a number of chemical categories as well as specific chemicals. Chemicals in these categories may be subject to reporting, and the category listing contains additional information such as Category codes (e.g., N010 for "mercury compounds") that may be useful to users of this book.

The book is organized with information on environmental information for each chemical as follows: (1) Clean Air Act (CAA); (2) Clean Water Act (CWA); (3) Resource Conservation and Recovery Act (RCRA); (4) Safe Drinking Water Act (SDWA)-3 lists; (5) Comprehensive Environmental Response, Compensation, and Liability Act (CERCLA), a law commonly known as Superfund; (6) Emergency Planning and Community Right-to-Know Act (EPCRA, also known as SARA Title III; (7) Marine Pollutants from DOT Hazardous Materials Transportation Act; (8) California's Proposition 65 carcinogens and reproductive toxins.

Every attempt was made to ensure the accuracy of the information in this book. It is recommended that this book be used as a guide.

USING THIS BOOK and SUMMARY OF FEDERAL STATUTES AND REGULATIONS

Each entry consists of five data fields:

1. EPA name or regulatory category. The first line of each entry contains the common name or regulatory category names used in EPA documents and regulatory lists. Whenever possible, the EPCRA Section 313 name was used as the primary name. This was done because Section 313 chemicals are reportable under these listed names.

2. Chemical Abstract Service (CAS) number. In a few cases this field may contain more than one CAS number. These additional identifiers appear in government-supplied documents and are being provided. Any incorrect CAS numbers found in regulatory lists were corrected and a notation containing the incorrect CAS number appears in the record next to the appropriate list(s).

4. Synonyms. Most chemicals are known by several widely used names. A few useful synonyms have been included from various environmental regulatory lists to assist users in their identification.

5. Regulatory lists. Regulatory lists and important government phone numbers and contacts appear following the summary of each major statute.

GENERAL SUMMARY OF FEDERAL ENVIRONMENTAL REGULATIONS

The brief descriptions in this section are intended solely for general information. Depending on the nature or scope of the activities at a particular facility, these summaries may or may not necessarily describe all applicable environmental requirements. Moreover, they do not constitute formal interpretations or clarifications of the statutes and regulations. For further information, readers should consult the Code of Federal Regulations (CFR) and other state or local regulatory agencies. EPA Hotline contacts are also provided for each major statute.

CLEAN AIR ACT

The Clean Air Act (CAA), first enacted in 1970, consists of six sections, known as titles, that direct EPA to establish national standards for ambient or "outside" air quality from a multiplicity of sources, and for EPA and states to implement, maintain, and enforce these standards through a variety of mechanisms. Under the Clean Air Act Amendments (CAAA) of 1990, many facilities will be required to obtain permits for the first time. State and local governments oversee, manage, and enforce many of the requirements of the CAAA. CAA regulations appear at 40 CFR, Parts 50-59.

Pursuant to Title I of the CAA, EPA has established national ambient air quality standards (NAAQS) to limit levels of "criteria pollutants," including carbon monoxide, lead, nitrogen dioxide, particulate matter, ozone, and sulfur dioxide. In order to assess NAAQS compliance, the nation is divided geographically into Air Quality Control Regions (AQCRs). AQCRs that meet NAAQS for a given pollutant are classified as "attainment" areas; those that do not meet NAAQS are classified as "non-attainment" areas. Non-attainment areas are subject to additional air quality-related restrictions. Under Section 110 of the CAA, each state must develop a State Implementation Plan (SIP) to identify sources of air pollution and to determine what reductions are required to meet federal air quality standards.

Title I also authorizes EPA to establish New Source Performance Standards (NSPS), which are nationally uniform emission standards for new stationary sources falling within particular industrial categories. NSPS are based on the pollution control technology available to that category of industrial source but allow the affected industry the flexibility to devise a cost-effective means of reducing emissions.

Under Title I, EPA establishes and enforces National Emission Standards for Hazardous Air Pollutants (NESHAPS), nationally uniform standards oriented towards controlling particular hazardous pollutants (HAPs). Title III of the CAA further directed EPA to develop regulations for these categories of sources. To date EPA has listed 174 categories and developed a schedule for the establishment of emission standards. The emission standards will be developed for both new and existing sources based on "maximum achievable control technology" (MACT). The MACT is defined as the control technology achieving the maximum degree of reduction in the emission of the HAPs.

Title II of the CAA pertains to mobile sources, such as cars, trucks, buses, and planes. Reformulated gasoline, automobile pollution control devices, and vapor recovery nozzles on gas pumps are a few of the mechanisms EPA uses to regulate mobile air emission sources.

Title III of the 1990 CAA defines and lists hazardous air pollutants (HAPs) and amends Section 112 of the CAA. This initial list of pollutants can be revised by the EPA Administrator "as appropriate."

Title IV establishes a sulfur dioxide emissions program designed to reduce the formation of acid rain. Reduction of sulfur dioxide releases will be obtained by granting to certain sources limited emissions allowances, which, beginning in 1995, will be set below previous levels of sulfur dioxide releases.

Title V of the CAAA of 1990 extensively amended the CAA to include strict regulation of volatile organic compound (VOC) emissions, air toxics, stratospheric ozone, acid rain, motor vehicles, and source permitting. CAAA also created a permit program for all "major sources" (and certain other sources) regulated under the CAA. One purpose of the operating permit is to include in a single document all air emissions requirements that apply to a given facility. States are developing the permit programs in accordance with guidance and regulations from EPA. Once a state program is approved by EPA, permits will be issued and monitored by that state.

Title VI is intended to protect stratospheric ozone by phasing out the manufacture of ozone-depleting chemicals and restricting their use and distribution. Production of Class I substances, including 15 kinds of chlorofluorocarbons (CFCs), will be phased out entirely by the year 2000, while certain hydrochlorofluorocarbons (HCFCs) will be phased out by 2030. CFCs and HCFCs, both used as propellants, have been banned from use in consumer products.

There are criminal penalties for Clean Air Act violations. Various provisions of the CAA can be enforced by EPA, states, and private citizens.

This book contains information from the following CAA lists:

* Hazardous Air Pollutants (Title I, Part A, Section 112) as amended 1990 (Source 42 USC 7412). This list provided for regulating at least 189 specific substances using technology-based standards that employ Maximum Achievable Control Technology (MACT) standards and possibly health-based standards if required at a later time. Section 112 of the CAA requires emission control by the EPA on a source-by-source basis. Therefore, the emission of substances on this list does not necessarily mean that a firm is subject to regulation.

* Regulated Toxic Substances and Threshold Quantities for Accidental Release Prevention. These appear as Accidental Release Prevention/Flammable substances, (Section 112[r], Table 3), TQ (threshold quantity) xxx in pounds and kilograms (Source 40 CFR, Section 68.130). The accidental release prevention regulations applies to stationary sources that have present more than a threshold quantity of a Section 112[r] regulated substance.

* CAA (1990) Public Law 101-549, Title VI, Protection of Stratospheric Ozone, Subpart A, Appendix A—Class I Controlled Substances (CFCs), Ozone depleting substances (Source: 40 CFR, Part 82). This appears as CLEAN AIR ACT: Stratospheric ozone protection (Title VI, Subpart A, Appendix A), Class I, Ozone Depletion Potential = xxx.

* Protection of Stratospheric Ozone, Subpart A, Appendix B—Class II Controlled Substances (HCFCs), Ozone depleting substances (Source: 40 CFR, Part 82). This appears as CLEAN AIR ACT: Stratospheric ozone protection (Title VI, Subpart A, Appendix B), Class II, Ozone Depletion Potential = xxx

Additional CAA Information

EPA's Control Technology Center at (919) 541-0800, provides general assistance and information on CAA standards. The Stratospheric Ozone Information Hotline, at (800) 296-1996, provides general information about regulations promulgated under Title VI of the CAA, and EPA's EPCRA Hotline, at (800) 535-0202, answers questions about accidental release prevention under CAA, Section 112(r). In addition, the Technology Transfer Network Bulletin Board System [modem access (919) 541-5742)] includes recent CAA rules, EPA guidance documents, and updates of EPA activities.

CLEAN WATER ACT

The purpose of the Clean Water Act (CWA) is to "restore and maintain the chemical, physical and biological integrity of the nation's waters." Pollutants regulated under the Clean Water Act (CWA) include "priority" pollutants, including various toxic pollutants; "conventional" pollutants, such as biochemical oxygen demand (BOD), total suspended solids (TSS), fecal coliform, oil and grease, and pH, and "non-conventional" pollutants, including any pollutant not identified as either conventional or priority.

The CWA regulates both direct and indirect discharges. The National Pollutant Discharge Elimination System (NPDES) program (CWA, Section 402) controls direct discharges into surface water and navigable waters. Direct discharges or "point source" discharges are from sources such as pipes and sewers. NPDES permits, issued by either EPA or an authorized state (EPA has presently authorized forty states to administer the NPDES program), contain industry-specific, technology-based and/or water quality-based limits, and establish pollutant monitoring and reporting requirements. A facility that intends to discharge into the nation's waters must obtain a permit prior to initiating its discharge. A permit applicant must provide quantitative analytical data identifying the types of pollutants present in the facility's effluent. The permit will then set forth the conditions and effluent limitations under which a facility may make a discharge.

A NPDES permit may also include discharge limits based on federal or state water quality criteria or standards that were designed to protect designated uses of surface waters, such as supporting aquatic life or recreation. These standards, unlike the technological standards, generally do not take into account technological feasibility or costs. Water quality criteria and standards vary from state to state and site to site, depending on the use classification of the receiving body of water. Most states follow EPA guidelines, which propose aquatic life and human health criteria for many of the 126 priority pollutants.

Storm Water Discharges

In 1987 the CWA was amended to require EPA to establish a program to address storm water discharges. In response, EPA promulgated the NPDES storm water permit application regulations. Storm water discharge associated with industrial activity means the discharge from any conveyance that is used for collecting and conveying storm water and that is directly related to manufacturing, processing or raw materials storage areas at an industrial plant [40 CFR 122.26(b)(14)]. These regulations require that facilities with the following storm water discharges apply for a NPDES permit: (1) a discharge associated with industrial activity; (2) a discharge from a large or medium municipal storm sewer system; or (3) a discharge that EPA or the state determines to contribute to a violation of a water quality standard or is a significant contributor of pollutants to waters of the United States.

The term "storm water discharge associated with industrial activity" means a storm water discharge from one of 11 categories of industrial activity defined at 40 CFR 122.26. Six of the categories are defined by SIC codes, while the other five are identified through narrative descriptions of the regulated industrial activity. If the primary SIC code of the facility is one of those identified in the regulations, the facility is subject to the storm water permit application requirements. If any activity at a facility is covered by one of the five narrative categories, storm water discharges from those areas where the activities occur are subject to storm water discharge permit application requirements.

Those facilities/activities that are subject to storm water discharge permit application requirements are identified below. To determine whether a particular facility falls within one of these categories, the regulations should be consulted.

Category i: Facilities subject to storm water effluent guidelines, new source performance standards, or toxic pollutant effluent standards.

Category ii: Facilities classified as SIC 24—lumber and wood products (except wood kitchen cabinets); SIC 26—paper and allied products (except paperboard containers and products); SIC 28—chemicals and allied products (except drugs and paints); SIC 29—petroleum refining; and SIC 311—leather tanning and finishing.

Category iii: Facilities classified as SIC 10—metal mining; SIC 12—coal mining; SIC 13—oil and gas extraction; and SIC 14—nonmetallic mineral mining.

Category iv: Hazardous waste treatment, storage, or disposal facilities.

Category v: Landfills, land application sites, and open dumps that receive or have received industrial wastes.

Category vi: Facilities classified as SIC 5015—used motor vehicle parts; and SIC 5093—automotive scrap and waste material recycling facilities.

Category vii: Steam electric power generating facilities.

Category viii: Facilities classified as SIC 40—railroad transportation; SIC 41—local passenger transportation; SIC 42—trucking and warehousing (except public warehousing and storage); SIC 43—U.S. Postal Service; SIC 44—water transportation; SIC 45—transportation by air; and SIC 5171—petroleum bulk storage stations and terminals.

Category ix: Sewage treatment works.

Category x: Construction activities except operations that result in the disturbance of less than five acres of total land area.

Category xi: Facilities classified as SIC 20—food and kindred products; SIC 21—tobacco products; SIC 22—textile mill products; SIC 23—apparel related products; SIC 2434—wood kitchen cabinets manufacturing; SIC 25—furniture and fixtures; SIC 265—paperboard containers and boxes; SIC 267—converted paper and paperboard products; SIC 27—printing, publishing, and allied industries; SIC 283—drugs; SIC 285—paints, varnishes, lacquer, enamels, and allied products; SIC 30—rubber and plastics; SIC 31—leather and leather products (except leather and tanning and finishing); SIC 323—glass products; SIC 34—fabricated metal products (except fabricated structural metal); SIC 35—industrial and commercial machinery and computer equipment; SIC 36—electronic and other electrical equipment and components; SIC 37—transportation equipment (except ship and boat building and repairing); SIC 38—measuring, analyzing, and controlling instruments; SIC 39—miscellaneous manufacturing industries; and SIC 4221-4225—public warehousing and storage.

Pretreatment Program

Another type of discharge that is regulated by the CWA is one that goes to a publicly owned treatment works (POTWs). The national pretreatment program (CWA, Section 307(b)) controls the indirect discharge of pollutants to POTWs by "industrial users." Facilities regulated under Section 307(b) must meet certain pretreatment standards. The goal of the pretreatment program is to protect municipal wastewater treatment plants from damage that may occur when hazardous, toxic, or other wastes are discharged into sewer systems and to protect the quality of sludge generated by these plants. Discharges to a POTW are regulated primarily by the POTW itself, rather than the state or EPA.

EPA has developed technology-based standards for industrial users of POTWs. Different standards apply to existing and new sources within each category. "Categorical" pretreatment standards applicable to an industry on a nationwide basis are developed by EPA. In addition, another kind of pretreatment standard, "local limits," are developed by the POTW in order to assist the POTW in achieving the effluent limitations in its NPDES permit.

Regardless of whether a state is authorized to implement either the NPDES or the pretreatment program, if it develops its own program, it may enforce more stringent requirements than Federal standards.

This book contains information from the following CWA lists:

* List of Hazardous Substances (Source: 40 CFR, Sections 116.4, Table 116.4A).

* Section 311 Hazardous Materials Discharge Reportable Quantities (RQs). This regulation establishes reportable quantities for substances designated as hazardous (see Section 116.4, above) and sets forth requirements for notification in the event of discharges into navigable waters (Source: 40 CFR, Section 117.3, amended at 60 FR 30937). Note: In order to avoid confusion with Section 304 CERCLA and EHS reportable quantities, the CWA field contains the statement "RQ (same as CERCLA)" unless otherwise noted.

* Section 307 List of Toxic Pollutants. These toxic pollutant classes or categories were originally identified in the 1976 consent decree between the U.S. EPA and the National Resources Defense Council and later incorporated into the 1977 amendments to the CWA (Source: 40 CFR, Section 401.15).

* Section 307 Priority Pollutant List. This list was developed from the List of Toxic Pollutants classes discussed above and includes substances with known toxic effects on human and aquatic life, and those known to be, or suspected of being, carcinogens, mutagens, or teratogens (Source 40 CFR, Part 423, Appendix A).

* Section 313 Priority Chemicals. Source: 57 FR 41331.

Additional CWA Information

EPA's Office of Water, at (202) 260-5700, will direct callers with questions about CWA to the appropriate EPA office. EPA also maintains a bibliographic database of Office of Water publications that can be accessed through the Ground Water Drinking Water resource center, at (202) 260-7786.

RESOURCE CONSERVATION AND RECOVERY ACT

The Resource Conservation and Recovery Act addresses solid and hazardous waste management activities. The Hazardous and Solid Waste Amendments (HSWA) of 1984 strengthened RCRA's waste management provisions and added Subtitle I, which governs underground storage tanks (USTs).

Regulations promulgated pursuant to Subtitle C of RCRA (40 CFR Parts 260-299) establish a "cradle-to-grave" system governing hazardous waste from the point of generation to disposal. RCRA hazardous wastes include specific materials listed in the regulations (commercial chemical products, designated with the code "P" or "U"; hazardous wastes from specific industries/sources, designated with the code "K"; or hazardous wastes from non-specific sources, designated with the code "K"; or hazardous wastes from non-specific sources, designated with the code "F"; or materials that exhibit a hazardous waste characteristic (ignitability, corrosivity, reactivity, or toxicity) and designated with the code "D".

Regulated entities that generate hazardous waste are subject to waste accumulation, manifesting, and recordkeeping standards. Facilities that treat, store, or dispose of hazardous waste must obtain a permit, either from the EPA or from a state agency that EPA has authorized to implement the permitting program. Subtitle C permits contain general facility standards such as contingency plans, emergency procedures, recordkeeping and reporting requirements, financial assurance mechanisms, and unit-specific standards. RCRA also contains provisions (40 CFR Part 264 Subpart S and Section 264.10) for conducting corrective actions that govern the cleanup of releases of hazardous waste or constituents from solid waste management units at RCRA-regulated facilities.

Although RCRA is a federal statute, many states implement the RCRA program. Currently, EPA has delegated its authority to implement various provisions of RCRA to 46 of the 50 states.

Most RCRA requirements are not industry specific but apply to any company that transports, treats, stores, or disposes of hazardous waste. Here are some important RCRA regulatory requirements:

Identification of Solid and Hazardous Wastes (40 CFR Part 261) lays out the procedure every generator should follow to determine whether the material created is considered a hazardous waste, solid waste, or is exempted from regulation.

Standards for Generators of Hazardous Waste (40 CFR Part 262) establishes the responsibilities of hazardous waste generators including obtaining an ID number, preparing a manifest, ensuring proper packaging and labeling, meeting standards for waste accumulation units, and recordkeeping and reporting requirements. Generators can accumulate hazardous waste for up to 90 days (or 180 days depending on the amount of waste generated) without obtaining a permit. Based on waste volume and toxicity, EPA defines three different levels, or classes, of generators: (a) Conditionally exempt small-quantity generators that generate hazardous waste totaling less than 220 lbs./100 kgs. per month; (b) Small-quantity generators that generate hazardous waste totaling between 220 lbs./1,000 kgs per month. This category of generator has 180 days to dispose of accumulated waste.; and (c) Large-quantity generators which generate hazardous waste totaling 2,200 lbs./1,000 kgs. or more per month. Large-quantity generators have 90 days to dispose of accumulated waste.

Land Disposal Restrictions (LDRs) are regulations prohibiting the disposal of hazardous waste on land without prior treatment. Under the LDRs (40 CFR 268), materials must meet land disposal restriction (LDR) treatment standards prior to placement in a RCRA land disposal unit (landfill, land treatment unit, waste pile, or surface impoundment). Wastes subject to the LDRs include solvents, electroplating wastes, heavy metals, and acids. Generators of waste subject to the LDRs must provide notification of such to the designated TSD facility to ensure proper treatment to disposal.

Used Oil Management Standards (40 CFR 279) impose management requirements affecting the storage, transportation, burning, processing, and re-refining of the used oil. For parties that merely generate used oil, regulations establish storage standards. For a party considered a used oil marketer (one who generates and sells off-specification used oil directly to a used oil burner), additional tracking and paperwork requirements must be satisfied.

Underground Storage Tanks (USTs) containing petroleum and hazardous substance are regulated under Subtitle I of RCRA. Subtitle I regulations (40 CFR Part 280) contain tank design and release detection requirements, as well as financial responsibility and corrective action standards for USTs. The UST program also establishes increasingly stringent standards, including upgrade requirements for existing tanks, that must be met by 1998.

Boilers and Industrial Furnaces (BIFs) that use or burn fuel containing hazardous waste must comply with strict design and operating standards. BIF regulations (40 CFR Part 266, Subpart H) address unit design, provide performance standards, require emissions monitoring, and restrict the type of waste that may be burned.

This book contains information from the following RCRA lists:

* Maximum Concentration of Contaminants for the Toxicity Characteristic with Regulatory levels in mg/L (Source: 40 CFR 261.24).

* RCRA Hazardous Constituents (Source: 40 CFR Part 261, Appendix VIII). Substances listed in Appendix VIII are have been shown, in scientific studies, to have carcinogenic, mutagenic, teratogenic, or toxic effects on humans and other life forms. This list also contains "P" and "U" RCRA waste codes. The words "waste number not listed" appear when no RCRA number is provided in Appendix VIII.

* Maximum Concentration of Contaminants for the Toxicity Characteristic (Source 40 CFR Section 261.24). Table I. These are listed with regulatory level in mg/L and "D" waste numbers representing the broad waste classes of ignitability, corrosivity, and reactivity.

* EPA Hazardous Waste code(s), or RCRA number, appears in its own field. Acute hazardous wastes from commercial chemical products are identified with the prefix "P." Nonacutely hazardous wastes from commercial chemical products are identified with the prefix "U."

* RCRA Universal Treatment Standards. Lists hazardous wastes that are banned from land disposal unless treated to meet standards established by the regulations. Treatment standard levels for wastewater (reported in mg/L) and nonwastewater (reported in mg/kg or mg/L TCLP (Toxicity Characteristic Leachability Procedure) have been provided (Source: 40 CFR Section 268.48 and revision, 61 FR 15654).

* RCRA Ground Water Monitoring List. Sets standards for owners and operators of hazardous waste treatment, storage, and disposal facilities, and contains test methods suggested by the EPA (see EPA Report SW 846) followed by the Practical Quantitation Limit (PQL)

shown in parentheses. The regulation applies only to the listed chemical, and although both the test methods and PQL are provided, they are advisory only (Source: 40 CFR Part 264, Appendix IX).

Additional RCRA Information

EPA's RCRA/Superfund/UST Hotline, at (800) 424-9346 or (800) 535-0202, responds to questions and distributes guidance regarding all RCRA regulations. The RCRA Hotline operates from 8:30 A.M. to 7:30 P.M., EST, excluding federal holidays.

SAFE DRINKING WATER ACT

The Safe Drinking Water Act (SDWA) mandates that EPA establish regulations to protect human health from contaminants in drinking water. The law authorizes EPA to develop national drinking water standards and to create a joint federal-state system to ensure compliance with these standards. The SDWA also directs EPA to protect underground sources of drinking water through the control of underground injection of liquid wastes.

EPA has developed primary and secondary drinking water standards under its SDWA authority. EPA and authorized states enforce the primary drinking water standards, which are contaminant-specific concentration limits that apply to certain public drinking water supplies. Primary drinking water standards consist of maximum contaminant level goals (MCLGs), which are non-enforceable health-based goals, and maximum contaminant levels (MCLs), which are enforceable limits set as close to MCLGs as possible, considering cost and feasibility of attainment.

The SDWA Underground Injection Control (UIC) program (40 CFR parts 144-148) is a permit program that protects underground sources of drinking water by regulating five classes of injection wells. UIC permits include design, operating, inspection, and monitoring requirements. Wells used to inject hazardous wastes must comply with RCRA corrective action standards in order to be granted a RCRA permit and must meet applicable RCRA land disposal restrictions standards. The UIC permit program is primarily state-enforced, since EPA has authorized all but a few states to administer the program.

The SDWA also provides for the federally implemented Sole Source Aquifer program, which prohibits federal funds from being expended on projects that may contaminate the sole or principal source of drinking water for a given area, and for a state-implemented Wellhead Protection program, designed to protect drinking water recharge areas.

This book contains information from the following RCRA lists:

* Maximum Contaminant Levels for Inorganic Chemicals. The maximum contaminant level for arsenic applies only to community water systems. Compliance with the MCL for arsenic is calculated pursuant to Section 141.23. (Source 40 CFR Section 141.11).

* Maximum Contaminant Level Goals (MCLG) for Organic Contaminants. (Source 40 CFR Part 141 and Section 141.50, amended 57 FR 31776).

* Maximum Contaminant Levels (MCL) for Organic Contaminants (Source: 40 CFR 141.61).

* Maximum Contaminant Level Goals (MCLG) for Inorganic Contaminants (Source: 40 CFR Part 141.51).

* Maximum Contaminant Levels (MCL) for Inorganic Contaminants (Source: 40 CFR Part 141.62).

* Secondary Maximum Contaminant Levels (SMCL). Federal advisory standards for the states concerning substances that effect physical characteristics (i.e., smell, taste, color, etc.) of public drinking water systems (Source: 40 CFR Section 143.3).

Additional SDWA Information

EPA's Safe Drinking Water Hotline, at (800) 426-4791, answers questions and distributes guidance pertaining to SDWA standards. The Hotline operates from 9:00 A.M. through 5:30 P.M., EST, excluding federal holidays.

SUPERFUND/CERCLA/EPCRA

COMPREHENSIVE ENVIRONMENTAL RESPONSE, COMPENSATION AND LIABILITY ACT

The Comprehensive Environmental Response, Compensation, and Liability Act (CERCLA), a law commonly known as Superfund, authorizes EPA to respond to releases, or threatened releases, of hazardous substances that may endanger public health, welfare, or the environment. CERCLA also enables EPA to force parties responsible for environmental contamination to clean it up or to reimburse the Superfund for response costs incurred by the EPA. The Superfund Amendments and Reauthorization Act (SARA) of 1986 revised various sections of CERCLA, extended the taxing authority for the Superfund, and created a free-standing law, SARA title III, also known as the Emergency Planning and Community Right-to-Know Act (EPCRA).

The CERCLA hazardous substance release reporting regulations (40 CFR Part 302) directs the person in charge of a facility to report to the National Response Center (NRC) any environmental release of a hazardous substance that exceeds a reportable quantity (RQ). Reportable quantities are defined and listed in 40 CFR 302.4. A release report may trigger a response by EPA or by one or more federal or state emergency response authorities.

EPA implements hazardous substance responses according to procedures outlined in the National Oil and Hazardous Substances Pollution Contingency Plan (NCP) (40 CFR Part 300). The NCP includes provisions for permanent cleanups, referred to as "removals." EPA generally takes remedial actions only at sites on the National Priorities List (NPL), which currently includes approximately 1300 sites. Both EPA and states can act at other sites; however, EPA provides responsible parties the opportunity to conduct removal and remedial actions and encourages community involvement throughout the Superfund response process.

This book contains information from the following CERCLA lists:

CERCLA Hazardous Substances ("RQ" Chemicals). From Consolidated List of Chemicals Subject to the Emergency Planning and Community Right-to-Know Act (EPCRA) and Section 112(r) of the Clean Air Act, as Amended. Source: EPA document 740-R-95-001 "Title III List of Lists"

Releases of CERCLA hazardous substances in quantities equal to or greater than their reportable quantity (RQ), are subject to reporting to the National Response Center under CERCLA. Such releases are also subject to state and local reporting under Section 304 of SARA Title

III (EPCRA). CERCLA hazardous substances and their reportable quantities are listed in 40 CFR Part 302, Table 302.4. RQs are shown in pounds and kilograms for chemicals that are CERCLA hazardous substances. For metals listed under CERCLA (antimony, arsenic, beryllium, cadmium, chromium, copper, lead, nickel, selenium, silver, thallium, and zinc), no reporting of releases of the solid for is required if the diameter of the pieces of solid metal released is 100 micrometers (0.004 inches) or greater. The RQs shown apply to smaller particles.

Additional CERCLA Information

EPA's RCRA/Superfund/UST Hotline, at (800) 424-9346, responds to questions and distributes guidance regarding all RCRA regulations. The RCRA Hotline operates from 8:30 A.M. to 7:30 P.M., EST, excluding federal holidays.

Emergency Planning and Community Right-to-Know Act

The Superfund Amendments and Reauthorization Act (SARA) of 1986 created the Emergency Planning and Community Right-to-Know Act (EPCRA, also known as SARA Title III), a statute designed to improve community access to information about chemical hazards and to facilitate the development of chemical emergency response plans by state and local governments. EPCRA required the establishment of state emergency response commissions (SERCs), responsible for coordinating certain emergency response activities and for appointing local emergency planning committees (LEPCs).

EPCRA and the EPCRA regulations (40 CFR Parts 350-372) establish four types of reporting obligations for facilities that store or manage specified chemicals:

* EPCRA Section 302 requires facilities to notify the SERC and LEPC of the presence of any "extremely hazardous substance" (the list of such substances in the 40 CFR Part 355, Appendices A and B), if it has such substance in excess of the substance's threshold planning quantity, and directs the facility to appoint an emergency response coordinator.

* EPCRA Section 304 requires the facility to notify the SERC and the LEPC in the event of a release exceeding the reportable quantity (RQ) of a CERCLA hazardous substance or an EPCRA extremely hazardous substance (EHS).

* EPCRA Sections 311 and 312 require a facility at which a hazardous chemical, as defined by the Occupational Safety and Health Act, is present in an amount exceeding a specified threshold to submit to the SERC, LEPC, and local fire department material safety data sheets (MSDSs) or lists of MSDSs and hazardous chemical inventory forms (also known as Tier I and II forms). This information helps the local government respond in the event of a spill release of the chemical.

* EPCRA Section 313 requires manufacturing facilities included in SIC codes 20 through 39, which have ten or more employees, and which manufacture, process, or use specified chemicals in amounts greater than threshold quantities, to submit an annual toxic chemical release report. This report, commonly known as the Form R, covers releases and transfers of toxic chemicals to various facilities and environmental media, and allows EPA to compile the National Toxic Release Inventory (TRI) database.

All information submitted pursuant to EPCRA regulations is publicly accessible, unless protected by a trade secret claim.

This book contains information from the following EPCRA lists:

* EPCRA Section 302 Extremely Hazardous Substances (EHS). From Consolidated List of Chemicals Subject to the Emergency Planning and Community Right-to-Know Act (EPCRA) and Section 112[r] of the Clean Air Act, as Amended (Source: EPA document 740-R-95-001 "Title III List of Lists"). The presence of extremely hazardous substances in quantities in excess of the Threshold Planning Quantity (TPQ), requires certain emergency planning activities to be conducted. The EHSs and their TPQs are listed in 40 CFR Part 355, Appendices A and B. For chemicals that are solids, there may be two TPQs given (e.g., 500/10,000). In these cases, the lower quantity applies for solids in powder form with particle size less than 100 microns or if the substance is in solution or in molten form. Otherwise, the higher quantity (10,000 pounds in the example) TPQ applies.

* EPCRA Section 304 Reportable Quantities (RQ). In the event of a release or spill exceeding the reportable quantity, facilities are required to notify state emergency response commissions (SERCs) and Local Emergency Planning Committees (LEPCs). From Consolidated List of Chemicals Subject to the Emergency Planning and Community Right-to-Know Act (EPCRA) and Section 112[r] of the Clean Air Act, as amended (Source: EPA document 740-R-95-001 "Title III List of Lists").

* EPCRA Section 313 Toxic Chemicals. From Consolidated List of Chemicals Subject to the Emergency Planning and Community Right-to-Know Act (EPCRA) and Section 112[r] of the Clean Air Act, as amended (Source: EPA document 740-R-95-001 "Title III List of Lists" and 59 FR 61432). Chemicals on this list are reportable under Section 313 and Section 6607 of the Pollution Prevention Act. Some chemicals are reportable by category under Section 313. Category codes needed for reporting are provided for the EPCRA Section 313 categories. Information and Federal Register references have been provided where a chemical is subject to an administrative stay and not reportable until further notice.

* Toxic Chemical Release Inventory Reporting Form R and Instructions, Revised 1995 Edition (March 1996), EPA document 745-K-96-001 was used for de minimus concentrations, toxic chemical categories and Toxic Chemical Category Codes. Missing from the category listing was "Chlorophenols," which has a de minimus concentration of 1.0%.

Additional EPCRA Information.

EPA's EPCRA Hotline at (800) 535-0202, answers questions and distributes guidance regarding the emergency planning and community right-to-know regulations. The EPCRA Hotline operates weekdays from 8:30 A.M. to 7:30 P.M., EST, excluding federal holidays.

This book also contains information from the following lists:

* Marine Pollutants from the Hazardous Materials Transportation Act (Source: 49 CFR Section 172.101, Appendix B).

* California's "Proposition 65" List of Hazardous Chemicals (Source: California Environmental Protection Agency, Office of Environmental Health Hazard Assessment). Chemicals known to cause cancer. Chemicals known to cause reproductive toxicity. Updated as of 7/1/95.

- A -

EPA NAME: A2213

CAS: 30558-43-1

SYNONYMS: ETHANIMIDOTHIOIC ACID, 2-(DIMETHYLAMINO)-N-HYDROXY-2-OXO-, METHYL ESTER

EPA HAZARDOUS WASTE NUMBER (RCRA No.): U394.
RCRA Section 261 Hazardous Constituents.
RCRA Land Ban Waste.
RCRA Universal Treatment Standards: Wastewater (mg/L), 0.003; Nonwastewater (mg/kg), 1.4.

EPA NAME: ABAMECTIN

CAS: 71751-41-2

SYNONYMS: AVERMECTIN B1; AVERMECTIN B(SUB1)

EPCRA Section 313 Form R de minimus concentration reporting level: 1.0%.

EPA NAME: ACENAPHTHENE

CAS: 83-32-3

SYNONYMS: ACENAPHTHYLENE, 1,2-DIHYDRO-; PERIETHYLENENAPHTHALENE

CLEAN WATER ACT: Section 307 Priority Pollutants; Section 307 Toxic Pollutants.
RCRA Land Ban Waste.
RCRA Universal Treatment Standards: Wastewater (mg/L), 0.059; Nonwastewater (mg/kg), 3.4.
RCRA Ground Water Monitoring List, Suggested methods (PQL μg/L): 8100(200); 8270(10).
EPCRA Section 304 Reportable Quantity (RQ): CERCLA, 100 lbs (45.4 kgs.).

EPA NAME: ACENAPHTHYLENE

CAS: 208-96-8

SYNONYMS: CYCLOPENTA(de)NAPHTHALENE

CLEAN WATER ACT: Section 307 Priority Pollutants.
RCRA Land Ban Waste.
RCRA Universal Treatment Standards: Wastewater (mg/L), 0.059; Nonwastewater (mg/kg), 3.4.
RCRA Ground Water Monitoring List, Suggested methods (PQL μg/L): 8100(200); 8270(10).
EPCRA Section 304 Reportable Quantity (RQ): CERCLA, 5,000 lbs. (2270 kgs.).

EPA NAME: ACEPHATE

CAS: 30560-19-1

SYNONYMS: ACETYLPHOSPHORAMIDOTHIOIC ACID O,S-DIMETHYL ESTER; O,S-DIMETHYL ACETYLPHOSPHORAMIDOTHIOATE; EPA PESTICIDE CHEMICAL CODE 103301; N-(METHOXY(METHYLTHIO)PHOSPHINOYL)ACETAMIDE; PHOSPHORAMIDOTHIOIC ACID, ACETYL-,O,S-DIMETHYL ESTER; PHOSPHORAMIDOTHIOIC ACID, N-ACETYL-,O,S-DIMETHYL

EPCRA Section 313 Form R de minimus concentration reporting level: 1.0%.

EPA NAME: ACETALDEHYDE

CAS: 75-07-0

SYNONYMS: ACETIC ALDEHYDE; ACETEHYDE; ACETICEHYDE; ETHANAL; ETHYL ALDEHYDE; ETHYL EHYDE

CLEAN AIR ACT: Hazardous Air Pollutants (Title I, Part A, Section 112); Accidental Release Prevention/Flammable substances (Section 112[r], Table 3), TQ = 10,000 lbs. (4540 kgs.).

CLEAN WATER ACT: Section 311 Hazardous Substances/RQ (same as CERCLA); Section 313 Priority Chemicals.

EPA HAZARDOUS WASTE NUMBER (RCRA No.): U001, D001.

RCRA Section 261 Hazardous Constituents.

EPCRA Section 304 Reportable Quantity (RQ): CERCLA, 1,000 lbs. (454 kgs.).

EPCRA Section 313 Form R de minimus concentration reporting level: 0.1%.

CALIFORNIA'S PROPOSITION 65: Carcinogen.

EPA NAME: ACETALDEHYDE, TRICHLORO-

CAS: 75-87-6

SYNONYMS: ANHYDROUS CHLORAL; CHLORAL; TRICHLOROACETALDEHYDE

EPA HAZARDOUS WASTE NUMBER (RCRA No.): U034.

RCRA Section 261 Hazardous Constituents as chloral.

EPCRA Section 304 Reportable Quantity (RQ): CERCLA, 5,000 lbs (2270 kgs.).

EPA NAME: ACETAMIDE

CAS: 60-35-5

SYNONYMS: ACETIC ACID AMIDE; ACETIMIDIC ACID; AMID KYSELINY OCTOVE; ETHANAMIDE; METHANECARBOXAMIDE; NCI-C02108

CLEAN AIR ACT: Hazardous Air Pollutants (Title I, Part A, Section 112).

EPCRA Section 304 Reportable Quantity (RQ): CERCLA, 100 lbs. (45.4 kgs.).

EPCRA Section 313 Form R de minimus concentration reporting level: 0.1%.

CALIFORNIA'S PROPOSITION 65: Carcinogen.

EPA NAME: ACETIC ACID

CAS: 64-19-7

SYNONYMS: ETHANOIC ACID; ETHYLIC ACID; GLACIAL ACETIC ACID; METHANE CARBOXYLIC ACID

CLEAN WATER ACT: Section 311 Hazardous Substances/RQ (same as CERCLA).

EPCRA Section 304 Reportable Quantity (RQ): CERCLA, 5,000 lbs. (2270 kgs.).

EPA NAME: ACETIC ACID (2,4-DICHLOROPHENOXY)-

[see 2,4-D]

CAS: 94-75-7

EPA NAME: ACETIC ACID, ETHENYL ESTER

[see VINYL ACETATE]

CAS: 108-05-4

EPA NAME: ACETIC ANHYDRIDE

CAS: 108-24-7

SYNONYMS: ACETIC ACID, ANHYDRIDE; ACETYL OXIDE; ETHANOIC ANHYDRIDE

CLEAN WATER ACT: Section 311 Hazardous Substances/RQ (same as CERCLA).

EPCRA Section 304 Reportable Quantity (RQ): CERCLA, 5,000 lbs. (2270 kgs.).

EPA NAME: ACETONE

CAS: 67-64-1

SYNONYMS: DIMETHYLFORMALDEHYDE; 2-PROPANONE

EPA HAZARDOUS WASTE NUMBER (RCRA No.): U002.

RCRA Section 261 Hazardous Constituents.

RCRA Land Ban Waste.

RCRA Universal Treatment Standards: Wastewater (mg/L), 0.28; Nonwastewater (mg/kg), 160.

RCRA Ground Water Monitoring List, Suggested methods (PQL µg/L): 8240(100).

EPCRA Section 304 Reportable Quantity (RQ): CERCLA, 5,000 lbs. (2270 kgs.).

EPCRA Section 313: Acetone deleted from this list, Federal Register, Vol. 60, No. 116, 6/16/95.

EPA NAME: ACETONE CYANOHYDRIN

[see 2-METHYLLACTONITRILE]

CAS: 75-86-5

EPA NAME: ACETONE THIOSEMICARBAZIDE

CAS: 1752-30-3

SYNONYMS: THIOSEMICARBAZONE ACETONE

EPCRA Section 302, Extremely Hazardous Substances: TPQ = 1,000/10,000 lbs. (454/4,540 kgs.).

EPCRA Section 304 Reportable Quantity (RQ): EHS, 1000 lbs. (454 kgs.).

EPA NAME: ACETONITRILE

CAS: 75-05-8

SYNONYMS: METHANECARBONITRILE; METHYL CYANIDE EK-488

CLEAN AIR ACT: Hazardous Air Pollutants (Title I, Part A, Section 112).

EPA HAZARDOUS WASTE NUMBER (RCRA No.): U003.

RCRA Section 261 Hazardous Constituents.

RCRA Land Ban Waste.

RCRA Universal Treatment Standards: Wastewater (mg/L), 5.6; Non-wastewater (mg/kg), 38.

RCRA Ground Water Monitoring List, Suggested methods (PQL µg/L): 8015(100).

EPCRA Section 304 Reportable Quantity (RQ): CERCLA, 5,000 lbs. (2270 kgs.).

EPCRA Section 313 Form R de minimus concentration reporting level: 1.0%.

EPA NAME: ACETOPHENONE

CAS: 98-86-2

SYNONYMS: ACETYL BENZENE; 1-PHENYL-; ETHANONE, 1-PHENYL-; PHENYL METHYL KETONE

CLEAN AIR ACT: Hazardous Air Pollutants (Title I, Part A, Section 112).

EPA HAZARDOUS WASTE NUMBER (RCRA No.): U004.

RCRA Section 261 Hazardous Constituents.

RCRA Land Ban Waste.

RCRA Universal Treatment Standards: Wastewater (mg/L), 0.010; on-wastewater (mg/kg), 9.7 (Note: listed as CAS 96-86-2).

RCRA Ground Water Monitoring List, Suggested methods (PQL µg/L): 8270(10).

EPCRA Section 304 Reportable Quantity (RQ): CERCLA, 5,000 lbs. (2270 kgs.).

EPCRA Section 313 Form R de minimus concentration reporting level: 1.0%.

EPA NAME: 2-ACETYLAMINOFLUORENE

CAS: 53-96-3

SYNONYMS: 2-AAF; ACETAMIDE, N-FLUOREN-2-YL-; ACETAMIDE, N-9H-FLUOREN-2-YL; 2-2-ACETYLAMIDOFLUORENE; 2-ACETYLAMINEFLUARONE; N-2-FLUORENYLACETAMIDE

CLEAN AIR ACT: Hazardous Air Pollutants (Title I, Part A, Section 112).

EPA HAZARDOUS WASTE NUMBER (RCRA No.): U005.

RCRA Section 261 Hazardous Constituents.

RCRA Land Ban Waste.

RCRA Universal Treatment Standards: Wastewater (mg/L), 0.059; Non-wastewater (mg/kg), 140.

RCRA Ground Water Monitoring List, Suggested methods (PQL µg/L): 8270(10).

EPCRA Section 304 Reportable Quantity (RQ): CERCLA, 1 lb. (0.454 kg.).
EPCRA Section 313 Form R de minimus concentration reporting level: 0.1%.
CALIFORNIA'S PROPOSITION 65: Carcinogen.

EPA NAME: ACETYL BROMIDE
CAS: 506-96-7
SYNONYMS: ACETIC BROMIDE

CLEAN WATER ACT: Section 311 Hazardous Substances/RQ (same as CERCLA).
EPCRA Section 304 Reportable Quantity (RQ): CERCLA, 5,000 lbs. (2270 kgs.).

EPA NAME: ACETYL CHLORIDE
CAS: 75-36-5
SYNONYMS: ACETIC ACID CHLORIDE; ACETIC CHLORIDE; ETHANOYL CHLORIDE

CLEAN WATER ACT: Section 311 Hazardous Substances/RQ (same as CERCLA).
EPA HAZARDOUS WASTE NUMBER (RCRA No.): U006, D001, D002.
RCRA Section 261 Hazardous Constituents.
EPCRA Section 304 Reportable Quantity (RQ): CERCLA, 5,000 lbs. (2270 kgs.).

EPA NAME: ACETYLENE
CAS: 74-86-2
SYNONYMS: ETHENE; ETHINE; ETHYNE

CLEAN AIR ACT: Section 112(r), Accidental Release Prevention/Flammable substances (Section 68.130) TQ = 10,000 lbs (4540 kgs.).
EPA HAZARDOUS WASTE NUMBER (RCRA No.): D001.

EPA NAME: ACETYLPHOSPHORAMIDOTHIOIC ACID O,S-DIMETHYL
[see ACEPHATE]
CAS: 30560-19-1

EPA NAME: 1-ACETYL-2-THIOREA
CAS: 591-08-2
SYNONYMS: ACETAMIDE, N-(AMINOTHIOXOMETHYL)-; ACETYL THIOUREA

EPA HAZARDOUS WASTE NUMBER (RCRA No.): P002.
RCRA Section 261 Hazardous Constituents.
EPCRA Section 304 Reportable Quantity (RQ): CERCLA, 1,000 lbs. (454 kgs.).
MARINE POLLUTANT (49CFR, Subchapter 172.101, Appendix B).

EPA NAME: ACIFLUORFEN, SODIUM SALT
CAS: 62476-59-9

SYNONYMS: ACIFLUORFEN; ACIFLUORFEN SODIUM; 5-(2-2-CHLORO-4-(TRIFLUOROMETHYL)PHENOXY)-2-NITROBENZOIC ACID, SODIUM SALT; EPA PESTICIDE CHEMICAL CODE 114402

EPCRA Section 313 Form R de minimus concentration reporting level: 1.0%.

CALIFORNIA'S PROPOSITION 65: Carcinogen.

EPA NAME: ACROLEIN

CAS: 107-02-8

SYNONYMS: ACROLEINE; ACRYLIC ALDEHYDE; ACRYLALDEHYDE; 2-PROPENAL; 2-PROPEN-1-ONE

CLEAN AIR ACT: Hazardous Air Pollutants (Title I, Part A, Section 112); Section 112[r], Accidental Release Prevention/Flammable substances (Section 68.130); TQ = 5,000 lbs (1275 kgs.).

CLEAN WATER ACT: Section 311 Hazardous Substances/RQ (same as CERCLA); Section 307 Priority Pollutants; Section 313 Priority Chemicals; Toxic Pollutant (Section 401.15).

EPA HAZARDOUS WASTE NUMBER (RCRA No.): P003.

RCRA Section 261 Hazardous Constituents.

RCRA Land Ban Waste.

RCRA Universal Treatment Standards: Wastewater (mg/L), 0.29; Nonwastewater, N/A.

RCRA Ground Water Monitoring List, Suggested methods (PQL μg/L): 8030(5); 8240(5).

EPCRA Section 302, Extremely Hazardous Substances: TPQ = 500 lbs. (228 kgs.).

EPCRA Section 304 Reportable Quantity (RQ): EHS/CERCLA, 1 lb. (0.454 kg.).

EPCRA Section 313 Form R de minimus concentration reporting level: 1.0%.

MARINE POLLUTANT (49CFR, Subchapter 172.101, Appendix B).

EPA NAME: ACRYLAMIDE

CAS: 79-06-1

SYNONYMS: ETHYLENECARBOXAMIDE; 2-PROPENAMIDE

CLEAN AIR ACT: Hazardous Air Pollutants (Title I, Part A, Section 112).

EPA HAZARDOUS WASTE NUMBER (RCRA No.): U007.

RCRA Section 261 Hazardous Constituents.

RCRA Land Ban Waste.

SAFE DRINKING WATER ACT: MCL, treatment technique; MCGL, zero; Regulated chemical (47 FR 9352).

RCRA Universal Treatment Standards: Wastewater (mg/L), 19; Nonwastewater (mg/kg), 23.

EPCRA Section 302, Extremely Hazardous Substances: TPQ = 1,000/10,000 lbs. (454/4,540 kgs.).

EPCRA Section 304 Reportable Quantity (RQ): EHS/CERCLA, 5,000 lbs. (2270 kgs.).

EPCRA Section 313 Form R de minimus concentration reporting level: 0.1%.

CALIFORNIA'S PROPOSITION 65: Carcinogen.

EPA NAME: ACRYLIC ACID

CAS: 79-10-7

SYNONYMS: ACROLEIC ACID; ACRYLIC ACID, GLACIAL; GLACIAL ACRYLIC ACID; 2-PROPENOIC ACID

CLEAN AIR ACT: Hazardous Air Pollutants (Title I, Part A, Section 112).

EPA HAZARDOUS WASTE NUMBER (RCRA No.): U008, D002.

RCRA Section 261 Hazardous Constituents.

EPCRA Section 304 Reportable Quantity (RQ): CERCLA, 5,000 lbs. (2270 kgs.).

EPCRA Section 313 Form R de minimus concentration reporting level: 1.0%.

EPA NAME: ACRYLONITRILE

CAS: 107-13-1

SYNONYMS: CYANOETHYLENE; FUMIGRAIN; 2-PROPENITRILE; 2-PROPENENITRILE; VINYL CYANIDE; VINYL CYANIDE, PROPENENITRILE; VENTOX

CLEAN AIR ACT: Hazardous Air Pollutants (Title I, Part A, Section 112); Section 112[r], Accidental Release Prevention/Flammable substances (Section 68.130), TQ = 20,000 lbs (9150 kgs.).

CLEAN WATER ACT: Section 307 Priority Pollutants; Section 311 Hazardous Substances/RQ (same as CERCLA); Section 313 Priority Chemicals.

EPA HAZARDOUS WASTE NUMBER (RCRA No.): U009.

RCRA Section 261 Hazardous Constituents.

RCRA Land Ban Waste.

RCRA Universal Treatment Standards: Wastewater (mg/L), 0.24; Nonwastewater (mg/kg), 84.

RCRA Ground Water Monitoring List, Suggested methods (PQL μg/L): 8030(5); 8240(5).

SAFE DRINKING WATER ACT: Priority List (55 FR 1470).

EPCRA Section 302, Extremely Hazardous Substances: TPQ = 10,000 lbs. (4,540 kgs.).

EPCRA Section 304 Reportable Quantity (RQ): EHS/CERCLA, 100 lbs. (45.4 kgs.).

EPCRA Section 313 Form R de minimus concentration reporting level: 0.1%.

CALIFORNIA'S PROPOSITION 65: Carcinogen.

EPA NAME: ACRYLYL CHLORIDE

CAS: 814-68-6

SYNONYMS: ACRYLIC ACID CHLORIDE; ACRYLOYL CHLORIDE; 2-PROPENOYL CHLORIDE

CLEAN AIR ACT: Section 112[r], Accidental Release Prevention/Flammable substances (Section 68.130), TQ = 5,000 lbs. (2270 kgs.).

EPA HAZARDOUS WASTE NUMBER (RCRA No.): D001.
EPCRA Section 302, Extremely Hazardous Substances: TPQ = 100 lbs. (45.4 kgs.).
EPCRA Section 304 Reportable Quantity (RQ): EHS, 100 lbs. (45.4 kgs.).

EPA NAME: ADIPIC ACID
CAS: 124-04-9
SYNONYMS: 1,4-BUTANEDICARBOXYLIC ACID; HEXANEDIOIC ACID; 1,6-HEXANEDIOIC ACID

CLEAN WATER ACT: Section 311 Hazardous Substances/RQ (same as CERCLA).
EPCRA Section 304 Reportable Quantity (RQ): CERCLA, 5,000 lbs. (2270 kgs.).

EPA NAME: ADIPONITRILE
[see also CYANIDE COMPOUNDS]
CAS: 111-69-3
SYNONYMS: 1,4-DICYANOBUTANE; HEXANEDINITRILE; HEXANEDIOIC ACID, DINITRILE

EPCRA Section 313: As cyanide compounds. Form R de minimus concentration reporting level: 1.0%.
EPCRA Section 302, Extremely Hazardous Substances: TPQ = 1,000 lbs. (454 kgs.).
EPCRA Section 304 Reportable Quantity (RQ): EHS, 1,000 lbs. (454 kgs.).
MARINE POLLUTANT (49CFR, Subchapter 172.101, Appendix B) as cyanide mixtures or solutions.

EPA NAME: ALACHLOR
CAS: 15972-60-8
SYNONYMS: ACETAMIDE,2-CHLORO-N-(2,6-DIETHYLPHENYL)-N-(METHOXYMETHYL)-; 2-CHLORO-N-(2,6-DIETHYLPHENYL)-N-(METHOXYMETHYL)ACETAMIDE; EPA PESTICIDE CHEMICAL CODE 090501

SAFE DRINKING WATER ACT: MCL, 0.002 mg/L; MGLC, zero; Regulated chemical (47 FR 9352).
EPCRA Section 304 Reportable Quantity (RQ): CERCLA, 1 lb. (0.454 kg.).
EPCRA Section 313 Form R de minimus concentration reporting level: 1.0%.
CALIFORNIA'S PROPOSITION 65: Carcinogen.

EPA NAME: ALDICARB
CAS: 116-06-3
SYNONYMS: EPA PESTICIDE CHEMICAL CODE 098301; PROPIONALDEHYDE, 2-METHYL-2-(METHYLTHIO)-,O-(METHYLCARBAMOYL)OXIME; PROPANAL, 2-METHYL-2-(METHYTHIO)-,O-((METHYLAMINO)CARBONYL)OXIME

EPA HAZARDOUS WASTE NUMBER (RCRA No.): P070.
RCRA Section 261 Hazardous Constituents.

SAFE DRINKING WATER ACT: MCL, 0.003 mg/L; MCLG 0.001 mg/L; Regulated chemical (47 FR 9352).
EPCRA Section 313 Form R de minimus concentration reporting level: 1.0%.
EPCRA Section 302, Extremely Hazardous Substances: TPQ = 100/10,000 lbs. (45.4/4,540 kgs.).
EPCRA Section 304 Reportable Quantity (RQ): EHS/CERCLA, 1 lb. (0.454 kg.).
MARINE POLLUTANT (49CFR, Subchapter 172.101, Appendix B).

EPA NAME: ALDICARB SULFONE

CAS: 1646-88-4
SYNONYMS: PROPANAL, 2-METHYL-2-(METHYLSULFONYL)-O-[(METHYLAMINO)CARBONYL]OXIME

EPA HAZARDOUS WASTE NUMBER (RCRA No.): P203.
RCRA Section 261 Hazardous Constituents.
RCRA Land Ban Waste.
RCRA Universal Treatment Standards: Wastewater (mg/L), 0.056; Nonwastewater (mg/kg), 0.28.
SAFE DRINKING WATER ACT: MCL, 0.002 mg/L; MCLG, 0.001 mg/L.
EPCRA Section 304 Reportable Quantity (RQ): CERCLA, 1 lb. (0.454 kg.).

EPA NAME: ALDRIN

CAS: 309-00-2
SYNONYMS: 1,4:5,8-DIMETHANONAPHTHALENE,1,2,3,4,10,10-HEXACHLORO-1,4,4a,5,8,8a-HEXAHYDRO-(1α,4α,4β,5α,8α,8β)-; HEXACHLOROHEXAHYDRO-ENDO-EXO-DIMETHANONAPHTHALENE; 1,2,3,4,10,10-HEXACHLORO-1,4,4A,5,8,8A-HEXAHYDRO-1,4,5,8-DIMETHANONAPHTHALENE; 1,2,3,4,10,10-HEXACHLORO-1,4,4A,5,8,8A-HEXAHYDRO-EXO-1,4-ENDO-5,8-DIMETHANONAPHTHALENE; 1,2,3,4,10-10-HEXACHLORO-1,4,4a,5,8,8a-HEXAHYDRO-1,4,5,8-ENDO,EXO-DIMETHANONAPHTHALENE

CLEAN WATER ACT: Section 311 Hazardous Substances/RQ (same as CERCLA); Section 307 Priority Pollutants; Section 313 Priority Chemicals; Toxic Pollutant (Section 401.15).
EPA HAZARDOUS WASTE NUMBER (RCRA No.): P004.
RCRA Section 261 Hazardous Constituents.
RCRA Land Ban Waste.
RCRA Universal Treatment Standards: Wastewater (mg/L), 0.021; Nonwastewater (mg/kg), 0.066.
RCRA Ground Water Monitoring List, Suggested methods (PQL μg/L): 8080(0.05); 8270(10).
EPCRA Section 302, Extremely Hazardous Substances: TPQ = 500/10,000 lbs. (227/4540 kgs.).
EPCRA Section 304 Reportable Quantity (RQ): EHS/CERCLA, 1 lb. (0.454 kg.).
EPCRA Section 313 Form R de minimus concentration reporting level: 1.0%.

MARINE POLLUTANT (49CFR, Subchapter 172.101, Appendix B), Severe pollutant.
CALIFORNIA'S PROPOSITION 65: Carcinogen.

EPA NAME: d-trans-ALLETHRIN
CAS: 28057-48-9
SYNONYMS: d-trans-CHRYSANTHEMIC ACID of ALLETHRONE; d-trans-CHRYSANTHEMUMONOCARBOXYLIC ESTER; EPA PESTICIDE CHEMICAL CODE 004003

EPCRA Section 313 Form R de minimus concentration reporting level: 1.0%.

EPA NAME: ALLYL ALCOHOL
CAS: 107-18-6
SYNONYMS: 2-PROPEN-1-OL; PROPENYL ALCOHOL; VINYL CARBINOL

CLEAN AIR ACT: Section 112[r], Accidental Release Prevention/Flammable substances (Section 68.130), TQ = 15,000 lbs. (5825 kgs.).
CLEAN WATER ACT: Section 311 Hazardous Substances/RQ (same as CERCLA).
EPA HAZARDOUS WASTE NUMBER (RCRA No.): P005.
EPCRA Section 302, Extremely Hazardous Substances: TPQ = 1,000 lbs. (454 kgs.).
EPCRA Section 304 Reportable Quantity (RQ): EHS/CERCLA, 100 lbs. (45.4 kgs.).
EPCRA Section 313 Form R de minimus concentration reporting level: 1.0%.

EPA NAME: ALLYLAMINE
CAS: 107-11-9
SYNONYMS: 3-AMINOPROPENE; 3-AMINO-1-PROPENE; 2-PROPEN-1-AMINE; 2-PROPENAMINE; 2-PROPENYLAMINE

CLEAN AIR ACT: Section 112[r], Accidental Release Prevention/Flammable substances (Section 68.130); TQ = 10,000 lbs (4540 kgs.).
EPCRA Section 302, Extremely Hazardous Substances: TPQ = 500 lbs. (228 kgs.).
EPCRA Section 304 Reportable Quantity (RQ): EHS, 500 lbs. (227 kgs.).
EPCRA Section 313 Form R de minimus concentration reporting level: 1.0%.
CALIFORNIA'S PROPOSITION 65: Carcinogen.

EPA NAME: ALLYL CHLORIDE
CAS: 107-05-1
SYNONYMS: 3-CHLOROPRENE; 3-CHLOROPROPYLENE; 3-PROPENE, 3-CHLORO; 1-PROPENE, 3-CHLORO; 3-PROPENYL CHLORIDE

CLEAN AIR ACT: Hazardous Air Pollutants (Title I, Part A, Section 112).

CLEAN WATER ACT: Section 311 Hazardous Substances/RQ (same as CERCLA); Section 313 Priority Chemicals.

Hazardous constituents (40 CFR/261, Appendix VIII), waste number not listed.

RCRA Land Ban Waste.

RCRA Universal Treatment Standards: Wastewater (mg/L), 0.036; Non-wastewater (mg/kg), 30.

RCRA Ground Water Monitoring List, Suggested methods (PQL μg/L): 8010(5); 8240(100).

EPCRA Section 304 Reportable Quantity (RQ): CERCLA, 1,000 lbs. (454 kgs.).

EPCRA Section 313 Form R de minimus concentration reporting level: 1.0%.

MARINE POLLUTANT (49CFR, Subchapter 172.101, Appendix B).

CALIFORNIA'S PROPOSITION 65: Carcinogen.

EPA NAME: ALUMINUM (fume or dust)

CAS: 7429-90-5

SYNONYMS: ALUMINUM, ELEMENTAL; ALUMINUM, METAL AND OXIDE

CLEAN WATER ACT: Section 313 Priority Chemicals.

EPA HAZARDOUS WASTE NUMBER (RCRA No.): D003.

SAFE DRINKING WATER ACT: SMLC, 0.05 to 0.2 mg/L; Priority List (55 FR 1470).

EPCRA Section 313 (Reportable only in a fume or dust form) Form R de minimus concentration reporting level: 1.0%.

EPA NAME: ALUMINUM OXIDE (fibrous forms)

CAS: 1344-28-1

SYNONYMS: A-ALUMINA; α-ALUMINA; ALUNDUM; ALUNDUM (Al2O3); ALUNDUM 600

EPCRA Section 313 (Reportable in fibrous forms only) Form R de minimus concentration reporting level: 1.0%.

EPA NAME: ALUMINUM PHOSPHIDE

CAS: 20859-73-8

SYNONYMS: ALUMINUM MONOPHOSPHIDE; EPA PESTICIDE CHEMICAL CODE 066501; PHOSTOXIN

EPA HAZARDOUS WASTE NUMBER (RCRA No.): P006.

RCRA Section 261 Hazardous Constituents.

EPCRA Section 302 Extremely Hazardous Substances: TPQ = 500 lbs (228 kgs.).

EPCRA Section 304 Reportable Quantity (RQ): EHS/CERCLA, 100 lbs. (45.4 kgs.).

EPCRA Section 313 Form R de minimus concentration reporting level: 1.0%.

EPA NAME: ALUMINUM SULFATE

CAS: 10043-01-3

SYNONYMS: ALUM; ALUMINUM ALUM, ALUMINUM SALT (3:2)

CLEAN WATER ACT: Section 311 Hazardous Substances/RQ (same as CERCLA).
EPCRA Section 304 Reportable Quantity (RQ): CERCLA, 5,000 lbs. (2270 kgs.).

EPA NAME: AMETRYN
CAS: 834-12-8
SYNONYMS: N-ETHYL-N'-(1-METHYLETHYL)-6-(METHYLTHIOL)-1,3,5,-TRIAZINE-2,4-DIAMINE; 1,3,5-TRIAZINE-2,4-DIAMINE,N-ETHYL-N'-(1-METHYLETHYL)-6-(METHYLTHIO)-
EPCRA Section 313 Form R de minimus concentration reporting level: 1.0%.

EPA NAME: 2-AMINOANTHRAQUINONE
CAS: 117-79-3
SYNONYMS: β-AMINOANTHRAQUINONE; 9,10-ANTHRACENEDIONE, 2-AMINO-
EPCRA Section 313 Form R de minimus concentration reporting level: 0.1%.
CALIFORNIA'S PROPOSITION 65: Carcinogen.

EPA NAME: 4-AMINOAZOBENZENE
CAS: 60-09-3
SYNONYMS: AMINOAZOBENZENE; p-AMINOAZOBENZENE; para-AMINOAZOBENZENE; C.I. SOLVENT YELLOW 1; 4-PHENYLAZO ANILINE
EPCRA Section 313 Form R de minimus concentration reporting level: 0.1%.
CALIFORNIA'S PROPOSITION 65: Carcinogen.

EPA NAME: 4-AMINOBIPHENYL
CAS: 92-67-1
SYNONYMS: p-AMINOBIPHENYL; 4-AMINODIPHENYL; p-AMINODIPHENYLE; [1,1'-BIPHENYL]-4-AMINE
CLEAN AIR ACT: Hazardous Air Pollutants (Title I, Part A, Section 112).
RCRA Section 261 Hazardous Constituents, waste number not listed.
RCRA Land Ban Waste.
RCRA Universal Treatment Standards: Wastewater (mg/L), 0.13; Nonwastewater (mg/kg), N/A.
RCRA Ground Water Monitoring List, Suggested methods (PQL μg/L): 8270(10).
EPCRA Section 304 Reportable Quantity (RQ): CERCLA, 1 lb. (0.454 kg.).
EPCRA Section 313 Form R de minimus concentration reporting level: 0.1%.
CALIFORNIA'S PROPOSITION 65: Carcinogen.

EPA NAME: 1-AMINO-2-METHYLANTHRAQUINONE
CAS: 82-28-0

SYNONYMS: 1-AMINO-2-METHYL-9,10-ANTHRACENEDIONE; 9,10-ANTHRACENEDIONE, 1-AMINO-2-METHYL-

EPCRA Section 313 Form R de minimus concentration reporting level: 0.1%.

CALIFORNIA'S PROPOSITION 65: Carcinogen.

EPA NAME: 5-(AMINOMETHYL)-3-ISOXAZOLOL

[see MUSCIMOL]

CAS: 2763-96-4

EPA NAME: AMINOPTERIN

CAS: 54-62-6

SYNONYMS: 4-AMINO-4-DEOXYPTEROYLGLUTAMATE; AMINOPTERIDINE; 4-AMINOPTEROYLGLUTAMIC ACID

EPCRA Section 302, Extremely Hazardous Substances: TPQ = 500/10,000 lbs. (227/4,540 kgs.).

EPCRA Section 304 Reportable Quantity (RQ): EHS, 500 lbs. (227 kgs.).

CALIFORNIA'S PROPOSITION 65: Reproductive toxin (female).

EPA NAME: 4-AMINOPYRIDINE

CAS: 504-24-5

SYNONYMS: AMINO-4-PYRIDINE; γ-AMINOPYRIDINE; p-AMINOPYRIDINE; AVITROL; 4-PYRIDINAMINE; PYRIDINE, 4-AMINO-; 4-PYRIDYLAMINE

EPA HAZARDOUS WASTE NUMBER (RCRA No.): P008.

RCRA Section 261 Hazardous Constituents.

EPCRA Section 302, Extremely Hazardous Substances: TPQ = 500/10,000 lbs. (227/4,540 kgs.).

EPCRA Section 304 Reportable Quantity (RQ): EHS/CERCLA, 1,000 lbs. (454 kgs.).

EPA NAME: AMITON

CAS: 78-53-5

SYNONYMS: DIETHYL-S-2-DIETHYLAMINOETHYL PHOSPHOROTHIOATE

EPCRA Section 302, Extremely Hazardous Substances: TPQ = 500 lbs. (227 kgs.).

EPCRA Section 304 Reportable Quantity (RQ): EHS, 500 lbs. (227 kgs.).

EPA NAME: AMITON OXALATE

CAS: 3734-97-2

SYNONYMS: 2-(2-DIETHYLAMINO)ETHYL)-O,O-DIETHYL ESTER, OXALATE (1:1)

EPCRA Section 302, Extremely Hazardous Substances: TPQ = 100/10,000 lbs. (45.4/4,540 kgs.).

EPCRA Section 304 Reportable Quantity (RQ): EHS, 100 lbs. (45.4 kgs.).

EPA NAME: AMITRAZ

CAS: 33089-61-1

SYNONYMS: N,N-BIS(2,4-XYLYLIMINOMETHYL)METHYLAMINE; 1,5-DI-(2,4-DIMETHYLPHENYL)-3-METHYL-1,3,5-TRIAZAPENTA-1,4-DIENE; N,N'-((METHYLIMINO)DIMETHYLIDYNE)BIS(2,4-XYLIDINE); EPA PESTICIDE CHEMICAL CODE 106201

EPCRA Section 313 Form R de minimus concentration reporting level: 1.0%.

EPA NAME: AMITROLE

CAS: 61-82-5

SYNONYMS: AMINOTRIAZOLE; 2-AMINOTRIAZOLE; 3-AMINOTRIAZOLE; 3-AMINO-S-TRIAZOLE; 3-AMINO-1,2,4-TRIAZOLE; 2-AMINO-1,3,4-TRIAZOLE; 3-AMINO-1H-1,2,4-TRIAZOLE; EPA PESTICIDE CHEMICAL CODE 004401

EPA HAZARDOUS WASTE NUMBER (RCRA No.): U011.

RCRA Section 261 Hazardous Constituents.

EPCRA Section 304 Reportable Quantity (RQ): CERCLA, 10 lbs. (4.54 kgs.).

EPCRA Section 313 Form R de minimus concentration reporting level: 0.1%.

CALIFORNIA'S PROPOSITION 65: Carcinogen.

EPA NAME: AMMONIA

CAS: 7664-41-7

SYNONYMS: SPIRIT OF HARTSHORN

CLEAN AIR ACT: Section 112[r], Accidental Release Prevention/Flammable substances (Section 68.130); (anhydrous) TQ = 10,000 lbs. (4540 kgs.); (concentration equal to or greater than 20%) TQ = 20,000 lbs. (9150 kgs.).

CLEAN WATER ACT: Section 311 Hazardous Substances/RQ (same as CERCLA); Section 313 Priority Chemicals.

EPCRA Section 302, Extremely Hazardous Substances: TPQ = 500 lbs. (228 kgs.).

EPCRA Section 304 Reportable Quantity (RQ): EHS/CERCLA, 100 lbs. (45.4 kgs.).

EPCRA Section 313 Form R de minimus concentration reporting level: 1.0% (includes anhydrous ammonia and aqueous ammonia from water dissociable ammonium salts and other sources; 10 % of total aqueous ammonia, and 100% of anhydrous forms of ammonia is reportable under this listing).

EPA NAME: AMMONIA (anhydrous)

CAS: 7664-41-7

SYNONYMS: ANHYDROUS AMMONIA

CLEAN AIR ACT: Section 112[r], Accidental Release Prevention/Flammable substances (Section 68.130); TQ = 10,000 lbs (4540 kgs.).

CLEAN WATER ACT: Section 311 Hazardous Substances/RQ (same as CERCLA).

EPCRA Section 302, Extremely Hazardous Substances: TPQ = 500 lbs. (228 kgs.).
EPCRA Section 304 Reportable Quantity (RQ): EHS/CERCLA, 100 lbs. (45.4 kgs.).
EPCRA Section 313 Form R de minimus concentration reporting level: 1.0% (100% of anhydrous forms is reportable under this listing).

EPA NAME: AMMONIA (conc. 20% or greater)
CAS: 7664-41-7; 1336-21-6 (25% solution in water)
SYNONYMS: AMMONIUM AMIDE; AMMONIUM HYDROXIDE; SPIRIT OF HARTSHORN

CLEAN AIR ACT: Section 112[r], Accidental Release Prevention/Flammable substances (Section 68.130); TQ = 20,000 lbs (9150 kgs.) (concentration equal to or greater than 20%).
CLEAN WATER ACT: Section 311 Hazardous Substances/RQ (same as CERCLA).
EPCRA Section 302, Extremely Hazardous Substances: TPQ = 500 lbs. (228 kgs.).
EPCRA Section 304 Reportable Quantity (RQ): EHS/CERCLA, 100 lbs. (45.4 kgs.).
EPCRA Section 313 Form R de minimus concentration reporting level: 1.0% (includes anhydrous ammonia and aqueous ammonia from water dissociable ammonium salts and other sources; 10% total aqueous ammonia is reportable under this listing).

EPA NAME: AMMONIUM ACETATE
[see also AMMONIA]
CAS: 631-61-8
SYNONYMS: ACETIC ACID, AMMONIUM SALT

CLEAN WATER ACT: Section 311 Hazardous Substances/RQ (same as CERCLA).
EPCRA Section 304 Reportable Quantity (RQ): CERCLA, 5,000 lbs. (2,270 kgs.).
EPCRA Section 313: Source of aqueous ammonia. Molecular weight 77.08. NH_3 equivalent weight 22.09.

EPA NAME: AMMONIUM BENZOATE
[see also AMMONIA]
CAS: 1863-63-4
SYNONYMS: BENZOIC ACID, AMMONIUM SALT

EPCRA Section 304 Reportable Quantity (RQ): CERCLA, 5,000 lbs. (2,270 kgs.).
CLEAN WATER ACT: Section 311 Hazardous Substances/RQ (same as CERCLA).
EPCRA Section 313: Source of aqueous ammonia. Molecular weight 139.15. NH_3 equivalent weight 12.24.

EPA NAME: AMMONIUM BICARBONATE
CAS: 1066-33-7
SYNONYMS: AMMONIUM HYDROGEN CARBONATE

CLEAN WATER ACT: Section 311 Hazardous Substances/RQ (same as CERCLA).

EPCRA Section 304 Reportable Quantity (RQ): CERCLA, 5,000 lbs. (2,270 kgs.).

EPCRA Section 313: Source of aqueous ammonia. Molecular weight 79.06. NH_3 equivalent weight 21.54.

EPA NAME: AMMONIUM BICHROMATE

[see CHROMIUM COMPOUNDS]

CAS: 7789-09-5

SYNONYMS: AMMONIUM DICHROMATE; AMMONIUM DICHROMATE(VI)

CLEAN WATER ACT: Section 311 Hazardous Substances/RQ (same as CERCLA); Section 307 Priority Pollutants as chromium compounds.

EPCRA Section 304 Reportable Quantity (RQ): CERCLA, 10 lbs. (4.54 kgs.).

EPCRA Section 313 Form R de minimus concentration reporting level 0.1% [chromium (VI) compounds]. Source of aqueous ammonia. Molecular weight 252.06. NH_3 equivalent weight 13.51.

EPA NAME: AMMONIUM BIFLUORIDE

CAS: 1341-49-7

SYNONYMS: AMMONIUM FLUORIDE $(NH_4)(HF_2)$; AMMONIUM HYDROGEN DIFLUORIDE

CLEAN WATER ACT: Section 311 Hazardous Substances/RQ (same as CERCLA).

EPCRA Section 304 Reportable Quantity (RQ): CERCLA, 100 lbs. (45.4 kgs.).

EPCRA Section 313: Source of aqueous ammonia. Molecular weight 57.04. NH_3 equivalent weight 29.86.

EPA NAME: AMMONIUM BISULFITE

[see also AMMONIA]

CAS: 10192-30-0

SYNONYMS: AMMONIUM HYDROGEN SULFITE

CLEAN WATER ACT: Section 311 Hazardous Substances/RQ (same as CERCLA).

EPCRA Section 304 Reportable Quantity (RQ): CERCLA, 5,000 lbs. (2270 kgs.).

EPCRA Section 313: Source of aqueous ammonia. Molecular weight 99.10. NH_3 equivalent weight 17.18.

EPA NAME: AMMONIUM CARBAMATE

[see also AMMONIA]

CAS: 1111-78-0

SYNONYMS: CARBAMIC ACID, AMMONIUM SALT; CARBAMIC ACID, MONOAMMONIUM SALT

CLEAN WATER ACT: Section 311 Hazardous Substances/RQ (same as CERCLA).

EPCRA Section 304 Reportable Quantity (RQ): CERCLA, 5,000 lbs. (2270 kgs.).

EPCRA Section 313: Source of aqueous ammonia. Molecular weight 78.07. NH_3 equivalent weight 21.81.

EPA NAME: AMMONIUM CARBONATE

[see also AMMONIA]

CAS: 506-87-6

SYNONYMS: CARBONIC ACID, AMMONIUM SALT; CARBONIC ACID, DIAMMONIUM SALT

CLEAN WATER ACT: Section 311 Hazardous Substances/RQ (same as CERCLA).

EPCRA Section 304 Reportable Quantity (RQ): CERCLA, 5,000 lbs. (2270 kgs.).

EPCRA Section 313: Source of aqueous ammonia. Molecular weight 96.09. NH_3 equivalent weight 35.45.

EPA NAME: AMMONIUM CHLORIDE

[see also AMMONIA]

CAS: 12125-02-9

SYNONYMS: AMMONIUM MURIATE; SAL AMMONIAC

CLEAN WATER ACT: Section 311 Hazardous Substances/RQ (same as CERCLA).

EPCRA Section 304 Reportable Quantity (RQ): CERCLA, 5,000 lbs. (2270 kgs.).

EPCRA Section 313: Source of aqueous ammonia. Molecular weight 53.49. NH_3 equivalent weight 31.84.

EPA NAME: AMMONIUM CHROMATE

[see also AMMONIA and CHROMIUM COMPOUNDS]

CAS: 7788-98-9

SYNONYMS: CHROMIC ACID, DIAMMONIUM SALT; DIAMMONIUM CHROMATE

CLEAN AIR ACT: Hazardous Air Pollutants (Title I, Part A, Section 112); as chromium compounds.

CLEAN WATER ACT: Section 311 Hazardous Substances/RQ (same as CERCLA); Section 307 Priority Pollutants.

EPCRA Section 304 Reportable Quantity (RQ): CERCLA, 10 lbs. (4.54 kgs.).

EPCRA Section 313: Form R de minimus concentration reporting level 0.1% [chromium (VI) compounds]. Source of aqueous ammonia. Molecular weight 152.07. NH_3 equivalent weight 22.40.

EPA NAME: AMMONIUM CITRATE, DIBASIC

[see also AMMONIA]

CAS: 3012-65-5

SYNONYMS: CITRIC ACID, DIAMMONIUM SALT; DIAMMONIUM CITRATE; 1,2,3-PROPANE TRICARBOXYLIC ACID, 2-HYDROXY-, AMMONIUM SALT

CLEAN WATER ACT: Section 311 Hazardous Substances/RQ (same as CERCLA).
EPCRA Section 304 Reportable Quantity (RQ): CERCLA, 5,000 lbs. (2270 kgs.).
EPCRA Section 313: Source of aqueous ammonia. Molecular weight 226.19. NH_3 equivalent weight 15.06.

EPA NAME: AMMONIUM FLUOBORATE
[see also AMMONIA]
CAS: 13826-83-0
SYNONYMS: AMMONIUM BOROFLUORIDE; AMMONIUM TETRAFLUOBORATE

CLEAN WATER ACT: Section 311 Hazardous Substances/RQ (same as CERCLA).
EPCRA Section 304 Reportable Quantity (RQ): CERCLA, 5,000 lbs. (2270 kgs.).
EPCRA Section 313: Source of aqueous ammonia. Molecular weight 104.84. NH_3 equivalent weight 16.24.

EPA NAME: AMMONIUM FLUORIDE
[see also AMMONIA]
CAS: 12125-01-8
CLEAN WATER ACT: Section 311 Hazardous Substances/RQ (same as CERCLA).
EPCRA Section 304 Reportable Quantity (RQ): CERCLA, 100 lbs. (45.4 kgs.).
EPCRA Section 313: Source of aqueous ammonia. Molecular weight 37.04. NH_3 equivalent weight 45.98.

EPA NAME: AMMONIUM HYDROXIDE
[see also AMMONIA]
CAS: 1336-21-6
CLEAN WATER ACT: Section 311 Hazardous Substances/RQ (same as CERCLA).
EPCRA Section 304 Reportable Quantity (RQ): CERCLA, 1,000 lbs. (454 kgs.).
EPCRA Section 313: Source of aqueous ammonia. Molecular weight 35.05. NH_3 equivalent weight 48.59.

EPA NAME: AMMONIUM NITRATE (solution)
[see also AMMONIA and NITRATE COMPOUNDS]
CAS: 6484-52-2
SYNONYMS: NITRIC ACID, AMMONIUM SALT

EPCRA Section 313: See ammonia and nitrate compounds. Form R, Toxic Chemical Category Code: N511 (nitrate compounds).
Note: This chemical is a source of aqueous ammonia and is in the water-dissociable nitrate compound category. Molecular weight 80.0. NH_3 equivalent weight 21.28.

EPA NAME: AMMONIUM OXALATE
[see also AMMONIA]

CAS: 5972-73-6

SYNONYMS: ETHANEDIOIC ACID, MONOAMMONIUM SALT, MONOHYDRATE

CLEAN WATER ACT: Section 311 Hazardous Substances/RQ (same as CERCLA).

EPCRA Section 304 Reportable Quantity (RQ): CERCLA, 5,000 lbs. (2270 kgs.).

EPCRA Section 313: Source of aqueous ammonia. Molecular weight 124.10. NH_3 equivalent weight 27.45.

EPA NAME: AMMONIUM OXALATE

[see also AMMONIA]

CAS: 6009-70-7

SYNONYMS: AMMONIUM OXALATE HYDRATE; AMMONIUM OXALATE, MONOHYDRATE; DIAMMONIUM OXALATE; ETHANEDIOIC ACID, DIAMMONIUM SALT, MONOHYDRATE

CLEAN WATER ACT: Section 311 Hazardous Substances/RQ (same as CERCLA).

EPCRA Section 304 Reportable Quantity (RQ): CERCLA, 5,000 lbs. (2270 kgs.).

EPCRA Section 313: Source of aqueous ammonia. Molecular weight 124.10. NH_3 equivalent weight 27.45.

EPA NAME: AMMONIUM OXALATE

[see also AMMONIA]

CAS: 14258-49-2

SYNONYMS: OXALIC ACID, AMMONIUM SALT; OXALIC ACID, DIAMMONIUM SALT; DIAMMONIUM OXALATE; ETHANEDIOIC ACID, AMMONIUM SALT

CLEAN WATER ACT: Section 311 Hazardous Substances/RQ (same as CERCLA).

EPCRA Section 304 Reportable Quantity (RQ): CERCLA, 5,000 lbs. (2270 kgs.).

EPCRA Section 313: Source of aqueous ammonia. Molecular weight 124.10. NH_3 equivalent weight 27.45.

EPA NAME: AMMONIUM PICRATE

[see also AMMONIA]

CAS: 131-74-8

SYNONYMS: PHENOL, 2,4,6-TRINITRO-, AMMONIUM SALT (9CI); PICRIC ACID, AMMONIUM SALT

EPA HAZARDOUS WASTE NUMBER (RCRA No.): P009.

RCRA Section 261 Hazardous Constituents.

EPCRA Section 304 Reportable Quantity (RQ): CERCLA, 10 lbs. (4.54 kgs.).

EPCRA Section 313: See ammonia.

EPA NAME: AMMONIUM SILICOFLUORIDE

[see also AMMONIA]

CAS: 16919-19-0

SYNONYMS: AMMONIUM FLUOROSILICATE

CLEAN WATER ACT: Section 311 Hazardous Substances/RQ (same as CERCLA).

EPCRA Section 304 Reportable Quantity (RQ): CERCLA, 1,000 lbs. (454 kgs.).

EPCRA Section 313: Source of aqueous ammonia. Molecular weight 178.15. NH_3 equivalent weight 19.12.

EPA NAME: AMMONIUM SULFAMATE

[see also AMMONIA]

CAS: 7773-06-0

SYNONYMS: AMMATE; AMMONIUM AMINOSULFONATE; AMMONIUM SULPHAMIDATE; AMMONIUM AMIDOSULPHATE; SULFAMIC ACID, MONOAMMONIUM SALT

CLEAN WATER ACT: Section 311 Hazardous Substances/RQ (same as CERCLA).

EPCRA Section 304 Reportable Quantity (RQ): CERCLA, 5,000 lbs. (2270 kgs.).

EPCRA Section 313: Source of aqueous ammonia. Molecular weight 114.12. NH_3 equivalent weight 14.92.

EPA NAME: AMMONIUM SULFATE (solution)

[see also AMMONIA]

CAS: 7783-20-1

SYNONYMS: AMMONIUM HYDROGEN SULFATE; SULFURIC ACID, DIAMMONIUM SALT

EPCRA Section 313: Source of aqueous ammonia. Molecular weight 132.13. NH_3 equivalent weight 25.78.

EPA NAME: AMMONIUM SULFIDE

[see also AMMONIA]

CAS: 12135-76-1

SYNONYMS: AMMONIUM BISULFIDE

CLEAN WATER ACT: Section 311 Hazardous Substances/RQ (same as CERCLA).

EPCRA Section 304 Reportable Quantity (RQ): CERCLA, 100 lbs. (45.4 kgs.).

EPCRA Section 313: Source of aqueous ammonia. Molecular weight 68.14. NH_3 equivalent weight 49.99.

EPA NAME: AMMONIUM SULFITE

[see also AMMONIA]

CAS: 10196-04-0

CLEAN WATER ACT: Section 311 Hazardous Substances/RQ (same as CERCLA).

EPCRA Section 304 Reportable Quantity (RQ): CERCLA, 5,000 lbs. (2270 kgs.).

EPCRA Section 313: Source of aqueous ammonia. Molecular weight 99.10. NH_3 equivalent weight 17.18 (based on CAS 10192-30-0).

EPA NAME: AMMONIUM TARTRATE
[see also AMMONIA]
CAS: 3164-29-2
SYNONYMS: AMMONIUM TARTRATE, DIAMMONIUM SALT DI-AMMONIUM TARTRATE

CLEAN WATER ACT: Section 311 Hazardous Substances/RQ (same as CERCLA).
EPCRA Section 304 Reportable Quantity (RQ): CERCLA, 5,000 lbs. (2270 kgs.).
EPCRA Section 313: See ammonia.

EPA NAME: AMMONIUM TARTRATE
[see also AMMONIA]
CAS: 14307-43-8
CLEAN WATER ACT: Section 311 Hazardous Substances/RQ (same as CERCLA).
EPCRA Section 304 Reportable Quantity (RQ): CERCLA, 5,000 lbs. (2270 kgs.).
EPCRA Section 313: See ammonia.

EPA NAME: AMMONIUM THIOCYANATE
[see also AMMONIA and CYANIDE COMPOUNDS]
CAS: 1762-95-4
SYNONYMS: AMMONIUM SULFOCYANATE; AMMONIUM RHODANATE; AMMONIUM RHODANIDE; AMMONIUM SULFOCYANIDE; THIOCYANIC ACID, AMMONIUM SALT

CLEAN WATER ACT: Section 311 Hazardous Substances/RQ (same as CERCLA).
EPCRA Section 304 Reportable Quantity (RQ): CERCLA, 5,000 lbs. (2270 kgs.).
EPCRA Section 313: (As cyanide compounds.) Form R de minimus concentration reporting level: 1.0%. Source of aqueous ammonia. Molecular weight 76.12. NH_3 equivalent weight 22.37. Form R, Toxic Chemical Category Code: N106 (as cyanide compounds).

EPA NAME: AMMONIUM VANADATE
CAS: 7803-55-6
SYNONYMS: AMMONIUM METAVANADATE; VANADIC ACID, AMMONIUM SALT; VANADATE (V031-), AMMONIUM

EPA HAZARDOUS WASTE NUMBER (RCRA No.): P119.
RCRA Section 261 Hazardous Constituents.
EPCRA Section 304 Reportable Quantity (RQ): CERCLA, 1,000 lbs. (454 kgs.).
EPCRA Section 313: See ammonia.

EPA NAME: AMPHETAMINE
CAS: 300-62-9

SYNONYMS: α-METHYLBENZENEETHANEAMINE

EPCRA Section 302, Extremely Hazardous Substances: TPQ = 1,000 lbs. (454 kgs.).
EPCRA Section 304 Reportable Quantity (RQ): EHS, 1 lb. (0.454 kg.).

EPA NAME: AMYL ACETATE
CAS: 628-63-7
SYNONYMS: ACETIC ACID, PENTYL ESTER; 1-PENTANOL ACETATE; PENTYL ACETATE

CLEAN WATER ACT: Section 311 Hazardous Substances/RQ (same as CERCLA).
EPCRA Section 304 Reportable Quantity (RQ): CERCLA, 5,000 lbs. (2270 kgs.).

EPA NAME: iso-AMYL ACETATE
CAS: 123-92-2
SYNONYMS: 1-BUTANOL, 3-METHYL-, ACETATE; iso-PENTYL ACETATE

CLEAN WATER ACT: Section 311 Hazardous Substances/RQ (same as CERCLA).
EPCRA Section 304 Reportable Quantity (RQ): CERCLA, 5,000 lbs. (2270 kgs.).

EPA NAME: sec-AMYL ACETATE
CAS: 626-38-0
SYNONYMS: 1-METHYLBUTYL ACETATE; PEAR OIL; 2-PENTANOL, ACETATE

CLEAN WATER ACT: Section 311 Hazardous Substances/RQ (same as CERCLA).
EPCRA Section 304 Reportable Quantity (RQ): CERCLA, 5,000 lbs. (2270 kgs.).

EPA NAME: tert-AMYL ACETATE
CAS: 625-16-1
SYNONYMS: BANANA OIL; 2-BUTANOL, 2-METHYL-, ACETATE; tert-PENTYL ACETATE

CLEAN WATER ACT: Section 311 Hazardous Substances/RQ (same as CERCLA).
EPCRA Section 304 Reportable Quantity (RQ): CERCLA, 5,000 lbs. (2270 kgs.).

EPA NAME: ANILAZINE
CAS: 101-05-3
SYNONYMS: 4,6-DICHLORO-N-(2-CHLOROPHENYL)-1,3,5-TRIAZIN-2-AMINE; 2,4-DICHLORO-6-(2-CHLOROANILINO)-1,3,5-TRIAZINE; 2,4-DICHLORO-6-(o-CHLOROANILINO)-s-TRIAZINE; EPA PESTICIDE CHEMICAL CODE 080811; 1,3,5-TRIAZIN-2-AMINE,4,6-DICHLORO-N-(2-CHLOROPHENYL)-

EPCRA Section 313 Form R de minimus concentration reporting level: 1.0%.

EPA NAME: ANILINE

CAS: 62-53-3

SYNONYMS: AMINOBENZENE; BENZENEAMINE; PHENYLAMINE

CLEAN AIR ACT: Hazardous Air Pollutants (Title I, Part A, Section 112).

CLEAN WATER ACT: Section 311 Hazardous Substances/RQ (same as CERCLA); Section 313 Priority Chemicals.

EPA HAZARDOUS WASTE NUMBER (RCRA No.): U012.

RCRA Section 261 Hazardous Constituents.

RCRA Land Ban Waste.

RCRA Universal Treatment Standards: Wastewater (mg/L), 0.81; Nonwastewater (mg/kg), 14.

RCRA Ground Water Monitoring List, Suggested methods (PQL μg/L): 8270(10).

EPCRA Section 302, Extremely Hazardous Substances: TPQ = 1,000 lbs. (454 kgs.).

EPCRA Section 304 Reportable Quantity (RQ): EHS/CERCLA, 5,000 lbs. (2270 kgs.).

EPCRA Section 313 Form R de minimus concentration reporting level: 1.0%.

CALIFORNIA'S PROPOSITION 65: Carcinogen.

EPA NAME: ANILINE, 2,4,6-TRIMETHYL-

CAS: 88-05-1

SYNONYMS: 2,4,6-TRIMETHYLANILINE; 2,4,6-TRIMETHYLBENZENAMINE

EPCRA Section 302, Extremely Hazardous Substances: TPQ = 500 lbs. (228 kgs.).

EPCRA Section 304 Reportable Quantity (RQ): EHS, 500 lbs. (227 kgs.).

EPA NAME: o-ANISIDINE

CAS: 90-04-0

SYNONYMS: ortho-AMINOANISOLE; BENZENAMINE, 2-METHOXY-; ortho-ANSIDINE

CLEAN AIR ACT: Hazardous Air Pollutants (Title I, Part A, Section 112).

EPCRA Section 304 Reportable Quantity (RQ): CERCLA, 100 lbs. (45.4 kgs.).

EPCRA Section 313 Form R de minimus concentration reporting level: 0.1%.

MARINE POLLUTANT (49CFR, Subchapter 172.101, Appendix B) as ortho-anisidines.

CALIFORNIA'S PROPOSITION 65: Carcinogen.

EPA NAME: p-ANISIDINE

CAS: 104-94-9

SYNONYMS: ANILINE, 4-METHOXY-; para-ANISIDINE; BENZENAMINE, 4-METHOXY-; para-AMINOANISOLE

EPCRA Section 313 Form R de minimus concentration reporting level: 1.0%.

EPA NAME: o-ANISIDINE HYDROCHLORIDE

CAS: 134-29-2

SYNONYMS: ortho-ANISIDINE HYDROCHLORIDE; BENZENAMINE, 2-METHOXY, HYDROCHLORIDE; 2-METHOXYANILINE HYDROCHLORIDE

EPCRA Section 313 Form R de minimus concentration reporting level: 0.1%.

MARINE POLLUTANT (49CFR, Subchapter 172.101, Appendix B) as ortho-anisidines.

CALIFORNIA'S PROPOSITION 65: Carcinogen.

EPA NAME: ANTHRACENE

[see also POLYCYCLIC AROMATIC COMPOUNDS]

CAS: 120-12-7

SYNONYMS: para-NAPHTHALENE

CLEAN WATER ACT: Section 307 Priority Pollutants; Section 313 Priority Chemicals.

RCRA Section 261 Hazardous Constituents, waste number not listed.

RCRA Land Ban Waste.

RCRA Universal Treatment Standards: Wastewater (mg/L), 0.059; Nonwastewater (mg/kg), 3.4.

RCRA Ground Water Monitoring List, Suggested methods (PQL µg/L): 8100(200); 8270(10).

EPCRA Section 304 Reportable Quantity (RQ): CERCLA, 5,000 lbs. (2270 kgs.).

EPCRA Section 313 Form R de minimus concentration reporting level: 1.0%.

Note: May contain POLYCYCLIC AROMATIC COMPOUNDS.

EPA NAME: ANTIMONY

[see also ANTIMONY COMPOUNDS]

CAS: 7440-36-0

SYNONYMS: ANTIMONY, ELEMENTAL

CLEAN WATER ACT: Section 307 Priority Pollutants; Section 313 Priority Chemicals; Toxic Pollutant (Section 401.15).

RCRA Section 261 Hazardous Constituents, waste number not listed.

RCRA Land Ban Waste.

SAFE DRINKING WATER ACT: MCL, 0.006 mg/L; MCLG, 0.006 mg/L.

RCRA Universal Treatment Standards: Wastewater (mg/L), 1.9; Nonwastewater (mg/L), TCLP.

RCRA Ground Water Monitoring List, Suggested methods (PQL µg/L): 6010(300); 7040(2,000); 7041(30).

EPCRA Section 304 Reportable Quantity (RQ): CERCLA, 5,000 lbs. (2270 kgs.). Only if the diameter of pieces of solid metal is equal to or greater than 0.004 in.

EPCRA Section 313 Form R de minimus concentration reporting level: 1.0%. Form R, Toxic Chemical Category Code: N010.

EPA NAME: ANTIMONY COMPOUNDS

CLEAN AIR ACT: Hazardous Air Pollutants (Title I, Part A, Section 112).

CLEAN WATER ACT: Toxic Pollutant (Section 401.15).

RCRA Section 261 Hazardous Constituents, waste number not listed, as antimony compounds, n.o.s.

SAFE DRINKING WATER ACT: MCL 0.006 mg/L; MCLG, 0.006 mg/L.

EPCRA Section 313: Includes any unique chemical substance that contains antimony as part of that chemical's infrastructure. Form R de minimus concentration reporting level: 0.1%. Form R, Toxic Chemical Category Code: N010.

EPA NAME: ANTIMONY PENTACHLORIDE

[see also ANTIMONY COMPOUNDS]

CAS: 7647-18-9

SYNONYMS: ANTIMONIC CHLORIDE; ANTIMONY CHLORIDE ($SbCl_5$)

CLEAN AIR ACT: Hazardous Air Pollutants (Title I, Part A, Section 112), as antimony compounds.

CLEAN WATER ACT: Section 311 Hazardous Substances/RQ (same as CERCLA); Section 307 Priority Pollutants; Section 313 Priority Chemicals; Section 311 Hazardous Substances/RQ (same as CERCLA); Toxic Pollutant (Section 401.15) as antimony compounds.

RCRA Section 261 Hazardous Constituents, waste number not listed, as antimony compounds, n.o.s.

EPCRA Section 304 Reportable Quantity (RQ): CERCLA, 1,000 lbs. (454 kgs.).

EPCRA Section 313 (as antimony compounds) Form R de minimus concentration reporting level: 0.1%. Form R, Toxic Chemical Category Code: N010.

EPA NAME: ANTIMONY PENTAFLUORIDE

[see also ANTIMONY COMPOUNDS]

CAS: 7783-70-2

SYNONYMS: ANTIMONY FLUORIDE (SbF_5); ANTIMONY(V) PENTAFLUORIDE; ANTIMONY(5+) PENTAFLUORIDE

CLEAN AIR ACT: Hazardous Air Pollutants (Title I, Part A, Section 112), as antimony compounds.

CLEAN WATER ACT: Toxic Pollutant (Section 401.15) as antimony compounds.

RCRA Section 261 Hazardous Constituents, waste number not listed, as antimony compounds, n.o.s.

EPCRA Section 302, Extremely Hazardous Substances: TPQ = 500 lbs. (227 kgs.).

EPCRA Section 304 Reportable Quantity (RQ): EHS, 500 lbs. (227 kgs.).

EPCRA Section 313 (as antimony compounds) Form R de minimus concentration reporting level: 0.1%. Form R, Toxic Chemical Category Code: N010.

EPA NAME: ANTIMONY POTASSIUM TARTRATE

[see also ANTIMONY COMPOUNDS]

CAS: 28300-74-5

SYNONYMS: ANTIMONATE(2-), BIS μ-2,3-DIHYDROXYBUTANEDIOATA (4-)-01,02:03,04DI-, DIPOTASSIUM, TRIHYDRATE, STEREOISOMER; ANTIMONYL POTASSIUM TARTRATE

CLEAN AIR ACT: Hazardous Air Pollutants (Title I, Part A, Section 112), as antimony compounds.

CLEAN WATER ACT: Section 311 Hazardous Substances/RQ (same as CERCLA); Section 307 Priority Pollutants; Section 313 Priority Chemicals; Toxic Pollutant (Section 401.15) as antimony compounds.

RCRA Section 261 Hazardous constituents (40 CFR/261, Appendix VIII), waste number not listed, as antimony compounds, n.o.s.

EPCRA Section 304 Reportable Quantity (RQ): CERCLA, 100 lbs. (45.4 kgs.).

EPCRA Section 313 (as antimony compounds) Form R de minimus concentration reporting level: 0.1%. Form R, Toxic Chemical Category Code: N010.

EPA NAME: ANTIMONY TRIBROMIDE

[see also ANTIMONY COMPOUNDS]

CAS: 7789-61-9

SYNONYMS: ANTIMONOUS BROMIDE; STIBINE, TRIBROMO-

CLEAN AIR ACT: Hazardous Air Pollutants (Title I, Part A, Section 112), as antimony compounds.

CLEAN WATER ACT: Section 311 Hazardous Substances/RQ (same as CERCLA); Section 307 Priority Pollutants; Section 313 Priority Chemicals; Toxic Pollutant (Section 401.15) as antimony compounds.

RCRA Section 261 Hazardous Constituents, waste number not listed, as antimony compounds, n.o.s.

EPCRA Section 304 Reportable Quantity (RQ): CERCLA, 1,000 lbs. (454 kgs.).

EPCRA Section 313 (as antimony compounds) Form R de minimus concentration reporting level: 0.1%. Form R, Toxic Chemical Category Code: N010.

EPA NAME: ANTIMONY TRICHLORIDE

[see also ANTIMONY COMPOUNDS]

CAS: 10025-91-9

SYNONYMS: ANTIMONY(III) CHLORIDE; ANTIMONY(III) TRICHLORIDE; STIBINE, TRICHLORO-

CLEAN AIR ACT: Hazardous Air Pollutants (Title I, Part A, Section 112), as antimony compounds.

CLEAN WATER ACT: Section 311 Hazardous Substances/RQ (same as CERCLA); Section 307 Priority Pollutants; Section 313 Priority Chemicals; Toxic Pollutant (Section 401.15) as antimony compounds.

RCRA Section 261 Hazardous Constituents, waste number not listed, as antimony compounds, n.o.s.

EPCRA Section 304 Reportable Quantity (RQ): CERCLA, 1,000 lbs. (454 kgs.).

EPCRA Section 313 (as antimony compounds) Form R de minimus concentration reporting level: 0.1%. Form R, Toxic Chemical Category Code: N010.

EPA NAME: ANTIMONY TRIFLUORIDE
[see also ANTIMONY COMPOUNDS]
CAS: 7783-56-4
SYNONYMS: ANTIMONY(III) FLUORIDE (1:3); STIBINE, TRIFLUORO-

CLEAN AIR ACT: Hazardous Air Pollutants (Title I, Part A, Section 112), as antimony compounds.
CLEAN WATER ACT: Section 311 Hazardous Substances/RQ (same as CERCLA); Section 307 Priority Pollutants; Section 313 Priority Chemicals; Toxic Pollutant (Section 401.15) as antimony compounds.
RCRA Section 261 Hazardous Constituents, waste number not listed, as antimony compounds, n.o.s.
EPCRA Section 304 Reportable Quantity (RQ): CERCLA, 1,000 lbs. (454 kgs.).
EPCRA Section 313 (as antimony compounds) Form R de minimus concentration reporting level: 0.1%. Form R, Toxic Chemical Category Code: N010.

EPA NAME: ANTIMONY TRIOXIDE
[see also ANTIMONY COMPOUNDS]
CAS: 1309-64-4
SYNONYMS: ANTIMONOUS OXIDE; ANTIMONY OXIDE (Sb_2O_3)

CLEAN AIR ACT: Hazardous Air Pollutants (Title I, Part A, Section 112), as antimony compounds.
CLEAN WATER ACT: Section 311 Hazardous Substances/RQ (same as CERCLA); Section 307 Priority Pollutants; Section 313 Priority Chemicals; Toxic Pollutant (Section 401.15) as antimony compounds.
RCRA Section 261 Hazardous Constituents, waste number not listed, as antimony compounds, n.o.s.
EPCRA Section 304 Reportable Quantity (RQ): CERCLA, 1,000 lbs. (454 kgs.).
EPCRA Section 313 (as antimony compounds) Form R de minimus concentration reporting level: 0.1%. Form R, Toxic Chemical Category Code: N010.
CALIFORNIA'S PROPOSITION 65: Carcinogen.

EPA NAME: ANTIMYCIN
CAS: 1397-94-0
SYNONYMS: ANTIMYCIN A; ANTIPIRICULLIN; VIROSIN

EPCRA Section 302, Extremely Hazardous Substances: TPQ = 1,000/10,000 lbs. (454/4,540 kgs.).
EPCRA Section 304 Reportable Quantity (RQ): EHS, 1000 lbs. (454 kgs.).

EPA NAME: ANTU
CAS: 86-88-4

SYNONYMS: ALPHANAPHTHYL THIOUREA; 1-NAPHTHYL THIOUREA; 1-(1-NAPHTHYL)-2-THIOUREA; α-NAPHTHYL-THIOUREA; THIOUREA, 1-NAPHTHALENYL-

EPA HAZARDOUS WASTE NUMBER (RCRA No.): P072.
RCRA Section 261 Hazardous Constituents.
EPCRA Section 302, Extremely Hazardous Substances: TPQ = 500/10,000 lbs. (227/4,540 kgs.).
EPCRA Section 304 Reportable Quantity (RQ): EHS/CERCLA, 100 lbs. (45.4 kgs.).

EPA NAME: ARAMITE
CAS: 140-57-8
SYNONYMS: SULFUROUS ACID, 2-CHLOROETHYL 2-[4-(1,1-DIMETHYLETHYL)PHENOXY]-1-METHYLETHYL ESTER

RCRA Section 261 Hazardous Constituents, waste number not listed.
RCRA Land Ban Waste.
RCRA Universal Treatment Standards: Wastewater (mg/L), 0.36; Non-wastewater N/A.
CALIFORNIA'S PROPOSITION 65: Carcinogen.

EPA NAME: AROCLOR 1016
[see also POLYCHLORINATED BIPHENYLS]
CAS: 12674-11-2
SYNONYMS: AROCHLOR 1016; PCB-1016; POLYCHLORINATED BIPHENYL (AROCLOR 1016)

CLEAN WATER ACT: Section 307 Priority Pollutants.
RCRA Ground Water Monitoring List. Suggested test method(s) (PQL µg/L): 8080(50); 8250(100).
EPCRA Section 304 Reportable Quantity (RQ): CERCLA, 1 lb. (0.454 kg.).
EPCRA Section 313 (as PCBs) Form R de minimus concentration reporting level: 0.1%.
MARINE POLLUTANT (49CFR, Subchapter 172.101, Appendix B). Severe pollutant.

EPA NAME: AROCLOR 1221
[see also POLYCHLORINATED BIPHENYLS]
CAS: 11104-28-2
SYNONYMS: AROCHLOR 1221; CHLORODIPHENYL (21%. Cl); PCB-1221; POLYCHLORINATED BIPHENYL (AROCLOR 1221)

CLEAN WATER ACT: Section 307 Priority Pollutants.
RCRA Ground Water Monitoring List. Suggested test method(s) (PQL µg/L): 8080(50); 8250(100).
EPCRA Section 304 Reportable Quantity (RQ): CERCLA, 1 lb. (0.454 kg.).
EPCRA Section 313 (as PCBs) Form R de minimus concentration reporting level: 0.1%.
MARINE POLLUTANT (49CFR, Subchapter 172.101, Appendix B). Severe pollutant.

EPA NAME: AROCLOR 1232
[see also POLYCHLORINATED BIPHENYLS]
CAS: 11141-16-5
SYNONYMS: AROCHLOR 1232; CHLORODIPHENYL (32% CI); PCB-1232; POLYCHLORINATED BIPHENYL (AROCLOR 1232)

CLEAN WATER ACT: Section 307 Priority Pollutants.
RCRA Ground Water Monitoring List. Suggested test method(s) (PQL µg/L): 8080(50); 8250(100).
EPCRA Section 304 Reportable Quantity (RQ): CERCLA, 1 lb. (0.454 kg.).
EPCRA Section 313 (as PCBs) Form R de minimus concentration reporting level: 0.1%.
MARINE POLLUTANT (49CFR, Subchapter 172.101, Appendix B). Severe pollutant.

EPA NAME: AROCLOR 1242
[see also POLYCHLORINATED BIPHENYLS]
CAS: 53469-21-9
SYNONYMS: AROCHLOR 1242; CHLORODIPHENYL (42% CI); PCB-1242; POLYCHLORINATED BIPHENYL (AROCLOR 1242)

CLEAN WATER ACT: Section 307 Priority Pollutants.
RCRA Ground Water Monitoring List. Suggested test method(s) (PQL µg/L): 8080(50); 8250(100).
EPCRA Section 304 Reportable Quantity (RQ): CERCLA, 1 lb. (0.454 kg.).
EPCRA Section 313 (as PCBs) Form R de minimus concentration reporting level: 0.1%.
MARINE POLLUTANT (49CFR, Subchapter 172.101, Appendix B). Severe pollutant.

EPA NAME: AROCLOR 1248
[see also POLYCHLORINATED BIPHENYLS]
CAS: 12672-29-6
SYNONYMS: AROCHLOR 1248; CHLORODIPHENYL (48% CI); PCB-1248; POLYCHLORINATED BIPHENYL (AROCLOR 1248)

CLEAN WATER ACT: Section 307 Priority Pollutants.
RCRA Ground Water Monitoring List. Suggested test method(s) (PQL µg/L): 8080(50); 8250(100).
EPCRA Section 304 Reportable Quantity (RQ): CERCLA, 1 lb. (0.454 kg.).
EPCRA Section 313 (as PCBs) Form R de minimus concentration reporting level: 0.1%.
MARINE POLLUTANT (49CFR, Subchapter 172.101, Appendix B). Severe pollutant.

EPA NAME: AROCLOR 1254
[see also POLYCHLORINATED BIPHENYLS]
CAS: 11097-69-1
SYNONYMS: AROCHLOR 1254; CHLORODIPHENYL (54% CI); PCB-1254; POLYCHLORINATED BIPHENYL (AROCLOR 1254)

CLEAN WATER ACT: Section 307 Priority Pollutants.
RCRA Ground Water Monitoring List. Suggested test method(s) (PQL μg/L): 8080(50); 8250(100).
EPCRA Section 304 Reportable Quantity (RQ): CERCLA, 1 lb. (0.454 kg.).
EPCRA Section 313 (as PCBs) Form R de minimus concentration reporting level: 0.1%.
MARINE POLLUTANT (49CFR, Subchapter 172.101, Appendix B). Severe pollutant.

EPA NAME: AROCLOR 1260
[see also POLYCHLORINATED BIPHENYLS]
CAS: 11096-82-5
SYNONYMS: AROCHLOR 1260; CHLORODIPHENYL (60%. Cl); PCB-1260; POLYCHLORINATED BIPHENYL (AROCLOR 1260)

CLEAN WATER ACT: Section 307 Priority Pollutants.
RCRA Ground Water Monitoring List. Suggested test method(s) (PQL μg/L): 8080(50); 8250(100).
EPCRA Section 304 Reportable Quantity (RQ): CERCLA, 1 lb. (0.454 kg.).
EPCRA Section 313 (as PCBs) Form R de minimus concentration reporting level: 0.1%.
MARINE POLLUTANT (49CFR, Subchapter 172.101, Appendix B). Severe pollutant.

EPA NAME: ARSENIC
[see also ARSENIC COMPOUNDS]
CAS: 7440-38-2
SYNONYMS: ARSENIC, ELEMENTAL; INORGANIC ARSENIC

CLEAN AIR ACT: Hazardous Air Pollutants (Title I, Part A, Section 112).
CLEAN WATER ACT: Section 307 Priority Pollutants; Section 313 Priority Chemicals; Toxic Pollutant (Section 401.15).
RCRA Maximum Concentration of Contaminants, Regulatory level, 5.0 mg/L.
RCRA Section 261 Hazardous Constituents, waste number not listed.
RCRA Land Ban Waste.
RCRA Universal Treatment Standards: Wastewater (mg/L), 1.4; Nonwastewater (mg/L), 5.0 TCLP.
RCRA Ground Water Monitoring List: Suggested methods (PQL μg/L): (total) 6010(500), 7060(10), 7061(20).
SAFE DRINKING WATER ACT: Regulated chemical (47 FR 9352); MCL, 0.05 mg/L (Section 141.11) applies only to community water systems.
EPCRA Section 304 Reportable Quantity (RQ): CERCLA, 1 lb. (0.454 kg.), no reporting required if diameter of metal is equal to or exceeds 0.004 in.
EPCRA Section 313 Form R de minimus concentration reporting level: 0.1%. Form R, Toxic Chemical Category Code: N020.
MARINE POLLUTANT (49CFR, Subchapter 172.101, Appendix B).
CALIFORNIA'S PROPOSITION 65: Carcinogen.

EPA NAME: ARSENIC ACID
[see also ARSENIC and ARSENIC COMPOUNDS]
CAS: 1327-52-2
SYNONYMS: ARSENIC ACID (H_3AsO_4)

CLEAN AIR ACT: Hazardous Air Pollutants (Title I, Part A, Section 112) as arsenic compounds.
CLEAN WATER ACT: Toxic Pollutant (Section 401.15) as arsenic and compounds.
RCRA Section 261 Hazardous Constituents as arsenic compounds, n.o.s.
EPCRA Section 304 Reportable Quantity (RQ): CERCLA, 1 lb. (0.454 kg.).
EPCRA Section 313 Form R de minimus concentration reporting level: 0.1%. Form R, Toxic Chemical Category Code: N020.
MARINE POLLUTANT (49CFR, Subchapter 172.101, Appendix B).

EPA NAME: ARSENIC ACID
[see also ARSENIC and ARSENIC COMPOUNDS]
CAS: 7778-39-4
SYNONYMS: ARSENIC ACID (H_3AsO_4); ortho-ARSENIC ACID

CLEAN AIR ACT: Hazardous Air Pollutants (Title I, Part A, Section 112) as arsenic compounds.
CLEAN WATER ACT: Toxic Pollutant (Section 401.15) as arsenic and compounds.
RCRA Section 261 Hazardous Constituents.
EPCRA Section 304 Reportable Quantity (RQ): CERCLA, 1 lb. (0.454 kg.).
EPA HAZARDOUS WASTE NUMBER (RCRA No.): P010.
EPCRA Section 313 Form R de minimus concentration reporting level: 0.1%. Form R, Toxic Chemical Category Code: N020.
MARINE POLLUTANT (49CFR, Subchapter 172.101, Appendix B).

EPA NAME: ARSENIC COMPOUNDS
[see also ARSENIC]
CLEAN AIR ACT: Hazardous Air Pollutants (Title I, Part A, Section 112) as arsenic compounds.
CLEAN WATER ACT: Toxic Pollutant (Section 401.15) as arsenic and compounds.
RCRA Section 261 Hazardous Constituents, waste number not listed.
EPCRA Section 304 Reportable Quantity (RQ): CERCLA, 1 lb. (0.454 kg.).
EPCRA Section 313: Includes any unique chemical substance that contains arsenic as part of that chemical's infrastructure. Form R de minimus concentration reporting level: (inorganics) 0.1%.; organics 1.0%. Form R, Toxic Chemical Category Code: N020.
MARINE POLLUTANT (49CFR, Subchapter 172.101, Appendix B) as arsenates, liquid, n.o.s.; arsenates, solid, n.o.s.; arsenical pesticides liquid, toxic, flammable, n.o.s.

EPA NAME: ARSENIC DISULFIDE
[see also ARSENIC and ARSENIC COMPOUNDS]

CAS: 1303-32-8

CLEAN AIR ACT: Hazardous Air Pollutants (Title I, Part A, Section 112) as arsenic compounds.

CLEAN WATER ACT: Section 311 Hazardous Substances/RQ (same as CERCLA); Section 313 Priority Chemicals; Toxic Pollutant (Section 401.15) as arsenic and compounds.

RCRA Section 261 Hazardous Constituents, waste number not listed, as arsenic compounds, n.o.s.

EPCRA Section 304 Reportable Quantity (RQ): CERCLA, 1 lb. (0.454 kg.).

EPCRA Section 313 Form R de minimus concentration reporting level: 0.1%. Form R, Toxic Chemical Category Code: N020.

EPA NAME: ARSENIC PENTOXIDE

[see also ARSENIC and ARSENIC COMPOUNDS]

CAS: 1303-28-2

SYNONYMS: ARSENIC OXIDE (As_2O_5); ARSENIC(V) OXIDE

CLEAN AIR ACT: Hazardous Air Pollutants (Title I, Part A, Section 112) as arsenic compounds.

CLEAN WATER ACT: Section 311 Hazardous Substances/RQ (same as CERCLA); Section 313 Priority Chemicals; Toxic Pollutant (Section 401.15) as arsenic and compounds.

EPA HAZARDOUS WASTE NUMBER (RCRA No.): P011.

RCRA Section 261 Hazardous Constituents.

EPCRA Section 302, Extremely Hazardous Substances: TPQ = 100/10,000 lbs. (45.4/4,540 kgs.), solid.

EPCRA Section 304 Reportable Quantity (RQ): EHS/CERCLA, 1 lb. (0.454 kg.).

EPCRA Section 313 Form R de minimus concentration reporting level: 0.1%. Form R, Toxic Chemical Category Code: N020.

EPA NAME: ARSENIC TRICHLORIDE

[see ARSENOUS TRICHLORIDE]

CAS: 7784-34-1

EPA NAME: ARSENIC TRIOXIDE

[see also ARSENIC and ARSENIC COMPOUNDS]

CAS: 1327-53-3

SYNONYMS: ARSENIC(III) OXIDE; ARSENOUS OXIDE; ARSENOUS OXIDE ANHYDRIDE; WHITE ARSENIC

CLEAN AIR ACT: Hazardous Air Pollutants (Title I, Part A, Section 112) as arsenic compounds.

CLEAN WATER ACT: Section 311 Hazardous Substances/RQ (same as CERCLA); Section 313 Priority Chemicals; Toxic Pollutant (Section 401.15) as arsenic and compounds.

EPA HAZARDOUS WASTE NUMBER (RCRA No.): P012.

RCRA Section 261 Hazardous Constituents.

EPCRA Section 302, Extremely Hazardous Substances: TPQ = 100/10,000 lbs. (45.4/4,540 kgs.), solid.

EPCRA Section 304 Reportable Quantity (RQ): EHS/CERCLA, 1 lb. (0.454 kg.).
EPCRA Section 313 Form R de minimus concentration reporting level: 0.1%. Form R, Toxic Chemical Category Code: N020.

EPA NAME: ARSENIC TRISULFIDE
[see also ARSENIC and ARSENIC COMPOUNDS]
CAS: 1303-33-9
SYNONYMS: ARSENIC SULFIDE (As_2S_3); ARSENIC YELLOW; DIARSENIC TRISULFIDE

CLEAN AIR ACT: Hazardous Air Pollutants (Title I, Part A, Section 112) as arsenic compounds.
CLEAN WATER ACT: Section 311 Hazardous Substances/RQ (same as CERCLA); Section 313 Priority Chemicals; Toxic Pollutant (Section 401.15) as arsenic and compounds.
RCRA Section 261 Hazardous Constituents, waste number not listed, as arsenic compounds, n.o.s.
EPCRA Section 304 Reportable Quantity (RQ): CERCLA, 1 lb. (0.454 kg.).
EPCRA Section 313 Form R de minimus concentration reporting level: 0.1%. Form R, Toxic Chemical Category Code: N020.

EPA NAME: ARSENOUS OXIDE
[see ARSENIC TRIOXIDE]
CAS: 1327-53-3

EPA NAME: ARSENOUS TRICHLORIDE
[see also ARSENIC and ARSENIC COMPOUNDS]
CAS: 7784-34-1
SYNONYMS: ARSENIC(III) CHLORIDE; ARSENIC CHLORIDE; ARSENIC(III) TRICHLORIDE; ARSENOUS CHLORIDE

CLEAN AIR ACT: Section 112[r], Accidental Release Prevention/Flammable substances (Section 68.130), TQ = 15,000 lbs (5825 kgs.); Hazardous Air Pollutants (Title I, Part A, Section 112) as arsenic compounds.
CLEAN WATER ACT: Section 311 Hazardous Substances/RQ (same as CERCLA); Toxic Pollutant (Section 401.15) as arsenic and compounds.
RCRA Section 261 Hazardous Constituents, waste number not listed, as arsenic compounds, n.o.s.
EPCRA Section 302, Extremely Hazardous Substances: TPQ = 500 lbs. (228 kgs.).
EPCRA Section 304 Reportable Quantity (RQ): EHS/CERCLA, 1 lb. (0.454 kg.).
EPCRA Section 313 Form R de minimus concentration reporting level: 0.1%. Form R, Toxic Chemical Category Code: N020.
MARINE POLLUTANT (49CFR, Subchapter 172.101, Appendix B).

EPA NAME: ARSINE
[see also ARSENIC and ARSENIC COMPOUNDS]

CAS: 7784-42-1
SYNONYMS: ARSENOUS HYDRIDE; HYDROGEN ARSENIDE

CLEAN AIR ACT: Section 112[r], Accidental Release Prevention/Flammable substances (Section 68.130), TQ = 1,000 lbs. (454 kgs.); Hazardous Air Pollutants (Title I, Part A, Section 112) as arsenic compounds.
CLEAN WATER ACT: Toxic Pollutant (Section 401.15).
RCRA Section 261 Hazardous Constituents, waste number not listed.
EPCRA Section 302, Extremely Hazardous Substances: TPQ = 100 lbs. (45.4 kgs.).
EPCRA Section 304 Reportable Quantity (RQ): EHS, 100 lbs. (45.4 kgs.).
EPCRA Section 313 Form R de minimus concentration reporting level: 0.1%. Form R, Toxic Chemical Category Code: N020.
MARINE POLLUTANT (49CFR, Subchapter 172.101, Appendix B).

EPA NAME: ASBESTOS (Friable)
CAS: 1332-21-4
SYNONYMS: WHITE ASBESTOS

CLEAN AIR ACT: Hazardous Air Pollutants (Title I, Part A, Section 112).
CLEAN WATER ACT: Section 307 Priority Pollutants; Section 313 Priority Chemicals; Toxic Pollutant (Section 401.15).
SAFE DRINKING WATER ACT: MCL, 7 million fibers/L (longer than 10 microns); MCLG, 7 million fibers/L (longer than 10 microns); Regulated chemical (47 FR 9352).
EPCRA Section 304 Reportable Quantity (RQ): CERCLA, 1 lb. (0.454 kg.).
EPCRA Section 313 (Reportable only in friable form) Form R de minimus concentration reporting level: 0.1%.
CALIFORNIA'S PROPOSITION 65: Carcinogen.

EPA NAME: ATRAZINE
CAS: 1912-24-9
SYNONYMS: 6-CHLORO-N-ETHYL-N'-(1-METHYLETHYL)-1,3,5-TRIAZINE-2,4-DIAMINE; EPA PESTICIDE CHEMICAL CODE 080803; 1,3,5-TRIAZINE-2,4-DIAMINE,6-CHLORO-N-ETHYL-N'-(1-METHYLETHYL)-; 2-CHLORO-4-ETHYLAMINEISOPROPYLAMINE-s-TRIAZINE

SAFE DRINKING WATER ACT: MCL, 0.003 mg/L; MCGL 0.003 mg/L; Regulated chemical (47 FR 9352).
EPCRA Section 313 Form R de minimus concentration reporting level: 0.1%.

EPA NAME: AURAMINE
[see C.I. SOLVENT YELLOW 34]
CAS: 492-80-8

EPA NAME: AVERMECTIN B1
[see ABAMECTIN]
CAS: 71751-41-2

EPA NAME: AZASERINE

CAS: 115-02-6

SYNONYMS: 1-AZASERINE; DIAZOACETATE (ESTER)-1-SERINE; 1-DIAZOACETATE (ESTER)-1-SERINE; o-DIAZOACETYL-1-SERINE; 1-SERINE DIAZOACETATE; 1-SERINE DIAZOACETATE (ESTER)

EPA HAZARDOUS WASTE NUMBER (RCRA No.): U015.

EPCRA Section 304 Reportable Quantity (RQ): CERCLA, 1 lb. (0.454 kg.).

CALIFORNIA'S PROPOSITION 65: Carcinogen.

EPA NAME: 1H-AZEPINE-1-CARBOTHIOIC ACID, HEXAHYDRO-S-ETHYL ESTER

[see MOLINATE]

CAS: 2212-67-1

EPA NAME: AZINPHOS-ETHYL

CAS: 2642-71-9

SYNONYMS: 3,4-DIHYDRO-4-OXO-3-BENZOTRIAZINYL-METHYL O,O-DIETHYL PHOSPHORODITHIOATE

EPCRA Section 302, Extremely Hazardous Substances: TPQ = 100/10,000 lbs. (45.4/4,540 kgs.).

EPCRA Section 304 Reportable Quantity (RQ): EHS, 100 lbs. (45.4 kgs.).

MARINE POLLUTANT (49CFR, Subchapter 172.101, Appendix B).

EPA NAME: AZINPHOS-METHYL

CAS: 86-50-0

SYNONYMS: AZINPHOS-METHYL (ISO); O,O-DIMETHYL-S-(4-OXO-1,2,3-BEZOTRIAZIN-3(4H)-YL METHYL)PHOSPHORODITHIOATE; GUSATHION; GUTHION; PHOSPHORODITHIOIC ACID, O,O-DIMETHYL S-((4-OXO-1,2,3-BENZOTRIAZIN-3(4H)-YL)METHYL)ESTER

CLEAN WATER ACT: Section 311 Hazardous Substances/RQ (same as CERCLA) as guthion.

EPCRA Section 302, Extremely Hazardous Substances: TPQ = 100/10,000 lbs. (45.4/4,540 kgs.).

EPCRA Section 304 Reportable Quantity (RQ): EHS/CERCLA, 1 lb. (0.454 kg.).

MARINE POLLUTANT (49CFR, Subchapter 172.101, Appendix B). Severe pollutant.

EPA NAME: AZIRIDINE

[see ETHYLENEIMINE]

CAS: 151-56-4

EPA NAME: AZIRIDINE, 2 METHYL

[see PROPYLENEIMINE]

CAS: 75-55-8

- B -

EPA NAME: BARBAN
CAS: 101-27-9
RCRA Land Ban Waste.
RCRA Universal Treatment Standards: Wastewater (mg/L), 0.056; Non-wastewater (mg/kg), 1.4.

EPA NAME: BARIUM
[see also BARIUM COMPOUNDS]
CAS: 7440-39-3
SYNONYMS: BARIUM, ELEMENTAL; BARIUM METAL

EPA HAZARDOUS WASTE NUMBER (RCRA No.): D005.
RCRA Toxicity Characteristic (Section 261.24), Maximum Concentration of Contaminants, regulatory level, 100.0 mg/L.
RCRA Section 261 Hazardous Constituents, waste number not listed.
RCRA Maximum Concentration Limit for Ground Water Protection (40 CFR/264.94), 1.0 mg/L.
RCRA Land Ban Waste.
RCRA Universal Treatment Standards: Wastewater (mg/L), 1.2; Non-wastewater (mg/L), 7.6 TCLP.
RCRA Ground Water Monitoring List, Suggested methods (PQL µg/L): 6010(20); 7080(1,000).
SAFE DRINKING WATER ACT: MCL, 2 mg/L; MCLG, 2 mg/L; Regulated chemical (47 FR 9352).
EPCRA Section 313 Form R de minimus concentration reporting level: 1.0%. Form R Toxic Chemical Category Code: N040.
MARINE POLLUTANT (49CFR, Subchapter 172.101, Appendix B).

EPA NAME: BARIUM COMPOUNDS
[see also BARIUM]
RCRA Section 261 Hazardous Constituents, as barium compounds, n.o.s., waste number not listed.
EPCRA Section 313: Includes any unique chemical substance that contains barium as part of that chemical's infrastructure. This category does not include barium sulfate (7727-43-7). Form R de minimus concentration reporting level: 0.1%. Form R Toxic Chemical Category Code: N040.
MARINE POLLUTANT (49CFR, Subchapter 172.101, Appendix B).

EPA NAME: BARIUM CYANIDE
[see also BARIUM COMPOUNDS and CYANIDE COMPOUNDS]
CAS No.: 542-62-1
SYNONYMS: BARIUM DICYANIDE

CLEAN WATER ACT: Section 311 Hazardous Substances/RQ (same as CERCLA); Section 313 Priority Chemicals.
EPA HAZARDOUS WASTE NUMBER (RCRA No.): P013.
RCRA Section 261 Hazardous Constituents.

EPCRA Section 304 Reportable Quantity (RQ): CERCLA, 10 lbs. (4.54 kgs.).

EPCRA Section 313: As barium compounds. Form R de minimus concentration reporting level: 0.1%. Form R Toxic Chemical Category Code: N040.

EPCRA Section 313 Form R de minimus concentration reporting level: barium compounds: 0.1%; cyanide compounds: 1.0%. Form R Toxic Chemical Category Code: N040 (barium); N106 (cyanide).

EPA NAME: BENDIOCARB

CAS: 22781-23-3

SYNONYMS: 2,2-DIMETHYL-1,3-BENZODIOXOL-4-YL-N-METHYLCARBAMATE; 2,2-DIMETHYL-1,3-BENZODIOXOL-4-OL METHYLCARBAMATE

EPA HAZARDOUS WASTE NUMBER (RCRA No.): U278.

RCRA Section 261 Hazardous Constituents.

RCRA Land Ban Waste.

RCRA Universal Treatment Standards: Wastewater (mg/L), 0.056; Nonwastewater (mg/kg), 1.4.

EPCRA Section 304 Reportable Quantity (RQ): CERCLA, 1 lb. (0.454 kg.).

EPA NAME: BENDIOCARB PHENOL

CAS: 22961-82-6

RCRA Land Ban Waste.

RCRA Universal Treatment Standards: Wastewater (mg/L), 0.056; Nonwastewater (mg/kg), 1.4.

EPCRA Section 304 Reportable Quantity (RQ): CERCLA, 1 lb. (0.454 kg.).

EPA NAME: BENEZENEAMINE, 2,6-DINITRO-N,N-DIPROPYL-4-(TRIFLUOROMETHYLANILINE)

[see TRIFLURALIN]

CAS: 1582-09-8

EPA NAME: BENFLURALIN

CAS: 1861-40-1

SYNONYMS: N-BUTYL-N-ETHYL-α,α,α-TRIFLUORO-2,6-DINITRO-p-TOLUIDINE; N-BUTYL-N-ETHYL-2,6-DINITRO-4-(TRIFLUROMETHYL)BENZENEAMINE; EPA PESTICIDE CHEMICAL CODE 084301

EPCRA Section 313 Form R de minimus concentration reporting level: 1.0%.

EPA NAME: BENOMYL

CAS: 17804-35-2

SYNONYMS: BENOMYL (ISO); CARBAMIC ACID, 1-(BUTYLAMINO)CARBONYL- 1H-BENZIMIDAZOL-2YL, METHYL ESTER

EPA HAZARDOUS WASTE NUMBER (RCRA No.): U271.

RCRA Section 261 Hazardous Constituents.

RCRA Land Ban Waste.
RCRA Universal Treatment Standards: Wastewater (mg/L), 0.056; Non-wastewater (mg/kg), 1.4.
EPCRA Section 304 Reportable Quantity (RQ): CERCLA, 1 lb. (0.454 kg.).
EPCRA Section 313 Form R de minimus concentration reporting level: 1.0%.
CALIFORNIA'S PROPOSITION 65: Reproductive toxin (male).

EPA NAME: BENZ[c]ACRIDINE
CAS: 225-51-4
SYNONYMS: 12-AZABENZ(a)ANTHRACENE

CLEAN WATER ACT: Section 307 Toxic Pollutants.
EPA HAZARDOUS WASTE NUMBER (RCRA No.): U016.
RCRA Section 261 Hazardous Constituents.
EPCRA Section 304 Reportable Quantity (RQ): CERCLA, 100 lbs. (45.5 kgs.).

EPA NAME: BENZAL CHLORIDE
CAS: 98-87-3
SYNONYMS: BENZENE, DICHLOROMETHYL; BENZYL DICHLORIDE; BENZYLIDENE CHLORIDE; CHLOROBENZAL; (DICHLOROMETHYL)BENZENE

EPA HAZARDOUS WASTE NUMBER (RCRA No.): U017.
RCRA Section 261 Hazardous Constituents.
RCRA Land Ban Waste.
RCRA Universal Treatment Standards: Wastewater (mg/L), 0.055; Non-wastewater (mg/kg), 6.0.
EPCRA Section 302 Extremely Hazardous Substances: TPQ = 500 lbs. (227 kgs.).
EPCRA Section 304 Reportable Quantity (RQ): EHS/CERCLA, 5,000 lbs. (2270 kgs.).
EPCRA Section 313 Form R de minimus concentration reporting level: 1.0%.

EPA NAME: BENZAMIDE
CAS: 55-21-0
SYNONYMS: BENZOIC ACID AMIDE

EPCRA Section 313 Form R de minimus concentration reporting level: 1.0%.

EPA NAME: BENZAMIDE, 3,5-DICHLORO-N-(1,1-DIMETHYL-2-PROPYNYL)
[see PRONAMIDE]
CAS: 23950-58-5

EPA NAME: BENZ[a]ANTHRACENE
[see also POLYCYCLIC AROMATIC COMPOUNDS]
CAS: 56-55-3

SYNONYMS: 1,2-BENZANTHRACENE; 1,2-BENZ(a)ANTHRACENE; 1,2-BENZO(a)ANTHRACENE

CLEAN WATER ACT: Section 307 Priority Pollutants; Section 307 Toxic Pollutants.
EPA HAZARDOUS WASTE NUMBER (RCRA No.): U018.
RCRA Section 261 Hazardous Constituents.
RCRA Land Ban Waste.
RCRA Universal Treatment Standards: Wastewater (mg/L), 0.059; Nonwastewater (mg/kg), 3.4.
RCRA Ground Water Monitoring List. Suggested test method(s) (PQL µg/L): 8100(200); 8270 (10).
EPCRA Section 304 Reportable Quantity (RQ): CERCLA, 10 lbs. (4.54 kgs.).
EPCRA Section 313 Form R de minimus concentration reporting level: 0.1%.
CALIFORNIA'S PROPOSITION 65: Carcinogen.

EPA NAME: BENZENEAMINE, 3-(TRIFLUOROMETHYL)-

CAS: 98-16-8
SYNONYMS: 3-(TRIFLUOROMETHYL)ANILINE; m-(TRIFLUOROMETHYL)BENZENAMINE

EPCRA Section 302 Extremely Hazardous Substances: TPQ = 500 lbs. (227 kgs.).
EPCRA Section 304 Reportable Quantity (RQ): EHS, 500 lbs. (227 kgs.).

EPA NAME: BENZENE

CAS: 71-43-2
SYNONYMS: BENZOL; BENZELENE

CLEAN AIR ACT: Hazardous Air Pollutants (Title I, Part A, Section 112) NOTE: Including benzene from gasoline.
CLEAN WATER ACT: Section 311 Hazardous Substances/RQ (same as CERCLA); Section 307 Priority Pollutants; Section 313 Priority Chemicals; Toxic Pollutant (Section 401.15).
EPA HAZARDOUS WASTE NUMBER (RCRA No.): D019.
RCRA Toxicity Characteristic (Section 261.24), Maximum Concentration of Contaminants, regulatory level, 0.5 mg/L.
RCRA Section 261 Hazardous Constituents.
RCRA Land Ban Waste.
RCRA Universal Treatment Standards: Wastewater (mg/L), 0.14; Nonwastewater (mg/kg), 10.
RCRA Ground Water Monitoring List. Suggested test method(s) (PQL µg/L): 8020(2); 8240(5).
SAFE DRINKING WATER ACT: MCL, 0.005 mg/L; MCLG, zero; Regulated chemical (47 FR 9352).
EPCRA Section 304 Reportable Quantity (RQ): CERCLA, 10 lbs. (4.54 kgs.).
EPCRA Section 313 Form R de minimus concentration reporting level: 0.1%.
MARINE POLLUTANT (49CFR, Subchapter 172.101, Appendix B).
CALIFORNIA'S PROPOSITION 65: Carcinogen.

EPA NAME: BENZENEACETIC ACID, 4-CHLORO-α-(4-CHLOROPHENYL)-α-HYDROXY-, ETHYL ESTER
[see CHLOROBENZILATE]
CAS: 510-15-6

EPA NAME: BENZENEAMINE, N-HYDROXY-N-NITROSO, AMMONIUM SALT
[see CUPFERRON]
CAS: 135-20-6

EPA NAME: BENZENEARSONIC ACID
CAS: 98-05-5
[see also ARSENIC and ARSENIC COMPOUNDS]
SYNONYMS: PHENYL ARSENIC ACID; PHENYLARSONIC ACID

CLEAN AIR ACT: Hazardous Air Pollutants (Title I, Part A, Section 112); List of high risk pollutants (Section 63.74) as arsenic compounds.
CLEAN WATER ACT: Toxic Pollutant (Section 401.15) as arsenic and compounds.
EPCRA Section 302 Extremely Hazardous Substances: TPQ = 10/10,000 lbs. (4.54/4540 kgs.).
RCRA Section 261 Hazardous Constituents, waste number not listed.
EPCRA Section 304 Reportable Quantity (RQ): EHS, 1 lb. (0.454 kg.).
EPCRA (Section 313): As an arsenic organic compound. Form R de minimus concentration reporting level: 1.0%. Form R, Toxic Chemical Category Code: N020.
MARINE POLLUTANT (49CFR, Subchapter 172.101, Appendix B) as arsenates, liquid, n.o.s.; arsenates, solid, n.o.s.; arsenical pesticides liquid, toxic, flammable, n.o.s.

EPA NAME: BENZENE, 1-(CHLOROMETHYL)-4-NITRO-
CAS: 100-14-1
SYNONYMS: p-NITROBENZYL CHLORIDE; α-CHLORO-p-NITROTOLUENE

EPCRA Section 302 Extremely Hazardous Substances: TPQ = 500/10,000 lbs. (227/4540 kgs.).
EPCRA Section 304 Reportable Quantity (RQ): EHS, 500 lbs. (227 kgs.).

EPA NAME: 1,3-BENZENEDICARBONITRILE, 2,4,6,6-TETRACHLORO-
[see CHLOROTHALONIL]
CAS: 1897-45-6

EPA NAME: BENZENE, 2,4-DICHLORO-1-(4-NITROPHENOXY)-
[see NITROFEN]
CAS: 1836-75-5

EPA NAME: BENZENE, 2,4-DIISOCYANATO-1-METHYL-
[see TOLUENE 2,4-DIISOCYANATE]
CAS: 584-84-9

EPA NAME: BENZENE, 1,3-DIISOCYANATO-2-METHYL-
[see TOLUENE-2,6-DIISOCYANATE]
CAS: 91-08-7

EPA NAME: BENZENE, 1,3-DIISOCYANATOMETHYL-
[see TOLUENEDIISOCYANATE (MIXED ISOMERS)]
CAS: 26471-62-5

EPA NAME: BENZENE-m-DIMETHYL-
[see m-XYLENE]
CAS: 108-38-3

EPA NAME: BENZENE-o-DIMETHYL
[see o-XYLENE]
CAS: 95-47-6

EPA NAME: BENZENE-p-DIMETHYL
[see p-XYLENE]
CAS: 106-42-3

EPA NAME: BENZENEETHANAMINE, α, α-DIMETHYL-
CAS: 122-09-8
SYNONYMS: α,α-DIMETHYLPHENETHYLAMINE

EPA HAZARDOUS WASTE NUMBER (RCRA No.): P046.
RCRA Section 261 Hazardous Constituents.
RCRA Ground Water Monitoring List. Suggested test method(s) (PQL μg/L): 8270(10).
EPCRA Section 304 Reportable Quantity (RQ): CERCLA, 5,000 lbs. (2270 kgs.).

EPA NAME: BENZENEMETHANOL, 4-CHLORO-α-(4-CHLOROPHENYL)-α-(TRICHLOROMETHYL)-
[see DICOFOL]
CAS: 115-32-2

EPA NAME: BENZENESULFONYL CHLORIDE
CAS: 98-09-9
SYNONYMS: BENZENESULFONIC ACID CHLORIDE

EPA HAZARDOUS WASTE NUMBER (RCRA No.): U020.
EPCRA Section 304 Reportable Quantity (RQ): CERCLA, 100 lbs. (45.4 kgs.).

EPA NAME: BENZENETHIOL
CAS: 108-98-5
SYNONYMS: MERCAPTOBENZENE; PHENYL MERCAPTAN; PHENYLTHIOL; THIOPHENOL

EPA HAZARDOUS WASTE NUMBER (RCRA No.): P014.
RCRA Section 261 Hazardous Constituents.
EPCRA Section 302 Extremely Hazardous Substances: TPQ = 500 lb. (227 kgs.).

EPCRA Section 304 Reportable Quantity (RQ): EHS/CERCLA, 100 lbs. (454 kgs.).

EPA NAME: BENZENE,1,1′-(2,2,2-TRICHLOROETHYLIDENE)BIS(4-METHOXY-)

[see METHOXYCHLOR]

CAS: 72-43-5

EPA NAME: BENZIDINE

CAS: 92-87-5

SYNONYMS: (1,1′-BIFENYL)-4,4′-DIAMINE; 4,4′-DIAMINO-1,1′-BIPHENYL

CLEAN AIR ACT: Hazardous Air Pollutants (Title I, Part A, Section 112).

CLEAN WATER ACT: Section 307 Priority Pollutants; Section 307 Toxic Pollutants; Section 313 Priority Chemicals.

EPA HAZARDOUS WASTE NUMBER (RCRA No.): U021.

RCRA Section 261 Hazardous Constituents.

EPCRA Section 304 Reportable Quantity (RQ): CERCLA, 1 lb. (0.454 kg.).

EPCRA Section 313 Form R de minimus concentration reporting level: 0.1%.

CALIFORNIA'S PROPOSITION 65: Carcinogen as benzidine and its salts; benzidine based dyes.

EPA NAME: BENZIMIDAZOLE, 4,5-DICHLORO-2-(TRIFLUOROMETHYL)-

CAS: 3615-21-2

SYNONYMS: CHLOROFLURAZOLE; 4,5-DICHLORO-2-TRIFLUOROMETHYLBENZIMIDAZOLE

EPCRA Section 302 Extremely Hazardous Substances: TPQ = 500/1,000 lbs. (227/455 kgs.).

EPCRA Section 304 Reportable Quantity (RQ): EHS, 500 lbs. (227 kgs.).

EPA NAME: BENZO(a)PHENANTHRENE

[see CHRYSENE]

CAS 218-01-9

EPA NAME: BENZO(rst)PENTAPHENE

[see DIBENZ[a,i]PYRENE]

CAS:189-55-9

EPA NAME: BENZO[b]FLUORANTHENE

[see also POLYCYCLIC AROMATIC COMPOUNDS]

CAS: 205-99-2

SYNONYMS: BENZ(e)ACEPHENANTHRYLENE; 2,3-BENZOFLUORANTHENE; BENZO(e)FLUORANTHENE

CLEAN WATER ACT: Section 307 Priority Pollutants; Section 307 Toxic Pollutants.

RCRA Section 261 Hazardous Constituents, waste number not listed.

RCRA Land Ban Waste.

RCRA Universal Treatment Standards: Wastewater NOTE: Difficult to distinguish from benzo(k)fluoranthene (mg/L), 0.11; Nonwastewater (mg/kg), 6.8.
RCRA Ground Water Monitoring List. Suggested test method(s) (PQL µg/L): 8100(200); 8270(10).
EPCRA Section 304 Reportable Quantity (RQ): CERCLA, 1 lb. (0.454 kg.).
EPCRA Section 313 Form R de minimus concentration reporting level: 0.1%.
CALIFORNIA'S PROPOSITION 65: Carcinogen.

EPA NAME: BENZO[j]FLUORANTHENE
[see also POLYCYCLIC AROMATIC COMPOUNDS]
CAS: 205-82-3
SYNONYMS: BENZ(j)FLUORANTHENE

RCRA Section 261 Hazardous Constituents, waste number not listed.
EPCRA Section 313: As polycyclic aromatic compounds. Form R de minimus concentration reporting level: 0.1%.
CALIFORNIA'S PROPOSITION 65: Carcinogen.

EPA NAME: BENZO[k]FLUORANTHENE
[see also POLYCYCLIC AROMATIC COMPOUNDS]
CAS: 207-08-9
SYNONYMS: BENZ(k)FLUORANTHENE

CLEAN WATER ACT: Section 307 Priority Pollutants.
RCRA Section 261 Hazardous Constituents, waste number not listed.
RCRA Land Ban Waste.
RCRA Universal Treatment Standards: Wastewater (mg/L), 0.11; Nonwastewater (mg/kg), 6.8. NOTE: Difficult to distinguish from benzo[b]fluoranthene.
RCRA Ground Water Monitoring List. Suggested test method(s) (PQL µg/L): 8100(200); 8270(10).
EPCRA Section 304 Reportable Quantity (RQ): CERCLA, 5,000 lbs. (2,270 kgs.).
EPCRA Section 313: As polycyclic aromatic compounds. Form R de minimus concentration reporting level: 0.1%.
CALIFORNIA'S PROPOSITION 65: Carcinogen.

EPA NAME: BENZOIC ACID
CAS: 65-85-0
SYNONYMS: BENZENECARBOXYLIC ACID

CLEAN WATER ACT: Section 311 Hazardous Substances/RQ (same as CERCLA).
EPCRA Section 304 Reportable Quantity (RQ): CERCLA, 5,000 lbs. (2270 kgs.).

EPA NAME: BENZOIC ACID, 3-AMINO-2,5-DICHLORO-
[see CHLORAMBEN]
CAS: 133-90-4

EPA NAME: BENZOIC ACID, 5-(2-CHLORO-4-(TRIFLUOROMETHYL)PHENOXY)-2-NITRO-2-ETHOXY-1-METHYL-2-OXOETHYL ESTER

[see LACTOFEN]

CAS: 77501-63-4

EPA NAME: BENZOIC TRICHLORIDE

CAS: 98-07-7

SYNONYMS: BENZENE, TRICHLOROMETHYL-; BENZOTRICHLORIDE; PHENYL CYANIDE

CLEAN AIR ACT: Hazardous Air Pollutants (Title I, Part A, Section 112).

EPA HAZARDOUS WASTE NUMBER (RCRA No.): U023.

RCRA Section 261 Hazardous Constituents.

RCRA Land Ban Waste.

EPCRA Section 302 Extremely Hazardous Substances: TPQ = 100 lbs. (45.4 kgs.).

EPCRA Section 304 Reportable Quantity (RQ): EHS/CERCLA, 10 lbs. (4.54 kgs.).

EPCRA Section 313 Form R de minimus concentration reporting level: 0.1%.

CALIFORNIA'S PROPOSITION 65: Carcinogen.

EPA NAME: BENZONITRILE

CAS: 100-47-0

SYNONYMS: BENZENENITRILE

CLEAN WATER ACT: Section 311 Hazardous Substances/RQ (same as CERCLA); Section 313 Priority Chemicals.

EPCRA Section 304 Reportable Quantity (RQ): CERCLA, 5,000 lbs. (2270 kgs.).

EPA NAME: BENZO[ghi]PERYLENE

CAS: 191-24-2

SYNONYMS: 1,12-BENZPERYLENE; BENZO[g]PERYLENE; 1,12-BENZOPERYLENE

CLEAN WATER ACT: Section 307 Priority Pollutants.

RCRA Land Ban Waste.

RCRA Universal Treatment Standards: Wastewater (mg/L), 0.0055; Nonwastewater (mg/kg), 1.8.

RCRA Ground Water Monitoring List. Suggested test method(s) (PQL µg/L): 8100(200); 8270(10).

EPCRA Section 304 Reportable Quantity (RQ): CERCLA, 5,000 lbs. (2270 kgs.).

EPA NAME: BENZO(a)PYRENE

[see also POLYCYCLIC AROMATIC COMPOUNDS]

CAS: 50-32-8

SYNONYMS: BENZO(d,e,f)CHRYSENE; 3,4-BENZ(a)PYRENE

CLEAN WATER ACT: Section 307 Priority Pollutants; Section 307 Toxic Pollutants.

EPA HAZARDOUS WASTE NUMBER (RCRA No.): U022.
RCRA Section 261 Hazardous Constituents.
RCRA Land Ban Waste.
RCRA Universal Treatment Standards: Wastewater (mg/L), 0.061; Nonwastewater (mg/kg), 3.4.
RCRA Ground Water Monitoring List. Suggested test method(s) (PQL μg/L): 8100(200); 8270(10).
SAFE DRINKING WATER ACT: MCL, 0.0002 mg/L; MCLG, zero.
EPCRA Section 304 Reportable Quantity (RQ): CERCLA, 1 lb. (0.454 kg.).
EPCRA Section 313: As polycyclic aromatic compounds. Form R de minimus concentration reporting level: 0.1%.
CALIFORNIA'S PROPOSITION 65: Carcinogen.

EPA NAME: p-BENZOQUINONE
[see QUINONE]
CAS: 106-51-4

EPA NAME: BENZOTRICHLORIDE
[see BENZOIC TRICHLORIDE]
CAS: 98-07-7

EPA NAME: BENZOYL CHLORIDE
CAS: 98-88-4
SYNONYMS: BENZALDEHYDE, α-CHLORO-; BENZOIC ACID, CHLORIDE; α-CHLOROBENZALDEHYDE

CLEAN WATER ACT: Section 311 Hazardous Substances/RQ (same as CERCLA); Section 313 Priority Chemicals.
EPCRA Section 304 Reportable Quantity (RQ): CERCLA, 1,000 lbs. (454 kgs.).
EPCRA Section 313 Form R de minimus concentration reporting level: 1.0%.

EPA NAME: BENZOYL PEROXIDE
CAS: 94-36-0
SYNONYMS: DIBENZOYL PEROXIDE; PEROXIDE, DIBENZOYL

EPCRA Section 313 Form R de minimus concentration reporting level: 1.0%.

EPA NAME: BENZYL CHLORIDE
CAS: 100-44-7
SYNONYMS: BENZENE, CHLOROMETHYL-BENZENE; CHLOROTOLUENE; α-CHLOROTOLUENE

CLEAN AIR ACT: Hazardous Air Pollutants (Title I, Part A, Section 112).
CLEAN WATER ACT: Section 311 Hazardous Substances/RQ (same as CERCLA); Section 313 Priority Chemicals.
EPA HAZARDOUS WASTE NUMBER (RCRA No.): P028.
RCRA Section 261 Hazardous Constituents.
EPCRA Section 302 Extremely Hazardous Substances: TPQ = 500 lbs. (227 kgs.).

EPCRA Section 304 Reportable Quantity (RQ): EHS/CERCLA, 100 lbs. (45.4 kgs.).
EPCRA Section 313 Form R de minimus concentration reporting level: 1.0%.
CALIFORNIA'S PROPOSITION 65: Carcinogen.

EPA NAME: BENZYL CYANIDE
[see also CYANIDE COMPOUNDS]
CAS: 140-29-0
SYNONYMS: 2-PHENYLACETONITRILE; PHENYLACETONITRILE

CLEAN AIR ACT: Hazardous Air Pollutants (Title I, Part A, Section 112) as cyanide compound.
EPCRA Section 302 Extremely Hazardous Substances: TPQ = 500 lbs. (227 kgs.).
EPCRA Section 304 Reportable Quantity (RQ): EHS, 500 lbs. (227 kgs.).
EPCRA Section 313 Form R de minimus concentration reporting level: 1.0%. Form R Toxic Chemical Category Code: N106 (cyanide compounds).
MARINE POLLUTANT (49CFR, Subchapter 172.101, Appendix B) as cyanide mixtures or solutions.

EPA NAME: BERYLLIUM
[see also BERYLLIUM COMPOUNDS]
CAS: 7440-41-7
SYNONYMS: BERYLLIUM-9; BERYLLIUM, ELEMENTAL; BERYLLIUM POWDER

CLEAN WATER ACT: Section 307 Priority Pollutants; Toxic Pollutant (Section 401.15).
EPA HAZARDOUS WASTE NUMBER (RCRA No.): P015.
RCRA Section 261 Hazardous Constituents.
RCRA Land Ban Waste.
RCRA Universal Treatment Standards: Wastewater (mg/L), 0.82; Nonwastewater (mg/L), 0.014 TCLP.
RCRA Ground Water Monitoring List. Suggested test method(s) (PQL µg/L): (total) 6010(3); 7090(50); 7091(2).
SAFE DRINKING WATER ACT: MCL, 0.004 mg/L; MCLG, 0.004 mg/L.
EPCRA Section 304 Reportable Quantity (RQ): CERCLA, 10 lbs. (4.54 kgs.). NOTE: No report required if the diameter of the pieces of solid metal is equal to or exceeds 0.004 in.
EPCRA Section 313 Form R de minimus concentration reporting level: 0.1%. Form R Toxic Chemical Category Code: N050.
CALIFORNIA'S PROPOSITION 65: Carcinogen.

EPA NAME: BERYLLIUM CHLORIDE
[see also BERYLLIUM and BERYLLIUM COMPOUNDS]
CAS: 7787-47-5
SYNONYMS: BERYLLIUM DICHLORIDE

CLEAN WATER ACT: Section 311 Hazardous Substances/RQ (same as CERCLA); Section 313 Priority Chemicals; Toxic Pollutant (Section 401.15).
RCRA Section 261 Hazardous Constituents, as beryllium compounds, n.o.s., waste number not listed.
EPCRA Section 304 Reportable Quantity (RQ): CERCLA, 1 lb. (0.454 kg.).
EPCRA Section 313 Form R de minimus concentration reporting level: 0.1%. Form R Toxic Chemical Category Code: N050.
CALIFORNIA'S PROPOSITION 65: Carcinogen.

EPA NAME: BERYLLIUM COMPOUNDS
[see also BERYLLIUM]
SYNONYMS: BERYLLIUM COMPOUNDS, N.O.S.

CLEAN AIR ACT: Hazardous Air Pollutants (Title I, Part A, Section 112).
CLEAN WATER ACT: Toxic Pollutant (Section 401.15) as beryllium and compounds.
RCRA Section 261 Hazardous Constituents, waste number not listed.
EPCRA Section 313: Includes any unique chemical substance that contains beryllium as part of that chemical's infrastructure. Form R de minimus concentration reporting level: inorganic compounds, 0.1%.; organic compounds, 1.0%. Form R Toxic Chemical Category Code: N050.
CALIFORNIA'S PROPOSITION 65: Carcinogen as beryllium compounds.

EPA NAME: BERYLLIUM FLUORIDE
[see also BERYLLIUM, BERYLLIUM COMPOUNDS and NITRATE COMPOUNDS]
CAS: 7787-49-7
SYNONYMS: BERYLLIUM DIFLUORIDE

CLEAN WATER ACT: Section 311 Hazardous Substances/RQ (same as CERCLA); Section 313 Priority Chemicals; Toxic Pollutant (Section 401.15).
RCRA Section 261 Hazardous Constituents, as beryllium compounds, n.o.s., waste number not listed.
EPCRA Section 304 Reportable Quantity (RQ): CERCLA, 1 lb. (0.454 kg.).
EPCRA Section 313 Form R de minimus concentration reporting level: 0.1%. Form R Toxic Chemical Category Code: N050.
CALIFORNIA'S PROPOSITION 65: Carcinogen.

EPA NAME: BERYLLIUM NITRATE
[see also BERYLLIUM COMPOUNDS and NITRATE COMPOUNDS]
CAS: 7787-55-5
SYNONYMS: BERYLLIUM NITRATE (HYDRATED); NITRIC ACID, BERYLLIUM SALT, TRIHYDRATE

CLEAN WATER ACT: Section 311 Hazardous Substances/RQ (same as CERCLA) as beryllium nitrate; Section 313 Priority Chemicals; Toxic Pollutant (Section 401.15).

RCRA Section 261 Hazardous Constituents, as beryllium compounds, n.o.s., waste number not listed.
EPCRA Section 304 Reportable Quantity (RQ): CERCLA, 1 lb. (0.454 kg.).
EPCRA Section 313 Form R de minimus concentration reporting level: 0.1% (beryllium); 1.0% water-dissociable nitrate compound; reportable only when in aqueous solution. Molecular weight 133.02. Form R Toxic Chemical Category Code: N050 (beryllium); N511 (nitrate).
CALIFORNIA'S PROPOSITION 65: Carcinogen.

EPA NAME: BERYLLIUM NITRATE
[see also BERYLLIUM and BERYLLIUM COMPOUNDS]
CAS: 13597-99-4
SYNONYMS: BERYLLIUM DINITRATE; NITRIC ACID, BERYLLIUM SALT

CLEAN WATER ACT: Section 311 Hazardous Substances/RQ (same as CERCLA); Toxic Pollutant (Section 401.15).
RCRA Section 261 Hazardous Constituents, as beryllium compounds, n.o.s., waste number not listed.
EPCRA Section 304 Reportable Quantity (RQ): CERCLA, 1 lb. (0.454 kg.).
EPCRA Section 313 Form R de minimus concentration reporting level: 0.1%. Form R Toxic Chemical Category Code: N050.
Note: This chemical is *not* listed by the USEPA in EPA 745-R-96-004 (Revised May 1996) as a water-dissociable nitrate compound.
CALIFORNIA'S PROPOSITION 65: Carcinogen.

EPA NAME: α-BHC
[see α-HEXACHLOROCYCLOHEXANE]
CAS: 319-84-6

EPA NAME: β-BHC
[see β-HEXACHLOROCYCLOHEXANE]
CAS: 319-85-7

EPA NAME: δ-BHC
[see δ- HEXACHLOROCYCLOHEXANE]
CAS: 319-86-8

EPA NAME: BICYCLO[2.2.1]HEPTANE-2-CARBONITRILE, 5-CHLORO-6-((((METHYAMINO)CARBONYL)OXY)IMINO)-, (1ST-(1-α,2-β,4-α,5-α,6 e))-
CAS: 15271-41-7
SYNONYMS: 2-exo-3-CHLORO-6-endo-CYANO-2-NORBORNANONE-o-(METHYLCARBOMOYL)OXIME2-CARBON ITRILE

EPCRA Section 302 Extremely Hazardous Substances: TPQ = 500/10,000 lbs. (227/4550 kgs.).
EPCRA Section 304 Reportable Quantity (RQ): EHS, 500 lbs. (227 kgs.).

EPA NAME: BIFENTHRIN
CAS: 82657-04-3

SYNONYMS: CYCLOPROPANECARBOXYLIC ACID,3-(2-CHLORO-3,3,3-TRIFLUORO-1-PROPENYL)-2,2-DIMETHYL-,(2-METHYL(1,1′-BIPHENYL)3-YL)METHYL ESTER,(Z)-; EPA PESTICIDE CHEMICAL CODE 128825

EPCRA Section 313 Form R de minimus concentration reporting level: 1.0%.

EPA NAME: 2,2′-BIOXIRANE
[see DIEPOXYBUTANE]
CAS: 1464-53-5

EPA NAME: BIPHENYL
CAS: 92-52-4
SYNONYMS: 1,1′-BIPHENYL; DIPHENYL; 1,1′-DIPHENYL

CLEAN AIR ACT: Hazardous Air Pollutants (Title I, Part A, Section 112).
EPCRA Section 304 Reportable Quantity (RQ): CERCLA, 100 lbs. (45.4 kgs.).
EPCRA Section 313 Form R de minimus concentration reporting level: 1.0%.

EPA NAME: BIS(2-CHLOROETHOXY)METHANE
CAS: 111-91-1
SYNONYMS: BIS(β-CHLORETHYL)FORMAL; BIS(2-CHLORETHYL)FORMAL; DICHLOROMETHOXY ETHANE; ETHANE, 1,1′-(METHYLENEBIS(OXY))BIS(2-CHLORO-

CLEAN WATER ACT: Section 307 Priority Pollutants.
EPA HAZARDOUS WASTE NUMBER (RCRA No.): U024.
RCRA Section 261 Hazardous Constituents.
RCRA Land Ban Waste.
RCRA Universal Treatment Standards: Wastewater (mg/L), 0.036; Nonwastewater (mg/kg), 7.2.
RCRA Ground Water Monitoring List. Suggested test method(s) (PQL μg/L): 8270(10).
EPCRA Section 304 Reportable Quantity (RQ): CERCLA, 1,000 lbs. (454 kgs.).
EPCRA Section 313 Form R de minimus concentration reporting level: 1.0%.

EPA NAME: BIS(2-CHLOROETHYL)ETHER
[see also HALOETHERS]
CAS: 111-44-4
SYNONYMS: BIS(2-CHLOROETHYL) ETHER; BIS(2-CHLOROETHOXY)METHANE; 2,2′-DICHLORO-DIETHYLETHER; DICHLOROETHYL ETHER; DI(2-CHLOROETHYL) ETHER; ETHANE, 1,1′-OXYBIS(2-CHLORO-; 1,1′-OXYBIS(2-CHLORO)ETHANE

CLEAN AIR ACT: Hazardous Air Pollutants (Title I, Part A, Section 112).
CLEAN WATER ACT: Section 307 Priority Pollutants; Section 313 Priority Chemicals.
EPA HAZARDOUS WASTE NUMBER (RCRA No.): U025.

RCRA Section 261 Hazardous Constituents.
RCRA Land Ban Waste.
RCRA Universal Treatment Standards: Wastewater (mg/L), 0.033; Non-wastewater (mg/kg), 6.0.
RCRA Ground Water Monitoring List. Suggested test method(s) (PQL µg/L): 8270(10).
EPCRA Section 302 Extremely Hazardous Substances: TPQ = 10,000 lbs. (4,550 kgs.).
EPCRA Section 304 Reportable Quantity (RQ): EHS/CERCLA, 10 lbs. (4.54 kgs.).
EPCRA Section 313 Form R de minimus concentration reporting level: 0.1%.
CALIFORNIA'S PROPOSITION 65: Carcinogen.

EPA NAME: BIS(CHLOROMETHYL)ETHER

[see also HALOETHERS]
CAS: 542-88-1
SYNONYMS: BIS(2-CHLOROMETHYL)ETHER; CHLOROMETHYL ETHER; DICHLOROMETHYL ETHER; METHANE, OXYBIS(CHLORO)

CLEAN AIR ACT: Hazardous Air Pollutants (Title I, Part A, Section 112); Accidental Release Prevention/Flammable substances, (Section 112[r], Table 3), TQ = 1,000 lbs. (454 kgs.).
EPA HAZARDOUS WASTE NUMBER (RCRA No.): P016.
RCRA Section 261 Hazardous Constituents.
EPCRA Section 302 Extremely Hazardous Substances: TPQ = 100 lbs. (45.4 kgs.).
EPCRA Section 304 Reportable Quantity (RQ): EHS/CERCLA, 10 lbs. (4.54 kgs.).
EPCRA Section 313 Form R de minimus concentration reporting level: 1.0%.
CALIFORNIA'S PROPOSITION 65: Carcinogen.

EPA NAME: BIS(2-CHLORO-1-METHYLETHYL)ETHER

CAS: 108-60-1
SYNONYMS: BIS(CHLOROMETHYL) ETHER; BIS(2-CHLOROISOPROPYL) ETHER; DICHLOROISOPROPYL ETHER; 2,2'-DICHLORO ISOPROPYL ETHER; PROPANE, 2,2'-OXYBIS(1-CHLORO-)

CLEAN WATER ACT: Section 307 Priority Pollutants.
EPA HAZARDOUS WASTE NUMBER (RCRA No.): U027.
RCRA Section 261 Hazardous Constituents.
RCRA Universal Treatment Standards: Wastewater (mg/L), 0.055; Non-wastewater (mg/kg), 7.2.
RCRA Ground Water Monitoring List. Suggested test method(s) (PQL µg/L): 8010 (100); 8270 (10).
EPCRA Section 304 Reportable Quantity (RQ): CERCLA, 1,000 lbs. (454 kgs.).
EPCRA Section 313 Form R de minimus concentration reporting level: 1.0%.

EPA NAME: BIS(CHLOROMETHYL)KETONE
CAS: 534-07-6
SYNONYMS: 1,3-DICHLOROACETONE; 1,3-DICHLORO-2-PROPANONE; 2-PROPANONE, 1,3-DICHLORO-

EPCRA Section 302 Extremely Hazardous Substances: TPQ = 10/10,000 lbs. (4.54/4540 kgs.).
EPCRA Section 304 Reportable Quantity (RQ): EHS/CERCLA, 10 lbs. (4.54 kgs.).

EPA NAME: BIS(2-ETHYLHEXYL)ADIPATE
CAS: 103-23-1
SYNONYMS: DI-(2-ETHYLHEXYL)ADIPATE; DI-2-ETHYLHEXYL ADIPATE; DIOCTYL ADIPATE

SAFE DRINKING WATER ACT: MCL, 0.4 mg/L; MCLG, 0.4 mg/L.
EPCRA Section 313: Deleted 7/31/96 (FR Vol. 61, No. 148 p.39891-39892).

EPA NAME: BIS(2-ETHYLHEXYL)PHTHALATE
[see DI(2-ETHYLHEXYL)PHTHALATE]
CAS: 117-81-7

EPA NAME: N,N'-BIS(1-METHYLETHYL)-6-METHYLTHIO-1,3,5-TRIAZINE-2,4-DIAMINE
[see PROMETHRYN]
CAS: 7287-19-6

EPA NAME: 1,4-BIS(METHYLISOCYANATE)CYCLOHEXANE
[see also DIISOCYANATES]
CAS: 10347-54-3
SYNONYMS: CYCLOHEXANE, 1,4-BIS(ISOCYANATOMETHYL)-; 1,4-CYCLOHEXANE BIS(METHYLISOCYANATE)

EPCRA Section 313 Form R de minimus concentration reporting level: 1.0%. Form R Toxic Chemical Category Code: N120.

EPA NAME: 1,3-BIS(METHYLISOCYANATE)CYCLOHEXANE
[see also DIISOCYANATES]
CAS: 38661-72-2
SYNONYMS: 1,3-BIS(ISOCYANATOMETHYL)CYCLOHEXANE; CYCLOHEXANE, 1,3-BIS(ISOCYANATOMETHYL)-; 1,3-CYCLOHEXANE BIS(METHYLISOCYANATE)

EPCRA Section 313 Form R de minimus concentration reporting level: 1.0%. Form R Toxic Chemical Category Code: N120.

EPA NAME: BIS(TRIBUTYLTIN) OXIDE
CAS: 56-35-9
SYNONYMS: 6-OXA-5,7-DISTANNAUNDECANE,5,5,7,7-TETRABUTYL-; EPA PESTICIDE CHEMICAL CODE 083001; TIN,BIS(TRIBUTYL)-,OXIDE

EPCRA Section 313 Form R de minimus concentration reporting level: 1.0%.

EPA NAME: BITOSCANATE
[see also DIISOCYANATES]
CAS: 4044-65-9
SYNONYMS: 1,4-DIISOTHIOCYANATOBENZENE

EPCRA Section 302 Extremely Hazardous Substances: TPQ = 500/10,000 lbs. (227/4,540 kgs.).
EPCRA Section 304 Reportable Quantity (RQ): EHS, 500 lbs. (227 kgs.).
EPCRA Section 313 Form R de minimus concentration reporting level: 1.0%. Form R Toxic Chemical Category Code: N120.

EPA NAME: BORANE, TRICHLORO-
[see BORON TRICHLORIDE]
CAS: 10294-34-5

EPA NAME: BORANE, TRIFLUORO-
[see BORON TRIFLUORIDE]
CAS: 7637-07-2

EPA NAME: BORON TRICHLORIDE
CAS: 10294-34-5
SYNONYMS: BORANE, TRICHLORO-

CLEAN AIR ACT: Accidental Release Prevention/Flammable substances (Section 112[r], Table 3), TQ = 5,000 lbs. (2270 kgs.).
EPCRA Section 302 Extremely Hazardous Substances: TPQ = 500 lbs. (227 kgs.).
EPCRA Section 304 Reportable Quantity (RQ): EHS, 500 lbs. (227 kgs.).
EPCRA Section 313 Form R de minimus concentration reporting level: 1.0%.

EPA NAME: BORON TRIFLUORIDE
CAS: 7637-07-2
SYNONYMS: BORANE, TRIFLUORO-; TRIFLUOROBORANE

CLEAN AIR ACT: Accidental Release Prevention/Flammable substances (Section 112[r], Table 3), TQ = 5,000 lbs. (2270 kgs.).
SAFE DRINKING WATER ACT: As boron, Priority List (55 FR 1470).
EPCRA Section 302 Extremely Hazardous Substances: TPQ = 500 lbs. (227 kgs.).
EPCRA Section 304 Reportable Quantity (RQ): EHS, 500 lbs. (227 kgs.).
EPCRA Section 313 Form R de minimus concentration reporting level: 1.0%.

EPA NAME: BORON TRIFLUORIDE COMPOUND with METHYL ETHER (1:1)
CAS: 353-42-4
SYNONYMS: BORON TRIFLUORIDE DIMETHYLETHERATE; BORON, TRIFLUORO[OXYBIS[METHANE]]-, (T-4)-

CLEAN AIR ACT: Accidental Release Prevention/Flammable substances (Section 112[r], Table 3), TQ = 15,000 lbs. (6810 kgs.).
SAFE DRINKING WATER ACT: As boron, Priority List (55 FR 1470).

EPCRA Section 302 Extremely Hazardous Substances: TPQ = 1,000 lbs. (454 kgs.).
EPCRA Section 304 Reportable Quantity (RQ): EHS, 1000 lbs. (454 kgs.).

EPA NAME: BROMACIL
CAS: 314-40-9
SYNONYMS: 5-BROMO-6-METHYL-3-(1-METHYLPROPYL)-2,4(1H,3H)-PYRIMIDINEDIONE; 5-BROMO-3-sec-BUTYL-6-METHYLURACIL

SAFE DRINKING WATER ACT: Priority List (55 FR 1470).
EPCRA Section 313 Form R de minimus concentration reporting level: 1.0%.

EPA NAME: BROMACIL, LITHIUM SALT
CAS: 53404-19-6
SYNONYMS: 5-BROMO-3-sec-BUTYL-6-METHYLPYRIMIDINE-2,4(1H,3H)-DIONE, LITHIUM SALT; 2,4-(1H,3H)-PYRIMIDINE-DIONE, 5-BROMO-6-METHYL-3(1-METHYLPROPYL), LITHIUM SALT; EPA PESTICIDE CHEMICAL CODE 012302; 2,4(1H,3H)-PYRIMIDINEDIONE,5-BROMO-6-METHYL-3-(1-METHYLPROPYL)-,LITHIUM SALT

EPCRA Section 313 Form R de minimus concentration reporting level: 1.0%.

EPA NAME: BROMADIOLONE
CAS: 28772-56-7
SYNONYMS: 3-(3-(4′-BROMO(1,1′-BIPHENYL)-4-YL)3-HYDROXY-1-PHENYLPROPYL)-4-HYDROXY-2H-1-BENZOPYRAN-2-ONE

EPCRA Section 302 Extremely Hazardous Substances: TPQ = 100/10,000 lbs. (45.4/4,540 kgs.).
EPCRA Section 304 Reportable Quantity (RQ): EHS, 100 lbs. (45.4 kgs.).

EPA NAME: BROMINE
CAS: 7726-95-6
SYNONYMS: BROMINE, ELEMENTAL; EPA PESTICIDE CHEMICAL CODE 008701

CLEAN AIR ACT: Accidental Release Prevention/Flammable substances, (Section 112[r]), TQ = 10,000 lbs. (4540 kgs.).
EPCRA Section 302 Extremely Hazardous Substances: TPQ = 500 lbs. (227 kgs.).
EPCRA Section 304 Reportable Quantity (RQ): EHS, 500 lbs. (227 kgs.).
EPCRA Section 313 Form R de minimus concentration reporting level: 1.0%.

EPA NAME: BROMOACETONE
CAS: 598-31-2
SYNONYMS: BROMO-2-PROPANONE; 2-PROPANONE, 1-BROMO

EPCRA Section 304 Reportable Quantity (RQ): CERCLA, 1,000 lbs. (454 kgs.).
EPA HAZARDOUS WASTE NUMBER (RCRA No.): P017.
RCRA Section 261 Hazardous Constituents.

EPA NAME: 1-BROMO-1-(BROMOMETHYL)-1,3-PROPANE-DICARBONITRILE
CAS: 35691-65-7
SYNONYMS: 2-BROMO-2-(BROMOETHYL)PENTANEDINITRILE; 1,2-DIBROMO-2,4-DICYANOBUTANE; EPA PESTICIDE CHEMICAL CODE 111001

EPCRA Section 313 Form R de minimus concentration reporting level: 1.0%.

EPA NAME: BROMOCHLORODIFLUOROMETHANE
CAS: 353-59-3
SYNONYMS: CHLORODIFLUOROBROMOMETHANE; HALON 1211

CLEAN AIR ACT: Stratospheric ozone protection (Title VI, Subpart A, Appendix A), Class I, Ozone Depletion Potential = 3.0.
EPCRA Section 313 Form R de minimus concentration reporting level: 1.0%.

EPA NAME: O-(4-BROMO-2-CHLOROPHENYL)-O-ETHYL-S-PROPYLPHOSPHOROTHIOATE
[see PROFENOFOS]
CAS: 41198-08-7

EPA NAME: BROMOFORM
CAS: 75-25-2
SYNONYMS: METHENYL TRIBROMIDE; METHANE, TRIBROMO-; METHYL TRIBROMIDE; TRIBROMOMETHANE

CLEAN AIR ACT: Hazardous Air Pollutants (Title I, Part A, Section 112).
CLEAN WATER ACT: Section 307 Priority Pollutants; Section 313 Priority Chemicals.
EPA HAZARDOUS WASTE NUMBER (RCRA No.): U225.
RCRA Section 261 Hazardous Constituents.
RCRA Land Ban Waste.
RCRA Universal Treatment Standards: Wastewater (mg/L), 0.63; Non-wastewater (mg/kg), 15.
RCRA Ground Water Monitoring List. Suggested test method(s) (PQL µg/L): 8010(2); 8240(5).
SAFE DRINKING WATER ACT: Priority List (55 FR 1470).
EPCRA Section 304 Reportable Quantity (RQ): CERCLA, 100 lbs. (45.4 kgs.).
EPCRA Section 313 Form R de minimus concentration reporting level: 1.0%.
MARINE POLLUTANT (49CFR, Subchapter 172.101, Appendix B).
CALIFORNIA'S PROPOSITION 65: Carcinogen.

EPA NAME: BROMOMETHANE

CAS: 74-83-9

SYNONYMS: METHANE, BROMO-; METHYL BROMIDE; MONO-BROMOMETHANE

CLEAN AIR ACT: Hazardous Air Pollutants (Title I, Part A, Section 112); Stratospheric ozone protection (Title VI, Subpart A, Appendix A), Class I, Ozone Depletion Potential = 0.7.

CLEAN WATER ACT: Section 307 Priority Pollutants; Section 313 Priority Chemicals.

EPA HAZARDOUS WASTE NUMBER (RCRA No.): U029.

RCRA Section 261 Hazardous Constituents.

RCRA Land Ban Waste.

RCRA Universal Treatment Standards: Wastewater (mg/L), 0.11; Non-wastewater (mg/kg), 15.

RCRA Ground Water Monitoring List. Suggested test method(s) (PQL μg/L): 8010(20); 8240(10).

SAFE DRINKING WATER ACT: Priority List (55 FR 1470).

EPCRA Section 302 Extremely Hazardous Substances: TPQ = 1,000 lbs. (454 kgs.).

EPCRA Section 304 Reportable Quantity (RQ): EHS/CERCLA, 1,000 lbs. (454 kgs.).

EPCRA Section 313 Form R de minimus concentration reporting level: 1.0%.

MARINE POLLUTANT (49CFR, Subchapter 172.101, Appendix B). Only as methyl bromide and ethylene dibromide mixture, liquid.

CALIFORNIA'S PROPOSITION 65: Reproductive toxin, as a structural fumigant.

EPA NAME: 5-BROMO-6-METHYL-3-(1-METHYLPROPYL)-2,4(1H,3H)-PYRIMIDINEDIONE

[see BROMACIL]

CAS: 314-40-9

EPA NAME: 2-BROMO-2-NITROPROPANE-1,3-DIOL

CAS: 52-51-7

SYNONYMS: 2-BROMO-2-NITRO-1,3-PROPANEDIOL; BRONOPOL; EPA PESTICIDE CHEMICAL CODE 216400

EPCRA Section 313 Form R de minimus concentration reporting level: 1.0%.

EPA NAME: 4-BROMOPHENYL PHENYL ETHER

CAS: 101-55-3

SYNONYMS: BENZENE, 1-BROMO-4-PHENOXY-; BENZENE, 2-BROMO-4-PHENOXY-

CLEAN WATER ACT: Section 307 Priority Pollutants.

EPA HAZARDOUS WASTE NUMBER (RCRA No.): U030.

RCRA Section 261 Hazardous Constituents.

RCRA Land Ban Waste.

RCRA Universal Treatment Standards: Wastewater (mg/L), 0.055; Non-wastewater (mg/kg), 15.

RCRA Ground Water Monitoring List. Suggested test method(s) (PQL μg/L): 8270(10).

EPCRA Section 304 Reportable Quantity (RQ): CERCLA, 100 lbs. (45.4 kgs.).

EPA NAME: BROMOTRIFLUORETHYLENE

CAS: 598-73-2

SYNONYMS: BROMOTRIFLUORETHENE; ETHENE, BROMOTRIFLUORO-; TRIFLUOROBROMOETHYLENE; TRIFLUOROVINYLBROMIDE

CLEAN AIR ACT: Accidental Release Prevention/Flammable substances (Section 112[r], Table 3), TQ = 10,000 lbs. (4540 kgs.).

EPA NAME: BROMOTRIFLUROMETHANE

CAS: 75-63-8

SYNONYMS: FLUOROCARBON 1301; HALON 1301; METHANE, BROMOTRIFLUORO-; TRIFLUOROBROMOMETHANE

CLEAN AIR ACT: Stratospheric ozone protection (Title VI, Subpart A, Appendix A), Class I, Ozone Depletion Potential = 10.0.

EPCRA Section 313 Form R de minimus concentration reporting level: 1.0%.

EPA NAME: BROMOXYNIL

CAS: 1689-84-5

SYNONYMS: BENSONITRILE, 3,5-DIBROMO-4-HYDROXY-; 3,5-DIBROMO-4-HYDROXYBENZONITRILE; EPA PESTICIDE CHEMICAL CODE 035301; 4-HYDROXY-3,5-DIBROMOBENZONITRILE

EPCRA Section 313 Form R de minimus concentration reporting level: 1.0%.

MARINE POLLUTANT (49CFR, Subchapter 172.101, Appendix B).

CALIFORNIA'S PROPOSITION 65: Reproductive toxin.

EPA NAME: BROMOXYNIL OCTANOATE

CAS: 1689-99-2

SYNONYMS: 3,5-DIBROMO-4-OCTANOYLOXY-BENZONITRILE; EPA PESTICIDE CHEMICAL CODE 035302; OCTANOIC ACID, 2,6-DIBROMO-4-CYANOPHENYL ESTER

EPCRA Section 313 Form R de minimus concentration reporting level: 1.0%.

EPA NAME: BRONOPOL

[see 2-BROMO-2-NITROPROPANE-1,3-DIOL]

CAS: 52-51-7

EPA NAME: BRUCINE

[see also STRYCHNINE AND SALTS]

CAS: 357-57-3

SYNONYMS: DIMETHOXY STRYCHNINE; STRYCHNIDIN-10-ONE, 2,3-DIMETHOXY-

EPA HAZARDOUS WASTE NUMBER (RCRA No.): P018.
RCRA Section 261 Hazardous Constituents.
EPCRA Section 304 Reportable Quantity (RQ): CERCLA, 100 lbs. (45.4 kgs.).
EPCRA Section 313 Form R de minimus concentration reporting level: 1.0%.

EPA NAME: 1,3-BUTADIENE
CAS: 106-99-0
SYNONYMS: BUTADIENE; BUTA-1,3-DIENE; VINYLETHYLENE

CLEAN AIR ACT: Hazardous Air Pollutants (Title I, Part A, Section 112); Accidental Release Prevention/Flammable substances (Section 112[r], Table 3), TQ = 10,000 lbs. (4540 kgs.).
EPCRA Section 304 Reportable Quantity (RQ): CERCLA, 10 lbs. (4.54 kgs.).
EPCRA Section 313 Form R de minimus concentration reporting level: 0.1%.

EPA NAME: 1,3-BUTADIENE, 2-METHYL
[see ISOPRENE]
CAS: 78-79-5

EPA NAME: BUTANE
CAS: 106-97-8
SYNONYMS: n-BUTANE; METHYLETHYL METHANE

CLEAN AIR ACT: Accidental Release Prevention/Flammable substances (Section 112[r], Table 3), TQ = 10,000 lbs. (4540 kgs.).

EPA NAME: BUTANE, 2-METHYL-
[see ISOPENTANE]
CAS: 78-78-4

EPA NAME: 2-BUTENAL
[see CROTONALDEHYDE]
CAS: 4170-30-3

EPA NAME: 2-BUTENAL, (E)
[see CROTONALDEHYDE, (E)]
CAS: 123-73-9

EPA NAME: BUTENE
CAS: 25167-67-3
SYNONYMS: α-BUTYLENE; BUTYLENE; ETHYLETHYLENE

CLEAN AIR ACT: Accidental Release Prevention/Flammable substances (Section 112[r], Table 3), TQ = 10,000 lbs. (4540 kgs.).

EPA NAME: 2-BUTENE-cis
CAS: 590-18-1
SYNONYMS: 2-BUTENE, (z)-

CLEAN AIR ACT: Accidental Release Prevention/Flammable substances (Section 112[r], Table 3), TQ = 10,000 lbs. (4540 kgs.).

EPA NAME: 2-BUTENE, (E)-
CAS: 624-64-6
SYNONYMS: (E)-, 2-BUTENE; 2-BUTENE, trans; trans-2-BUTENE

CLEAN AIR ACT: Accidental Release Prevention/Flammable substances (Section 112[r], Table 3), TQ = 10,000 lbs. (4540 kgs.).

EPA NAME: 2-BUTENE, trans
[see 2-BUTENE, (E)]
CAS: 624-64-6

EPA NAME: 1-BUTENE
CAS: 106-98-9
SYNONYMS: BUTENE-1; α-BUTYLENE; BUTYLENE; ETHYL-ETHYLENE

CLEAN AIR ACT: Accidental Release Prevention/Flammable substances (Section 112[r], Table 3), TQ = 10,000 lbs. (4540 kgs.).

EPA NAME: 2-BUTENE
CAS: 107-01-7
SYNONYMS: β-BUTYLENE; PSEUDOBUTYLENE

CLEAN AIR ACT: Accidental Release Prevention/Flammable substances (Section 112[r], Table 3), TQ = 10,000 lbs. (4540 kgs.).

EPA NAME: 2, BUTENE, 1,4-DICHLORO-
[see 1,4-DICHLORO-2-BUTENE]
CAS: 764-41-0

EPA NAME: 1-BUTEN-3-YNE
[see VINYL ACETYLENE]
CAS: 689-97-4

EPA NAME: 2,4-D BUTOXYETHYL ESTER
[see 2,4-D BUTOXYETHYL ESTER in D section)
CAS: 1929-73-3

EPA NAME: BUTYL ACETATE
CAS: 123-86-4
SYNONYMS: ACETIC ACID, n-BUTYL ESTER; ACETIC ACID, BUTYL ESTER

CLEAN WATER ACT: Section 311 Hazardous Substances/RQ (same as CERCLA).
EPCRA Section 304 Reportable Quantity (RQ): CERCLA, 5,000 lbs. (2270 kgs.).

EPA NAME: iso-BUTYL ACETATE
CAS: 110-19-0
SYNONYMS: ACETIC ACID, ISOBUTYL ESTER; ACETIC ACID, 2-METHYLPROPYL ESTER; 2-METHYLPROPYL ACETATE

CLEAN WATER ACT: Section 311 Hazardous Substances/RQ (same as CERCLA).
EPCRA Section 304 Reportable Quantity (RQ): CERCLA, 5,000 lbs. (2270 kgs.).

EPA NAME: sec-BUTYL ACETATE
CAS: 105-46-4
SYNONYMS: ACETIC ACID, 2-BUTOXY ESTER; 1-METHYL PROPYL ACETATE

CLEAN WATER ACT: Section 311 Hazardous Substances/RQ (same as CERCLA).
EPCRA Section 304 Reportable Quantity (RQ): CERCLA, 5,000 lbs. (2270 kgs.).

EPA NAME: tert-BUTYL ACETATE
CAS: 540-88-5
SYNONYMS: ACETIC ACID, 1,1-DIMETHYLETHYL ESTER (9CI); ACETIC ACID, tert-BUTYL ESTER-

CLEAN WATER ACT: Section 311 Hazardous Substances/RQ (same as CERCLA).
EPCRA Section 304 Reportable Quantity (RQ): CERCLA, 5,000 lbs. (2270 kgs.).

EPA NAME: BUTYL ACRYLATE
CAS: 141-32-2
SYNONYMS: ACRYLIC ACID, BUTYL ESTER; ACRYLIC ACID n-BUTYL ESTER; n-BUTYL ACRYLATE; 2-PROPENOIC ACID, BUTYL ESTER

EPCRA Section 313 Form R de minimus concentration reporting level: 1.0%.

EPA NAME: n-BUTYL ALCOHOL
CAS: 71-36-3
SYNONYMS: 1-BUTANOL; BUTAN-1-OL; BUTYRIC ALCOHOL; BUTYL ALCOHOL

EPA HAZARDOUS WASTE NUMBER (RCRA No.): U031.
RCRA Section 261 Hazardous Constituents.
RCRA Land Ban Waste.
RCRA Universal Treatment Standards: Wastewater (mg/L), 5.6; Nonwastewater (mg/kg), 2.6.
EPCRA Section 304 Reportable Quantity (RQ): CERCLA, 5,000 lbs. (2270 kgs.).
EPCRA Section 313 Form R de minimus concentration reporting level: 1.0%.

EPA NAME: sec-BUTYL ALCOHOL
CAS: 78-92-2
SYNONYMS: 2-BUTANOL; 2-BUTYL ALCOHOL; BUTAN-2-OL; sec-BUTANOL

EPCRA Section 313 Form R de minimus concentration reporting level: 1.0%.

EPA NAME: tert-BUTYL ALCOHOL

CAS: 75-65-0

SYNONYMS: tert-BUTANOL; 2-METHYL-,2-PROPANOL; 2-METHYL-PROPAN-2-OL; 2-PROPANOL, 2-METHYL-

EPCRA Section 313 Form R de minimus concentration reporting level: 1.0%.

EPA NAME: BUTYLAMINE

CAS: 109-73-9

SYNONYMS: 1-AMINOBUTANE; 1-BUTANAMINE; N-BUTYLAMINE; MONOBUTYLAMINE

CLEAN WATER ACT: Section 311 Hazardous Substances/RQ (same as CERCLA).

EPCRA Section 304 Reportable Quantity (RQ): CERCLA, 1,000 lbs. (454 kgs.).

EPA NAME: iso-BUTYLAMINE

CAS: 78-81-9

SYNONYMS: 1-AMINO-2-METHYLPROPANE; 2-METHYLPROPYLAMINE; 1-PROPANAMINE, 2-METHYL-

CLEAN WATER ACT: Section 311 Hazardous Substances/RQ (same as CERCLA).

EPCRA Section 304 Reportable Quantity (RQ): CERCLA, 1,000 lbs. (454 kgs.).

EPA NAME: sec-BUTYLAMINE

CAS: 513-49-5

SYNONYMS: sec-BUTYLAMINE (s-)

CLEAN WATER ACT: Section 311 Hazardous Substances/RQ (same as CERCLA).

EPCRA Section 304 Reportable Quantity (RQ): CERCLA, 1,000 lbs. (454 kgs.).

EPA NAME: sec-BUTYLAMINE

CAS: 13952-84-6

SYNONYMS: 2-BUTANAMINE; PROPYLAMINE, 1-METHYL; (RS)-sec-BUTYLAMINE

CLEAN WATER ACT: Section 311 Hazardous Substances/RQ (same as CERCLA).

EPCRA Section 304 Reportable Quantity (RQ): CERCLA, 1,000 lbs. (454 kgs.).

EPA NAME: tert-BUTYLAMINE

CAS: 75-64-9

SYNONYMS: 1,1-DIMETHYLETHYLAMINE; 2-PROPANEAMINE, 2-METHYL-

CLEAN WATER ACT: Section 311 Hazardous Substances/RQ (same as CERCLA).
EPCRA Section 304 Reportable Quantity (RQ): CERCLA, 1,000 lbs. (454 kgs.).

EPA NAME: BUTYLATE
CAS: 2008-41-5
RCRA Land Ban Waste.
RCRA Universal Treatment Standards: Wastewater (mg/L), 0.003; Non-wastewater (mg/kg), 1.4.
EPCRA Section 304 Reportable Quantity (RQ): CERCLA, 1 lb. (0.454 kg.).

EPA NAME: BUTYL BENZYL PHTHALATE
CAS: 85-68-7
SYNONYMS: 1,2-BENZENEDICARBOXYLIC ACID, BUTYL PHENYLMETHYL ESTER; n-BENZYL BUTYL PHTHALATE

CLEAN WATER ACT: Section 307 Priority Pollutants; Section 313 Priority Chemicals.
RCRA Section 261 Hazardous Constituents, waste number not listed.
RCRA Land Ban Waste.
RCRA Universal Treatment Standards: Wastewater (mg/L), 0.017; Non-wastewater (mg/kg), 28.
RCRA Ground Water Monitoring List. Suggested test method(s) (PQL µg/L): 8060(5); 8270(10).
EPCRA Section 304 Reportable Quantity (RQ): CERCLA, 100 lbs. (45.4 kgs.).
MARINE POLLUTANT (49CFR, Subchapter 172.101, Appendix B).

EPA NAME: α-BUTYL-α-(4-CHLOROPHENYL)-1H-1,2,4-TRIAZOLE-1-PROPANENITRILE
[see MYCLOBUTANIL]
CAS: 88671-89-0

EPA NAME: 1,2-BUTYLENE OXIDE
CAS: 106-88-7
SYNONYMS: OXIRANE, ETHYL-; 1,2-EPOXYBUTANE; 2-ETHYLOXIRANE; PROPYL OXIRANE

CLEAN AIR ACT: Hazardous Air Pollutants (Title I, Part A, Section 112).
EPCRA Section 304 Reportable Quantity (RQ): CERCLA, 100 lbs. (45.4 kgs.).
EPCRA Section 313 Form R de minimus concentration reporting level: 1.0%.

EPA NAME: BUTYLETHYLCARBAMOTHIOIC ACID S-PROPYL ESTER
[see PEBULATE]
CAS: 1114-71-2

EPA NAME: N-BUTYL-N-ETHYL-2,6-DINITRO-4-(TRIFLUOROMETHYL)BENZENEAMINE
[see BENFLURALIN]
CAS: 1861-40-1

EPA NAME: n-BUTYL PHTHALATE
[see DIBUTYL PHTHALATE]
CAS: 84-74-2

EPA NAME: 1-BUTYNE
CAS: 107-00-6
SYNONYMS: ETHYL ACETYLENE; ETHYLETHYNE

CLEAN AIR ACT: Accidental Release Prevention/Flammable substances (Section 112[r], Table 3), TQ = 10,000 lbs. (4540 kgs.).

EPA NAME: BUTYRALDEHYDE
CAS: 123-72-8
SYNONYMS: BUTANAL; n-BUTYRALDEHYDE; BUTYL ALDEHYDE

EPCRA Section 313 Form R de minimus concentration reporting level: 1.0%.

EPA NAME: BUTYRIC ACID
CAS: 107-92-6
SYNONYMS: BUTANOIC ACID; n-BUTANOIC ACID; 1-PROPANECARBOXYIC ACID

CLEAN WATER ACT: Section 311 Hazardous Substances/RQ (same as CERCLA).
EPCRA Section 304 Reportable Quantity (RQ): CERCLA, 5,000 lbs. (2270 kgs.).

EPA NAME: iso-BUTYRIC ACID
CAS: 79-31-2
SYNONYMS: ISOBUTANOIC ACID; METHYLPROPIONIC ACID, 2-; PROPANE-2-CARBOXYLIC ACID; PROPANOIC ACID, 2-METHYL-

CLEAN WATER ACT: Section 311 Hazardous Substances/RQ (same as CERCLA), as isomer of butyric acid.
EPCRA Section 304 Reportable Quantity (RQ): CERCLA, 5,000 lbs. (2270 kgs.).

- C -

EPA NAME: CACODYLIC ACID
[see also ARSENIC AND ARSENIC COMPOUNDS]
CAS: 75-60-5
SYNONYMS: HYDROXYDIMETHYLARSINE OXIDE; DIMETHYLARSINIC ACID

EPA HAZARDOUS WASTE NUMBER (RCRA No.): U136.
RCRA Section 261 Hazardous Constituents.
EPCRA Section 304 Reportable Quantity (RQ): CERCLA, 1 lb. (0.454 kg.).
EPCRA Section 313 Form R (as organic arsenic compound) de minimus concentration reporting level: 1.0%. Form R, Toxic Chemical Category Code: N020.
MARINE POLLUTANT (49CFR, Subchapter 172.101, Appendix B) as dimethylarsinic acid.

EPA NAME: CADMIUM
[see also CADMIUM COMPOUNDS]
CAS: 7440-43-9
SYNONYMS: CADMIUM, ELEMENTAL

CLEAN AIR ACT: Hazardous Air Pollutants (Title I, Part A, Section 112). Includes any unique chemical substance that contains cadmium as part of that chemical's infrastructure.
CLEAN WATER ACT: Section 307 Priority Pollutants; Section 313 Priority Chemicals; Toxic Pollutant (Section 401.15).
EPA HAZARDOUS WASTE NUMBER (RCRA No.): D006.
RCRA Section 261 Hazardous Constituents, waste number not listed.
RCRA Toxicity Characteristic (Section 261.24), Maximum Concentration of Contaminants, regulatory level, 1.0 mg/L.
RCRA Land Ban Waste.
RCRA Universal Treatment Standards: Wastewater (mg/L), 0.69; Nonwastewater (mg/L), 0.19 TCLP.
RCRA Ground Water Monitoring List. Suggested test method(s) (PQL µg/L): 6010(40); 7130(50); 7131(1).
SAFE DRINKING WATER ACT: MCL, 0.005 mg/L; MCLG, 0.005 mg/L; Regulated chemical (47 FR 9352); Priority List (55 FR 1470).
EPCRA Section 304 Reportable Quantity (RQ): CERCLA, 10 lbs. (4.54 kgs.). Note: No release report required if diameter of pieces is equal to or exceeds 0.004 in.
EPCRA Section 313 Form R de minimus concentration reporting level: 0.1%. Form R Toxic Chemical Category Code: N078.
MARINE POLLUTANT (49CFR, Subchapter 172.101, Appendix B): Severe pollutant, as cadmium compounds.
CALIFORNIA'S PROPOSITION 65: Carcinogen.

EPA NAME: CADMIUM ACETATE
[see also CADMIUM and CADMIUM COMPOUNDS]

CAS: 543-90-8
SYNONYMS: ACETIC ACID, CADMIUM SALT; CADMIUM(II) ACETATE

CLEAN AIR ACT: Hazardous Air Pollutants (Title I, Part A, Section 112). Includes any unique chemical substance that contains cadmium as part of that chemical's infrastructure.
CLEAN WATER ACT: Section 311 Hazardous Substances/RQ (same as CERCLA); Section 313 Priority Chemicals.
EPCRA Section 304 Reportable Quantity (RQ): CERCLA, 10 lbs. (4.54 kgs.).
EPCRA Section 313 Form R de minimus concentration reporting level: 1.0%. Form R Toxic Chemical Category Code: N078.
MARINE POLLUTANT (49CFR, Subchapter 172.101, Appendix B): Severe pollutant, as cadmium compounds.
CALIFORNIA'S PROPOSITION 65: Carcinogen.

EPA NAME: CADMIUM BROMIDE
[see also CADMIUM and CADMIUM COMPOUNDS]
CAS: 7789-42-6
SYNONYMS: CADMIUM DIBROMIDE

CLEAN AIR ACT: Hazardous Air Pollutants (Title I, Part A, Section 112). Includes any unique chemical substance that contains cadmium as part of that chemical's infrastructure.
CLEAN WATER ACT: Section 311 Hazardous Substances/RQ (same as CERCLA); Section 313 Priority Chemicals.
EPCRA Section 304 Reportable Quantity (RQ): CERCLA, 10 lbs. (4.54 kgs.).
EPCRA Section 313 Form R de minimus concentration reporting level: 0.1%. Form R Toxic Chemical Category Code: N078.
MARINE POLLUTANT (49CFR, Subchapter 172.101, Appendix B): Severe pollutant, as cadmium compounds.
CALIFORNIA'S PROPOSITION 65: Carcinogen.

EPA NAME: CADMIUM CHLORIDE
[see also CADMIUM and CADMIUM COMPOUNDS]
CAS: 10108-64-2
SYNONYMS: CADMIUM DICHLORIDE

CLEAN AIR ACT: Hazardous Air Pollutants (Title I, Part A, Section 112). Includes any unique chemical substance that contains cadmium as part of that chemical's infrastructure.
CLEAN WATER ACT: Section 311 Hazardous Substances/RQ (same as CERCLA); Section 313 Priority Chemicals.
EPCRA Section 304 Reportable Quantity (RQ): CERCLA, 10 lbs. (4.54 kgs.).
EPCRA Section 313 Form R de minimus concentration reporting level: 0.1%. Form R Toxic Chemical Category Code: N078.
MARINE POLLUTANT (49CFR, Subchapter 172.101, Appendix B): Severe pollutant, as cadmium compounds.
CALIFORNIA'S PROPOSITION 65: Carcinogen.

EPA NAME: CADMIUM COMPOUNDS
[see also CADMIUM]
CLEAN AIR ACT: Hazardous Air Pollutants (Title I, Part A, Section 112). Includes any unique chemical substance that contains cadmium as part of that chemical's infrastructure.
CLEAN WATER ACT: Toxic Pollutant (Section 401.15).
RCRA Section 261 Hazardous Constituents, waste number not listed.
EPCRA Section 313: Includes any unique chemical substance that contains cadmium as part of that chemical's infrastructure. Form R de minimus concentration reporting level: (inorganic compounds: 0.1%.; organic compounds: 1.0%). Form R Toxic Chemical Category Code: N078.
MARINE POLLUTANT (49CFR, Subchapter 172.101, Appendix B): Severe pollutant, as cadmium compounds.
CALIFORNIA'S PROPOSITION 65: Carcinogen.

EPA NAME: CADMIUM OXIDE
[see also CADMIUM and CADMIUM COMPOUNDS]
CAS: 1306-19-0
CLEAN AIR ACT: Hazardous Air Pollutants (Title I, Part A, Section 112). Includes any unique chemical substance that contains cadmium as part of that chemical's infrastructure.
EPCRA Section 302 Extremely Hazardous Substances: TPQ = 10/10,000 lbs. (4.54/4,540 kgs.).
EPCRA Section 304 Reportable Quantity (RQ): EHS, 100 lbs. (45.4 kgs.).
EPCRA Section 313 Form R de minimus concentration reporting level: 0.1%. Form R Toxic Chemical Category Code: N078.
MARINE POLLUTANT (49CFR, Subchapter 172.101, Appendix B): Severe pollutant, as cadmium compounds.
CALIFORNIA'S PROPOSITION 65: Carcinogen.

EPA NAME: CADMIUM STEARATE
[see also CADMIUM and CADMIUM COMPOUNDS]
CAS: 2223-93-0
SYNONYMS: STEARIC ACID, CADMIUM SALT
CLEAN AIR ACT: Hazardous Air Pollutants (Title I, Part A, Section 112). Includes any unique chemical substance that contains cadmium as part of that chemical's infrastructure.
EPCRA Section 302 Extremely Hazardous Substances: TPQ = 1,000/10,000 lbs. (454/4,540 kgs.).
EPCRA Section 304 Reportable Quantity (RQ): EHS, 1000 lbs. (454 kgs.).
EPCRA Section 313 Form R de minimus concentration reporting level: 1.0%. Form R Toxic Chemical Category Code: N078.
MARINE POLLUTANT (49CFR, Subchapter 172.101, Appendix B): Severe pollutant, as cadmium compounds.
CALIFORNIA'S PROPOSITION 65: Carcinogen.

EPA NAME: CALCIUM ARSENATE
[see also ARSENIC and ARSENIC COMPOUNDS]
CAS: 7778-44-1

SYNONYMS: ARSENIC ACID, CALCIUM SALT (2:3); TRICALCIUM ORTHOARSENITE

CLEAN AIR ACT: List of high risk pollutants (Section 63.74) as arsenic compounds.

CLEAN WATER ACT: Section 311 Hazardous Substances/RQ (same as CERCLA); Section 313 Priority Chemicals; Toxic Pollutant (Section 401.15) as arsenic and compounds.

RCRA Section 261 Hazardous Constituents, waste number not listed.

EPCRA Section 302 Extremely Hazardous Substances: TPQ = 500/10,000 lbs. (227/4,540 kgs.).

EPCRA Section 304 Reportable Quantity (RQ): EHS/CERCLA, 1 lb. (0.454 kg.).

EPCRA Section 313 Form R de minimus concentration reporting level: 0.1%. Form R, Toxic Chemical Category Code: N020.

MARINE POLLUTANT (49CFR, Subchapter 172.101, Appendix B), listed by name; also listed as calcium arsenate and calcium arsenite, mixtures, solid.

CALIFORNIA'S PROPOSITION 65: Carcinogen.

EPA NAME: CALCIUM ARSENITE

[see also ARSENIC and ARSENIC COMPOUNDS]

CAS: 52740-16-6

SYNONYMS: ARSENOUS ACID, CALCIUM SALT; ARSONIC ACID, CALCIUM SALT(1:1)

CLEAN AIR ACT: List of high risk pollutants (Section 63.74) as arsenic compounds.

CLEAN WATER ACT: Section 311 Hazardous Substances/RQ (same as CERCLA); Section 313 Priority Chemicals; Toxic Pollutant (Section 401.15) as arsenic and compounds.

RCRA Section 261 Hazardous Constituents, waste number not listed.

EPCRA Section 304 Reportable Quantity (RQ): CERCLA, 1 lb. (0.454 kg.).

EPCRA Section 313 Form R de minimus concentration reporting level: 0.1%. Form R, Toxic Chemical Category Code: N020.

MARINE POLLUTANT (49CFR, Subchapter 172.101, Appendix B) as arsenates, liquid, n.o.s.; arsenates, solid, n.o.s.; arsenical pesticides liquid, toxic, flammable, n.o.s.; also listed as calcium arsenate and calcium arsenite, mixtures, solid.

CALIFORNIA'S PROPOSITION 65: Carcinogen.

EPA NAME: CALCIUM CARBIDE

CAS: 75-20-7

SYNONYMS: CARBIDE, ACETYLENOGEN

CLEAN WATER ACT: Section 311 Hazardous Substances/RQ (same as CERCLA).

EPCRA Section 304 Reportable Quantity (RQ): CERCLA, 10 lbs. (4.54 kgs.).

EPA NAME: CALCIUM CHROMATE

[see also CHROMIUM and CHROMIUM COMPOUNDS]

CAS: 13765-19-0
SYNONYMS: CHROMIC ACID, CALCIUM SALT (1:1)

CLEAN AIR ACT: Hazardous Air Pollutants (Title I, Part A, Section 112).
CLEAN WATER ACT: Section 311 Hazardous Substances/RQ (same as CERCLA); Section 313 Priority Chemicals; Toxic Pollutant (Section 401.15).
EPA HAZARDOUS WASTE NUMBER (RCRA No.): U032.
RCRA Section 261 Hazardous Constituents.
EPCRA Section 304 Reportable Quantity (RQ): CERCLA, 10 lbs. (4.54 kgs.).
EPCRA Section 313 Form R de minimus concentration reporting level: 0.1%. Form R Toxic Chemical Category Code: N090.

EPA NAME: CALCIUM CYANAMIDE
[see also CYANIDES and CYANIDE COMPOUNDS]
CAS: 156-62-7
SYNONYMS: CYANAMIDE, CALCIUM SALT (1:1)

CLEAN AIR ACT: Hazardous Air Pollutants (Title I, Part A, Section 112).
EPCRA Section 304 Reportable Quantity (RQ): CERCLA, 1,000 lbs. (454 kgs.).
EPCRA Section 313 Form R de minimus concentration reporting level: 1.0%.
MARINE POLLUTANT (49CFR, Subchapter 172.101, Appendix B) as cyanide mixtures, cyanide solutions or cyanides, inorganic, n.o.s.

EPA NAME: CALCIUM CYANIDE
[see also CYANIDES and CYANIDE COMPOUNDS]
CAS: 592-01-8
SYNONYMS: CALCYANIDE

CLEAN AIR ACT: Hazardous Air Pollutants (Title I, Part A, Section 112).
CLEAN WATER ACT: Section 311 Hazardous Substances/RQ (same as CERCLA); Section 313 Priority Chemicals.
EPA HAZARDOUS WASTE NUMBER (RCRA No.): P021.
RCRA Section 261 Hazardous Constituents.
RCRA Land Ban Waste.
EPCRA Section 304 Reportable Quantity (RQ): CERCLA, 10 lbs. (4.54 kgs.).
MARINE POLLUTANT (49CFR, Subchapter 172.101, Appendix B) as cyanide mixtures, cyanide solutions or cyanides, inorganic, n.o.s.

EPA NAME: CALCIUM DODECYLBENZENESULFONATE
CAS: 26264-06-2
SYNONYMS: BENZENESULFONIC ACID, DODECYL-, CALCIUM SALT; DODECYLBENZENESULFONIC ACID, CALCIUM SALT

CLEAN WATER ACT: Section 311 Hazardous Substances/RQ (same as CERCLA).
EPCRA Section 304 Reportable Quantity (RQ): CERCLA, 1,000 lbs. (454 kgs.).

EPA NAME: CALCIUM HYPOCHLORITE

CAS: 7778-54-3

SYNONYMS: CALCIUM CHLOROHYDROCHLORITE; HYPOCHLOROUS ACID, CALCIUM SALT

CLEAN WATER ACT: Section 311 Hazardous Substances/RQ (same as CERCLA).

EPCRA Section 304 Reportable Quantity (RQ): CERCLA, 10 lbs. (4.54 kgs.).

EPA NAME: CAMPHECHLOR

[see TOXAPHENE]

CAS: 8001-35-2

EPA NAME: CAMPHENE, OCTACHLORO-

[see TOXAPHENE]

CAS: 8001-35-2

EPA NAME: CANTHARIDIN

CAS: 56-25-7

SYNONYMS: CANTHARIDINE; 2,3-DIMETHYL-7-OXABICYCLO(2,2,1)HEPTANE-2,3-DICARBOXYLIC ANHYDRIDE

EPCRA Section 302 Extremely Hazardous Substances: TPQ = 100/10,000 lbs. (454/4,540 kgs.).

EPCRA Section 304 Reportable Quantity (RQ): EHS, 100 lbs. (45.4 kgs.).

EPA NAME: CAPROLACTUM

CAS: 105-60-2

SYNONYMS: 2H-AZEPIN-2-ONE,HEXAHYDRO; ε-CAPROLACTAM; σ-CAPROLACTAM; 1,6-HEXANOLACTAM; HEXAHYDRO-2H-AZEPINE-2-ONE

CLEAN AIR ACT: Hazardous Air Pollutants (Title I, Part A, Section 112).

EPCRA Section 304 Reportable Quantity (RQ): CERCLA, 5,000 lbs. (2,270 kgs.).

EPA NAME: CAPTAN

CAS: 133-06-2

SYNONYMS: 4-CYCLOHEXENE-1,2-DICARBOXIMIDE,N-[(TRICHLOROMETHYL)MERCAPTO; H-ISOINDOLE-1,3(2H)-DIONE,3a,4,7,7a-TETRAHYDRO-2-[(TRICHLOROMETHYL)-THIO]-

CLEAN AIR ACT: Hazardous Air Pollutants (Title I, Part A, Section 112).

CLEAN WATER ACT: Section 311 Hazardous Substances/RQ (same as CERCLA); Section 313 Priority Chemicals.

EPCRA Section 304 Reportable Quantity (RQ): CERCLA, 10 lbs. (4.54 kgs.).

EPCRA Section 313 Form R de minimus concentration reporting level: 1.0%.

CALIFORNIA'S PROPOSITION 65: Carcinogen.

EPA NAME: CARBACHOL CHLORIDE
CAS: 51-83-2
SYNONYMS: 2-((AMINOCARBONYL)OXY)-N,N,N-TRIMETHYLETHANAMINIUM CHLORIDE CARBAMOYLCHOLINE CHLORIDE

EPCRA Section 302 Extremely Hazardous Substances: TPQ = 500/10,000 lbs. (227/4,540 kgs.).
EPCRA Section 304 Reportable Quantity (RQ): EHS, 500 lbs. (227 kgs.).

EPA NAME: CARBAMIC ACID, DIETHYLTHIO-, S-(P-CHLOROBENZYL) ESTER
[see THIOBENCARB]
CAS: 28249-77-6

EPA NAME: CARBAMIC ACID, ETHYL ESTER
[see URETHANE]
CAS: 51-79-6

EPA NAME: CARBAMIC ACID, METHYL-, O-(((2,4-DIMETHYL-1,3-DITHIOLAN-2-YL)METHYLENE)AMINO)-
[see TRIPATE]
CAS: 26419-73-8

EPA NAME: CARBAMODITHIOIC ACID, 1,2-ETHANEDIYLBIS-, MANGANESE SALT
[see MANEB]
CAS: 12427-38-2

EPA NAME: CARBAMODITHIOIC ACID, 1,2-ETHANEDIYLBIS-, ZINC COMPLEX
[see ZINEB]
CAS: 12122-67-7

EPA NAME: CARBAMODITHIOIC ACID, BIS(1-METHYLETHYL)-, S-(2,3-DICHLORO-2-PROPENYL) ESTER
[see DIALLATE]
CAS 2303-16-4

EPA NAME: CARBARYL
CAS: 63-25-2
SYNONYMS: CARBARYL (ISO); CARBARYL (SEVIN); α-NAPHTHALENOL METHYLCARBAMATE; 1-NAPHTHOL N-METHYLCARBAMATE; 1-NAPHTHYL N-METHYLCARBAMATE; SEVIN

CLEAN AIR ACT: Hazardous Air Pollutants (Title I, Part A, Section 112).
CLEAN WATER ACT: Section 311 Hazardous Substances/RQ (same as CERCLA); Section 313 Priority Chemicals.
RCRA Land Ban Waste.
RCRA Universal Treatment Standards: Wastewater (mg/L), 0.006; Nonwastewater (mg/kg), 0.14.

EPCRA Section 304 Reportable Quantity (RQ): CERCLA, 100 lbs. (45.4 kgs.).
EPCRA Section 313 Form R de minimus concentration reporting level: 1.0%.
MARINE POLLUTANT (49CFR, Subchapter 172.101, Appendix B).

EPA NAME: CARBENZADIM
CAS: 10605-21-7
RCRA Land Ban Waste.
RCRA Universal Treatment Standards: Wastewater (mg/L), 0.056; Nonwastewater (mg/kg), 1.4.
EPCRA Section 304 Reportable Quantity (RQ): CERCLA, 1 lb. (0.454 kg.).

EPA NAME: CARBOFURAN
CAS: 1563-66-2
SYNONYMS: 7-BENZOFURANOL, 2,3-DIHYDRO-2,2-DIMETHYL-, METHYLCARBAMATE; 2,3-DIHYDRO-2,2-DIMETHYL-7-BENZOFURANOLMETHYLCARBAMATE; FURADAN

CLEAN WATER ACT: Section 311 Hazardous Substances/RQ (same as CERCLA).
RCRA Land Ban Waste.
RCRA Universal Treatment Standards: Wastewater (mg/L), 0.006; Nonwastewater (mg/kg), 0.14.
SAFE DRINKING WATER ACT: MCL, 0.04 mg/L; MCLG, 0.04 mg/L; Regulated chemical (47 FR 9352).
EPCRA Section 302 Extremely Hazardous Substances: TPQ = 10/10,000 lbs. (4.54/4,540 kgs.).
EPCRA Section 304 Reportable Quantity (RQ): EHS/CERCLA, 10 lbs. (4.54 kgs.).
EPCRA Section 313 Form R de minimus concentration reporting level: 1.0%.
MARINE POLLUTANT (49CFR, Subchapter 172.101, Appendix B).

EPA NAME: CARBOFURAN PHENOL
CAS: 1563-38-8
RCRA Land Ban Waste.
RCRA Universal Treatment Standards: Wastewater (mg/L), 0.056; Nonwastewater (mg/L), 1.4.
EPCRA Section 304 Reportable Quantity (RQ): CERCLA, 1 lb. (0.454 kg.).

EPA NAME: CARBON DISULFIDE
CAS: 75-15-0
SYNONYMS: CARBON BISULFIDE

CLEAN AIR ACT: Hazardous Air Pollutants (Title I, Part A, Section 112); Accidental Release Prevention/Flammable substances, (Section 112[r], Table 3), TQ = 20,000 lbs. (9080 kgs.).
CLEAN WATER ACT: Section 311 Hazardous Substances/RQ (same as CERCLA); Section 313 Priority Chemicals.

EPA HAZARDOUS WASTE NUMBER (RCRA No.): P022.
RCRA Section 261 Hazardous Constituents.
RCRA Land Ban Waste.
RCRA Universal Treatment Standards: Wastewater (mg/L), 3.8; Non-wastewater (mg/L), 4.8 TCLP.
RCRA Ground Water Monitoring List. Suggested test method(s) (PQL µg/L): 8240(5).
EPCRA Section 302 Extremely Hazardous Substances: TPQ = 10,000 lbs. (4,540 kgs.).
EPCRA Section 304 Reportable Quantity (RQ): EHS/CERCLA, 100 lbs. (45.4 kgs.).
EPCRA Section 313 Form R de minimus concentration reporting level: 1.0%.
MARINE POLLUTANT (49CFR, Subchapter 172.101, Appendix B) as carbon bisulphide.
CALIFORNIA'S PROPOSITION 65: Reproductive toxin (male, female).

EPA NAME: CARBONIC DIFLUORIDE
CAS: 353-50-4
SYNONYMS: CARBONYL FLUORIDE; CARBON OXYFLUORIDE; CARBONYL DIFLUORIDE

EPA HAZARDOUS WASTE NUMBER (RCRA No.): U033.
RCRA Section 261 Hazardous Constituents.
EPCRA Section 304 Reportable Quantity (RQ): CERCLA, 1,000 lbs. (454 kgs.).

EPA NAME: CARBONIC DICHLORIDE
[see PHOSGENE]
CAS: 75-44-5

EPA NAME: CARBONOCHLORIDIC ACID, METHYL ESTER
[see METHYL CHLOROCARBONATE]
CAS: 79-22-1

EPA NAME: CARBONOCHLORIDIC ACID, 1-METHYLETHYL ESTER
[see ISOPROPYL CHLOROFORMATE]
CAS: 108-23-6

EPA NAME: CARBONOCHLORIDIC ACID, PROPYL ESTER
[see PROPYL CHLOROFORMATE]
CAS: 109-61-5

EPA NAME: CARBON OXIDE SULFIDE (COS)
[see CARBONYL SULFIDE]
CAS: 463-58-1

EPA NAME: CARBON TETRACHLORIDE
CAS: 56-23-5
SYNONYMS: METHANE, TETRACHLORO-; TETRACHLOROMETHANE

CLEAN AIR ACT: Hazardous Air Pollutants (Title I, Part A, Section 112); Stratospheric ozone protection (Title VI, Subpart A, Appendix A), Class I, Ozone Depletion Potential = 1.1.
CLEAN WATER ACT: Section 311 Hazardous Substances/RQ (same as CERCLA); Section 307 Priority Pollutants; Section 313 Priority Chemicals; Toxic Pollutant (Section 401.15).
EPA HAZARDOUS WASTE NUMBER (RCRA No.): U211, D019.
RCRA Toxicity Characteristic (Section 261.24), Maximum Concentration of Contaminants, regulatory level, 0.5 mg/L.
RCRA Section 261 Hazardous Constituents.
RCRA Land Ban Waste.
RCRA Universal Treatment Standards: Wastewater (mg/L), 0.057; Non-wastewater (mg/kg), 6.0.
RCRA Maximum Concentration Limit for Ground Water Protection (Section 264.94): 8010(1); 8240(5).
SAFE DRINKING WATER ACT: MCL, 0.005 mg/L; MCLG, zero; Regulated chemical (47 FR 9352).
EPCRA Section 304 Reportable Quantity (RQ): CERCLA, 10 lbs. (4.54 kgs.).
EPCRA Section 313 Form R de minimus concentration reporting level: 0.1%.
CALIFORNIA'S PROPOSITION 65: Carcinogen.
MARINE POLLUTANT (49CFR, Subchapter 172.101, Appendix B).

EPA NAME: CARBONYL SULFIDE

CAS: 463-58-1
SYNONYMS: CARBON OXIDE SULFIDE (COS); CARBON OXYSULFIDE; OXYCARBON SULFIDE

CLEAN AIR ACT: Hazardous Air Pollutants (Title I, Part A, Section 112); Accidental Release Prevention/Flammable substances (Section 112[r], Table 3), TQ = 10,000 lbs. (4,540 kgs.).
EPCRA Section 304 Reportable Quantity (RQ): CERCLA, 100 lbs. (45.4 kgs.).
EPCRA Section 313 Form R de minimus concentration reporting level: 1.0%.

EPA NAME: CARBOPHENOTHION

CAS: 786-19-6
SYNONYMS: s-(4-CHLOROPHENYLTHIOMETHYL)DIETHYL PHOSPHOROTHIOLOTHIONATE; O,O-DIETHY-S-P-CHLOROPHENYLTHIOMETHYL DITHIOPHOSPHATE

EPCRA Section 302 Extremely Hazardous Substances: TPQ = 500 lbs. (227 kgs.).
EPCRA Section 304 Reportable Quantity (RQ): EHS, 500 lbs. (227 kgs.).
MARINE POLLUTANT (49CFR, Subchapter 172.101, Appendix B), severe pollutant.

EPA NAME: CARBOXIN

CAS: 5234-68-4

SYNONYMS: 5,6-DIHYDRO-2-METHYL-N-PHENYL-1,4-OXATHIIN-3-CARBOXAMIDE; 2,3-DIHYDRO-6-METHYL-1,4-OXATHIIN-5-CARBOXANILIDE; EPA PESTICIDE CHEMICAL CODE 090201; 1,4-OXATHIIN-3-CARBOXANILIDE,5,6-DIHYDRO-2-METHYL-

EPCRA Section 313 Form R de minimus concentration reporting level: 1.0%.

EPA NAME: CATECHOL
CAS: 120-80-9
SYNONYMS: 1,2-BENZENEDIOL; 1,2-DIHYDROXYBENZENE; o-HYDROXYPHENOL; PYROCATECHOL

CLEAN AIR ACT: Hazardous Air Pollutants (Title I, Part A, Section 112).
EPCRA Section 304 Reportable Quantity (RQ): CERCLA, 100 lbs. (45.4 kgs.).
EPCRA Section 313 Form R de minimus concentration reporting level: 1.0%.

EPA NAME: CFC-11
[see TRICHLOROFLUOROMETHANE]
CAS: 75-69-4

EPA NAME: CFC-12
[see DICHLORODIFLUOROMETHANE]
CAS: 75-71-8

EPA NAME: CFC-114
[see DICHLOROTETRAFLUOROETHANE]
CAS 76-14-2

EPA NAME: CFC-115
[see MONOCHLOROPENTAFLUOROETHANE]
CAS: 76-15-3

EPA NAME: CFC-13
[see CHLOROTRIFLUOROMETHANE]
CAS: 75-72-9

EPA NAME: CHINOMETHIONAT
CAS: 2439-01-2
SYNONYMS: CARBONIC ACID, DITHIO-, CYCLIC S,S-(6-METHYL-2,3-QUINOXALINEDIYL)ESTER; EPA PESTICIDE CHEMICAL CODE 054101; 6-METHYL-1,3-DITHIOLO[4,5-B]QUINOXALIN-2-ONE; 6-METHYL-2,3-QUINOXALIN DITHIOCARBONATE

EPCRA Section 313 Form R de minimus concentration reporting level: 1.0%.

EPA NAME: CHLORAMBEN
CAS: 133-90-4

SYNONYMS: 3-AMINO-2,5-DICHLOROBENZOIC ACID; BENZOIC ACID, 3-AMINO-2,5-DICHLORO-

CLEAN AIR ACT: Hazardous Air Pollutants (Title I, Part A, Section 112).
EPCRA Section 304 Reportable Quantity (RQ): CERCLA, 100 lbs. (45.4 kgs.).
EPCRA Section 313 Form R de minimus concentration reporting level: 1.0%.

EPA NAME: CHLORAMBUCIL
CAS: 305-03-3
SYNONYMS: BENZENEBUTANOIC ACID, 4-[BIS(2-CHLOROETHYL)AMINO]-; 4-(BIS(2-CHLOROETHYL)AMINO)BENZENEBUTANOIC ACID; N,N-DI-2-CHLOROETHYL-γ-p-AMINOPHENYLBUTYRIC ACID

EPA HAZARDOUS WASTE NUMBER (RCRA No.): U035.
RCRA Section 261 Hazardous Constituents.
EPCRA Section 304 Reportable Quantity (RQ): CERCLA, 10 lbs. (4.54 kgs.).
CALIFORNIA'S PROPOSITION 65: Carcinogen; reproductive toxin.

EPA NAME: CHLORDANE
CAS: 57-74-9
SYNONYMS: CHLORDAN; 4,7-METHANO-1H-INDENE, 1,2,4,5,6,7,8,8-OCTACHLORO-2,3,3A,4,7,7A-HEXAHYDRO-; 4,7-METHANOINDAN, 1,2,3,4,5,6,7,8,8-OCTACHLORO-2,3,3a,4,7,7a-HEXAHYDRO-; 1,2,4,5,6,7,8,8-OCTACHLORO-2,3,3A,4,7,7A-HEXAHYDRO-4,7-METHANO-1H-INDENE; TRICHLOR

CLEAN AIR ACT: Hazardous Air Pollutants (Title I, Part A, Section 112).
CLEAN WATER ACT: Section 311 Hazardous Substances/RQ (same as CERCLA); Section 307 Priority Pollutants; Section 313 Priority Chemicals; Toxic Pollutant (Section 401.15) as technical mixture and metabolites.
EPA HAZARDOUS WASTE NUMBER (RCRA No.): U036; D020.
RCRA Toxicity Characteristic (Section 261.24), Maximum Concentration of Contaminants, regulatory level, 0.03 mg/L.
RCRA Section 261 Hazardous Constituents.
RCRA Land Ban Waste.
RCRA Universal Treatment Standards: Wastewater (mg/L), (α and γ isomers) 0.0033; Nonwastewater (mg/kg), 0.26.
RCRA Ground Water Monitoring List. Suggested test method(s) (PQL μg/L): 8080(0.1); 8250(10).
SAFE DRINKING WATER ACT: MCL, 0.002 mg/L; MCLG, zero; Regulated chemical (47 FR 9352).
EPCRA Section 302 Extremely Hazardous Substances: TPQ = 1,000 lbs. (454 kgs.).
EPCRA Section 304 Reportable Quantity (RQ): EHS/CERCLA, 1 lb. (0.454 kg.).

EPCRA Section 313 Form R de minimus concentration reporting level: 1.0%.

MARINE POLLUTANT (49CFR, Subchapter 172.101, Appendix B), severe pollutant.

CALIFORNIA'S PROPOSITION 65: Carcinogen.

EPA NAME: CHLORDANE (TECHNICAL MIXTURE AND METABOLITES)

[see CHLORDANE]

CAS: 57-74-9

EPA NAME: CHLORENDIC ACID

CAS: 115-28-6

SYNONYMS: BICYCLO(2.2.1)HEPT-5-ENE-2,3-DICARBOXYLIC ACID, 1,4,5,6,7,7-HEXACHLORO-; 1,4,5,6,7,7-HEXACHLORO-5-NORBORNENE-2,3-DICARBOXYLIC ACID;

EPCRA Section 313 Form R de minimus concentration reporting level: 0.1%.

CALIFORNIA'S PROPOSITION 65: Carcinogen.

EPA NAME: CHLORFENVINFOS

CAS: 470-90-6

SYNONYMS: BENZYL ALCOHOL,2,4-DICHLORO-α-(CHLOROMETHYLENE)-, DIETHYL PHOSPHATE; PHOSPHORIC ACID, 2-CHLORO-1-(2,4-DICHLOROPHENYL)ETHENYL DIETHYL ESTER

EPCRA Section 302 Extremely Hazardous Substances: TPQ = 500 lbs. (227 kgs.).

EPCRA Section 304 Reportable Quantity (RQ): EHS, 500 lbs. (227 kgs.).

MARINE POLLUTANT (49CFR, Subchapter 172.101, Appendix B).

EPA NAME: CHLORIMURON ETHYL

CAS: 90982-32-4

SYNONYMS: ETHYL-2-[[[(4-CHLORO-6-METHOXYPYRIMIDIN-2-YL)-CARBONYL]-AMINO]SULFONYL]BENZOATE

EPCRA Section 313 Form R de minimus concentration reporting level: 1.0%.

EPA NAME: CHLORINATED BENZENES

CLEAN WATER ACT: Toxic Pollutant (Section 401.15), other than dichlorobenzenes.

RCRA Section 261 Hazardous Constituents, waste number not listed.

EPA NAME: CHLORINATED ETHANES

CLEAN WATER ACT: Toxic Pollutant (Section 401.15) including 1,2 dichloroethane (107-06-2), 1,1,1-trichloroethane (71-55-6), and hexachloroethane (67-72-1).

RCRA Section 261 Hazardous Constituents, waste number not listed.

EPA NAME: CHLORINATED NAPHTHALENES

CLEAN WATER ACT: Toxic Pollutant (Section 401.15).

RCRA Section 261 Hazardous Constituents, waste number not listed.

EPA NAME: CHLORINE

CAS: 7782-50-5

SYNONYMS: CHLORINE MOLECULAR (C12); DIATOMIC CHLORINE; MOLECULAR CHLORINE

CLEAN AIR ACT: Hazardous Air Pollutants (Title I, Part A, Section 112); Accidental Release Prevention/Flammable substances (Section 112[r], Table 3), TQ = 2,500 lbs. (1135 kgs.).

CLEAN WATER ACT: Section 311 Hazardous Substances/RQ (same as CERCLA); Section 313 Priority Chemicals.

SAFE DRINKING WATER ACT: SMCL, 250 mg/L; Priority List (55 FR 1470).

EPCRA Section 302 Extremely Hazardous Substances: TPQ = 100 lbs. (45.4 kgs.).

EPCRA Section 304 Reportable Quantity (RQ): EHS/CERCLA, 10 lbs. (4.54 kgs.).

EPCRA Section 313 Form R de minimus concentration reporting level: 1.0%.

MARINE POLLUTANT (49CFR, Subchapter 172.101, Appendix B).

EPA NAME: CHLORINE DIOXIDE

CAS: 10049-04-4

SYNONYMS: CHLORINE OXIDE (ClO2); CHLORINE(IV) OXIDE

CLEAN AIR ACT: Accidental Release Prevention/Flammable substances (Section 112[r], Table 3), TQ = 1,000 lbs. (454 kgs.).

SAFE DRINKING WATER ACT: Priority List (55 FR 1470).

EPCRA Section 313 Form R de minimus concentration reporting level: 1.0%.

EPA NAME: CHLORINE MONOXIDE

[see CHLORINE OXIDE]

CAS: 7791-21-1

EPA NAME: CHLORINE OXIDE

CAS: 7791-21-1

SYNONYMS: CHLORINE MONOXIDE; DICHLORINE OXIDE (Cl2O)

CLEAN AIR ACT: Accidental Release Prevention/Flammable substances (Section 112[r], Table 3), TQ = 10,000 lbs. (4,540 kgs.).

EPA NAME: CHLORINE OXIDE (ClO_2)

[see CHLORINE DIOXIDE]

CAS: 10049-04-4

EPA NAME: CHLORMEPHOS

CAS: 24934-91-6

SYNONYMS: S-(CHLOROMETHYL)-O,O-DIETHYL PHOSPHORODITHIOATE

EPCRA Section 302 Extremely Hazardous Substances: TPQ = 500 lbs. (227 kgs.).
EPCRA Section 304 Reportable Quantity (RQ): EHS, 500 lbs. (227 kgs.).
MARINE POLLUTANT (49CFR, Subchapter 172.101, Appendix B).

EPA NAME: CHLORMEQUAT CHLORIDE
CAS: 999-81-5
SYNONYMS: 2-CHLORAETHYLTRIMETHYLAMMONIUM CHLORIDE; 2-CHLORO-N,N,N-TRIMETHYLAMMONIUM CHLORIDE

EPCRA Section 302 Extremely Hazardous Substances: TPQ = 100/10,000 lbs. (45.4/4,540 kgs.).
EPCRA Section 304 Reportable Quantity (RQ): EHS, 100 lbs. (45.4 kgs.).

EPA NAME: CHLORNAPHAZINE
CAS: 494-03-1
SYNONYMS: 2-BIS(2-CHLOROETHYL)AMINONAPHTHALENE; N,N-BIS(2-CHLOROETHYL)-2-NAPHTHYLAMINE; BIS(2-CHLOROETHYL)-β-NAPHTHYLAMINE; 2-NAPHTHALENAMINE, N,N-BIS(2-CHLOROETHYL)-

EPA HAZARDOUS WASTE NUMBER (RCRA No.): U026.
RCRA Section 261 Hazardous Constituents.
EPCRA Section 304 Reportable Quantity (RQ): CERCLA, 100 lbs. (45.4 kgs.).
CALIFORNIA'S PROPOSITION 65: Carcinogen.

EPA NAME: CHLOROACETALDEHYDE
CAS: 107-20-0
SYNONYMS: ACETALDEHYDE, CHLORO-; 2-CHLORO-1-ETHANAL

EPA HAZARDOUS WASTE NUMBER (RCRA No.): P023.
RCRA Section 261 Hazardous Constituents.
EPCRA Section 304 Reportable Quantity (RQ): CERCLA, 1,000 lbs. (454 kgs.).

EPA NAME: CHLOROACETIC ACID
CAS: 79-11-8
SYNONYMS: ACETIC ACID, CHLORO-

CLEAN AIR ACT: Hazardous Air Pollutants (Title I, Part A, Section 112).
EPCRA Section 302 Extremely Hazardous Substances: TPQ = 100/10,000 lbs. (45.4/4,540 kgs.).
EPCRA Section 304 Reportable Quantity (RQ): EHS/CERCLA, 100 lbs. (45.4 kgs.).
EPCRA Section 313 Form R de minimus concentration reporting level: 1.0%.

EPA NAME: 2-CHLOROACETOPHENONE CHLOROALKYL ESTERS
CAS: 532-27-4

SYNONYMS: α-CHLOROACETOPHENONE; 2-CHLORO-1-PHENYLETHANONE; 2-CHLOROACETOPHENONE; CHLOROMETHYL PHENYL KETONE

CLEAN AIR ACT: Hazardous Air Pollutants (Title I, Part A, Section 112).
EPCRA Section 304 Reportable Quantity (RQ): CERCLA, 100 lbs. (45.4 kgs.).
EPCRA Section 313 Form R de minimus concentration reporting level: 1.0%.

EPA NAME: 1-(3-CHLORALLYL)-3,5,7-TRIAZA-1-AZONIA-ADAMANTANE CHLORIDE

CAS: 4080-31-3
SYNONYMS: 1-(3-CHLORO-2-PROPENYL)-3,5,7-TRIAZA-1-AZONIATRICYCLO(3.3.1)DECANE CHLORIDE; EPA PESTICIDE CHEMICAL CODE 017901; 3,5,7-TRIAZA-1-AZONIATRICYCLODECANE-1-(3-CHLORO-2-PROPENYL)-,CHLORIDE

EPCRA Section 313 Form R de minimus concentration reporting level: 1.0%.

EPA NAME: p-CHLOROANILINE

[see also CHLORINATED BENZENES]
CAS: 106-47-8
SYNONYMS: 1-AMINO-4-CHLOROBENZENE; BENZENEAMINE, 4-CHLORO-; para-CHLOROANILINE

EPA HAZARDOUS WASTE NUMBER (RCRA No.): P024.
RCRA Section 261 Hazardous Constituents.
RCRA Land Ban Waste.
RCRA Universal Treatment Standards: Wastewater (mg/L), 0.46; Nonwastewater (mg/kg), 16.
RCRA Ground Water Monitoring List. Suggested test method(s) (PQL μg/L): 8270(20).
EPCRA Section 304 Reportable Quantity (RQ): CERCLA, 1,000 lbs. (454 kgs.).
EPCRA Section 313 Form R de minimus concentration reporting level: 1.0%.
CALIFORNIA'S PROPOSITION 65: Carcinogen.

EPA NAME: CHLOROBENZENE

[see also CHLORINATED BENZENES]
CAS: 108-90-7
SYNONYMS: BENZENE, CHLORO-; CHLOROBENZENE (MONO); MONOCHLORBENZENE

CLEAN AIR ACT: Hazardous Air Pollutants (Title I, Part A, Section 112).
CLEAN WATER ACT: Section 311 Hazardous Substances/RQ (same as CERCLA); Section 307 Priority Pollutants; Section 313 Priority Chemicals.
EPA HAZARDOUS WASTE NUMBER (RCRA No.): U037; D021.
RCRA Toxicity Characteristic (Section 261.24), Maximum Concentration of Contaminants, regulatory level, 100 mg/L.

RCRA Section 261 Hazardous Constituents.
RCRA Land Ban Waste.
RCRA Universal Treatment Standards: Wastewater (mg/L), 0.057; Non-wastewater (mg/kg), 6.0.
RCRA Ground Water Monitoring List. Suggested test method(s) (PQL μg/L): 8010(2); 8020(2); 8240(5).
SAFE DRINKING WATER ACT: MCL, 0.1 mg/L; MCLG, 0.1 mg/L; Regulated chemical (47 FR 9352).
EPCRA Section 304 Reportable Quantity (RQ): CERCLA, 100 lbs. (45.4 kgs.).
EPCRA Section 313 Form R de minimus concentration reporting level: 1.0%.

EPA NAME: CHLOROBENZILATE
CAS: 510-15-6
SYNONYMS: BENZENEACETIC ACID, 4-CHLORO-α-(4-CHLOROPHENYL)-α-HYDROXY-, ETHYL ESTER; BENZILIC ACID, 4,4′-DICHLORO-, ETHYL ESTER; ETHYL 4,4′-DICHLOROBENZILATE

CLEAN AIR ACT: Hazardous Air Pollutants (Title I, Part A, Section 112).
EPA HAZARDOUS WASTE NUMBER (RCRA No.): U038.
RCRA Section 261 Hazardous Constituents.
RCRA Land Ban Waste.
RCRA Universal Treatment Standards: Wastewater (mg/L), 0.10; Non-wastewater (mg/kg), N/A.
RCRA Ground Water Monitoring List. Suggested test method(s) (PQL μg/L): 8270(10).
EPCRA Section 304 Reportable Quantity (RQ): CERCLA, 10 lbs. (4.54 kgs.).
EPCRA Section 313 Form R de minimus concentration reporting level: 1.0%.
MARINE POLLUTANT (49CFR, Subchapter 172.101, Appendix B).
CALIFORNIA'S PROPOSITION 65: Carcinogen.

EPA NAME: 2-(4-((6-CHLORO-2-BENZOXAZOLYLEN)OXY)PHENOXY)PROPANOIC ACID, ETHYL ESTER
[see FENOXAPROP ETHYL]
CAS: 66441-23-4

EPA NAME: 2-CHLORO-N-(2-CHLOROETHYL)-N-METHYLETHANAMINE
[see NITROGEN MUSTARD]
CAS: 51-75-2

EPA NAME: p-CHLORO-m-CRESOL
CAS: 59-50-7
SYNONYMS: 4-CHLORO-m-CRESOL; para-CHLORO-meta-CRESOL; CHLOROCRESOL; PARACHLOROMETA CRESOL

CLEAN WATER ACT: Section 307 Priority Pollutants; Section 313 Priority Chemicals.

EPA HAZARDOUS WASTE NUMBER (RCRA No.): U039.
RCRA Section 261 Hazardous Constituents.
RCRA Land Ban Waste.
RCRA Universal Treatment Standards: Wastewater (mg/L), 0.018; Nonwastewater (mg/kg), 14.
RCRA Ground Water Monitoring List. Suggested test method(s) (PQL μg/L): 8040(5); 8270(20).
EPCRA Section 304 Reportable Quantity (RQ): CERCLA, 5,000 lbs. (2270 kgs.).

EPA NAME: CHLORODIBROMOMETHANE
CAS: 124-48-1
SYNONYMS: DIBROMOCHLOROMETHANE

CLEAN WATER ACT: Section 307 Priority Pollutants.
SAFE DRINKING WATER ACT: Priority List (55 FR 1470) as dibromochloromethane.
RCRA Land Ban Waste.
RCRA Universal Treatment Standards: Wastewater (mg/L), 0.057; Nonwastewater (mg/kg), 15.
RCRA Ground Water Monitoring List. Suggested test method(s) (PQL μg/L): 8010(1); 8240(5).
EPCRA Section 304 Reportable Quantity (RQ): CERCLA, 100 lbs. (45.4 kgs.).
CALIFORNIA'S PROPOSITION 65: Carcinogen.

EPA NAME: 1-CHLORO-1,1-DIFLUOROETHANE
[see also CHLORINATED ETHANES]
CAS: 75-68-3; 65762-25-6
SYNONYMS: CHLORODIFLUOROETHANE; DIFLUOROMONOCHLOROETHANE; DIFLUORO-1-CHLOROETHANE; ETHANE, 1-CHLORO-1,1-DIFLUORO-; FREON 142b; HCFC-142b

CLEAN AIR ACT: Stratospheric ozone protection (Title VI, Subpart A, Appendix B), Class II, Ozone Depletion Potential = 0.06.
EPCRA Section 313 Form R de minimus concentration reporting level: 1.0%.

EPA NAME: CHLORODIFLUOROMETHANE
CAS: 75-45-6; 73666-77-0
SYNONYMS: DIFLUOROCHLOROMETHANE; DIFLUOROMONOCHLOROMETHANE; FREON 22; HCFC-22; MONOCHLORODIFLUOROMETHANE

CLEAN AIR ACT: Stratospheric ozone protection (Title VI, Subpart A, Appendix B), Class II, Ozone Depletion Potential = 0.05.
EPCRA Section 313 Form R de minimus concentration reporting level: 1.0%.

EPA NAME: 5-CHLORO-3-(1,1-DIMETHYLETHYL)-6-METHYL-2,4(1H,3H)-PYRIMIDINEDIONE
[see TERBACIL]

EPA NAME: CHLOROETHANE
[see also CHLORINATED ETHANES]
CAS: 75-00-3
SYNONYMS: ETHANE, CHLORO-; ETHYL CHLORIDE; MONO-CHLOROETHANE

CLEAN AIR ACT: Hazardous Air Pollutants (Title I, Part A, Section 112); Accidental Release Prevention/Flammable substances (Section 112[r], Table 3), TQ = 10,000 lbs. (4540 kgs.).
CLEAN WATER ACT: Section 307 Priority Pollutants; Section 313 Priority Chemicals; Section 307 Toxic Pollutants as chlorinated ethanes.
RCRA Land Ban Waste.
RCRA Universal Treatment Standards: Wastewater (mg/L), 0.27; Non-wastewater (mg/kg), 6.0.
RCRA Ground Water Monitoring List. Suggested test method(s) (PQL µg/L): 8010(5); 8240(10).
SAFE DRINKING WATER ACT: Priority List (55 FR 1470).
EPCRA Section 304 Reportable Quantity (RQ): CERCLA, 100 lbs. (45.4 kgs.).
EPCRA Section 313 Form R de minimus concentration reporting level: 1.0%.
CALIFORNIA'S PROPOSITION 65: Carcinogen.

EPA NAME: CHLOROETHANOL
CAS: 107-07-3
SYNONYMS: 2-CHLOROETHANOL; ETHYLENE CHLORHY-DRIN; ETHANOL, 2-CHLORO-; ETHYLENE CHLOROHYD-RINE

EPCRA Section 302 Extremely Hazardous Substances: TPQ = 500 lbs. (227 kgs.).
EPCRA Section 304 Reportable Quantity (RQ): EHS, 500 lbs. (227 kgs.).

EPA NAME: CHLOROETHYL CHLOROFORMATE
CAS: 627-11-2
SYNONYMS: CARBONOCHLORIDIC ACID-2-CHLOROETHYL ESTER; 2-CHLOROETHYL CHLOROCARBONATE

EPCRA Section 302 Extremely Hazardous Substances: TPQ = 1,000 lbs. (454 kgs.).
EPCRA Section 304 Reportable Quantity (RQ): EHS, 1000 lbs. (454 kgs.).

EPA NAME: 6-CHLORO-N-ETHYL-N′-(1-METHYLETHYL)-1,3,5-TRIAZINE-2,4-DIAMINE
[see ATRAZINE]
CAS: 1912-24-9

EPA NAME: 2-CHLOROETHYL VINYL ETHER
[see also HALOETHERS]
CAS: 110-75-8
SYNONYMS: (2-CHLOROETHOXY)ETHENE; ETHENE, (2-CHLOROETHOXY)-; VINYL-2-CHLOROETHYL ETHER

CLEAN WATER ACT: Section 307 Priority Pollutants as 2-chloroethyl vinyl ether (mixed).
EPA HAZARDOUS WASTE NUMBER (RCRA No.): U042.
RCRA Section 261 Hazardous Constituents.
RCRA Land Ban Waste.
RCRA Universal Treatment Standards: Wastewater (mg/L), 0.062; Nonwastewater (mg/kg), N/A.
EPCRA Section 304 Reportable Quantity (RQ): CERCLA, 1,000 lbs. (454 kgs.).

EPA NAME: CHLOROFORM

CAS: 67-66-3
SYNONYMS: FREON 20; METHANE, TRICHLORO-; TRICHLOROMETHANE

CLEAN AIR ACT: Hazardous Air Pollutants (Title I, Part A, Section 112); Accidental Release Prevention/Flammable substances (Section 112[r], Table 3), TQ = 20,000 lbs. (9080 kgs.).
CLEAN WATER ACT: Section 311 Hazardous Substances/RQ (same as CERCLA); Section 307 Priority Pollutants; Section 313 Priority Chemicals; Toxic Pollutant (Section 401.15).
EPA HAZARDOUS WASTE NUMBER (RCRA No.): U044; D022.
RCRA Toxicity Characteristic (Section 261.24), Maximum Concentration of Contaminants, regulatory level, 6.0 mg/L.
RCRA Section 261 Hazardous Constituents.
RCRA Land Ban Waste.
RCRA Universal Treatment Standards: Wastewater (mg/L), 0.046; Nonwastewater (mg/kg), 6.0.
RCRA Ground Water Monitoring List. Suggested test method(s) (PQL µg/L): 8010(0.5); 8240(5).
SAFE DRINKING WATER ACT: Priority List (55 FR 1470).
EPCRA Section 302 Extremely Hazardous Substances: TPQ = 10,000 lbs. (4,540 kgs.).
EPCRA Section 304 Reportable Quantity (RQ): EHS/CERCLA, 10 lbs. (4.54 kgs.).
EPCRA Section 313 Form R de minimus concentration reporting level: 0.1%.
CALIFORNIA'S PROPOSITION 65: Carcinogen.

EPA NAME: CHLOROMETHANE

CAS: 74-87-3
SYNONYMS: METHANE, CHLORO; MONOCHLOROMETHANE

CLEAN AIR ACT: Hazardous Air Pollutants (Title I, Part A, Section 112); Accidental Release Prevention/Flammable substances (Section 112[r], Table 3), TQ = 10,000 lbs. (4540 kgs.).
CLEAN WATER ACT: Section 307 Priority Pollutants; Section 313 Priority Chemicals.
EPA HAZARDOUS WASTE NUMBER (RCRA No.): U045.
RCRA Section 261 Hazardous Constituents.
RCRA Land Ban Waste.
SAFE DRINKING WATER ACT: Priority List (55 FR 1470).

RCRA Universal Treatment Standards: Wastewater (mg/L), 0.19; Non-wastewater (mg/kg), 30.
RCRA Ground Water Monitoring List. Suggested test method(s) (PQL µg/L): 8010(1); 8240(10).
EPCRA Section 304 Reportable Quantity (RQ): CERCLA, 100 lbs. (45.4 kgs.).
EPCRA Section 313 Form R de minimus concentration reporting level: 1.0%.

EPA NAME: 2-CHLORO-N-(((4-METHOXY-6-METHYL-1,3,5-TRIAZIN-2-YL)AMINO)CARBONYL)BENZENESULFONAMIDE
[see CHLORSULFURON]
CAS: 64902-72-3

EPA NAME: CHLOROMETHYL METHYL ETHER
[see also HALOETHERS]
CAS: 107-30-2
SYNONYMS: DIMETHYLCHLOROETHER; METHANE, CHLOROMETHOXY-; METHYL CHLOROMETHYL ETHER; MONOCHLOROMETHYL METHYL ETHER

CLEAN AIR ACT: Hazardous Air Pollutants (Title I, Part A, Section 112); Accidental Release Prevention/Flammable substances (Section 112[r], Table 3), TQ = 5,000 lbs. (2270 kgs.).
EPA HAZARDOUS WASTE NUMBER (RCRA No.): U046.
RCRA Section 261 Hazardous Constituents.
EPCRA Section 302 Extremely Hazardous Substances: TPQ = 100 lbs. (45.4 kgs.).
EPCRA Section 304 Reportable Quantity (RQ): EHS/CERCLA, 10 lbs. (4.54 kgs.).
EPCRA Section 313 Form R de minimus concentration reporting level: 0.1%.
CALIFORNIA'S PROPOSITION 65: Carcinogen.

EPA NAME: CHLOROMETHYL ETHER
[see BIS(CHLOROMETHYL)ETHER]
CAS: 542-88-1

EPA NAME: 3-CHLORO-2-METHYL-1-PROPENE
CAS: 563-47-3
SYNONYMS: 3-CHLORO-2-METHYLPROP-1-ENE; 3-CHLORO-2-METHYLPROPENE; METHALLYL CHLORIDE; 1-PROPENE, 3-CHLORO-2-METHYL-

EPCRA Section 313 Form R de minimus concentration reporting level: 0.1%.
CALIFORNIA'S PROPOSITION 65: Carcinogen.

EPA NAME: 4-CHLORO-5-(METHYLAMINO)-2-[3-(TRIFLUOROMETHYL)PHENYL]-3(2H)-PYRIDAZINONE
[see NORFLURAZON]

CAS: 27314-13-2

EPA NAME: 4-CHLORO-α-(1-METHYLETHYL)BENZENE-ACETIC ACID CYANO(3-PHENOXYPHENYL)METHYL ESTER

[see FENVALERATE]

CAS: 51630-58-1

EPA NAME: 2-CHLORO-N-(1-METHYLETHYL)-N-PHENYL-ACETAMIDE

[see PROPACHLOR]

CAS: 1918-16-7

EPA NAME: (4-CHLORO-2-METHYLPHENOXY) ACETATE SODIUM SALT

[see METHOXONE, SODIUM SALT]

CAS: 3653-48-3

EPA NAME: (4-CHLORO-2-METHYLPHENOXY) ACETIC ACID

[see METHOXONE]

CAS: 94-74-6

EPA NAME: 2-CHLORONAPHTHALENE

[see also CHLORONATED NAPHTHALENE]

CAS: 91-58-7

SYNONYMS: β-CHLORONAPHTHALENE; NAPHTHALENE, 2-CHLORO-

CLEAN WATER ACT: Section 307 Priority Pollutants.

EPA HAZARDOUS WASTE NUMBER (RCRA No.): U047.

RCRA Section 261 Hazardous Constituents.

RCRA Land Ban Waste.

RCRA Universal Treatment Standards: Wastewater (mg/L), 0.055; Non-wastewater (mg/kg), 5.6.

RCRA Ground Water Monitoring List. Suggested test method(s) (PQL μg/L): 8120(10); 8270(10).

EPCRA Section 304 Reportable Quantity (RQ): CERCLA, 5,000 lbs. (2270 kgs.).

EPA NAME: CHLOROPHACIONONE

CAS: 3691-35-8

SYNONYMS: 2-(α-p-CHLOROPHENYLACETYL)INDANE-1,3-DIONE; ROZOL

EPCRA Section 302 Extremely Hazardous Substances: TPQ = 100/10,000 lbs. (45.4/4,540 kgs.).

EPCRA Section 304 Reportable Quantity (RQ): EHS, 100 lbs. (45.4 kgs.).

EPA NAME: 2-CHLOROPHENOL

[see also CHLOROPHENOLS]

CAS: 95-57-8

SYNONYMS: PHENOL, 2-CHLORO; PHENOL, o-CHLORO; o-CHLOROPHENOL

CLEAN WATER ACT: Section 307 Priority Pollutants; Section 313 Priority Chemicals; Toxic Pollutant (Section 401.15).
EPCRA Section 304 Reportable Quantity (RQ): CERCLA, 100 lbs. (45.4 kgs.).
EPA HAZARDOUS WASTE NUMBER (RCRA No.): U048.
RCRA Section 261 Hazardous Constituents.
RCRA Land Ban Waste.
RCRA Universal Treatment Standards: Wastewater (mg/L), 0.44; Nonwastewater (mg/kg), 5.7.
RCRA Ground Water Monitoring List. Suggested test method(s) (PQL µg/L): 8040(5); 8270(10).
EPCRA Section 313 Form R de minimus concentration reporting level: 0.1%. Form R Toxic Chemical Category Code: N084.
MARINE POLLUTANT (49CFR, Subchapter 172.101, Appendix B).

EPA NAME: CHLOROPHENOLS
[see also CHLORINATED PHENOLS]
CAS: 25167-80-0
CLEAN WATER ACT: Toxic Pollutant (Section 401.15) includes chlorinated cresols.
RCRA Section 261 Hazardous Constituents, waste number not listed.

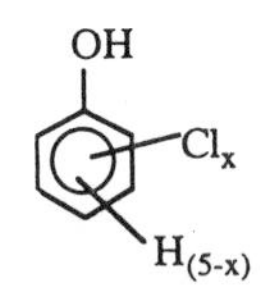

EPCRA Section 313 Form R de minimus concentration reporting level: 0.1%. Form R Toxic Chemical Category Code: N084.
MARINE POLLUTANT (49CFR, Subchapter 172.101, Appendix B).

EPA NAME: (1-(4-CHLOROPHENOXY)-3,3-DIMETHYL-1-(1H-1,2,4-TRIAZOL-1-YL)-2-BUTANONE
[see TRIADIMEFON]
CAS: 43121-43-3

EPA NAME: α-(2-CHLOROPHENYL)-α-4-CHLOROPHENYL-5-PYRIMIDINEMETHANOL
[see FENARIMOL]
CAS: 60168-88-9

EPA NAME: p-CHLOROPHENYL ISOCYANATE
CAS: 104-12-1
SYNONYMS: 1-CHLORO-4-ISOCYANATO-; 4-CHLOROPHENYL ISOCYANATE; ISOCYANIC ACID, p-CHLOROPHENYL ESTER

EPCRA Section 313 Form R de minimus concentration reporting level: 1.0%.

EPA NAME: 4-CHLOROPHENYL PHENYL ETHER
CAS: 7005-72-3
SYNONYMS: BENZENE, 1-CHLORO-4-PHENOXY-

CLEAN WATER ACT: Section 307 Priority Pollutants.

RCRA Ground Water Monitoring List. Suggested test method(s) (PQL µg/L): 8270(10).
EPCRA Section 304 Reportable Quantity (RQ): CERCLA, 5,000 lbs. (2270 kgs.).

EPA NAME: CHLOROPICRIN
CAS: 76-06-2
SYNONYMS: CHLOROPICRINE; METHANE, TRICHLORONITRO-; NITROTRICHLOROMETHANE; TRICHLORONITROMETHANE

SAFE DRINKING WATER ACT: Priority List (55 FR 1470).
EPCRA Section 313 Form R de minimus concentration reporting level: 1.0%.

EPA NAME: CHLOROPRENE
CAS: 126-99-8
SYNONYMS: 1,3-BUTADIENE, 2-CHLORO-; 2-CHLORO-1,3-BUTADIENE; 2-CHLOROBUTA-1,3-DIENE; β-CHLOROPRENE; NEOPRENE

CLEAN AIR ACT: Hazardous Air Pollutants (Title I, Part A, Section 112).
RCRA Section 261 Hazardous Constituents, waste number not listed.
RCRA Land Ban Waste.
RCRA Universal Treatment Standards: Wastewater (mg/L), 0.057; Nonwastewater (mg/kg), 0.28.
RCRA Ground Water Monitoring List. Suggested test method(s) (PQL µg/L): 8010(50); 8240(5).
EPCRA Section 304 Reportable Quantity (RQ): CERCLA, 100 lbs. (45.4 kgs.).
EPCRA Section 313 Form R de minimus concentration reporting level: 1.0%.

EPA NAME: 3-CHLOROPROPIONITRILE
CAS: 542-76-7
SYNONYMS: β-CHLOROPROPIONITRILE; PROPIONITRILE, 3-CHLORO-

EPA HAZARDOUS WASTE NUMBER (RCRA No.): P027.
RCRA Section 261 Hazardous Constituents.
EPCRA Section 302 Extremely Hazardous Substances: TPQ = 1,000 lbs. (454 kgs.).
EPCRA Section 304 Reportable Quantity (RQ): EHS/CERCLA, 1,000 lbs. (454 kgs.).
EPCRA Section 313 Form R de minimus concentration reporting level: 1.0%.

EPA NAME: 2-CHLOROPROPYLENE
CAS: 557-98-2
SYNONYMS: 2-CHLORO-1-PROPENE; 1-PROPENE, 2-CHLORO-; 2-PROPENYL CHLORIDE

CLEAN AIR ACT: Accidental Release Prevention/Flammable substances (Section 112[r], Table 3), TQ = 10,000 lbs. (4540 kgs.).
MARINE POLLUTANT (49CFR, Subchapter 172.101, Appendix B).

EPA NAME: 1-CHLOROPROPYLENE
CAS: 590-21-6
SYNONYMS: CHLOROPRENE; 1-CHLORO-1-PROPENE; 1-PROPENE, 1-CHLORO-; PROPENYL CHLORIDE

CLEAN AIR ACT: Accidental Release Prevention/Flammable substances (Section 112[r], Table 3), TQ = 10,000 lbs. (4540 kgs.).
MARINE POLLUTANT (49CFR, Subchapter 172.101, Appendix B).

EPA NAME: 2-[4-[(6-CHLORO-2-QUINOXALINYL)OXY]PHENOXY]PROPANOIC ACID ETHYL ESTER
[see QUIZALOFOP-ETHYL]
CAS: 76578-14-8

EPA NAME: CHLOROSULFONIC ACID
CAS: 7790-94-5
SYNONYMS: CHLOROSULFURIC ACID; 4-CHLORO-o-TOLUIDINE, HYDROCHLORIDE; CHLOROSULPHONIC ACID; para-CHLORO-ortho-TOLUIDINE HYDROCHLORIDE

CLEAN WATER ACT: Section 311 Hazardous Substances/RQ (same as CERCLA).
EPCRA Section 304 Reportable Quantity (RQ): CERCLA, 1,000 lbs. (454 kgs.).

EPA NAME: CHLOROTETRAFLUOROETHANE
CAS: 63938-10-3
SYNONYMS: 1-CHLORO-1,1,2,2-TETRAFLUOROETHANE; MONOCHLOROTETRAFLUOROETHANE

EPCRA Section 313 Form R de minimus concentration reporting level: 1.0%.

EPA NAME: 1-CHLORO-1,1,2,2-TETRAFLUOROETHANE
[see also CHLORINATED ETHANES]
CAS: 354-25-6
SYNONYMS: ETHANE, 2-CHLORO-1,1,2,2-TETRAFLUORO-; FREON 124a; HCFC-124a; 1,1,2,2-TETRAFLUORO-2-CHLOROETHANE

EPCRA Section 313 Form R de minimus concentration reporting level: 1.0%.

EPA NAME: 2-CHLORO-1,1,1,2-TETRAFLUOROETHANE
[see also CHLOROETHANES]
CAS: 2837-89-0
SYNONYMS: 1-CHLORO-1,2,2,2-TETRAFLUOROETHANE; ETHANE, 2-CHLORO-1,1,1,2-TETRAFLUORO-; FREON 124; HCFC-124; PROPELLANT 124; R 124; REFRIGERANT 124; 1,1,1,2-TETRAFLUORO-2-CHLOROETHANE

CLEAN AIR ACT: Stratospheric ozone protection (Title VI, Subpart A, Appendix B), Class II, Ozone Depletion Potential = 0.02.

EPCRA Section 313 Form R de minimus concentration reporting level: 1.0%.

EPA NAME: CHLOROTHALONIL

[see also CHLORINATED BENZENES]

CAS: 1897-45-6

SYNONYMS: 1,3-BENZENEDICARBONITRILE,2,4,6,6-TETRACHLORO-; 1,3-DICYANOTETRACHLOROBENZENE; 2,4,5,6-TETRACHLORO-1,3-BENZENEDICARBONITRILE

EPCRA Section 313 Form R de minimus concentration reporting level: 1.0%.

CALIFORNIA'S PROPOSITION 65: Carcinogen.

EPA NAME: p-CHLORO-o-TOLUIDINE

CAS: 95-69-2

SYNONYMS: BENZENAMINE, 4-CHLORO-2-METHYL; 4-CHLORO-2-METHYLBENZENAMINE; 5-CHLORO-2-AMINOTOLUENE; para-CHLORO-ortho-TOLUIDINE

EPCRA Section 313 Form R de minimus concentration reporting level: 0.1%.

CALIFORNIA'S PROPOSITION 65: Carcinogen.

EPA NAME: 4-CHLORO-o-TOLUIDINE, HYDROCHLORIDE

CAS: 3165-93-3

SYNONYMS: 2-AMINO-5-CHLOROTOLUENE HYDROCHLORIDE; BENZENEAMINE, 4-CHLORO-2-METHYL-, HYDROCHLORIDE

EPA HAZARDOUS WASTE NUMBER (RCRA No.): U049.

RCRA Section 261 Hazardous Constituents.

EPCRA Section 304 Reportable Quantity (RQ): CERCLA, 100 lbs. (45.4 kgs.).

EPA NAME: 4-CHLORO-6-(TRICHLOROMETHYL)PYRIDINE

[see NITRAPYRIN]

CAS: 1929-82-4

EPA NAME: 2-CHLORO-1,1,1-TRIFLUOROETHANE

CAS: 75-88-7

SYNONYMS: CHLORO-1,1,1-TRIFLUOROETHANE; CHLOROTRIFLUOROETHANE; ETHANE, 2-CHLORO-1,1,1-TRIFLUORO-; FREON 133a; HCFC-133a

CLEAN AIR ACT: Stratospheric ozone protection (Title VI, Subpart A, Appendix B), Class II, Ozone Depletion Potential = listed as "reserved."

EPCRA Section 313 Form R de minimus concentration reporting level: 1.0%.

EPA NAME: CHLOROTRIFLUOROMETHANE
CAS: 75-72-9
SYNONYMS: CFC-13; FREON-13; MONOCHLOROTRIFLUORO-METHANE; TRIFLUOROCHLOROMETHANE; TRIFLUORO-MONOCHLOROMETHANE

CLEAN AIR ACT: Stratospheric ozone protection (Title VI, Subpart A, Appendix A), Class I, Ozone Depletion Potential = 1.0.
EPCRA Section 313 Form R de minimus concentration reporting level: 1.0%.

EPA NAME: 5-(2-2-CHLORO-4-(TRIFLUOROMETH-YL)PHENOXY)-2-NITROBENZOIC ACID, SODIUM SALT
[see ACIFLUOREN, SODIUM SALT]
CAS: 62476-59-9

EPA NAME: 5-(2-CHLORO-4-(TRIFLUOROMETH-YL)PHENOXY)-N-METHYLSULFONYL)-2-NITROBENZ-AMIDE
[see FOMESAFEN]
CAS: 72178-02-0

EPA NAME: 5-(2-CHLORO-4-(TRIFLUOROMETH-YL)PHENOXY)-2-NITRO-2-ETHOXY-1-METHYL-2-OXO-ETHYL ESTER
Note: This name appears in EPA 740-R-95-001. Corrected in EPA 745-K-96-001 as: BENZOIC ACID, 5-(2-CHLORO-4-(TRIFLUORO-METHYL)PHENOXY)-2-NITRO-2-ETHOXY-1-METHYL-2-OX-OETHYL ESTER.
[see LACTOFEN]
CAS: 77501-63-4

EPA NAME: N-[2-CHLORO-4-(TRIFLUOROMETHYL) PHEN-YL]-DL-VALINE(+)-CYANO(3-PHENOXYLPHEN-YL)METHYL ESTER
[see FLUVALINATE]
CAS: 69409-94-5

EPA NAME: 3-CHLORO-1,1,1-TRIFLUOROPROPANE
CAS: 460-35-5
SYNONYMS: 1-CHLORO-3,3,3-TRIFLUOROPROPANE; FREON 253fb; HCFC 253fb

EPCRA Section 313 Form R de minimus concentration reporting level: 1.0%.

EPA NAME: 3-(2-CHLORO-3,3,3-TRIFLUORO-1-PROPE-NYL)-2,2-DIMETHYLCYCLOPROPANECARBOXYLIC ACID CYANO(3-PHENOXYPHENYL)METHYL ESTER
[see CYHALOTHRIN]
CAS: 68085-85-8

EPA NAME: CHLOROXURON

CAS: 1982-47-4

SYNONYMS: N'-(4-(4-CHLOROPHENOXY)PHENYL)-N,N-DIMETHYLUREA

EPCRA Section 302 Extremely Hazardous Substances: TPQ = 500/10,000 lbs. (227/4,540 kgs.).

EPCRA Section 304 Reportable Quantity (RQ): EHS, 500 lbs. (227 kgs.).

EPA NAME: CHLORPYRIFOS

CAS: 2921-88-2

SYNONYMS: CHLORPYRIFOS-ETHYL; DURSBAN; O,O-DIMETHYL O-(3,5,6-TRICHLORO-2-PYRIDINYL)PHOSPHOROTHIOATE

CLEAN WATER ACT: Section 311 Hazardous Substances/RQ (same as CERCLA).

EPCRA Section 304 Reportable Quantity (RQ): CERCLA, 1 lb. (0.454 kg.).

MARINE POLLUTANT (49CFR, Subchapter 172.101, Appendix B), severe pollutant

EPA NAME: CHLORPYRIFOS—METHYL

CAS: 5598-13-0

SYNONYMS: O,O-DIMETHYL-O-(3,5,6-TRICHLORO-2-PYRIDYL)PHOSPHOROTHIOATE; DURSBAN METHYL

EPCRA Section 313 Form R de minimus concentration reporting level: 1.0%.

EPA NAME: CHLORSULFURON

CAS: 64902-72-3

SYNONYMS: BENZENESULFONAMIDE, 2-CHLORO-N-(((4-METHOXY-6-METHYL-1,3,5-TRIAZIN-2-YL)AMINO)CARBONYL); 2-CHLORO-N-(((4-METHOXY-6-METHYL-1,3,5-TRIAZIN-2-YL)AMINO)CARBONYL)BENZENESULF ONAMIDE; EPA PESTICIDE CHEMICAL CODE 118601

EPCRA Section 313 Form R de minimus concentration reporting level: 1.0%.

EPA NAME: CHLORTHIOPHOS

CAS: 21923-23-9

SYNONYMS: O,O-(DIETHYL-O-2,4,5-DICHLORO(METHYLTHIO)PHENYL THIONOPHOSPHATE

EPCRA Section 302 Extremely Hazardous Substances: TPQ = 500 lbs. (227 kgs.).

EPCRA Section 304 Reportable Quantity (RQ): EHS, 500 lbs. (227 kgs.).

MARINE POLLUTANT (49CFR, Subchapter 172.101, Appendix B).

EPA NAME: CHROMIC ACETATE

[see also CHROMIUM AND CHROMIUM COMPOUNDS]

CAS: 1066-30-4

SYNONYMS: ACETIC ACID, CHROMIUM(3+) SALT; CHROMIUM(III) ACETATE; CHROMIUM TRIACETATE

CLEAN AIR ACT: Hazardous Air Pollutants (Title I, Part A, Section 112) as chromium compounds.
CLEAN WATER ACT: Section 311 Hazardous Substances/RQ (same as CERCLA); Section 313 Priority Chemicals; Section 307 Toxic Pollutants.
EPCRA Section 304 Reportable Quantity (RQ): CERCLA, 1,000 lbs. (454 kgs.).
EPCRA Section 313 Form R de minimus concentration reporting level: Chromium III compounds: 1.0%. Form R Toxic Chemical Category Code: N090.

EPA NAME: CHROMIC ACID
[see also CHROMIUM and CHROMIUM COMPOUNDS]
CAS: 7738-94-5
SYNONYMS: CHROMIC(VI) ACID

CLEAN AIR ACT: Hazardous Air Pollutants (Title I, Part A, Section 112) as chromium compounds.
CLEAN WATER ACT: Section 307 Toxic Pollutants
EPCRA Section 304 Reportable Quantity (RQ): CERCLA, 10 lbs. (4.54 kgs.).
EPCRA Section 313 Form R de minimus concentration reporting level: Chromium VI compounds: 0.1%. Form R Toxic Chemical Category Code: N090.

EPA NAME: CHROMIC ACID
[see also CHROMIUM and CHROMIUM COMPOUNDS]
CAS: 11115-74-5
SYNONYMS: CHROMIC ACID, ESTER; PURATRONIC CHROMIUM TRIOXIDE

CLEAN AIR ACT: Hazardous Air Pollutants (Title I, Part A, Section 112) as chromium compounds.
CLEAN WATER ACT: Section 311 Hazardous Substances/RQ (same as CERCLA); Section 313 Priority Chemicals; Section 307 Toxic Pollutants.
EPCRA Section 304 Reportable Quantity (RQ): CERCLA, 10 lbs. (4.54 kgs.).

EPA NAME: CHROMIC CHLORIDE
[see also CHROMIUM and CHROMIUM COMPOUNDS]
CAS: 10025-73-7
SYNONYMS: CHROMIUM(III) CHLORIDE (1:3); CHROMIUM TRICHLORIDE; TRICHLOROCHROMIUM

CLEAN AIR ACT: Hazardous Air Pollutants (Title I, Part A, Section 112) as chromium compounds.
CLEAN WATER ACT: Section 307 Toxic Pollutants.
EPCRA Section 302 Extremely Hazardous Substances: TPQ = 1/10,000 lbs. (0.454/4,540 kgs.).

EPCRA Section 304 Reportable Quantity (RQ): EHS, 1 lb. (0.454 kg.).
EPCRA Section 313 Form R de minimus concentration reporting level: Chromium III compounds: 1.0%. Form R Toxic Chemical Category Code: N090.

EPA NAME: CHROMIC SULFATE
[see also CHROMIUM and CHROMIUM COMPOUNDS]
CAS: 10101-53-8
SYNONYMS: CHROMIUM(III) SULFATE; SULFURIC ACID, CHROMIUM(3+) SALT (3:2)

CLEAN AIR ACT: Hazardous Air Pollutants (Title I, Part A, Section 112) as chromium compounds.
CLEAN WATER ACT: Section 311 Hazardous Substances/RQ (same as CERCLA); Section 313 Priority Chemicals; Section 307 Toxic Pollutants.
EPCRA Section 304 Reportable Quantity (RQ): CERCLA, 1,000 lbs. (454 kgs.).
EPCRA Section 313 Form R de minimus concentration reporting level: Chromium III compounds: 1.0%. Form R Toxic Chemical Category Code: N090.

EPA NAME: CHROMIUM
[see also CHROMIUM and CHROMIUM COMPOUNDS]
CAS: 7440-47-3
SYNONYMS: CHROMIUM METAL

CLEAN AIR ACT: Hazardous Air Pollutants (Title I, Part A, Section 112) as chromium compounds.
CLEAN WATER ACT: Section 307 Priority Pollutants; Section 313 Priority Chemicals; Toxic Pollutant (Section 401.15).
EPA HAZARDOUS WASTE NUMBER (RCRA No.): D007.
RCRA Section 261 Hazardous Constituents, waste number not listed.
RCRA Toxicity Characteristic (Section 261.24), Maximum Concentration of Contaminants, regulatory level, 5.0 mg/L.
RCRA Land Ban Waste.
RCRA Universal Treatment Standards: Wastewater (mg/L), 2.77; Nonwastewater (mg/kg), 0.86 as chromium (total).
RCRA Ground Water Monitoring List. Suggested test method(s) (PQL µg/L): (total) 6010(70); 7190(500); 7191(10).
SAFE DRINKING WATER ACT: MCL, 0.1 mg/L; MCLG, 0.1 mg/L; Regulated chemical (47 FR 9352).
EPCRA Section 304 Reportable Quantity (RQ): CERCLA, 5,000 lbs. (2270 kgs.).
EPCRA Section 313 Form R de minimus concentration reporting level: 0.1%. Form R Toxic Chemical Category Code: N090.

EPA NAME: CHROMIUM COMPOUNDS
[see also CHROMIUM]
CLEAN AIR ACT: Hazardous Air Pollutants (Title I, Part A, Section 112).
CLEAN WATER ACT: Toxic Pollutant (Section 401.15); Section 307 Toxic Pollutants as chromium and compounds.

RCRA Section 261 Hazardous Constituents, waste number not listed.

EPCRA Section 313: Includes any unique chemical substances that contains chromium as part of that chemical's infrastructure. Form R de minimus concentration reporting level: Chromium VI compounds: 0.1%; Chromium(III) compounds: 1.0%. Form R Toxic Chemical Category Code: N090.

CALIFORNIA'S PROPOSITION 65: Carcinogen as chromium (hexavalent compounds).

EPA NAME: CHROMOUS CHLORIDE

[see also CHROMIUM and CHROMIUM COMPOUNDS]

CAS: 10049-05-5

SYNONYMS: CHROMIC(II) CHLORIDE; CHROMIUM CHLORIDE; CHROMIUM DICHLORIDE

CLEAN AIR ACT: Hazardous Air Pollutants (Title I, Part A, Section 112) as chromium compounds.

CLEAN WATER ACT: Section 311 Hazardous Substances/RQ (same as CERCLA); Section 313 Priority Chemicals; Section 307 Toxic Pollutants.

EPCRA Section 304 Reportable Quantity (RQ): CERCLA, 1,000 lbs. (454 kgs.).

EPA NAME: d-trans-CHRYSANTHEMIC ACID of ALLETHRONE

[see d-trans-ALLETHRIN]

CAS: 28057-48-9

EPA NAME: CHRYSENE

[see also POLYCYCLIC AROMATIC COMPOUNDS]

CAS: 218-01-9

SYNONYMS: BENZO(A)PHENANTHRENE; 1,2-BENZOPHENANTHRENE; 1,2-BENZPHENANTHRENE; BENZ(A)PHENANTHRENE; 1,2,5,6-DIBENZONAPHTHALENE

CLEAN WATER ACT: Section 307 Priority Pollutants; Section 307 Toxic Pollutants as polynuclear aromatic hydrocarbons.

EPA HAZARDOUS WASTE NUMBER (RCRA No.): U050.

RCRA Section 261 Hazardous Constituents.

RCRA Land Ban Waste.

RCRA Universal Treatment Standards: Wastewater (mg/L), 0.059; Nonwastewater (mg/kg), 3.4.

RCRA Ground Water Monitoring List. Suggested test method(s) (PQL µg/L): 8100(200); 8270(10).

EPCRA Section 304 Reportable Quantity (RQ): CERCLA, 100 lbs. (45.4 kgs.).

EPCRA Section 313 Form R de minimus concentration reporting level: 0.1%. Form R Toxic Chemical Category Code: N590.

CALIFORNIA'S PROPOSITION 65: Carcinogen.

EPA NAME: C.I. ACID GREEN 3

CAS: 4680-78-8

SYNONYMS: GUINEA GREEN; GUINEA GREEN b

EPCRA Section 313 Form R de minimus concentration reporting level: 1.0%.

EPA NAME: C.I. ACID RED 114

CAS: 6459-94-5

SYNONYMS: C.I. ACID RED 114, DISODIUM SALT; DISODIUM 8-((3,3′-DIMETHYL-4′-(4-(4-METHYLPHENYLSULPHONYLOXY)PHENYLAZO)(1,1′-BIPHENYL)-4-YL)AZO)-7-HYDROXY-NAPHTHALENE-1,3-DISULPHONATE

EPCRA Section 313 Form R de minimus concentration reporting level: 0.1%.

EPA NAME: C.I. BASIC GREEN 4

CAS: 569-64-2

SYNONYMS: MALACHITE GREEN; MALACHITE GREEN CHLORIDE; VICTORIA GREEN B

EPCRA Section 313 Form R de minimus concentration reporting level: 1.0%.

EPA NAME: C.I. BASIC RED 1

CAS: 989-38-8

SYNONYMS: RHODAMINE 6G; RHODAMINE 6G EXTRA BASE

EPCRA Section 313 Form R de minimus concentration reporting level: 1.0%.

EPA NAME: C.I. DIRECT BLACK 38

CAS: 1937-37-7

SYNONYMS: DIRECT BLACK 38; 2,7-NAPHTHALENEDISULFONIC ACID, 4-AMINO-3-[[4′-[(2,4-DIAMINOPHENYL)AZO][1,1′-BIPHENYL]-4-YL]AZO]-5-HYDROXY-6-(PHENYLAZO)-, DISODIUM SALT

EPCRA Section 313 Form R de minimus concentration reporting level: 0.1%.

CALIFORNIA'S PROPOSITION 65: Carcinogen.

EPA NAME: C.I. DIRECT BLUE 6

CAS: 2602-46-2

SYNONYMS: DIRECT BLUE 6; 2,7-NAPHTHALENEDISULFONIC, 3,3′-[[1,1′-BIPHENYL]-4,4′=DIYBIS(AZO)BIS[5-AMINO-4-HYDROXYTETRASODIUM SALT

EPCRA Section 313 Form R de minimus concentration reporting level: 0.1%.

CALIFORNIA'S PROPOSITION 65: Carcinogen.

EPA NAME: C.I. DIRECT BLUE 218

CAS: 28407-37-6

SYNONYMS: COPPER, (μ-((TETRAHYDROGEN, 3,3′-((3,3′-DIHYDROXY-4,4′-BIPHENYLENE)BIS(AZO)BIS(5-AMINO-4-HYDROXY-2,7-NAPHTHALENEDISULFONATO))(8-)))DI-TETRA-SODIUM SALT

EPCRA Section 313 Form R de minimus concentration reporting level: 0.1%.

EPA NAME: C.I. DIRECT BROWN 95
[see also COPPER COMPOUNDS]
CAS: 16071-86-6
SYNONYMS: DIRECT BROWN 95; COPPER, [DIHYDROGEN 5-[[4′-[[2,6-DIHYDROXY-3-[(2-HYDROXY-5-SULFOPHENYL)AZO]PHENYL]AZO]-4-BIPHENYLY L]AZO]-2-HYDROXYBENZOATO(4-)]-, DISODIUM; CUPRATE(2-), [5-[[4′-[[2,6-DIHYDROXY-3-[(2-HYDROXY-5-SULFOPHENYL) AZO] PHENYL]AZO][1,1′-HYDROXY-5-SULFOPHENYL)AZO][1,1′-BIPHENYL]-4-YL]AZO]-2-HYDROXYBANZOATO(4-)]-, DISODIUM; CUPRATE(2-), [5[[4′-[[2,6-DIHYDROXY-3-[(2-HYDROXY-5-SULFOPHENYL)AZO]PHENYL]AZO] C-4-BIPHENYLYL]AZO][1,1′BIPHENYL]-4-YL]AZO]-2-HYDROXYBENZOATO(4-)]-, DISODIUM; CUPRATE(2-), [5-[[4′-[[2,6-DIHYDROXY-3-[(2-HYDROXY-5-SULFOPHENYL)AZO]PHENYL]AZO][1,1′-HYDROXY-5-SULFOPHENYL)AZO][1,1′-BIPHENYL]-4-YL]AZO]-2-HYDROXYBANZOATO(4-)]-, DISODIUM

CLEAN WATER ACT: Toxic Pollutant (Section 401.15) as copper and compounds.
RCRA Ground Water Monitoring List. Suggested test method(s) (PQL μg/L): 6010(60); 7210 (200). Note: All species in the ground water that contain copper are included.
EPCRA Section 313 Form R de minimus concentration reporting level: 0.1%. Form R Toxic Chemical Category Code: N100.
CALIFORNIA'S PROPOSITION 65: Carcinogen.

EPA NAME: C.I. DISPERSE YELLOW 3
CAS: 2832-40-8
SYNONYMS: ACETAMIDE, N-[4-[(2-HYDROXY-5-METHYLPHENYL)AZO]PHENYL-; DISPERSE YELLOW 3

EPCRA Section 313 Form R de minimus concentration reporting level: 1.0%.

EPA NAME: C.I. FOOD RED 5
CAS: 3761-53-3
SYNONYMS: ACIDAL PONCEAU G; FOOD RED 5; 3-HYDROXY-4-(2,4-XYLYLAZO)-2,7-NAPHTHALENEDISULFONIC ACID, DI-SODIUM SALT; 2,7-NAPHTHALENEDISULFONIC ACID, 4-[(2,4-DIMETHYLPHENYL)AZO]-3-HYDROXY-, DISODIUM SALT; PONCEAU MX

EPCRA Section 313 Form R de minimus concentration reporting level: 0.1%.

EPA NAME: C.I. FOOD RED 15

CAS: 81-88-9

SYNONYMS: 11411 RED; D AND C RED NO. 19; FOOD RED 15; RHODAMINE B; RHODAMINE B 500

EPCRA Section 313 Form R de minimus concentration reporting level: 1.0%.

CALIFORNIA'S PROPOSITION 65: Carcinogen.

EPA NAME: C.I. SOLVENT ORANGE 7

CAS: 3118-97-6

SYNONYMS: AF RED NO. 5; JAPAN RED 5; SOLVENT ORANGE 7; SUDAN II; SUDAN ORANGE; SUDAN ORANGE RPA

EPCRA Section 313 Form R de minimus concentration reporting level: 1.0%.

EPA NAME: C.I. SOLVENT YELLOW 3

CAS: 97-56-3

SYNONYMS: o-AMINOAZOTOLUENE; ortho-AMINOAZOTOLUENE; BENZENAMINE, 2-METHYL-4-[(2-METHYLPHENYL)AZO]-; 2′,3-DIMETHYL-4-AMINOAZOBENZENE; 4-(o-TOLYLAZO)-o-TOLUIDINE; SOLVENT YELLOW 3

EPCRA Section 313 Form R de minimus concentration reporting level: 1.0%.

CALIFORNIA'S PROPOSITION 65: Carcinogen.

EPA NAME: C.I. SOLVENT YELLOW 14

CAS: 842-07-9

SYNONYMS: BENZENEAZO-β-NAPHTHOL; SOLVENT YELLOW 14; SUDAN I; SUDAN ORANGE R

EPCRA Section 313 Form R de minimus concentration reporting level: 1.0%.

EPA NAME: C.I. SOLVENT YELLOW 34

CAS: 492-80-8

SYNONYMS: AURAMINE; AURAMINE BASE; BENZENEAMINE, 4,4′-CABONIMIDOYLBIS[N-DIMETHYL-; SOLVENT YELLOW 34

EPA HAZARDOUS WASTE NUMBER (RCRA No.): U014.

RCRA Section 261 Hazardous Constituents.

EPCRA Section 304 Reportable Quantity (RQ): CERCLA, 100 lbs. (45.4 kgs.).

EPCRA Section 313 Form R de minimus concentration reporting level: 0.1%.

CALIFORNIA'S PROPOSITION 65: Carcinogen.

EPA NAME: C.I. VAT YELLOW 4

CAS: 128-66-5

SYNONYMS: VAT YELLOW 4

EPCRA Section 313 Form R de minimus concentration reporting level: 1.0%.

EPA NAME: COBALT

[see also COBALT COMPOUNDS]

CAS: 7440-48-4

CLEAN AIR ACT: Hazardous Air Pollutants (Title I, Part A, Section 112) as cobalt compounds.

RCRA Ground Water Monitoring List. Suggested test method(s) (PQL µg/L): (total) 6010(70); 7200(500); 7201(10).

EPCRA Section 313 Form R de minimus concentration reporting level: 1.0%. Form R Toxic Chemical Category Code: N096.

CALIFORNIA'S PROPOSITION 65: Carcinogen as cobalt metal powder.

EPA NAME: COBALT CARBONYL

[see also COBALT and COBALT COMPOUNDS]

CAS: 10210-68-1

SYNONYMS: COBALT TETRACARBONYL; DICOBALT CARBONYL

CLEAN AIR ACT: Hazardous Air Pollutants (Title I, Part A, Section 112) as cobalt compounds.

EPCRA Section 302 Extremely Hazardous Substances: TPQ = 10/10,000 lbs. (4.54/4,540 kgs.).

EPCRA Section 304 Reportable Quantity (RQ): EHS, 10 lbs. (4.54 kgs.).

EPCRA Section 313 Form R de minimus concentration reporting level: 1.0%. Form R Toxic Chemical Category Code: N096.

EPA NAME: COBALT COMPOUNDS

[see also COBALT]

CLEAN AIR ACT: Hazardous Air Pollutants (Title I, Part A, Section 112). Note: Includes any unique chemical substance that contains cobalt as part of that chemical's infrastructure.

EPCRA Section 313: Includes any unique chemical substance that contains cobalt as part of that chemical's infrastructure. Form R de minimus concentration reporting level: 1.0%. Form R Toxic Chemical Category Code: N096.

EPA NAME: COBALT, ((2,2′-(1,2-ETHANEDIYLBIS(NITRILOMETHYLIDYNE))BIS(6-FLUOROPHENOLATO))(2)-

[see also COBALT and COBALT COMPOUNDS]

CAS: 62207-76-5

SYNONYMS: BIS(3-FLUOROSALICYLALDEHYDE)ETHYLENEDIIMINE-COBALT; COBALT(II), N,N′-ETHYLENEBIS(3-FLUOROSALICYLIDENEIMINATO)-

CLEAN AIR ACT: Hazardous Air Pollutants (Title I, Part A, Section 112) as cobalt compounds.

EPCRA Section 313: As cobalt compounds. Form R de minimus concentration reporting level: 1.0%. Form R Toxic Chemical Category Code: N096.

EPCRA Section 302 Extremely Hazardous Substances: TPQ = 100/10,000 lbs. (45.4/4,540 kgs.).

EPCRA Section 304 Reportable Quantity (RQ): EHS, 100 lbs. (45.4 kgs.).

EPA NAME: COBALTOUS BROMIDE

[see also COBALT and COBALT COMPOUNDS]

CAS: 7789-43-7

SYNONYMS: COBALT BROMIDE; COBALT(II) BROMIDE

CLEAN AIR ACT: Hazardous Air Pollutants (Title I, Part A, Section 112) as cobalt compounds.

CLEAN WATER ACT: Section 311 Hazardous Substances/RQ (same as CERCLA); Section 313 Priority Chemicals.

EPCRA Section 304 Reportable Quantity (RQ): CERCLA, 1,000 lbs. (454 kgs.).

EPCRA Section 313 Form R de minimus concentration reporting level: 1.0%. Form R Toxic Chemical Category Code: N096.

EPA NAME: COBALTOUS FORMATE

[see also COBALT and COBALT COMPOUNDS]

CAS: 544-18-3

SYNONYMS: COBALT FORMATE; COBALT(II) FORMATE

CLEAN AIR ACT: Hazardous Air Pollutants (Title I, Part A, Section 112).

CLEAN WATER ACT: Section 311 Hazardous Substances/RQ (same as CERCLA); Section 313 Priority Chemicals.

EPCRA Section 304 Reportable Quantity (RQ): CERCLA, 1,000 lbs. (454 kgs.).

EPCRA Section 313: As cobalt compounds. Form R de minimus concentration reporting level: 1.0%. Form R Toxic Chemical Category Code: N096.

EPA NAME: COBALTOUS SULFAMATE

[see also COBALT and COBALT COMPOUNDS]

CAS: 14017-41-5

SYNONYMS: COBALT SULFAMATE; COBALT(II) SULFAMATE

CLEAN AIR ACT: Hazardous Air Pollutants (Title I, Part A, Section 112).

CLEAN WATER ACT: Section 311 Hazardous Substances/RQ (same as CERCLA); Section 313 Priority Chemicals.

EPCRA Section 304 Reportable Quantity (RQ): CERCLA, 1,000 lbs. (454 kgs.).

EPCRA Section 313: As cobalt compounds. Form R de minimus concentration reporting level: 1.0%. Form R Toxic Chemical Category Code: N096.

EPA NAME: COKE OVEN EMISSIONS

CLEAN AIR ACT: Hazardous Air Pollutants (Title I, Part A, Section 112).

CALIFORNIA'S PROPOSITION 65: Carcinogen.

EPCRA Section 304 Reportable Quantity (RQ): CERCLA, 1 lb. (0.454 kg.).

EPA NAME: COLCHICINE

CAS: 64-86-8

SYNONYMS: N-(5,6,7,9)-TETRAHYDRO-1,2,3,10-TETRAMETHOXY-9-OXOBENZO(a)HEPTALEN-7-YL)-ACETAMIDE

CLEAN AIR ACT: Hazardous Air Pollutants (Title I, Part A, Section 112).

EPCRA Section 302 Extremely Hazardous Substances: TPQ = 10/10,000 lb. (4.54/4,540 kg.).

EPCRA Section 304 Reportable Quantity (RQ): EHS, 10 lbs. (4.54 kgs.).

CALIFORNIA'S PROPOSITION 65: Carcinogen; Reproductive toxin (male).

EPA NAME: COPPER

[see also COPPER COMPOUNDS]

CAS: 7440-50-8

SYNONYMS: COPPER, ELEMENTAL

CLEAN WATER ACT: Section 307 Priority Pollutants; Section 313 Priority Chemicals; Toxic Pollutant (Section 401.15).

RCRA Ground Water Monitoring List. Suggested test method(s) (PQL μg/L): (total) 6010(60); 7210 (200).

SAFE DRINKING WATER ACT: MCL, 1.3 mg/L; MCLG, 1 mg/L; SMLC, 1.0 mg/L; Regulated chemical (47 FR 9352).

EPCRA Section 304 Reportable Quantity (RQ): CERCLA, 5,000 lbs. (2270 kgs.).

EPCRA Section 313 Form R de minimus concentration reporting level: 1.0%. Form R Toxic Chemical Category Code: N100.

EPA NAME: COPPER COMPOUNDS

[see also COPPER]

CLEAN WATER ACT: Toxic Pollutant (Section 401.15) as copper and compounds.

RCRA Ground Water Monitoring List. Suggested test method(s) (PQL μg/L): 6010(60); 7210 (200). Note: All species in the ground water that contain copper are included.

EPCRA Section 313: Includes any unique chemical substance that contains copper as part of that chemical's infrastructure. This category does not include copper phthalocyanide compounds that are substituted with only hydrogen, and/or chlorine, and/or bromine. Form R de minimus concentration reporting level: 1.0%. Form R Toxic Chemical Category Code: N100.

EPA NAME: COPPER CYANIDE

[see also COPPER, COPPER COMPOUNDS, and CYANIDE COMPOUNDS]

CAS: 544-92-3

SYNONYMS: COPPER(II) CYANIDE; CUPRIC CYANIDE

CLEAN AIR ACT: Hazardous Air Pollutants (Title I, Part A, Section 112) as cyanide compounds.

CLEAN WATER ACT: Toxic Pollutant (Section 401.15).
EPA HAZARDOUS WASTE NUMBER (RCRA No.): P029
RCRA Ground Water Monitoring List. Suggested test method(s) (PQL μg/L): 6010(60); 7210 (200). Note: All species in the ground water that contain copper are included.
EPCRA Section 304 Reportable Quantity (RQ): CERCLA, 10 lbs. (4.54 kgs.).
EPCRA Section 313 Form R de minimus concentration reporting level: 1.0%. Form R Toxic Chemical Category Code: N100 (copper).
EPCRA Section 313 Form R de minimus concentration reporting level: 1.0%. Form R Toxic Chemical Category Code: N016 (cyanide).
MARINE POLLUTANT (49CFR, Subchapter 172.101, Appendix B).

EPA NAME: COUMAPHOS

CAS: 56-72-4
SYNONYMS: 3-CHLORO-7-HYDROXY-4-METHYL-COUMARIN O,O-DIETHYL PHOSPHOROTHIOATE; CO-RAL; PHOSPHOROTHIOIC ACID, O-(3-CHLORO-4-METHYL-2-OXO-2H-1-BENZOPYRAN-7-YL) O,O-DIETHYL ESTER

CLEAN WATER ACT: Section 311 Hazardous Substances/RQ (same as CERCLA).
EPCRA Section 302 Extremely Hazardous Substances: TPQ = 100/10,000 lbs. (455/4,540 kgs.).
EPCRA Section 304 Reportable Quantity (RQ): EHS/CERCLA, 10 lbs. (4.54 kgs.).
MARINE POLLUTANT (49CFR, Subchapter 172.101, Appendix B): Severe pollutant.

EPA NAME: COUMATETRALYL

CAS: 5836-29-3
SYNONYMS: 4-HYDROXY-3-(1,2,3,4-TETRAHYDRO-1-NAPTHALENYL)-2H-1-BENZOPYRAN-2-ONE(9CI)

EPCRA Section 302 Extremely Hazardous Substances: TPQ = 500/10,000 lbs. (227/4,540 kgs.).
EPCRA Section 304 Reportable Quantity (RQ): EHS, 500 lbs. (227 kgs.).

EPA NAME: CREOSOTE

CAS: 8001-58-9
SYNONYMS: COAL TAR CREOSOTE; CREOSOTE OIL

EPA HAZARDOUS WASTE NUMBER (RCRA No.): U051.
RCRA Section 261 Hazardous Constituents.
RCRA Land Ban Waste.
EPCRA Section 304 Reportable Quantity (RQ): CERCLA, 1 lb. (0.454 kg.).
EPCRA Section 313 Form R de minimus concentration reporting level: 0.1%.
MARINE POLLUTANT (49CFR, Subchapter 172.101, Appendix B).
CALIFORNIA'S PROPOSITION 65: Carcinogen as creosotes.

EPA NAME: p-CRESIDINE

CAS: 120-71-8

SYNONYMS: BENZENEAMINE, 2-METHOXY-5-METHYL-; 5-METHYL-o-ANISIDINE; para-CRESIDINE

EPCRA Section 313 Form R de minimus concentration reporting level: 0.1%.

CALIFORNIA'S PROPOSITION 65: Carcinogen.

EPA NAME: m-CRESOL

CAS: 108-39-4

SYNONYMS: BENZENE, 3-METHYL; 3-CRESOL; meta-CRESOL; m-CRESYLIC ACID; PHENOL, 3-METHYL-

CLEAN WATER ACT: Section 311 Hazardous Substances/RQ 100 lbs. (4.54 kgs.); Section 313 Priority Chemicals.

EPA HAZARDOUS WASTE NUMBER (RCRA No.): U052; D024.

RCRA Section 261 Hazardous Constituents.

RCRA Toxicity Characteristic (Section 261.24), Maximum Concentration of Contaminants, regulatory level, 200 mg/L. Note: if o-, m-, and p-Cresol concentrations cannot be differentiated, the total cresol (D026) concentration is used. The regulatory level of total cresol is 200 mg/L.

RCRA Land Ban Waste.

RCRA Universal Treatment Standards: Wastewater (mg/L), 0.77; Nonwastewater (mg/kg), 5.6. Note: Difficult to distinguish from p-cresol.

RCRA Ground Water Monitoring List. Suggested test method(s) (PQL µg/L): 8270(10).

EPCRA Section 304 Reportable Quantity (RQ): CERCLA, 100 lbs. (45.4 kgs.).

EPCRA Section 313 Form R de minimus concentration reporting level: 1.0%.

MARINE POLLUTANT (49CFR, Subchapter 172.101, Appendix B).

EPA NAME: o-CRESOL

CAS: 95-48-7

SYNONYMS: BENZENE, 2-METHYL; 2-CRESOL; CRESOL-o-; ortho-CRESOL; CRESOL-ortho; 2-METHYLPHENOL; PHENOL, 2-METHYL; o-TOLUOL

CLEAN AIR ACT: Hazardous Air Pollutants (Title I, Part A, Section 112).

CLEAN WATER ACT: Section 311 Hazardous Substances/RQ/RQ 100 lbs. (4.54 kgs.); Section 313 Priority Chemicals.

EPA HAZARDOUS WASTE NUMBER (RCRA No.): U052; D023.

RCRA Toxicity Characteristic (Section 261.24), Maximum Concentration of Contaminants, regulatory level, 200. Note: if o-, m-, and p-Cresol concentrations cannot be differentiated, the total cresol (D026) concentration is used. The regulatory level of total cresol is 200 mg/L.

RCRA Section 261 Hazardous Constituents.

RCRA Land Ban Waste.

RCRA Universal Treatment Standards: Wastewater (mg/L), 0.11; Nonwastewater (mg/kg), 5.6.

RCRA Ground Water Monitoring List. Suggested test method(s) (PQL µg/L): 8270(10).
EPCRA Section 302 Extremely Hazardous Substances: TPQ = 1,000/10,000 lbs. (454/4,540 kgs.).
EPCRA Section 304 Reportable Quantity (RQ): EHS/CERCLA, 100 lbs. (45.4 kgs.).
EPCRA Section 313 Form R de minimus concentration reporting level: 1.0%.
MARINE POLLUTANT (49CFR, Subchapter 172.101, Appendix B).

EPA NAME: p-CRESOL
CAS: 106-44-5
SYNONYMS: 4-CRESOL; para-CRESOL; CRESOL-PARA; PHENOL, 4-METHYL

CLEAN AIR ACT: Hazardous Air Pollutants (Title I, Part A, Section 112).
CLEAN WATER ACT: Section 311 Hazardous Substances/RQ/RQ 100 lbs. (4.54 kgs.); Section 313 Priority Chemicals.
EPA HAZARDOUS WASTE NUMBER (RCRA No.): U052; D025.
RCRA Toxicity Characteristic (Section 261.24), Maximum Concentration of Contaminants, regulatory level, 200 mg/L. Note: if o-, m-, and p-Cresol concentrations cannot be differentiated, the total cresol (D026) concentration is used. The regulatory level of total cresol is 200 mg/L.
RCRA Section 261 Hazardous Constituents.
RCRA Land Ban Waste.
RCRA Universal Treatment Standards: Wastewater (mg/L), 0.77; Nonwastewater (mg/kg), 5.6. NOTE: Difficult to distinguish from m-cresol.
RCRA Ground Water Monitoring List. Suggested test method(s) (PQL µg/L): 8270(10).
EPCRA Section 304 Reportable Quantity (RQ): CERCLA, 100 lbs. (45.4 kgs.).
EPCRA Section 313 Form R de minimus concentration reporting level: 1.0%.
MARINE POLLUTANT (49CFR, Subchapter 172.101, Appendix B).

EPA NAME: CRESOL (MIXED ISOMERS)
CAS: 1319-77-3
SYNONYMS: CRESOLS (ALL ISOMERS); CRESOLS AND CRESYLIC ACIDS, MIXED; CRESOL ISOMERS; CRESOLS (o-; m-; p-); CRESYLIC ACID; HYDROXYTOLUENE; METHYLPHENOL; PHENOL, METHYL-

CLEAN AIR ACT: Hazardous Air Pollutants (Title I, Part A, Section 112).
CLEAN WATER ACT: Section 311 Hazardous Substances/RQ 100 lbs. (4.54 kgs.); Section 313 Priority Chemicals.
EPA HAZARDOUS WASTE NUMBER (RCRA No.): U052; D026.

RCRA Toxicity Characteristic (Section 261.24), Maximum Concentration of Contaminants, regulatory level, 200 mg/L. Note: if o-, m-, and p-Cresol concentrations cannot be differentiated, the total cresol (D026) concentration is used. The regulatory level of total cresol is 200 mg/L.
RCRA Section 261 Hazardous Constituents.
RCRA Land Ban Waste.
EPCRA Section 304 Reportable Quantity (RQ): CERCLA, 100 lbs. (45.4 kgs.).
EPCRA Section 313 Form R de minimus concentration reporting level: 1.0%.
MARINE POLLUTANT (49CFR, Subchapter 172.101, Appendix B) as cresols (o-; m; p-).

EPA NAME: CRIMIDINE
CAS: 535-89-7
SYNONYMS: CASTRIX; 2-CHLORO-4-METHYL-6-DIMETHYLAMINOPYRIMIDINE

EPCRA Section 302 Extremely Hazardous Substances: TPQ = 100/10,000 lbs. (4.54/4,540 kgs.).
EPCRA Section 304 Reportable Quantity (RQ): EHS, 100 lbs. (45.4 kgs.).

EPA NAME: CROTONALDEHYDE
CAS: 4170-30-3
SYNONYMS: 2-BUTENAL; 2-BUTENAL PROPYLENE ALDEHYDE; CROTONIC ALDEHYDE; PROPYLENE ALDEHYDE

CLEAN AIR ACT: Accidental Release Prevention/Flammable substances (Section 112[r], Table 3), TQ = 20,000 lbs. (9080 kgs.).
CLEAN WATER ACT: Section 311 Hazardous Substances/RQ (same as CERCLA).
EPA HAZARDOUS WASTE NUMBER (RCRA No.): U053.
RCRA Section 261 Hazardous Constituents.
EPCRA Section 302 Extremely Hazardous Substances: TPQ = 1,000 lbs. (454 kgs.).
EPCRA Section 304 Reportable Quantity (RQ): EHS/CERCLA, 100 lbs. (45.4 kgs.).
EPCRA Section 313 Form R de minimus concentration reporting level: 1.0%.
MARINE POLLUTANT (49CFR, Subchapter 172.101, Appendix B).

EPA NAME: CROTONALDEHYDE, (E)
CAS: 123-73-9
SYNONYMS: 2-BUTENAL, (e); (E)-2-BUTENAL; CROTONIC ALDEHYDE; ETHYLENE DIPROPIONATE (8CI); 3-METHYLACROLEINE

CLEAN AIR ACT: Accidental Release Prevention/Flammable substances (Section 112[r], Table 3), TQ = 20,000 lbs. (9080 kgs.).
EPCRA Section 302 Extremely Hazardous Substances: TPQ = 1,000 lbs. (454 kgs.).

EPCRA Section 304 Reportable Quantity (RQ): EHS/CERCLA, 100 lbs. (45.4 kgs.).
EPCRA Section 313 Form R de minimus concentration reporting level: 1.0%.
MARINE POLLUTANT (49CFR, Subchapter 172.101, Appendix B).

EPA NAME: CUMENE
CAS: 98-82-8
SYNONYMS: BENZENE, (1-METHYLETHYL-)-; CUMOL; ISOPROPYL BENZENE; 1-METHYLETHYL BENZENE

CLEAN AIR ACT: Hazardous Air Pollutants (Title I, Part A, Section 112).
EPA HAZARDOUS WASTE NUMBER (RCRA No.): U055.
RCRA Section 261 Hazardous Constituents.
EPCRA Section 304 Reportable Quantity (RQ): CERCLA, 5,000 lbs. (2270 kgs.).
EPCRA Section 313 Form R de minimus concentration reporting level: 1.0%.
MARINE POLLUTANT (49CFR, Subchapter 172.101, Appendix B).

EPA NAME: CUMENE HYDROPEROXIDE
CAS: 80-15-9
SYNONYMS: CUMYL HYDROPEROXIDE; α,α-DIMETHYLBENZYL HYDROPEROXIDE; HYDROPEROXIDE, 1-METHYL-1-PHENYLETHYL-

EPA HAZARDOUS WASTE NUMBER (RCRA No.): U096.
RCRA Section 261 Hazardous Constituents.
EPCRA Section 304 Reportable Quantity (RQ): CERCLA, 10 lbs. (4.54 kgs.).
EPCRA Section 313 Form R de minimus concentration reporting level: 1.0%.

EPA NAME: CUPFERRON
CAS: 135-20-6
SYNONYMS: BENZENEAMINE,N-HYDROXY-N-NITROSO, AMMONIUM SALT; HYDROXYLAMINE,N-NITROSO-N-PHENYL-,AMMONIUM SALT

EPCRA Section 313 Form R de minimus concentration reporting level: 0.1%.
CALIFORNIA'S PROPOSITION 65: Carcinogen.

EPA NAME: CUPRIC ACETATE
[see also COPPER and COPPER COMPOUNDS]
CAS: 142-71-2
SYNONYMS: ACETIC ACID, COPPER(2+) SALT; COPPER ACETATE

CLEAN WATER ACT: Section 311 Hazardous Substances/RQ (same as CERCLA); Section 313 Priority Chemicals; Toxic Pollutant (Section 401.15) as copper and compounds.

RCRA Ground Water Monitoring List. Suggested test method(s) (PQL μg/L): 6010(60); 7210 (200). Note: All species in the ground water that contain copper are included.
EPCRA Section 304 Reportable Quantity (RQ): CERCLA, 100 lbs. (45.4 kgs.).
EPCRA Section 313 Form R de minimus concentration reporting level: 1.0%. Form R Toxic Chemical Category Code: N100 as copper compounds.

EPA NAME: CUPRIC ACETOARSENITE

[see also ARSENIC and ARSENIC COMPOUNDS; COPPER and COPPER COMPOUNDS]
CAS: 12002-03-8
SYNONYMS: C.I. PIGMENT GREEN 21; COPPER ACETOARSENITE; CUPRIC ACETOARSENITE; PARIS GREEN; VIENNA GREEN

CLEAN AIR ACT: Hazardous Air Pollutants (Title I, Part A, Section 112); List of high risk pollutants (Section 63.74) as arsenic compounds.
CLEAN WATER ACT: Section 311 Hazardous Substances/RQ 1 lb. (0.454 kg.); Toxic Pollutant (Section 401.15) as copper and compounds; Section 313 Priority Chemicals.
RCRA Ground Water Monitoring List. Suggested test method(s) (PQL μg/L): 6010(60); 7210 (200). Note: All species in the ground water that contain copper are included.
EPCRA Section 302 Extremely Hazardous Substances: TPQ = 500/10,000 lbs. (227/4,540 kgs.).
EPCRA Section 304 Reportable Quantity (RQ): EHS/CERCLA, 1 lb. (0.454 kg.).
EPCRA Section 313 Form R de minimus concentration reporting level: 1.0%. Form R Toxic Chemical Category Code: N100 as copper compounds: N020 as arsenic compounds.
MARINE POLLUTANT (49CFR, Subchapter 172.101, Appendix B) as arsenates, liquid, n.o.s.; arsenates, solid, n.o.s.; arsenical pesticides liquid, toxic, flammable, n.o.s.

EPA NAME: CUPRIC CHLORIDE

[see also COPPER and COPPER COMPOUNDS]
CAS: 7447-39-4
SYNONYMS: COPPER CHLORIDE; COPPER(II) CHLORIDE (1:2)

CLEAN WATER ACT: Section 311 Hazardous Substances/RQ (same as CERCLA); Section 313 Priority Chemicals; Toxic Pollutant (Section 401.15) as copper and compounds.
RCRA Ground Water Monitoring List. Suggested test method(s) (PQL μg/L): 6010(60); 7210 (200). Note: All species in the ground water that contain copper are included.
EPCRA Section 304 Reportable Quantity (RQ): CERCLA, 10 lbs. (4.54 kgs.).
EPCRA Section 313 Form R de minimus concentration reporting level: 1.0%. Form R Toxic Chemical Category Code: N100 as copper compounds.

MARINE POLLUTANT (49CFR, Subchapter 172.101, Appendix B).

EPA NAME: CUPRIC NITRATE

[see also COPPER and COPPER COMPOUNDS]

CAS: 3251-23-8

SYNONYMS: COPPER NITRATE; COPPER(II) NITRATE; NITRIC ACID, COPPER(II) SALT

CLEAN WATER ACT: Section 311 Hazardous Substances/RQ (same as CERCLA); Section 313 Priority Chemicals; Toxic Pollutant (Section 401.15) as copper and compounds.

RCRA Ground Water Monitoring List. Suggested test method(s) (PQL μg/L): 6010(60); 7210 (200). Note: All species in the ground water that contain copper are included.

EPCRA Section 304 Reportable Quantity (RQ): CERCLA, 100 lbs. (45.4 kgs.).

EPCRA Section 313 Form R de minimus concentration reporting level: 1.0%. Form R Toxic Chemical Category Code: N100 as copper compounds.

EPA NAME: CUPRIC OXALATE

CAS: 5893-66-3

SYNONYMS: COPPER OXALATE

CLEAN WATER ACT: Section 311 Hazardous Substances/RQ (same as CERCLA); Section 313 Priority Chemicals; Toxic Pollutant (Section 401.15). as copper and compounds.

EPCRA Section 304 Reportable Quantity (RQ): CERCLA, 100 lbs. (45.4 kgs.).

RCRA Ground Water Monitoring List. Suggested test method(s) (PQL μg/L): 6010(60); 7210 (200) Note: All species in the ground water that contain copper are included.

EPCRA Section 313 Form R de minimus concentration reporting level: 1.0%. Form R Toxic Chemical Category Code: N100 as copper compounds.

EPA NAME: CUPRIC SULFATE

[see also COPPER and COPPER COMPOUNDS]

CAS: 7758-98-7

SYNONYMS: COPPER SULFATE; COPPER(II) SULFATE; SULFURIC ACID, COPPER(2+) SALT (1:1); SULFURIC ACID COPPER(II) SALT (1:1)

CLEAN WATER ACT: Section 311 Hazardous Substances/RQ (same as CERCLA); Section 313 Priority Chemicals; Toxic Pollutant (Section 401.15) as copper and compounds.

RCRA Ground Water Monitoring List. Suggested test method(s) (PQL μg/L): 6010 (60); 7210 (200). Note: All species in the ground water that contain copper are included.

EPCRA Section 304 Reportable Quantity (RQ): CERCLA, 10 lbs. (4.54 kgs.).

EPCRA Section 313 Form R de minimus concentration reporting level: 1.0%. Form R Toxic Chemical Category Code: N100 as copper compounds.

EPA NAME: CUPRIC SULFATE, AMMONIATED

[see also COPPER and COPPER COMPOUNDS]

CAS: 10380-29-7

SYNONYMS: COPPER(2+), TETRAAMINE-, SULFATE (1:1), MONOHYDRATE; COPPER AMMONIUM SULFATE; TETRA-AMINE COPPER SULFATE

CLEAN WATER ACT: Section 311 Hazardous Substances/RQ (same as CERCLA); Section 313 Priority Chemicals; Toxic Pollutant (Section 401.15) as copper and compounds.

RCRA Ground Water Monitoring List. Suggested test method(s) (PQL µg/L): 6010 (60); 7210 (200). Note: All species in the ground water that contain copper are included.

EPCRA Section 304 Reportable Quantity (RQ): CERCLA, 100 lbs. (45.4 kgs.).

EPCRA Section 313 Form R de minimus concentration reporting level: 1.0%. Form R Toxic Chemical Category Code: N100 as copper compounds.

EPA NAME: CUPRIC TARTRATE

[see also COPPER and COPPER COMPOUNDS]

CAS: 815-82-7

SYNONYMS: BUTANEDIOIC ACID, 2,3-DIHYDROXY-[R-(R*,R*)]-, COPPER(2+) SALT (1:1)

CLEAN WATER ACT: Section 311 Hazardous Substances/RQ (same as CERCLA); Section 313 Priority Chemicals; Toxic Pollutant (Section 401.15). as copper and compounds.

RCRA Ground Water Monitoring List. Suggested test method(s) (PQL µg/L): 6010(60); 7210 (200) Note: All species in the ground water that contain copper are included.

EPCRA Section 304 Reportable Quantity (RQ): CERCLA, 100 lbs. (45.4 kgs.).

EPCRA Section 313 Form R de minimus concentration reporting level: 1.0%. Form R Toxic Chemical Category Code: N100 as copper compounds.

EPA NAME: CYANAZINE

[see also CYANIDES and CYANIDE COMPOUNDS]

CAS: 21725-46-2

SYNONYMS: BLADEX; 2-CHLORO-4-((1-CYANO-1-METHYLETHYL)AMINO)-6-(ETHYLAMINO)-s-TRIAZINE; EPA PESTICIDE CHEMICAL CODE 100101

SAFE DRINKING WATER ACT: Priority List (55 FR 1470).

EPCRA Section 313 Form R de minimus concentration reporting level: 1.0%. Form R Toxic Chemical Category Code: N106.

MARINE POLLUTANT (49CFR, Subchapter 172.101, Appendix B) as cyanide mixtures, cyanide solutions or cyanides, inorganic, n.o.s.

CALIFORNIA'S PROPOSITION 65: Reproductive toxin.

EPA NAME: CYANIDE
[see also CYANIDES and CYANIDE COMPOUNDS]
CAS: 57-12-5
EPA HAZARDOUS WASTE NUMBER (RCRA No.): P030
CLEAN WATER ACT: Section 313 Priority Chemicals.
RCRA Land Ban Waste.
RCRA Universal Treatment Standards: Wastewater (mg/L), 1.2; Nonwastewater (mg/kg), 590; RCRA Universal Treatment Standards as cyanides (total): Wastewater (mg/L), 0.86; Nonwastewater (mg/kg), 30 as cyanides (amenable) Note: Both Cyanides (Total) and Cyanides (amenable) for nonwastewaters are to be analyzed using Method 9010 or 9012, found in "Test Methods for Evaluating Solid Waste, Physical/Chemical Methods," EPA Publication SW-846, as incorporated by reference in 40 CFR 260.11, with a sample size of 10 grams and a distillation time of one hour and 15 minutes.
SAFE DRINKING WATER ACT: MCL, 0.2 mg/L; MCLG, 0.2 mg/L as free cyanide; Regulated chemical (47 FR 9352).
MARINE POLLUTANT (49CFR, Subchapter 172.101, Appendix B) as cyanides, inorganic, n.o.s.

EPA NAME: CYANIDE COMPOUNDS
[see also CYANIDES]
CLEAN AIR ACT: Hazardous Air Pollutants (Title I, Part A, Section 112).
CLEAN WATER ACT: Section 307 Priority Pollutants as cyanide, total.
EPA HAZARDOUS WASTE NUMBER (RCRA No.): P030 as cyanides soluble salts and complexes, n.o.s.
RCRA Section 261 Hazardous Constituents as cyanides, soluble salts, and complexes, n.o.s.
EPCRA Section 313: X^+CN^- where $X = H^+$ or any other group where a formal dissociation may occur. For example, KCN or $Ca(CN)_2$. Form R de minimus concentration reporting level: 1.0%. Form R Toxic Chemical Category Code: N016.
MARINE POLLUTANT (49CFR, Subchapter 172.101, Appendix B) as cyanide mixtures, cyanide solutions or cyanides, inorganic, n.o.s.

EPA NAME: CYANIDES (SOLUBLE SALTS AND COMPLEXES)
[see also CYANIDE COMPOUNDS]
CAS: 57-12-5
SYNONYMS: CYANIDE ANION; CYANIDE, COMPLEXED
CLEAN AIR ACT: Hazardous Air Pollutants (Title I, Part A, Section 112).
CLEAN WATER ACT: Section 307 Priority Pollutants as cyanide, total; Toxic Pollutant (Section 401.15).
EPA HAZARDOUS WASTE NUMBER (RCRA No.): P030.
RCRA Section 261 Hazardous Constituents.
RCRA Land Ban Waste.
RCRA Universal Treatment Standards: Wastewater (mg/L), 1.2 (total); 0.86 (amenable); Nonwastewater (mg/kg), 590 (total); 30 (amenable).

RCRA Ground Water Monitoring List. Suggested test method(s) (PQL μg/L): 9010(40).

SAFE DRINKING WATER ACT: MCL, 0.2 mg/L; MCLG, 0.2 mg/L; Regulated chemical (47 FR 9352).

EPCRA Section 304 Reportable Quantity (RQ): CERCLA, 10 lbs. (4.54 kgs.).

EPCRA Section 313 Form R de minimus concentration reporting level: 1.0%. Form R Toxic Chemical Category Code: N016.

MARINE POLLUTANT (49CFR, Subchapter 172.101, Appendix B) as cyanides, inorganic, n.o.s.

EPA NAME: CYANOGEN

[see also CYANIDES and CYANIDE COMPOUNDS]

CAS: 460-19-5

SYNONYMS: CARBON NITRIDE; ETHANEDINITRILE; OXALONITRILE

CLEAN AIR ACT: Accidental Release Prevention/Flammable substances (Section 112[r] Table 3), TQ = 10,000 lbs. (4540 kgs.).

EPA HAZARDOUS WASTE NUMBER (RCRA No.): P031.

RCRA Section 261 Hazardous Constituents.

EPCRA Section 304 Reportable Quantity (RQ): EHS/CERCLA, 100 lbs. (45.4 kgs.).

EPCRA Section 313 Form R de minimus concentration reporting level: 1.0%. Form R Toxic Chemical Category Code: N106.

MARINE POLLUTANT (49CFR, Subchapter 172.101, Appendix B) as cyanides, inorganic, n.o.s.

EPA NAME: CYANOGEN BROMIDE

[see also CYANIDES and CYANIDE COMPOUNDS]

CAS: 506-68-3

SYNONYMS: BROMINE CYANIDE; CYANOGEN MONOBROMIDE

EPA HAZARDOUS WASTE NUMBER (RCRA No.): U246.

RCRA Section 261 Hazardous Constituents.

EPCRA Section 302 Extremely Hazardous Substances: TPQ = 500/10,000 lbs. (227/4,540 kgs.).

EPCRA Section 304 Reportable Quantity (RQ): CERCLA, 1,000 lbs. (454 kgs.).

EPCRA Section 313 Form R de minimus concentration reporting level: 1.0%. Form R Toxic Chemical Category Code: N106.

MARINE POLLUTANT (49CFR, Subchapter 172.10).

EPA NAME: CYANOGEN CHLORIDE

[see also CYANIDES and CYANIDE COMPOUNDS]

CAS: 506-77-4

SYNONYMS: CHLORINE CYANIDE; CYANOGEN CHLORIDE ((CN)CL)

CLEAN AIR ACT: Accidental Release Prevention/Flammable substances (Section 112[r] Table 3), TQ = 10,000 lbs. (4540 kgs.).

CLEAN WATER ACT: Section 311 Hazardous Substances/RQ (same as CERCLA); Section 313 Priority Chemicals.
EPA HAZARDOUS WASTE NUMBER (RCRA No.): P033.
RCRA Section 261 Hazardous Constituents.
SAFE DRINKING WATER ACT: Priority List (55 FR 1470).
EPCRA Section 304 Reportable Quantity (RQ): CERCLA, 10 lbs. (4.54 kgs.).
EPCRA Section 313 Form R de minimus concentration reporting level: 1.0%. Form R Toxic Chemical Category Code: N106.
MARINE POLLUTANT (49CFR, Subchapter 172.101, Appendix B).

EPA NAME: CYANOGEN IODIDE

[see also CYANIDES and CYANIDE COMPOUNDS]
CAS: 506-78-5
SYNONYMS: IODINE CYANIDE

EPCRA Section 302 Extremely Hazardous Substances: TPQ = 1,000/10,000 lbs. (454/4,540 kgs.).
EPCRA Section 304 Reportable Quantity (RQ): EHS, 1000 lbs. (454 kgs.).
EPCRA Section 313 Form R de minimus concentration reporting level: 1.0%. Form R Toxic Chemical Category Code: N106.
MARINE POLLUTANT (49CFR, Subchapter 172.101, Appendix B) as cyanides, inorganic, n.o.s.

EPA NAME: CYANOPHOS

CAS: 2636-26-2
SYNONYMS: O-(4-CYANOPHENYL) O,O-DIMETHYL PHOSPHOROTHIOATE; O,O-DIMETHYL-O-4-CYANOPHENYL-PHOSPHOROTHIOATE

EPCRA Section 302 Extremely Hazardous Substances: TPQ = 1,000 lbs. (454 kgs.).
EPCRA Section 304 Reportable Quantity (RQ): EHS, 1000 lbs. (454 kgs.).
EPCRA Section 313 Form R de minimus concentration reporting level: 1.0%. Form R Toxic Chemical Category Code: N106.
MARINE POLLUTANT (49CFR, Subchapter 172.101, Appendix B).

EPA NAME: CYANURIC FLUORIDE

[see also CYANIDES and CYANIDE COMPOUNDS]
CAS: 675-14-9
SYNONYMS: 2,4,6-TRIFLUORO-s-TRIAZINE

EPCRA Section 302 Extremely Hazardous Substances: TPQ = 100 lbs. (45.4 kgs.).
EPCRA Section 304 Reportable Quantity (RQ): EHS, 100 lbs. (45.4 kgs.).
EPCRA Section 313 Form R de minimus concentration reporting level: 1.0%. Form R Toxic Chemical Category Code: N106.
MARINE POLLUTANT (49CFR, Subchapter 172.101, Appendix B) as cyanides, inorganic, n.o.s.

EPA NAME: CYCLOATE
CAS: 1134-23-2
SYNONYMS: EPA PESTICIDE CHEMICAL CODE 041301; S-ETHYL CYCLOHEXYLETHYLCARBAMOTHIOATE

RCRA Land Ban Waste.
RCRA Universal Treatment Standards: Wastewater (mg/L), 0.003; Non-wastewater (mg/kg), 1.4.
EPCRA Section 304 Reportable Quantity (RQ): CERCLA, 1 lb. (0.454 kg.).
EPCRA Section 313 Form R de minimus concentration reporting level: 1.0%.

EPA NAME: 2,5-CYCLOHEXADIENE-1,4-DIONE, 2,3,5-TRIS(1-AZIRIDINYL)-
[see TRIAZIQUONE]
CAS: 68-76-8

EPA NAME: CYCLOHEXANAMINE
[see CYCLOHEXYLAMINE]
CAS: 108-91-8

EPA NAME: CYCLOHEXANE
CAS: 110-82-7
SYNONYMS: BENZENE, HEXAHYDRO; HEXAHYDROBENZENE

CLEAN WATER ACT: Section 311 Hazardous Substances/RQ (same as CERCLA); Section 313 Priority Chemicals.
EPA HAZARDOUS WASTE NUMBER (RCRA No.): U056.
RCRA Section 261 Hazardous Constituents.
EPCRA Section 304 Reportable Quantity (RQ): CERCLA, 1,000 lbs. (454 kgs.).
EPCRA Section 313 Form R de minimus concentration reporting level: 1.0%.

EPA NAME: 1,4-CYCLOHEXANE DIISOCYANATE
[see also DIISOCYANATES]
CAS: 2556-36-7
SYNONYMS: CYCLOHEXANE, 1,4-DIISOCYANATE; 1,4-CYCLOHEXANE DIISOCYANATO-

EPCRA Section 313 Form R de minimus concentration reporting level: 1.0%. Form R Toxic Chemical Category Code: N120.

EPA NAME: 2,5-CYCLOHEXANE, 1,2,3,4,5,6-HEXACHLORO-, (1α,2α,3β,4α,5α,6β)-
[see LINDANE]
CAS: 58-89-9

EPA NAME: CYCLOHEXANOL
CAS: 108-93-0
SYNONYMS: HEXAHYDROPHENOL

EPCRA Section 313 Form R de minimus concentration reporting level: 1.0%.

EPA NAME: CYCLOHEXANONE

CAS: 108-94-1

SYNONYMS: CYCLOHEXYL KETONE; KETOHEXAMETHYLENE

EPA HAZARDOUS WASTE NUMBER (RCRA No.): U057.

RCRA Section 261 Hazardous Constituents.

RCRA Land Ban Waste.

RCRA Universal Treatment Standards: Wastewater (mg/L), 0.36; Non-wastewater (mg/L), 0.75 TCLP.

EPCRA Section 304 Reportable Quantity (RQ): CERCLA, 5,000 lbs. (2270 kgs.).

EPA NAME: CYCLOHEXIMIDE

CAS: 66-81-9

SYNONYMS: 3(2-(3,5-DIMETHYL-2-OXOCYCLOHEXYL)-2-HYDROXYETHYL)GLUTARIMIDE; 2,6-PIPERIDINEDIONE, 4-[2-3,5-DIMETHYL-2-OXOCYCLOHEXYL)-2-HYDROXY-ETHYL]-,[IS-[1α(S*),3α,5β]-

EPCRA Section 302 Extremely Hazardous Substances: TPQ = 100/10,000 lbs. (45.4/4,540 kgs.).

EPCRA Section 304 Reportable Quantity (RQ): EHS, 100 lbs. (45.4 kgs.).

CALIFORNIA'S PROPOSITION 65: Reproductive toxin.

EPA NAME: CYCLOHEXYLAMINE

CAS: 108-91-8

SYNONYMS: CYCLOHEXANAMINE; HEXAHYDROBENZENAMINE

CLEAN AIR ACT: Accidental Release Prevention/Flammable substances (Section 112[r], Table 3), TQ = 15,000 lbs. (6810 kgs.).

EPCRA Section 302 Extremely Hazardous Substances: TPQ = 10,000 lbs. (4,540 kgs.).

EPCRA Section 304 Reportable Quantity (RQ): EHS, 10,000 lbs. (4540 kgs.).

EPA NAME: 2-CYCLOHEXYL-4,6-DINITROPHENOL

CAS: 131-89-5

SYNONYMS: 6-CYCLOHEXYL-2,4-DINITROPHENOL; DINITRO-o-CYCLOHEXYLPHENOL; DINEX

EPA HAZARDOUS WASTE NUMBER (RCRA No.): P034.

RCRA Section 261 Hazardous Constituents.

EPCRA Section 304 Reportable Quantity (RQ): CERCLA, 100 lbs. (45.4 kgs.).

EPA NAME: CYCLOPHOSPHAMIDE

CAS: 50-18-0 (anhydrous); 6055-19-2 (hydrated)

SYNONYMS: 2-(BIS(2-CHLOROETHYL)AMINO)-1-OXA-3-AZA-2-PHOSPHOCYCLOHEXANE 2-OXIDE MONOHYDRATE; 2H-1,3,2-OXAZAPHOSPHORIN-2-AMINE, N,N-BIS(2-CHLOROETHYL)TETRAHYDRO-, 2-OXIDE

EPA HAZARDOUS WASTE NUMBER (RCRA No.): U058.

RCRA Section 261 Hazardous Constituents.

EPCRA Section 304 Reportable Quantity (RQ): CERCLA, 10 lbs. (4.54 kgs.).

CALIFORNIA'S PROPOSITION 65: Carcinogen; Reproductive toxin (male).

EPA NAME: CYCLOPROPANE

CAS: 75-19-4

SYNONYMS: TRIMETHYLENE

CLEAN AIR ACT: Accidental Release Prevention/Flammable substances (Section 112[r], Table 3), TQ = 10,000 lbs. (4540 kgs.).

EPA NAME: CYFLUTHRIN

CAS: 68359-37-5

SYNONYMS: 3-(2,2-DICHLOROETHENYL)-2,2-DIMETHYL-CYCLOPROPANECARBOXYLIC ACID, CYANO(4-FLUORO-3-PHENOXYPHENYL)METHYL ESTER; POLY(OXY-1,2-ETHANEDIYL),α-[2-[BIS(2-AMINOETHYL)METHYL-AMMONIO]-ETHYL]-ω-HYDROXY-, N,N′-DICOCOACYL DERIVATIVES, METHYL SULFATES

EPCRA Section 313 Form R de minimus concentration reporting level: 1.0%.

EPA NAME: CYHALOTHRIN

CAS: 68085-85-8

SYNONYMS: 3-(CHLORO-3,3,3-TRIFLUORO-1-PROPENYL)-2,2-DIMETHYLCYCLOPROPANECARBOXYLIC ACID CYANO(3-PHENOXYPHENYL)METHYL ESTER; α-CYANO-3-PHENOXYBENZYL 3-(2-CHLORO-3,3,3-TRIFLUOROPROP-1-ENYL)-2,2-DIMETHYLCYCLOPROPANECARBOXYLATE; CYCLOPROPANECARBOXYLIC ACID, 3-(CHLORO-3,3,3-TRIFLUORO-1-PROPENYL)-2,2-DIMETHYL-,CYANO(3-PHENOXYPHENYL)METHYL ESTER

EPCRA Section 313 Form R de minimus concentration reporting level: 1.0%.

- D -

EPA NAME: 2,4-D

CAS: 94-75-7

SYNONYMS: ACETIC ACID (2,4-DICHLOROPHENOXY)-; 2,4-D ACID; DICHLOROPHENOXYACETIC ACID; 2,4-DICHLORPHENOXYACETIC ACID; 2,4-DICHLOROPHENOXYACETIC ACID, SALTS AND ESTERS; 2,4-D, SALTS AND ESTERS

CLEAN AIR ACT: Hazardous Air Pollutants (Title I, Part A, Section 112).

CLEAN WATER ACT: Section 311 Hazardous Substances/RQ (same as CERCLA); Section 313 Priority Chemicals. Reportable Quantity (RQ): CERCLA, 100 lbs. (45.5 kgs).

EPA HAZARDOUS WASTE NUMBER (RCRA No.): U240, D016.

RCRA Section 261 Hazardous Constituents.

RCRA Toxicity Characteristic (Section 261.24), Maximum Concentration of Contaminants, regulatory level, 10.0 mg/L.

RCRA Land Ban Waste.

RCRA Universal Treatment Standards: Wastewater (mg/L), 0.72; Nonwastewater (mg/kg), 10.

RCRA Ground Water Monitoring List. Suggested test method(s) (PQL µg/L): 8150(10).

SAFE DRINKING WATER ACT: MCL, 0.1 mg/L; MCGL, 0.07 mg/L; Regulated chemical (47 FR 9352) as 2,4-D.

EPCRA Section 304 Reportable Quantity (RQ): CERCLA, 100 lbs. (45.4 kgs.).

CERCLA/SARA 313: Form R de minimus concentration reporting level: 1.0%.

MARINE POLLUTANT (49CFR, Subchapter 172.101, Appendix B).

CALIFORNIA'S PROPOSITION 65: Carcinogen.

EPA NAME: 2,4-D ACID

[see 2,4-D]

CAS: 94-75-7

EPA NAME: DAUNOMYCIN

CAS: 20830-81-3

SYNONYMS: ACETYLADRIAMYCIN; 5,12-NAPHTHACENEDIONE, 8-ACETYL-10-((3-AMINO-2,3,6-TRIDEOXY-α-L-LYXO-HEXOPYRANOSYL)OXY)-7,8,9,10-TETRAHYDRO-6,8,11-TRIHYDROXY-8-(HYDROXYACETYL)-1-METHOXY HYDROCHLORIDE

EPA HAZARDOUS WASTE NUMBER (RCRA No.): U059.

RCRA Section 261 Hazardous Constituents.

EPCRA Section 304 Reportable Quantity (RQ): CERCLA, 10 lbs. (4.54 kgs.).

CALIFORNIA'S PROPOSITION 65: Carcinogen.

EPA NAME: DAZOMET

CAS: 533-74-4

SYNONYMS: 3,5-DIMETHYL-1,2,3,5-TETRAHYDRO-1,3,5-THIADIAZINETHIONE-2; EPA PESTICIDE CHEMICAL CODE 035602; TETRAHYDRO-3,5-DIMETHYL-2H-1,3,5-THIADIAZINE-2-THIONE; TETRAHYDRO-2H-3,5-DIMETHYL-1,3,5-THIADIAZINE-2-THIONE; 2-THIO-3,5-DIMETHYLTETRAHYDRO-1,3,5-THIADIAZINE

EPA HAZARDOUS WASTE NUMBER (RCRA No.): U366.

RCRA Section 261 Hazardous Constituents.

EPCRA Section 304 Reportable Quantity (RQ): CERCLA, 1 lb. (0.454 kg.).

EPCRA Section 313 Form R de minimus concentration reporting level: 1.0%.

EPA NAME: DAZOMET, SODIUM SALT

CAS: 53404-60-7

SYNONYMS: TETRAHYDRO-3,5-DIMETHYL-2H-1,3,5-THIADIAZINE-2-THIONE, ION(1-), SODIUM; 2H-1,3,5-THIADIAZINE-2-THIONE, TETRAHYDRO-3,5-DIMETHYL-, ION(1-), SODIUM

EPCRA Section 313 Form R de minimus concentration reporting level: 1.0%.

EPA NAME: 2,4-DB

CAS: 94-82-6

SYNONYMS: BUTANOIC ACID, 4-(2,4-DICHLOROPHENOXY)-; 2,4-D BUTYRIC ACID; 4-(2,4-DICHLOROPHENOXY)BUTYRIC ACID; EPA PESTICIDE CHEMICAL CODE 030801

EPCRA Section 313 Form R de minimus concentration reporting level: 1.0%.

EPA NAME: DBCP

[see 1,2-DIBROMO-3-CHLOROPROPANE]

CAS: 96-12-8

EPA NAME: 2,4-D BUTOXYETHYL ESTER

CAS: 1929-73-3

SYNONYMS: BUTOXYETHYL 2,4-DICHLOROPHENOXYACETATE; BUTOXYETHYL 2,4-D; EPA PESTICIDE CHEMICAL CODE 030053; 2,4-D ESTER; 2,4-DICHLOROPHENOXYACETIC ACID BUTOXYETHYL ESTER

CLEAN WATER ACT: Section 311 Hazardous Substances/RQ (same as CERCLA).

EPA HAZARDOUS WASTE NUMBER (RCRA No.): U240.

RCRA Section 261 Hazardous Constituents.

SAFE DRINKING WATER ACT: Regulated chemical (47 FR 9352) as 2,4-D.

EPCRA Section 304 Reportable Quantity (RQ): CERCLA, 100 lbs. (45.4 kgs.).

EPCRA Section 313 Form R de minimus concentration reporting level: 0.1%.

EPA NAME: 2,4-D BUTYL ESTER

CAS: 94-80-4

SYNONYMS: ACETIC ACID, (2,4-DICHLOROPHENOXY)-, BUTYL ESTER; BUTYL 2,4-D; n-BUTYL 2,4-D ESTER; 2,4-D ESTERS; 2,4-DICHLOROPHENOXYACETIC ACID, BUTYL ESTER

CLEAN WATER ACT: Section 311 Hazardous Substances/RQ (same as CERCLA).

EPA HAZARDOUS WASTE NUMBER (RCRA No.): U240.

RCRA Section 261 Hazardous Constituents.

SAFE DRINKING WATER ACT: Regulated chemical (47 FR 9352) as 2,4-D.

EPCRA Section 304 Reportable Quantity (RQ): CERCLA, 100 lbs. (45.4 kgs.).

EPCRA Section 313 Form R de minimus concentration reporting level: 0.1%.

EPA NAME: 2,4-D sec-BUTYL ESTER

CAS: 94-79-1

SYNONYMS: ACETIC ACID, (2,4-DICHLOROPHENOXY), sec-BUTYL ESTER; ACETIC ACID, (2,4-DICHLOROPHENOXY), 1-METHYL PROPYL ESTER; sec-BUTYL, 2,4-D; sec-BUTYL, 2,4-D ESTER; 2,4-D ESTERS; 1-METHYL PROPYL 2,4-D

CLEAN WATER ACT: Section 311 Hazardous Substances/RQ (same as CERCLA).

EPA HAZARDOUS WASTE NUMBER (RCRA No.): U240.

RCRA Section 261 Hazardous Constituents.

EPCRA Section 304 Reportable Quantity (RQ): CERCLA, 100 lbs. (45.4 kgs.).

EPA NAME: 2,4-D CHLOROCROTYL ESTER

CAS: 2971-38-2

SYNONYMS: ACETIC ACID, (2,4-DICHLOROPHENOXY)-, 4-CHLORO-2-BUTENYL ESTER; 2,4-D α-CHLOROCROTYL ESTER; 2,4-D CHLOROCROTYL ESTER; 2,4-D ESTERS; 2,4-DICHLOROPHENOXYACETIC ACID, 4-CHLOROCROTONYL ESTER

CLEAN WATER ACT: Section 311 Hazardous Substances/RQ (same as CERCLA).

EPA HAZARDOUS WASTE NUMBER (RCRA No.): U240.

RCRA Section 261 Hazardous Constituents.

SAFE DRINKING WATER ACT: Regulated chemical (47 FR 9352) as 2,4-D.

EPCRA Section 304 Reportable Quantity (RQ): CERCLA, 100 lbs. (45.4 kgs.).

EPCRA Section 313 Form R de minimus concentration reporting level: 0.1%.

EPA NAME: DDD

CAS: 72-54-8

SYNONYMS: BENZENE, 1,1′-(2,2-DICHLOROETHYLIDENE)BIS[4-CHLORO-; 1,1-DICHLORO-2,2-BIS(p-CHLORO-PHENYL) ETHANE; 4,4′-DDD; DDD, p,p′; p,p′-DDD; DICHLORODIPHENYL-DICHLOROETHANE; TDE; p,p-TDE; TETRACHLORODIPHENYLETHANE

CLEAN WATER ACT: Section 311 Hazardous Substances/RQ (same as CERCLA); Section 307 Priority Pollutants.

EPA HAZARDOUS WASTE NUMBER (RCRA No.): U060.

RCRA Section 261 Hazardous Constituents.

RCRA Land Ban Waste.

RCRA Universal Treatment Standards: Wastewater (mg/L), 0.023; Non-wastewater (mg/kg), 0.087.

RCRA Ground Water Monitoring List. Suggested test method(s) (PQL μg/L): 8080(0.1); 8270(10).

EPCRA Section 304 Reportable Quantity (RQ): CERCLA, 1 lb. (0.454 kg.).

CALIFORNIA'S PROPOSITION 65: Carcinogen.

EPA NAME: DDE

[see also CHLORINATED BENZENES]

CAS: 72-55-9

SYNONYMS: BENZENE, 1,1′-(DICHLOROETHENYLIDENE)[4-CHLORO-; 2,2-BIS(p-CHLOROPHENYL)-1,1-DICHLOROETHYLENE; 4,4-DDE; p,p′-DDE p,p-DDX; DICHLORODIPHENYLDICHLOROETHYLENE; p,p′-DICHLORODIPHENYLDICHLOROETHYLENE; (1,1′-DICHLOROETHENYLIDENE)BIS(4-CHLOROBENZENE)

CLEAN WATER ACT: Section 307 Priority Pollutants; Section 313 Priority Chemicals.

RCRA Section 261 Hazardous Constituents, waste number not listed.

RCRA Land Ban Waste.

RCRA Universal Treatment Standards: Wastewater (mg/L), 0.031; Non-wastewater (mg/kg), 0.087.

RCRA Ground Water Monitoring List. Suggested test method(s) (PQL μg/L): 8080(0.05); 8270(10).

EPCRA Section 304 Reportable Quantity (RQ): CERCLA, 1 lb. (0.454 kg.).

CALIFORNIA'S PROPOSITION 65: Carcinogen.

EPA NAME: DDE

CAS: 3547-04-4

SYNONYMS: 1,1-BIS(p-CHLOROPHENYL)-; 2,2-BIS(p-CHLOROPHENYL)ETHANE; p,p′-DICHLORODIPHENYL ETHANE

CLEAN AIR ACT: Hazardous Air Pollutants (Title I, Part A, Section 112).

EPCRA Section 304 Reportable Quantity (RQ): CERCLA, 5,000 lbs. (2270 kgs.).

EPA NAME: DDT

CAS: 50-29-3

SYNONYMS: BENZENE, 1,1′-(2,2,2-TRICHLOROETHYLIDENE)BIS(4-CHLORO); p,p′-DDT; 4,4′-DDT; DICHLORODIPHENYLTRICHLOROETHANE; p,p′-DICHLORODIPHENYLTRICHLOROETHANE; 4,4′-DICHLORODIPHENYLTRICHLOROETHANE; TRICHLORO-BIS(4-CHLOROPHENYL)ETHANE; 1,1,1-TRICHLORO-2,2-BIS(P-CHLOROPHENYL)ETHANE; 1,1,1-TRICHLORO-2,2-DI(4-CHLOROPHENYL)-ETHANE

CLEAN WATER ACT: Section 311 Hazardous Substances/RQ (same as CERCLA); Section 307 Priority Pollutants; Section 313 Priority Chemicals; Toxic Pollutant (Section 401.15).

EPA HAZARDOUS WASTE NUMBER (RCRA No.): U061.

RCRA Section 261 Hazardous Constituents.

RCRA Land Ban Waste.

RCRA Universal Treatment Standards: Wastewater (mg/L), 0.0039; Nonwastewater (mg/kg), 0.087.

RCRA Ground Water Monitoring List. Suggested test method(s) (PQL μg/L): 8080(0.1); 8270(10).

EPCRA Section 304 Reportable Quantity (RQ): CERCLA, 1 lb. (0.454 kg.).

MARINE POLLUTANT (49CFR, Subchapter 172.101, Appendix B): Severe pollutant.

CALIFORNIA'S PROPOSITION 65: Carcinogen.

EPA NAME: DECABORANE(14)

CAS: 17702-41-9

SYNONYMS: BORON HYDRIDE; DECARBORON TETRADECAHYDRIDE

EPCRA Section 302 Extremely Hazardous Substances: TPQ = 500/10,000 lbs. (227/4,540 kgs.).

EPCRA Section 304 Reportable Quantity (RQ): EHS, 500 lbs. (227 kgs.).

EPA NAME: DECABROMODIPHENYL OXIDE

[ssee also HALOETHERS]

CAS: 1163-19-5

SYNONYMS: BENZENE, 1,1′-OXYBIS[2,3,4,5,6-PENTABROMO-; DECABROMOBIPHENYL ETHER

EPCRA Section 313 Form R de minimus concentration reporting level: 1.0%.

EPA NAME: DEF

[see S,S,S-TRIBUTYLTRITHIOPHOSPHATE]

CAS: 78-48-8

EPA NAME: DEHP
[see DI(2-ETHYLHEXYL)PHTHALATE]
CAS: 117-81-7

EPA NAME: DEMETON
CAS: 8065-48-3
SYNONYMS: DEMETON O + DEMETON S; O,O-DIETHYL-2-ETHYLMERCAPTOETHYL THIOPHOSPHATE, DIETHOXY-THIOPHOSPHORIC ACID; PHOSPHOROTHIOIC ACID, O,O-DIETHYL O-2-(ETHYLTHIO)ETHYL ESTER, mixed with O,O-DIETHYL S-2-(ETHYLTHIO)ETHYL PHOSPHOROTHIOATE; SYSTOX

EPCRA Section 302 Extremely Hazardous Substances: TPQ = 500 lbs. (227 kgs.).
EPCRA Section 304 Reportable Quantity (RQ): EHS, 500 lbs. (227 kgs.).

EPA NAME: DEMETON-s-METHYL
CAS: 919-86-8
SYNONYMS: O,O-DIMETHYL-S-(2-ETHTHIOETHYL)PHOS-PHOROTHIOATE; METHYL-s-DEMETON

EPCRA Section 302 Extremely Hazardous Substances: TPQ = 500 lbs. (227 kgs.).
EPCRA Section 304 Reportable Quantity (RQ): EHS, 500 lbs. (227 kgs.).

EPA NAME: DESMEDIPHAM
CAS: 13684-56-5
SYNONYMS: CARBAMIC ACID, (3-(((PHENYLAMINO)CAR-BONYL)OXY)PHENYL)-,ETHYL ESTER; EPA PESTICIDE CHEMICAL CODE 104801; ETHYL (3-(((PHENYLAMI-NO)CARBONYL)OXY)PHENYL)CARBAMATE

EPCRA Section 313 Form R de minimus concentration reporting level: 1.0%.

EPA NAME: 2,4-D ESTERS
[see 2,4-D BUTOXYETHYL ESTER]
CAS: 1929-73-3

EPA NAME: 2,4-D ESTERS
[see 2,4-D BUTYL ESTER]
CAS: 94-80-4

EPA NAME: 2,4-D ESTERS
[see 2,4-D sec-BUTYL ESTER)
CAS: 94-79-1

EPA NAME: 2,4-D ESTERS
[see 2,4-D CHLOROCROTYL ESTER]
CAS: 2971-38-2

EPA NAME: 2,4-D ESTERS
[see 2,4-D ISOPROPYL ESTER]

CAS: 94-11-1

EPA NAME: 2,4-D ESTERS
[see 2,4-D ISOOCTYL ESTER
CAS: 25168-26-7

EPA NAME: 2,4-D ESTERS
[see 2,4-D METHYL ESTER]
CAS: 1928-38-7

EPA NAME: 2,4-D ESTERS
[see 2,4-D PROPYL ESTER]
CAS: 1928-61-6

EPA NAME: 2,4-D ESTERS
[see 2,4-D PROPYLENE GLYCOL BUTYL ESTER]
CAS: 1320-18-9

EPA NAME: 2,4-D ESTERS
CAS: 53467-11-1
SYNONYMS: 2,4-DICHLOROPHENOXYACETIC ACID ESTER; POLY[OXY(METHYL)-1,2-ETHANEDIYL],α-(2,4-DICHLOROPHENOXY)ACETYL-.ω.-BUTOXY-

CLEAN WATER ACT: Section 311 Hazardous Substances/RQ (same as CERCLA).
EPA HAZARDOUS WASTE NUMBER (RCRA No.): U240.
RCRA Section 261 Hazardous Constituents.
SAFE DRINKING WATER ACT: Regulated chemical (47 FR 9352) as 2,4-D.
EPCRA Section 304 Reportable Quantity (RQ): CERCLA, 100 lbs. (45.4 kgs.).

EPA NAME: 2,4-D 2-ETHYLHEXYL ESTER
CAS: 1928-43-4
SYNONYMS: ACETIC ACID, (2,4-DICHLOROPHENOXY)-,2-BUTOXYETHYL ESTER; EPA PESTICIDE CHEMICAL CODE 030063; 2-ETHYLHEXYL(2,4-DICHLOROPHENOXY)ACETATE; 2,4-D 2-ETHYLHEXYL ESTER

CLEAN WATER ACT: Section 311 Hazardous Substances/RQ (same as CERCLA).
EPCRA Section 304 Reportable Quantity (RQ): CERCLA, 100 lbs. (45.4 kgs.).
EPCRA Section 313 Form R de minimus concentration reporting level: 0.1%.

EPA NAME: 2,4-D 2-ETHYL-4-METHYLPENTYL ESTER
CAS: 53404-37-8
SYNONYMS: ACETIC ACID, (2,4-DICHLOROPHENOXY)-,2-ETHYL-4-METHYLPENTYL ESTER; 2,4-DICHLOROPHENOXYACETIC ACID ISOOCTYL(2-ETHYL-4-METHYLPENTYL) ESTER; EPA PESTICIDE CHEMICAL CODE 030064

CLEAN WATER ACT: Section 311 Hazardous Substances/RQ (same as CERCLA).

EPCRA Section 304 Reportable Quantity (RQ): CERCLA, 100 lbs. (45.4 kgs.).

EPCRA Section 313 Form R de minimus concentration reporting level: 0.1%.

EPA NAME: DIALLATE

CAS: 2303-16-4

SYNONYMS: BIS(1-METHYLETHYL) CARBAMOTHIOIC ACID, S-(2,3-DICHLORO-2-PROPENYL)ESTER; CARBAMODITHIOIC ACID, BIS(1-METHYLETHYL)-, S-(2,3-DICHLORO-2-PROPENYL)ESTER; DICHLOROALLYLDIISOPROPYLTHIOCARBAMATE

EPA HAZARDOUS WASTE NUMBER (RCRA No.): U062.

RCRA Section 261 Hazardous Constituents.

RCRA Ground Water Monitoring List. Suggested test method(s) (PQL µg/L): 8270(10).

EPCRA Section 304 Reportable Quantity (RQ): CERCLA, 100 lbs. (45.4 kgs.).

EPCRA Section 313 Form R de minimus concentration reporting level: 1.0%.

MARINE POLLUTANT (49CFR, Subchapter 172.101, Appendix B).

EPA NAME: DIALIFOR

CAS: 10311-84-9

SYNONYMS: S-(2-CHLORO-1-(1,3-DIHYDRO-1,3-DIOXO-2H-ISOINDOL-2-YL)ETHYL)-O,O-DIETHYL PHOSPHORODITHIOATE; DIALIFOS; PHOSPHORODITHIOIC ACID-S-(2-CHLORO-1-PHTHALIMIDOETHYL)-O,)DIETHYL ESTER

EPCRA Section 302 Extremely Hazardous Substances: TPQ = 100/10,000 lbs. (45.4/4,540 kgs.).

EPCRA Section 304 Reportable Quantity (RQ): EHS, 100 lbs. (45.4 kgs.).

MARINE POLLUTANT (49CFR, Subchapter 172.101, Appendix B).

EPA NAME: 2,4-DIAMINOSOLE

CAS: 615-05-4

SYNONYMS: 1,3-BENZENEDIAMINE, 4-METHOXY-; 1,3-DIAMINO-4-METHOXYBENZENE; 2,4-DIAMINOPHENYL METHYL ETHER; p-METHOXY-m-PHENYLENEDIAMINE

EPCRA Section 313 Form R de minimus concentration reporting level: 0.1%.

CALIFORNIA'S PROPOSITION 65: Carcinogen.

EPA NAME: 2,4-DIAMINOSOLE, SULFATE

CAS: 39156-41-7

SYNONYMS: 1,3-BENZENEDIAMINE, 4-METHOXY SULFATE; 2,4-DIAMINOANISOLE SULPHATE; 4-METHOXY-1,3-BENZENEDIAMINE SULPHATE

EPCRA Section 313 Form R de minimus concentration reporting level: 0.1%.

CALIFORNIA'S PROPOSITION 65: Carcinogen.

EPA NAME: 4,4′-DIAMINOPHENYL ETHER

CAS: 101-80-4

SYNONYMS: BENZENAMINE, 4,4′-OXYBIS-; BIS(4-AMINOPHENYL) ETHER; 4,4′-OXYDIANILINE

EPCRA Section 313 Form R de minimus concentration reporting level: 0.1%.

CALIFORNIA'S PROPOSITION 65: Carcinogen.

EPA NAME: DIAMINOTOLUENE

CAS: 496-72-0

SYNONYMS: 1,2-BENZENEDIAMINE, 4-METHYL; 3,4-DIAMINOTOLUENE; 3,4-TOLUYLENEDIAMINE; TOLUENE-3,4-DIAMINE

EPA HAZARDOUS WASTE NUMBER (RCRA No.): U221.

RCRA Section 261 Hazardous Constituents.

EPCRA Section 304 Reportable Quantity (RQ): CERCLA, 10 lbs. (4.54 kgs.).

EPA NAME: DIAMINOTOLUENE

CAS: 823-40-5

SYNONYMS: 1,3-BENZENEDIAMINE, 2-METHYL-; 2,6-DIAMINOTOLUENE; 2-METHYL-1,2-BENZENEDIAMINE

EPA HAZARDOUS WASTE NUMBER (RCRA No.): U221.

RCRA Section 261 Hazardous Constituents.

EPCRA Section 304 Reportable Quantity (RQ): CERCLA, 10 lbs. (4.54 kgs.).

EPA NAME: 2,4-DIAMINOTOLUENE

CAS: 95-80-7

SYNONYMS: 1,3-BENZENEDIAMINE, 4-METHYL; 2,4-DIAMINO-1-METHYLBENZENE; DIAMINOTOLUENE, 2,4-; 2,4-DIAMINOTOLUENE; TOLUENE-2,4-DIAMINE

EPA HAZARDOUS WASTE NUMBER (RCRA No.): U221.

RCRA Section 261 Hazardous Constituents.

EPCRA Section 304 Reportable Quantity (RQ): CERCLA, 10 lbs. (4.54 kgs.).

EPCRA Section 313 Form R de minimus concentration reporting level: 1.0%.

EPA NAME: DIAMINOTOLUENE (MIXED ISOMERS)

CAS: 25376-45-8

SYNONYMS: BENZENEDIAMINE, AR-METHYL-; DIAMINOTOLUENE; TOLUENE-AR,AR′-DIAMINE; TOLUENEDIAMINE; TOLYLENEDIAMINE

EPA HAZARDOUS WASTE NUMBER (RCRA No.): U221.

RCRA Section 261 Hazardous Constituents.

EPCRA Section 304 Reportable Quantity (RQ): CERCLA, 10 lbs. (4.54 kgs.).

EPCRA Section 313 Form R de minimus concentration reporting level: 1.0%.

CALIFORNIA'S PROPOSITION 65: Carcinogen.

EPA NAME: o-DIANISIDINE DIHYDROCHLORIDE

[see 3,3'-DIMETHOXYBENZIDINE DIHYDROCHLORIDE]

CAS: 20325-40-0

EPA NAME: o-DIANISIDINE HYDROCHLORIDE

[see 3,3'-DIMETHOXYBENZIDINE HYDROCHLORIDE]

CAS: 111984-09-9

EPA NAME: DIAZINON

CAS: 333-41-5

SYNONYMS: DIAZINON (ISO); O,O-DIETHYL O-2-ISOPROPYL-6-METHYLPYRIMIDIN-4-YLPHOSPHOROTHIONATE; O,O-DIETHYL O-(6-METHYL-2-(1-METHYLETHYL)-4-PYRIMIDINYL) PHOSPHORTHIOATE; PHOSPHOROTHIOIC ACID, O,O-DIETHYL O-(2-ISOPROPYL-6-METHYL-4-PYRIMIDINYL) ESTER; EPA PESTICIDE CHEMICAL CODE 057801; PHOSPHOROTHIOIC ACID, O,O-DIETHYL O-(6-METHYL-2-(1-METHYLETHYL)-4-PYRIMIDINYL) ESTER

CLEAN WATER ACT: Section 311 Hazardous Substances/RQ (same as CERCLA).

EPCRA Section 304 Reportable Quantity (RQ): CERCLA, 1 lb. (0.454 kg.).

EPCRA Section 313 Form R de minimus concentration reporting level: 1.0%.

MARINE POLLUTANT (49CFR, Subchapter 172.101, Appendix B): Severe pollutant

---

EPA NAME: DIAZOMETHANE

CAS: 334-88-3

SYNONYMS: AZIMETHYLENE; DIAZIRINE; DIAZONIUM METHYLIDE; METHANE, DIAZO

CLEAN AIR ACT: Hazardous Air Pollutants (Title I, Part A, Section 112)

EPCRA Section 304 Reportable Quantity (RQ): CERCLA, 100 lbs. (45.4 kgs.).

EPCRA Section 313 Form R de minimus concentration reporting level: 1.0%.

EPA NAME: DIBENZ(a,h)ACRIDINE

[see also POLYCYCLIC AROMATIC COMPOUNDS]

CAS: 226-36-8

SYNONYMS: DIBENZ(a,d)ACRIDINE

RCRA Section 261 Hazardous Constituents, waste number not listed.

EPCRA Section 313 Form R de minimus concentration reporting level: 0.1%. Form R Toxic Chemical Category Code: N590.
CALIFORNIA'S PROPOSITION 65: Carcinogen.

EPA NAME: DIBENZ(a,j)ACRIDINE
[see also POLYCYCLIC AROMATIC COMPOUNDS]
CAS: 224-42-0
SYNONYMS: 3,4,5,6-DIBENZACRIDINE; DIBENZO(a,j)ACRIDINE

RCRA Section 261 Hazardous Constituents, waste number not listed.
EPCRA Section 313 Form R de minimus concentration reporting level: 0.1%. Form R Toxic Chemical Category Code: N590.
CALIFORNIA'S PROPOSITION 65: Carcinogen.

EPA NAME: DIBENZ[a,h]ANTHRACINE
[see also POLYCYCLIC AROMATIC COMPOUNDS]
CAS: 53-70-3
SYNONYMS: 1,2,5,6-DIBENZANTHRACENE; DIBENZO[a,h] ANTHRACINE

CLEAN WATER ACT: Section 307 Priority Pollutants; Section 313 Priority Chemicals; Section 307 Toxic Pollutants, as polynuclear aromatic hydrocarbons.
EPA HAZARDOUS WASTE NUMBER (RCRA No.): U063.
RCRA Section 261 Hazardous Constituents.
RCRA Land Ban Waste.
RCRA Universal Treatment Standards: Wastewater (mg/L), 0.055; Nonwastewater (mg/kg), 8.2.
RCRA Ground Water Monitoring List. Suggested test method(s) (PQL µg/L): 8100(200); 8270(10).
EPCRA Section 304 Reportable Quantity (RQ): CERCLA, 1 lb. (0.454 kg.).
EPCRA Section 313 Form R de minimus concentration reporting level: 0.1%. Form R Toxic Chemical Category Code: N590.
CALIFORNIA'S PROPOSITION 65: Carcinogen.

EPA NAME: 7H-DIBENZO(c,g)CARBAZOLE
[see also POLYCYCLIC AROMATIC COMPOUNDS]
CAS: 194-59-2
SYNONYMS: DIBENZO(c,g)CARBAZOLE

RCRA Section 261 Hazardous Constituents, waste number not listed.
EPCRA Section 313 Form R de minimus concentration reporting level: 0.1%. Form R Toxic Chemical Category Code: N590.
CALIFORNIA'S PROPOSITION 65: Carcinogen.

EPA NAME: DIBENZO(a,e)FLUORANTHENE
[see also POLYCYCLIC AROMATIC COMPOUNDS]
CAS: 5385-75-1
SYNONYMS: DIBENZ(a,e)ACEANTHRYLENE

CLEAN WATER ACT: Section 307 Toxic Pollutants as polynuclear aromatic hydrocarbons.

EPCRA Section 313 Form R de minimus concentration reporting level: 0.1%. Form R Toxic Chemical Category Code: N590.

EPA NAME: DIBENZOFURAN

CAS: 132-64-9

SYNONYMS: 2,2′-BIPHENYLYLEME OXIDE; DIBENZO[b,d]FURAN

CLEAN AIR ACT: Hazardous Air Pollutants (Title I, Part A, Section 112) as dibenzofurans.

RCRA Ground Water Monitoring List. Suggested test method(s) (PQL µg/L): 8270(10).

EPCRA Section 304 Reportable Quantity (RQ): CERCLA, 100 lbs. (45.4 kgs.).

EPCRA Section 313 Form R de minimus concentration reporting level: 1.0%. Form R Toxic Chemical Category Code: N590.

EPA NAME: DIBENZO(a,e)PYRENE

[see also POLYCYCLIC AROMATIC COMPOUNDS]

CAS: 192-65-4

SYNONYMS: 1,2:4,5-DIBENZOPYRENE; NAPTHOL(1,2,3,4-def)CHRYSENE

CLEAN WATER ACT: Section 307 Toxic Pollutants as polynuclear aromatic hydrocarbons.

RCRA Section 261 Hazardous Constituents, waste number not listed.

RCRA Land Ban Waste.

RCRA Universal Treatment Standards: Wastewater (mg/L), 0.061; Nonwastewater (mg/kg), N/A.

EPCRA Section 313 Form R de minimus concentration reporting level: 0.1%. Form R Toxic Chemical Category Code: N590.

CALIFORNIA'S PROPOSITION 65: Carcinogen.

EPA NAME: DIBENZO(a,h)PYRENE

[see also POLYCYCLIC AROMATIC COMPOUNDS]

CAS: 189-64-0

SYNONYMS: DIBENZO(b,def)CHRYSENE; 1,2,6,7-DIBENZOPYRENE

CLEAN WATER ACT: Section 307 Toxic Pollutants as polynuclear aromatic hydrocarbons.

RCRA Section 261 Hazardous Constituents, waste number not listed.

EPCRA Section 313 Form R de minimus concentration reporting level: 0.1%. Form R Toxic Chemical Category Code: N590.

CALIFORNIA'S PROPOSITION 65: Carcinogen.

EPA NAME: DIBENZO(a,l)PYRENE

[see also POLYCYCLIC AROMATIC COMPOUNDS]

CAS: 191-30-0

SYNONYMS: DIBENZO(d,e,f,p)CHRYSENE

CLEAN WATER ACT: Section 307 Toxic Pollutants as polynuclear aromatic hydrocarbons.

EPCRA Section 313 Form R de minimus concentration reporting level: 0.1%. Form R Toxic Chemical Category Code: N590.
CALIFORNIA'S PROPOSITION 65: Carcinogen.

EPA NAME: DIBENZ[a,i]PYRENE
[see also POLYCYCLIC AROMATIC COMPOUNDS]
CAS: 189-55-9
SYNONYMS: BENZO(rst)PENTAPHENE; 3,4,9,10-DIBENZOPYRENE; DIBENZO(a,i)PYRENE; DIBENZO(b,h)PYRENE; 1,2,7,8-DIBENZOPYRENE; DIBENZO-3,4,5,9,10-PYRENE; 3,4:9,10-DIBENZOPYRENE; 3,4:9,10-DIBENZPYRENE; DIBENZPYRENE; 3,4,9,10-DIBENZPYRENE; 1,2:7,8-DIBENZPYRENE

CLEAN WATER ACT: Section 307 Priority Pollutants; Section 307 Toxic Pollutants as polynuclear aromatic hydrocarbons.
EPA HAZARDOUS WASTE NUMBER (RCRA No.): U064.
RCRA Section 261 Hazardous Constituents.
EPCRA Section 304 Reportable Quantity (RQ): CERCLA, 10 lbs. (4.54 kgs.).
EPCRA Section 313 Form R de minimus concentration reporting level: 0.1%. Form R Toxic Chemical Category Code: N590.
CALIFORNIA'S PROPOSITION 65: Carcinogen.

EPA NAME: DIBORANE
CAS: 19287-45-7
SYNONYMS: DIBORANE (6); DIBORANE HEXANHYDRIDE

CLEAN AIR ACT: Accidental Release Prevention/Flammable substances (Section 112[r], Table 3), TQ=2,500 lbs. (1135 kgs.).
EPCRA Section 302 Extremely Hazardous Substances: TPQ=100 lbs. (45.4 kgs.).
EPCRA Section 304 Reportable Quantity (RQ): EHS, 100 lbs. (45.4 kgs.).

EPA NAME: 1,2-DIBROMO-3-CHLOROPROPANE
CAS: 96-12-8
SYNONYMS: 3-CHLORO-1,2-DIBROMOPROPANE; DBCP; DIBROMOCHLOROPROPANE; PROPANE, 1,2-DIBROMO-3-CHLORO-

CLEAN AIR ACT: Hazardous Air Pollutants (Title I, Part A, Section 112).
EPA HAZARDOUS WASTE NUMBER (RCRA No.): U066.
RCRA Section 261 Hazardous Constituents.
RCRA Land Ban Waste.
RCRA Universal Treatment Standards: Wastewater (mg/L), 0.11; Nonwastewater (mg/kg), 15.
RCRA Ground Water Monitoring List. Suggested test method(s) (PQL µg/L): 8010(100); 8240(5); 8270(10).
SAFE DRINKING WATER ACT: MCL, 0.0002 mg/L; MCLG, zero; Regulated chemical (47 FR 9352).
EPCRA Section 304 Reportable Quantity (RQ): CERCLA, 1 lb. (0.454 kg.).
EPCRA Section 313 Form R de minimus concentration reporting level: 0.1%.

CALIFORNIA'S PROPOSITION 65: Carcinogen; Reproductive toxin (male).

EPA NAME: 1,2-DIBROMOETHANE

CAS: 106-93-4

SYNONYMS: ETHANE, 1,2-DIBROMO-; ETHYLENE DIBROMIDE; 1,2-ETHYLENE DIBROMIDE

CLEAN AIR ACT: Hazardous Air Pollutants (Title I, Part A, Section 112); Accidental Release Prevention/Flammable substances (Section 112[r], Table 3), TQ=20,000 lbs. (9080 kgs.).

CLEAN WATER ACT: Section 311 Hazardous Substances/RQ (same as CERCLA); Section 313 Priority Chemicals.

EPA HAZARDOUS WASTE NUMBER (RCRA No.): U067.

RCRA Section 261 Hazardous Constituents.

RCRA Land Ban Waste.

RCRA Universal Treatment Standards: Wastewater (mg/L), 0.028; Non-wastewater (mg/kg), 15.

RCRA Ground Water Monitoring List. Suggested test method(s) (PQL µg/L): 8010(10); 8240(5).

SAFE DRINKING WATER ACT: MCL, 0.00005 mg/L; MCGL, zero.

EPCRA Section 304 Reportable Quantity (RQ): CERCLA, 1 lb. (0.454 kg.).

EPCRA Section 313 Form R de minimus concentration reporting level: 0.1%.

CALIFORNIA'S PROPOSITION 65: Carcinogen.

EPA NAME: 3,5-DIBROMO-4-HYDROXYBENZONITRILE

[see BROMOXYNIL]

CAS: 1689-84-5

EPA NAME: 2,2-DIBROMO-3-NITRILOPROPIONAMIDE

CAS: 10222-01-2

SYNONYMS: 2,2-DIBROMO-2-CARBAMOYLACETONITRILE; EPA PESTICIDE CHEMICAL CODE 101801

EPCRA Section 313 Form R de minimus concentration reporting level: 1.0%.

Note: Subject to an administrative stay under EPCRA Section 313; not reportable until stay is lifted. See 10/27/95 (60 FR 54949).

EPA NAME: DIBROMOTETRAFLUOROETHANE

CAS: 124-73-2

SYNONYMS: ETHANE, 1,2-DIBROMOTETRAFLUORO; FREON 114B2; HALON 2402

CLEAN AIR ACT: Stratospheric ozone protection (Title VI, Subpart A, Appendix A), Class I, Ozone Depletion Potential=6.0.

EPCRA Section 313 Form R de minimus concentration reporting level: 1.0%.

EPA NAME: DIBUTYL PHTHALATE

CAS: 84-74-2

SYNONYMS: 1,2-BENZENEDICARBOXYLIC ACID, DIBUTYL ESTER; n-BUTYL PHTHALATE; DIBUTYL O-PHTHALATE; DI-n-BUTYL PHTHALATE

CLEAN AIR ACT: Hazardous Air Pollutants (Title I, Part A, Section 112).
CLEAN WATER ACT: Section 307 Priority Pollutants; Section 313 Priority Chemicals.
EPA HAZARDOUS WASTE NUMBER (RCRA No.): U069.
RCRA Section 261 Hazardous Constituents.
RCRA Land Ban Waste.
RCRA Section 261 Hazardous Constituents.
RCRA Universal Treatment Standards: Wastewater (mg/L), 0.057; Nonwastewater (mg/kg), 28.
RCRA Ground Water Monitoring List. Suggested test method(s) (PQL μg/L): 8060(5); 8270(10).
EPCRA Section 304 Reportable Quantity (RQ): CERCLA, 10 lbs. (4.55 kgs.).
EPCRA Section 313 Form R de minimus concentration reporting level: 1.0%.
MARINE POLLUTANT (49CFR, Subchapter 172.101, Appendix B).

EPA NAME: DICAMBA
CAS: 1918-00-9
SYNONYMS: BENZOIC ACID, 3,6-DICHLORO-2-METHOXY-; DICAMBRA; 3,6-DICHLORO-2-METHOXYBENZOIC ACID; 3,6-DICHLORO-2-METHOXYBENZOIC ACID; EPA PESTICIDE CHEMICAL CODE 029801

CLEAN WATER ACT: Section 311 Hazardous Substances/RQ (same as CERCLA).
SAFE DRINKING WATER ACT: Priority List (55 FR 1470).
EPCRA Section 304 Reportable Quantity (RQ): CERCLA, 1,000 lbs. (454 kgs.).
EPCRA Section 313 Form R de minimus concentration reporting level: 1.0%.

EPA NAME: DICHLOBENIL
CAS: 1194-65-6
SYNONYMS: BENZONITRILE, 2,6-DICHLORO-; 2,6-DICHLOROBENZONITRILE

CLEAN WATER ACT: Section 311 Hazardous Substances/RQ (same as CERCLA).
EPCRA Section 304 Reportable Quantity (RQ): CERCLA, 100 lbs. (45.4 kgs.).

EPA NAME: DICHLONE
CAS: 117-80-6
SYNONYMS: PHYGON; DICHLORONAPHTHOQUINONE; 1,4-NAPHTHALENEDIONE, 2,3-DICHLORO-

CLEAN WATER ACT: Section 311 Hazardous Substances/RQ (same as CERCLA).

EPCRA Section 304 Reportable Quantity (RQ): CERCLA, 1 lb. (0.454 kg.).

EPA NAME: DICHLORAN

CAS: 99-30-9

SYNONYMS: BENZENAMINE, 2,6-DICHLORO-4-NITRO-; 2,6-DICHLORO-4-NITROANILINE; EPA PESTICIDE CHEMICAL CODE 031301

EPCRA Section 313 Form R de minimus concentration reporting level: 1.0%.

EPA NAME: o-DICHLOROBENZENE

[see also CHLORINATED BENZENES]

CAS: 95-50-1

SYNONYMS: BENZENE, 1,2-DICHLORO-; DICHLOROBENZENE, (ortho); 1,2-DICHLOROBENZENE; o-DICHLOROBENZOL; ORTHODICHLOROBENZENE

CLEAN WATER ACT: Section 311 Hazardous Substances/RQ (same as CERCLA); Section 307 Priority Pollutants; Section 313 Priority Chemicals; Toxic Pollutant (Section 401.15).

EPA HAZARDOUS WASTE NUMBER (RCRA No.): U070.

RCRA Section 261 Hazardous Constituents.

RCRA Land Ban Waste.

RCRA Universal Treatment Standards: Wastewater (mg/L), 0.088; Nonwastewater (mg/kg), 6.0.

RCRA Ground Water Monitoring List. Suggested test method(s) (PQL μg/L): 8010(2); 8020(5); 8120(10); 8270(10).

SAFE DRINKING WATER ACT: MCL, 0.6 mg/L; MCLG, 0.6 mg/L; Regulated chemical (47 FR 9352) as dichlorobenzene.

EPCRA Section 304 Reportable Quantity (RQ): CERCLA, 100 lbs. (45.4 kgs.).

EPCRA Section 313 Form R de minimus concentration reporting level: 1.0%.

RCRA Land Ban Waste.

MARINE POLLUTANT (49CFR, Subchapter 172.101, Appendix B).

EPA NAME: 1,2-DICHLOROBENZENE

[see o-DICHLOROBENZENE]

CAS: 95-50-1

EPA NAME: 1,3-DICHLOROBENZENE

[see also CHLORINATED BENZENES]

CAS: 541-73-1

SYNONYMS: BENZENE, 1,3-DICHLORO-; m-DICHLOROBENZENE; METADICHLOROBENZENE

CLEAN WATER ACT: Section 311 Hazardous Substances/RQ (same as CERCLA); Section 307 Priority Pollutants; Section 313 Priority Chemicals; Toxic Pollutant (Section 401.15).

EPA HAZARDOUS WASTE NUMBER (RCRA No.): U071.

RCRA Section 261 Hazardous Constituents.

RCRA Land Ban Waste.
RCRA Universal Treatment Standards: Wastewater (mg/L), 0.036; Nonwastewater (mg/kg), 6.0.
RCRA Ground Water Monitoring List. Suggested test method(s) (PQL μg/L): 8010(5); 8020(5); 8120(10); 8270(10).
SAFE DRINKING WATER ACT: Regulated chemical (47 FR 9352) as dichlorobenzene; Priority List (55 FR 1470).
EPCRA Section 304 Reportable Quantity (RQ): CERCLA, 100 lbs. (45.4 kgs.).
EPCRA Section 313 Form R de minimus concentration reporting level: 1.0%.
MARINE POLLUTANT (49CFR, Subchapter 172.101, Appendix B).

EPA NAME: 1,4-DICHLOROBENZENE

[see also CHLORINATED BENZENES]
CAS: 106-46-7
SYNONYMS: BENZENE, 1,4-DICHLORO-; p-DICHLOROBENZENE; PARADICHLOROBENZENE

CLEAN AIR ACT: Hazardous Air Pollutants (Title I, Part A, Section 112).
CLEAN WATER ACT: Section 311 Hazardous Substances/RQ (same as CERCLA); Section 307 Priority Pollutants; Section 313 Priority Chemicals; Toxic Pollutant (Section 401.15).
EPA HAZARDOUS WASTE NUMBER (RCRA No.): U072; D027.
RCRA Section 261 Hazardous Constituents.
RCRA Toxicity Characteristic (Section 261.24), Maximum Concentration of Contaminants, regulatory level, 7.5 mg/L.
RCRA Land Ban Waste.
RCRA Universal Treatment Standards: Wastewater (mg/L), 0.090; Nonwastewater (mg/kg), 6.0.
RCRA Ground Water Monitoring List. Suggested test method(s) (PQL μg/L): 8010(2); 8020(5); 8120(15); 8270(10).
SAFE DRINKING WATER ACT: MCL, 0.075 mg/L; MCLG, 0.075 mg/L; Regulated chemical (47 FR 9352) as dichlorobenzene.
EPCRA Section 304 Reportable Quantity (RQ): CERCLA, 100 lbs. (45.4 kgs.).
EPCRA Section 313 Form R de minimus concentration reporting level: 0.1%.
RCRA Land Ban Waste.
MARINE POLLUTANT (49CFR, Subchapter 172.101, Appendix B).
CALIFORNIA'S PROPOSITION 65: Carcinogen.

EPA NAME: DICHLOROBENZENE (MIXED ISOMERS)

[see also CHLORINATED BENZENES]
CAS: 25321-22-6
SYNONYMS: BENZENE, DICHLORO-; DICHLOROBENZENE; DICHLORICIDE

CLEAN WATER ACT: Section 311 Hazardous Substances/RQ (same as CERCLA); Section 307 Priority Pollutants; Section 313 Priority Chemicals; Toxic Pollutant (Section 401.15).
RCRA Section 261 Hazardous Constituents, waste number not listed.

SAFE DRINKING WATER ACT: Regulated chemical (47 FR 9352).
EPCRA Section 304 Reportable Quantity (RQ): CERCLA, 100 lbs. (45.4 kgs.).
EPCRA Section 313 Form R de minimus concentration reporting level: 0.1%.
MARINE POLLUTANT (49CFR, Subchapter 172.101, Appendix B).

EPA NAME: 3,3′-DICHLOROBENZIDINE
CAS: 91-94-1
SYNONYMS: BENZIDINE, 3,3′-DICHLORO-; DICHLOROBENZIDINE; 3,3′-DICHLOROBENZIDENE

CLEAN AIR ACT: Hazardous Air Pollutants (Title I, Part A, Section 112).
CLEAN WATER ACT: Section 307 Priority Pollutants; Section 313 Priority Chemicals; Toxic Pollutant (Section 401.15).
EPA HAZARDOUS WASTE NUMBER (RCRA No.): U073.
RCRA Section 261 Hazardous Constituents.
RCRA Ground Water Monitoring List. Suggested test method(s) (PQL μg/L): 8270(20).
EPCRA Section 304 Reportable Quantity (RQ): CERCLA, 1 lb. (0.454 kg.).
EPCRA Section 313 Form R de minimus concentration reporting level: 0.1%.
CALIFORNIA'S PROPOSITION 65: Carcinogen.

EPA NAME: 3,3′-DICHLOROBENZIDINE DIHYDROCHLORIDE
CAS: 612-83-9
SYNONYMS: BENZIDINE, 3,3′-DICHLORO-, DIHYDROCHLORIDE; (1,1′BIPHENYL)-4,4′-DIAMINE, 3,3′-DICHLORO-, DIHYDROCHLORIDE

EPCRA Section 313 Form R de minimus concentration reporting level: 0.1%.

EPA NAME: 3,3′-DICHLOROBENZIDINE SULFATE
CAS: 64969-34-2
SYNONYMS: (1,1′-BIPHENYL)-4,4′-DIAMINE, 3,3′-DICHLORO-, SULFATE (1:2)

EPCRA Section 313 Form R de minimus concentration reporting level: 0.1%.

EPA NAME: DICHLOROBROMOMETHANE
CAS: 75-27-4
SYNONYMS: BROMODICHLOROMETHANE; METHANE, BROMODICHLORO-

CLEAN WATER ACT: Section 307 Priority Pollutants; Section 313 Priority Chemicals.
RCRA Land Ban Waste.
RCRA Universal Treatment Standards: Wastewater (mg/L), 0.35; Nonwastewater (mg/kg), 15.

RCRA Ground Water Monitoring List. Suggested test method(s) (PQL µg/L): 8010(1); 8240(5).

SAFE DRINKING WATER ACT: Priority List (55 FR 1470) as bromodichloromethane.

EPCRA Section 304 Reportable Quantity (RQ): CERCLA, 5,000 lbs. (2270 kgs.).

EPCRA Section 313 Form R de minimus concentration reporting level: 1.0%.

CALIFORNIA'S PROPOSITION 65: Carcinogen.

EPA NAME: 1,4-DICHLORO-2-BUTENE

CAS: 764-41-0

SYNONYMS: 2, BUTENE, 1,4-DICHLORO-; 1,4-DICHLORO-2-BUTENE; 1,4-DICHLOROBUTENE-2

EPA HAZARDOUS WASTE NUMBER (RCRA No.): U074.

RCRA Section 261 Hazardous Constituents.

EPCRA Section 304 Reportable Quantity (RQ): CERCLA, 1 lb. (0.454 kg.).

EPCRA Section 313 Form R de minimus concentration reporting level: 1.0%.

CALIFORNIA'S PROPOSITION 65: Carcinogen.

EPA NAME: trans-1,4-DICHLORO-2-BUTENE

CAS: 110-57-4

SYNONYMS: 2-BUTYLENE DICHLORIDE; trans-1,4-DICHLOROBUTENE

RCRA Ground Water Monitoring List. Suggested test method(s) (PQL µg/L): 8240(5).

EPCRA Section 302 Extremely Hazardous Substances: TPQ = 500 lbs. (227 kgs.).

EPCRA Section 304 Reportable Quantity (RQ): EHS, 500 lbs. (227 kgs.).

EPCRA Section 313 Form R de minimus concentration reporting level: 1.0%.

EPA NAME: trans-1,4-DICHLOROBUTENE

[see trans-1,4-DICHLORO-2-BUTENE]

CAS: 110-57-4

EPA NAME: 4,6-DICHLORO-N-(2-CHLOROPHENYL)-1,3,5-TRIAZIN-2-AMINE

[see ANILAZINE]

CAS: 101-05-3

EPA NAME: 1,2-DICHLORO-1,1-DIFLUOROETHANE

CAS: 1649-08-7

SYNONYMS: 1,2-DICHLORO-2,2-DIFLUOROETHANE; ETHANE, 1,2-DICHLORO-1,1-DIFLUORO-; HCFC-132b

CLEAN AIR ACT: Stratospheric ozone protection (Title VI, Subpart A, Appendix B), Class II, Ozone Depletion Potential = listed as "reserved."

EPCRA Section 313 Form R de minimus concentration reporting level: 1.0%.

EPA NAME: DICHLORODIFLUOROMETHANE

CAS: 75-71-8

SYNONYMS: CFC-12; DIFLUORODICHLOROMETHANE; FREON 12; METHANE, DICHLORODIFLUORO-

CLEAN AIR ACT: Stratospheric ozone protection (Title VI, Subpart A, Appendix A), Class I, Ozone Depletion Potential = 1.0.
EPA HAZARDOUS WASTE NUMBER (RCRA No.): U075.
RCRA Section 261 Hazardous Constituents.
RCRA Land Ban Waste.
RCRA Universal Treatment Standards: Wastewater (mg/L), 0.23; Nonwastewater (mg/kg), 7.2.
RCRA Ground Water Monitoring List. Suggested test method(s) (PQL μg/L): 8010(10); 8240(5).
SAFE DRINKING WATER ACT: Priority List (55 FR 1470).
EPCRA Section 304 Reportable Quantity (RQ): CERCLA, 5,000 lbs. (2270 kgs.).
EPCRA Section 313 Form R de minimus concentration reporting level: 1.0%.

EPA NAME: 1,1-DICHLOROETHANE

[see ETHYLIDENE DICHLORIDE]

CAS: 75-34-3

EPA NAME: 1,2-DICHLOROETHANE

CAS: 107-06-2

SYNONYMS: DICHLOROETHYLENE; ETHANE, 1,2-DICHLORO-; ETHYLENE CHLORIDE; ETHYLENE DICHLORIDE; 1,2-ETHYLENE DICHLORIDE

CLEAN AIR ACT: Hazardous Air Pollutants (Title I, Part A, Section 112).
CLEAN WATER ACT: Section 311 Hazardous Substances/RQ (same as CERCLA); Section 307 Priority Pollutants; Section 313 Priority Chemicals; Toxic Pollutant (Section 401.15).
EPA HAZARDOUS WASTE NUMBER (RCRA No.): U077, D028.
RCRA Section 261 Hazardous Constituents.
RCRA Toxicity Characteristic (Section 261.24), Maximum Concentration of Contaminants, regulatory level, 0.5 mg/L.
RCRA Land Ban Waste.
RCRA Universal Treatment Standards: Wastewater (mg/L), 0.21; Nonwastewater (mg/kg), 6.0.
RCRA Ground Water Monitoring List. Suggested test method(s) (PQL μg/L): 8010(0.5); 8240(5).
SAFE DRINKING WATER ACT: MCL, 0.005 mg/L; MCLG, zero; Regulated chemical (47 FR 9352).
EPCRA Section 304 Reportable Quantity (RQ): CERCLA, 100 lbs. (45.4 kgs.).
MARINE POLLUTANT (49CFR, Subchapter 172.101, Appendix B).
CALIFORNIA'S PROPOSITION 65: Carcinogen.

EPA NAME: (3-(2,2-DICHLOROETHENYL)-2,2-DIMETHYL-CYCLOPROPANE CARBOXYLIC ACID, (3-PHENOXYPHENYL)METHYL ESTER)
[see PERMETHRIN]
CAS: 52645-53-1

EPA NAME: 3-(2,2-DICHLOROETHENYL)-2,2-DIMETHYLCYCLOPROPANECARBOXYLIC ACID, CYANO(4-FLUORO-3-PHENOXYPHENYL)METHYL ESTER
[see CYFLUTHRIN]
CAS: 68359-37-5

EPA NAME: DICHLOROETHYLENE
[see VINYLIDENE CHLORIDE]
CAS: 75-35-4

EPA NAME: 1,2-DICHLOROETHYLENE
CAS: 156-60-5
SYNONYMS: trans-ACETYLENE DICHLORIDE; ETHENE, 1,2-DICHLORO-, (E)-; ETHENE, trans-1,2-DICHLORO-; 1,2-trans-DICHLOROETHYLENE; 1,2-DICHLOROETHYLENE-trans-

CLEAN WATER ACT: Section 307 Priority Pollutants; Section 313 Priority Chemicals.
EPA HAZARDOUS WASTE NUMBER (RCRA No.): U079.
RCRA Section 261 Hazardous Constituents.
RCRA Land Ban Waste.
RCRA Universal Treatment Standards: Wastewater (mg/L), 0.054; Nonwastewater (mg/kg), 30.
RCRA Ground Water Monitoring List. Suggested test method(s) (PQL μg/L): 8010(1); 8240(5).
SAFE DRINKING WATER ACT: MCL, 0.1 mg/L; MCLG, 0.1 mg/L; Regulated chemical (47 FR 9352).
EPCRA Section 304 Reportable Quantity (RQ): CERCLA, 1,000 lbs. (454 kgs.).

EPA NAME: 1,2-DICHLOROETHYLENE
CAS: 540-59-0
SYNONYMS: ACETYLENE DICHLORIDE; trans-ACETYLENE DICHLORIDE; cis & trans-1,2-DICHLOROETHENE; ETHYLENE, 1,2-DICHLORO-

CLEAN WATER ACT: Section 307 Priority Pollutants; Section 313 Priority Chemicals; Toxic Pollutant (Section 401.15).
EPCRA Section 313 Form R de minimus concentration reporting level: 1.0%.

EPA NAME: DICHLOROETHYL ETHER
[see BIS(2-CHLOROETHYL)ETHER]
CAS: 111-44-4

EPA NAME: 1,1-DICHLORO-1-FLUOROETHANE
CAS: 1717-00-6

SYNONYMS: ETHANE, 1,1-DICHLORO-1-FLUORO-; FREON 141; HCFC-141b

CLEAN AIR ACT: Stratospheric ozone protection (Title VI, Subpart A, Appendix B), Class II, Ozone Depletion Potential = 0.12.
EPCRA Section 313 Form R de minimus concentration reporting level: 1.0%.

EPA NAME: DICHLOROFLUOROMETHANE

CAS: 75-43-4; 39289-28-6
SYNONYMS: DICHLOROMONOFLUOROMETHANE; FLUORODICHLOROMETHANE; FREON F 21; HCFC-21; METHANE, DICHLOROFLUORO-

CLEAN AIR ACT: Stratospheric ozone protection (Title VI, Subpart A, Appendix B), Class II, Ozone Depletion Potential = listed as "reserved."
EPCRA Section 313 Form R de minimus concentration reporting level: 1.0%.

EPA NAME: DICHLOROISOPROPYL ETHER

[see BIS(2-CHLORO-1-METHYLETHYL) ETHER]
CAS: 108-60-1

EPA NAME: DICHLOROMETHANE

CAS: 75-09-2
SYNONYMS: METHANE, DICHLORO-; METHYLENE CHLORIDE

CLEAN AIR ACT: Hazardous Air Pollutants (Title I, Part A, Section 112).
CLEAN WATER ACT: Section 307 Priority Pollutants.
EPA HAZARDOUS WASTE NUMBER (RCRA No.): U080.
RCRA Section 261 Hazardous Constituents.
RCRA Land Ban Waste.
SAFE DRINKING WATER ACT: Regulated chemical (47 FR 9352); MCL, 0.005 mg/L; MCLG, zero.
RCRA Universal Treatment Standards: Wastewater (mg/L), 0.089; Nonwastewater (mg/kg), 30.
RCRA Ground Water Monitoring List. Suggested test method(s) (PQL μg/L): 8010(5); 8240(5).
SAFE DRINKING WATER ACT: MCL, 0.005mg/L; MCLG, zero; Regulated chemical (47 FR 9352).
EPCRA Section 304 Reportable Quantity (RQ): CERCLA, 1,000 lbs. (454 kgs.).
EPCRA Section 313 Form R de minimus concentration reporting level: 1.0%.
CALIFORNIA'S PROPOSITION 65: Carcinogen.

EPA NAME: 3,6-DICHLORO-2-METHOXYBENZOIC ACID

[see DICAMBA]
CAS: 1918-00-9

EPA NAME: 3,6-DICHLORO-2-METHOXYBENZOIC ACID, SODIUM SALT)

[see SODIUM DICAMBA]

CAS: 1982-69-0

EPA NAME: DICHLOROMETHYLETHER
[see BIS(CHLOROMETHYL)ETHER]
CAS: 542-88-1

EPA NAME: 3-[2,4-DICHLORO-5-(1-METHYLETHOXY)PHENYL]-5-(1,1-DIMETHYLETHYL)-1,3,4-OXADIAZOL-2(3H)-ONE
[see OXYDIAZON]
CAS: 19666-30-9

EPA NAME: DICHLOROMETHYLPHENYLSILANE
CAS: 149-74-6
SYNONYMS: METHYLPHENYLDICHLOROSILANE; PHENYLMETHYLDICHLOROSILANE

EPCRA Section 302 Extremely Hazardous Substances: TPQ = 1,000 lbs. (454 kgs.).
EPCRA Section 304 Reportable Quantity (RQ): EHS, 1000 lbs. (454 kgs.).

EPA NAME: 2,6-DICHLORO-4-NITROANILINE
[see DICHLORAN]
CAS: 99-30-9

EPA NAME: DICHLOROPENTAFLUOROPROPANE
CAS: 127564-92-5
SYNONYMS: PROPANE, DICHLOROPENTAFLUORO-

EPCRA Section 313 Form R de minimus concentration reporting level: 1.0%.

EPA NAME: 1,1-DICHLORO-1,2,2,3,3-PENTAFLUOROPROPANE
CAS: 13474-88-9
SYNONYMS: HCFC-225cc; PROPANE, 1,1-DICHLORO-1,2,2,3,3-PENTAFLUORO-

EPCRA Section 313 Form R de minimus concentration reporting level: 1.0%.

EPA NAME: 1,1-DICHLORO-1,2,3,3,3-PENTAFLUOROPROPANE
CAS: 111512-56-2
SYNONYMS: HCFC-225eb; PROPANE, 1,1-DICHLORO-1,2,3,3,3-PENTAFLUORO-

EPCRA Section 313 Form R de minimus concentration reporting level: 1.0%.

EPA NAME: 1,2-DICHLORO-1,1,2,3,3-PENTAFLUOROPROPANE
CAS: 422-44-6

SYNONYMS: HCFC-225bb; PROPANE, 1,2-DICHLORO-1,1,2,3,3-PENTAFLUORO-

EPCRA Section 313 Form R de minimus concentration reporting level: 1.0%.

EPA NAME: 1,2-DICHLORO-1,1,3,3,3-PENTAFLUOROPROPANE

CAS: 431-86-7

SYNONYMS: HCFC-225da; PROPANE, 1,2-DICHLORO-1,1,2,3,3-PENTAFLUORO-

EPCRA Section 313 Form R de minimus concentration reporting level: 1.0%.

EPA NAME: 1,3-DICHLORO-1,1,2,2,3-PENTAFLUOROPROPANE

CAS: 507-55-1

SYNONYMS: HCFC-225cb; PROPANE, 1,3-DICHLORO-1,1,2,2,3-PENTAFLUORO-

CLEAN AIR ACT: Stratospheric ozone protection (Title VI, Subpart A, Appendix B), Class II, Ozone Depletion Potential = listed as "reserved."

EPCRA Section 313 Form R de minimus concentration reporting level: 1.0%.

EPA NAME: 1,3-DICHLORO-1,1,2,3,3-PENTAFLUOROPROPANE

CAS: 136013-79-1

SYNONYMS: HCFC-225ea; PROPANE, 1,3-DICHLORO-1,1,2,3,3-PENTAFLUORO-

EPCRA Section 313 Form R de minimus concentration reporting level: 1.0%.

EPA NAME: 2,2-DICHLORO-1,1,1,3,3-PENTAFLUOROPROPANE

CAS: 128903-21-9

SYNONYMS: HCFC-225aa; 1,1,1,3,3-PENTAFLUORO-2,2-DICHLOROPROPANE

EPCRA Section 313 Form R de minimus concentration reporting level: 1.0%.

EPA NAME: 2,3-DICHLORO-1,1,1,2,3-PENTAFLUOROPROPANE

CAS: 422-48-0

SYNONYMS: HCFC-225ba; PROPANE, 2,3-DICHLORO-1,1,1,2,3-PENTAFLUORO-

EPCRA Section 313 Form R de minimus concentration reporting level: 1.0%.

EPA NAME: 3,3-DICHLORO-1,1,1,2,2-PENTAFLUOROPROPANE

CAS: 422-56-0

SYNONYMS: 1,1-DICHLORO-2,2,3,3,3-PENTAFLUOROPROPANE; HCFC 225ca

CLEAN AIR ACT: Stratospheric ozone protection (Title VI, Subpart A, Appendix B), Class II, Ozone Depletion Potential = listed as "reserved."

EPCRA Section 313 Form R de minimus concentration reporting level: 1.0%.

EPA NAME: DICHLOROPHENE

CAS: 97-23-4

SYNONYMS: DICHLOROPHEN; 2,2'-METHYLENEBIS(4-CHLOROPHENOL); PHENOL, 2,2'-METHYLENEBIS(4-CHLORO-; DI(5-CHLORO-2-HYDROXYPHENYL)METHANE; EPA PESTICIDE CHEMICAL CODE 055001

EPCRA Section 313 Form R de minimus concentration reporting level: 1.0%. Form R Toxic Chemical Category Code: N084.

EPA NAME: 2,4-DICHLOROPHENOL

CAS: 120-83-2

SYNONYMS: 4,6-DICHLOROPHENOL; PHENOL, 2,4-DICHLORO-

CLEAN WATER ACT: Section 307 Toxic Pollutants; Section 313 Priority Chemicals.

EPA HAZARDOUS WASTE NUMBER (RCRA No.): U081.

RCRA Section 261 Hazardous Constituents.

RCRA Land Ban Waste.

RCRA Universal Treatment Standards: Wastewater (mg/L), 0.044; Nonwastewater (mg/kg), 14.

RCRA Ground Water Monitoring List. Suggested test method(s) (PQL μg/L): 8040(5); 8270(10).

EPCRA Section 304 Reportable Quantity (RQ): CERCLA, 100 lbs. (45.4 kgs.).

EPCRA Section 313 Form R de minimus concentration reporting level: 1.0%. Form R Toxic Chemical Category Code: N084.

Note: Threshold determinations should be made individually and separately from the chlorophenols category.

MARINE POLLUTANT (49CFR, Subchapter 172.101, Appendix B).

EPA NAME: 2,6-DICHLOROPHENOL

[see also CHLOROPHENOLS]

CAS: 87-65-0

CLEAN WATER ACT: Section 307 Priority Pollutants; Section 313 Priority Chemicals.

EPA HAZARDOUS WASTE NUMBER (RCRA No.): U082.

RCRA Section 261 Hazardous Constituents.

RCRA Land Ban Waste.

RCRA Universal Treatment Standards: Wastewater (mg/L), 0.044; Nonwastewater (mg/kg), 14.

RCRA Ground Water Monitoring List. Suggested test method(s) (PQL μg/L): 8270(10).
EPCRA Section 304 Reportable Quantity (RQ): CERCLA, 100 lbs. (45.4 kgs.).
EPCRA Section 313 Form R de minimus concentration reporting level: 1.0%. Form R Toxic Chemical Category Code: N084.
MARINE POLLUTANT (49CFR, Subchapter 172.101, Appendix B).

EPA NAME: 2-[4-(2,4-DICHLOROPHENOXY)PHENOXY]PROPANOIC ACID, METHYL ESTER
[see DICLOFOP METHYL]
CAS: 51338-27-3

EPA NAME: DICHLOROPHENYLARSINE
CAS: 696-28-6
SYNONYMS: ARSINE, DICHLOROPHENYL-; ARSONOUS DICHLORIDE, PHENYL-; PHENYLARSINEDICHLORIDE; PHENYLARSONOUS DICHLORIDE; PHENYLDICHLOROARSINE

EPA HAZARDOUS WASTE NUMBER (RCRA No.): P036.
RCRA Section 261 Hazardous Constituents.
EPCRA Section 302 Extremely Hazardous Substances: TPQ = 500 lbs. (227 kgs.).
EPCRA Section 304 Reportable Quantity (RQ): EHS/CERCLA, 1 lb. (0.454 kg.).

EPA NAME: (3-(3,5-DICHLOROPHENYL)-5-ETHENYL-5-METHYL-2,4-OXAZOLIDINEDIONE)
[see VINCLOZOLIN]
CAS: 50471-44-8

EPA NAME: 2-(3,4-DICHLOROPHENYL)-4-METHYL-1,2,4-OXADIAZOLIDINE-3,5-DIONE
[see METHAZOLE]
CAS: 20354-26-1

EPA NAME: N-(3,4-DICHLOROPHENYL)PROPANAMIDE
[see PROPANIL]
CAS: 709-98-8

EPA NAME: 1-[2-(2,4-DICHLOROPHENYL)-2-(2-PROPENYLOXY)ETHYL]-1H-IMIDAZOLE
[see IMAZALIL]
CAS: 35554-44-0

EPA NAME: 1-[2-(2,4-DICHLOROPHENYL)-4-PROPYL-1,3-DIOXOLAN-2-YL]-METHYL-1H-1,2,4,-TRIAZOLE
[see PROPICONAZOLE]
CAS: 60207-90-1

EPA NAME: DICHLOROPROPANE
CAS: 26638-19-7

SYNONYMS: PROPANE, DICHLORO-; PROPYLENE DICHLORIDE

CLEAN WATER ACT: Section 311 Hazardous Substances/RQ (same as CERCLA); Toxic Pollutant (Section 401.15).
RCRA Section 261 Hazardous Constituents, as dichloropropane, n.o.s, waste number not listed.
EPCRA Section 304 Reportable Quantity (RQ): CERCLA, 1,000 lbs. (454 kgs.).
MARINE POLLUTANT (49CFR, Subchapter 172.101, Appendix B).

EPA NAME: DICHLOROPROPANE–DICHLOROPROPENE (MIXTURE)
CAS: 8003-19-8
SYNONYMS: D-D MIXTURE VIDDEN D; 1-PROPENE, 1,3-DICHLORO- mixed with 1,2-DICHLOROPROPANE

CLEAN WATER ACT: Section 311 Hazardous Substances/RQ (same as CERCLA); Toxic Pollutant (Section 401.15).
RCRA Section 261 Hazardous Constituents, waste number not listed.
EPCRA Section 304 Reportable Quantity (RQ): CERCLA, 100 lbs. (45.4 kgs.).
MARINE POLLUTANT (49CFR, Subchapter 172.101, Appendix B).

EPA NAME: 1,1-DICHLOROPROPANE
CAS: 78-99-9
SYNONYMS: PROPYLIDENE CHLORIDE; PROPANE, 1,1-DICHLORO-

CLEAN WATER ACT: Section 311 Hazardous Substances/RQ (same as CERCLA); Toxic Pollutant (Section 401.15).
RCRA Section 261 Hazardous Constituents, waste number not listed.
EPCRA Section 304 Reportable Quantity (RQ): CERCLA, 1,000 lbs. (454 kgs.).
MARINE POLLUTANT (49CFR, Subchapter 172.101, Appendix B).

EPA NAME: 1,2-DICHLOROPROPANE
CAS: 78-87-5
SYNONYMS: PROPANE, 1,2-DICHLORO; PROPYLENE CHLORIDE; PROPYLENE DICHLORIDE

CLEAN AIR ACT: Hazardous Air Pollutants (Title I, Part A, Section 112).
CLEAN WATER ACT: Section 311 Hazardous Substances/RQ (same as CERCLA); Section 307 Priority Pollutants; Section 313 Priority Chemicals.
EPA HAZARDOUS WASTE NUMBER (RCRA No.): U083.
RCRA Section 261 Hazardous Constituents.
RCRA Universal Treatment Standards: Wastewater (mg/L), 0.85; Nonwastewater (mg/kg), 18.
RCRA Ground Water Monitoring List. Suggested test method(s) (PQL µg/L): 8010(0.5); 8240(5).
SAFE DRINKING WATER ACT: MCL, 0.005 mg/L; MCLG, zero; Regulated chemical (47 FR 9352); Priority List (55 FR 1470).

EPCRA Section 304 Reportable Quantity (RQ): CERCLA, 1,000 lbs. (454 kgs.).
EPCRA Section 313 Form R de minimus concentration reporting level: 1.0%.
MARINE POLLUTANT (49CFR, Subchapter 172.101, Appendix B).
CALIFORNIA'S PROPOSITION 65: Carcinogen.

EPA NAME: 1,3-DICHLOROPROPANE
CAS: 142-28-9
SYNONYMS: TRIMETHYLENE DICHLORIDE

CLEAN WATER ACT: Section 311 Hazardous Substances/RQ (same as CERCLA).
SAFE DRINKING WATER ACT: Priority List (55 FR 1470).
EPCRA Section 304 Reportable Quantity (RQ): CERCLA, 1,000 lbs. (454 kgs.).
MARINE POLLUTANT (49CFR, Subchapter 172.101, Appendix B).
CALIFORNIA'S PROPOSITION 65: Carcinogen.

EPA NAME: DICHLOROPROPENE
CAS: 26952-23-8
SYNONYMS: DICHLOROPROPYLENE; 1-PROPENE, DICHLORO-

CLEAN WATER ACT: Section 311 Hazardous Substances/RQ (same as CERCLA); Toxic Pollutant (Section 401.15).
RCRA Section 261 Hazardous Constituents, waste number not listed.
SAFE DRINKING WATER ACT: Priority List (55 FR 1470).
EPCRA Section 304 Reportable Quantity (RQ): CERCLA, 100 lbs. (45.4 kgs.).
MARINE POLLUTANT (49CFR, Subchapter 172.101, Appendix B).
CALIFORNIA'S PROPOSITION 65: Carcinogen.

EPA NAME: 1,3-DICHLOROPROPENE
[see 1,3-DICHLOROPROPYLENE]
CAS: 542-75-6

EPA NAME: trans-1,3-DICHLOROPROPENE
CAS: 10061-02-6
SYNONYMS: (E)-1,3-DICHLOROPROPENE; trans-1,3-DICHLOROPROPYLENE; 1-PROPENE, 1,3-DICHLORO-, (E)-

RCRA Land Ban Waste.
RCRA Universal Treatment Standards: Wastewater (mg/L), 0.036; Nonwastewater (mg/kg), 18.
RCRA Ground Water Monitoring List. Suggested test method(s) (PQL µg/L): 8010(5); 8240(5).
EPCRA Section 313 Form R de minimus concentration reporting level: 0.1%.

EPA NAME: 2,3-DICHLOROPRENE
CAS: 78-88-6
SYNONYMS: 2,3-DICHLORO-1-PROPENE; 2,3-DICHLOROPROPYLENE; PROPENE, 2,3-DICHLORO; 1-PROPENE, 2,3-DICHLORO-

CLEAN WATER ACT: Section 311 Hazardous Substances/RQ (same as CERCLA).

EPCRA Section 304 Reportable Quantity (RQ): CERCLA, 100 lbs. (45.4 kgs.).

EPCRA Section 313 Form R de minimus concentration reporting level: 1.0%.

EPA NAME: 2,2-DICHLOROPROPIONIC ACID

CAS: 75-99-0

SYNONYMS: DALAPON; PROPANOIC ACID, 2,2-DICHLORO-

CLEAN WATER ACT: Section 311 Hazardous Substances/RQ (same as CERCLA).

SAFE DRINKING WATER ACT: MCL, 0.2 mg/L; MCLG, 0.2 mg/L.

EPCRA Section 304 Reportable Quantity (RQ): CERCLA, 5,000 lbs. (2270 kgs.).

EPA NAME: 1,3-DICHLOROPROPYLENE

CAS: 542-75-6

SYNONYMS: 1,3-DICHLORO-1-PROPENE; 1,3-DICHLOROPROPENE; 1,3-DICHLORO-2-PROPENE; PROPENE, 1,3-DICHLORO-

CLEAN AIR ACT: Hazardous Air Pollutants (Title I, Part A, Section 112).

CLEAN WATER ACT: Section 311 Hazardous Substances/RQ (same as CERCLA); Section 307 Priority Pollutants; Section 313 Priority Chemicals.

EPA HAZARDOUS WASTE NUMBER (RCRA No.): U084.

RCRA Section 261 Hazardous Constituents.

EPCRA Section 304 Reportable Quantity (RQ): CERCLA, 100 lbs. (45.4 kgs.).

EPCRA Section 313 Form R de minimus concentration reporting level: 0.1%.

MARINE POLLUTANT (49CFR, Subchapter 172.101, Appendix B).

CALIFORNIA'S PROPOSITION 65: Carcinogen.

EPA NAME: DICHLOROSILANE

CAS: 4109-96-0

SYNONYMS: SILANE, DICHLORO-

CLEAN AIR ACT: Accidental Release Prevention/Flammable substances (Section 112[r], Table 3), TQ = 10,000 lbs. (4540 kgs.).

EPA NAME: DICHLOROTETRAFLUOROETHANE

CAS: 76-14-2

SYNONYMS: CFC-114; 1,2-DICHLORO-1,1,2,2-TETRAFLUOROETHANE; FREON 114; 1,1,2,2-TETRAFLUORO-1,2-DICHLOROETHANE

CLEAN AIR ACT: Stratospheric ozone protection (Title VI, Subpart A, Appendix A), Class I, Ozone Depletion Potential = 1.0.

EPCRA Section 313 Form R de minimus concentration reporting level: 1.0%.

EPA NAME: DICHLOROTRIFLUROETHANE
CAS: 34077-87-7
SYNONYMS: ETHANE, DICHLOROTRIFLUORO-; HCFC-123

CLEAN AIR ACT: Stratospheric ozone protection (Title VI, Subpart A, Appendix A), Class II, Ozone Depletion Potential = 0.02.
EPCRA Section 313 Form R de minimus concentration reporting level: 1.0%.

EPA NAME: DICHLORO-1,1,2-TRIFLUOROETHANE
CAS: 90454-18-5
SYNONYMS: ETHANE, DICHLORO-1,1,2-TRIFLUORO-; HCFC-123

CLEAN AIR ACT: Stratospheric ozone protection (Title VI, Subpart A, Appendix A), Class II, Ozone Depletion Potential = 0.02.
EPCRA Section 313 Form R de minimus concentration reporting level: 1.0%.

EPA NAME: 1,1-DICHLORO-1,2,2-TRIFLUOROETHANE
CAS: 812-04-4
SYNONYMS: ETHANE, 1,1-DICHLORO-1,2,2-TRIFLUORO-; HCFC-123b

EPCRA Section 313 Form R de minimus concentration reporting level: 1.0%.

EPA NAME: 1,2-DICHLORO-1,1,2-TRIFLUOROETHANE
CAS: 354-23-4
SYNONYMS: ETHANE, 1,2-DICHLORO-1,1,2-TRIFLUORO-; HCFC-123a; R-123a; 1,1,2-TRIFLUORO-1,2-DICHLOROETHANE

EPCRA Section 313 Form R de minimus concentration reporting level: 1.0%.

EPA NAME: 2,2-DICHLORO-1,1,1-TRIFLUOROETHANE
CAS: 306-83-2
SYNONYMS: ETHANE, 2,2-DICHLORO-1,1,1-TRIFLUORO-; FREON 123; HCFC-123; R-123; 1,1,1-TRIFLUORO-2,2-DICHLOROETHANE

CLEAN AIR ACT: Stratospheric ozone protection (Title VI, Subpart A, Appendix B), Class II, Ozone Depletion Potential = 0.02.
EPCRA Section 313 Form R de minimus concentration reporting level: 1.0%.

EPA NAME: DICHLORVOS
CAS: 62-73-7
SYNONYMS: DIMETHYL 2,2-DICHLOROETHENYL PHOSPHATE; O,O-DIMETHYL 2,2-DICHLOROVINYL PHOSPHATE; DIMETHYL O,O-DICHLOROVINYL-2,2-PHOSPHATE; PHOSPHORIC ACID, 2-DICHLOROETHENYL DIMETHYL ESTER; PHOSPHORIC ACID, 2,2-DICHLOROVINYL DIMETHYL ESTER

CLEAN AIR ACT: Hazardous Air Pollutants (Title I, Part A, Section 112).
CLEAN WATER ACT: Section 311 Hazardous Substances/RQ (same as CERCLA); Section 313 Priority Chemicals.
EPCRA Section 302 Extremely Hazardous Substances: TPQ = 1,000 lbs. (454 kgs.).
EPCRA Section 304 Reportable Quantity (RQ): EHS/CERCLA, 10 lbs. (4.54 kgs.).
EPCRA Section 313 Form R de minimus concentration reporting level: 1.0%.
MARINE POLLUTANT (49CFR, Subchapter 172.101, Appendix B).
CALIFORNIA'S PROPOSITION 65: Carcinogen.

EPA NAME: DICLOFOP METHYL
CAS: 51338-27-3
SYNONYMS: 2-(4-(2,4-DICHLOROPHENOXY)PHENOXY)PROPANOIC ACID, METHYL ESTER; EPA PESTICIDE CHEMICAL CODE 110902; HOELON; METHYL 2-(4-(2,4-DICHLOROPHENOXY)PHENOXY)PROPIONATE

EPCRA Section 313 Form R de minimus concentration reporting level: 1.0%.

EPA NAME: DICOFOL
CAS: 115-32-2
SYNONYMS: BENZENEMETHANOL,4-CHLORO-α-(4-CHLOROPHENYL)-α-(TRICHLOROMETHYL)-; 1,1-BIS(P-CHLOROPHENYL)-2,2,2-TRICHLOROETHANOL; DTMC; KELTHANE; 2,2,2-TRICHLORO-1,1-BIS(4-CHLOROPHENYL)ETHANOL; 2,2,2-TRICHLORO-1,1-BIS(P-CHLOROPHENYL)ETHANOL

CLEAN WATER ACT: Section 311 Hazardous Substances/RQ (same as CERCLA); Section 313 Priority Chemicals.
EPCRA Section 304 Reportable Quantity (RQ): CERCLA, 10 lbs. (4.54 kgs.).
EPCRA Section 313 Form R de minimus concentration reporting level: 1.0%.

EPA NAME: DICROTOPHOS
CAS: 141-66-2
SYNONYMS: BIDRIN; 3-HYDROXY-N,N-DIMETHYL-, cis-,DIMETHYL PHOSPHATE; CROTONAMIDE, 3-HYDROXY-N-N-DIMETHYL-, DIMETHYLPHOSPHATE, (E)-; DIDRIN; PHOSPHORIC ACID, 3-(DIMETHYLAMINO)-1-METHYL-3-OXO-1-PROPENYL DIMETHYL ESTER, (E)-

EPCRA Section 302 Extremely Hazardous Substances: TPQ = 100 lbs. (45.4 kgs.).
EPCRA Section 304 Reportable Quantity (RQ): EHS, 100 lbs. (45.4 kgs.).
MARINE POLLUTANT (49CFR, Subchapter 172.101, Appendix B).

EPA NAME: DICYCLOPENTADIENE
CAS: 77-73-6

SYNONYMS: BICYCLOPENTADIENE; 4,7-METHANO-1H-INDENE, 3A,4,7,7A-TETRAHYDRO-

EPCRA Section 313 Form R de minimus concentration reporting level: 1.0%.

EPA NAME: DIELDRIN

CAS: 60-57-1

SYNONYMS: ALVIT; 2,7:3,6-DIMETHANONAPHTH[2,3B] OXIRENE,3,4,5,6,9,9-HEXACHLORO-1A,2,2A,3,6,6A,7,7A-OCTAHYDRO-(1A α,2β,2Aα,3β,6β,6Aα,7β,7A α); 1,2,3,4,10,10-HEXACHLORO-6,7-EPOXY-1,4,4A,5,6,7,8,8A-OCTAHYDRO-1,4-endo-exo-5, 8-DI-METHANONAPHTHALENE; 3,4,5,6,9,9-HEXACHLORO-1A,2,2A,3,6,6A,7,7A-OCTAHYDRO-2,7:3,6-DIMETHANO

CLEAN WATER ACT: Section 311 Hazardous Substances/RQ (same as CERCLA); Section 307 Priority Pollutants; Section 313 Priority Chemicals; Toxic Pollutant (Section 401.15).

EPA HAZARDOUS WASTE NUMBER (RCRA No.): P037.

RCRA Section 261 Hazardous Constituents.

RCRA Land Ban Waste.

RCRA Universal Treatment Standards: Wastewater (mg/L), 0.017; Nonwastewater (mg/kg), 0.13.

RCRA Ground Water Monitoring List. Suggested test method(s) (PQL μg/L): 8080(0.05); 8270(10).

EPCRA Section 304 Reportable Quantity (RQ): CERCLA, 1 lb. (0.454 kg.).

MARINE POLLUTANT (49CFR, Subchapter 172.101, Appendix B).

CALIFORNIA'S PROPOSITION 65: Carcinogen.

EPA NAME: DIEPOXYBUTANE

CAS: 1464-53-5

SYNONYMS: 1,1′-BI[ETHYLENE OXIDE]; BIOXIRANE; 2,2′-BIOXIRANE; 1,3-BUTADIENE DIEPOXIDE; BUTADIENE DIEPOXIDE; BUTADIENE DIOXIDE; BUTANE DIEPOXIDE; BUTANE, 1,2:3,4-DIEPOXY-; DIOXYBUTADIENE

EPA HAZARDOUS WASTE NUMBER (RCRA No.): U085.

RCRA Section 261 Hazardous Constituents.

EPCRA Section 302 Extremely Hazardous Substances: TPQ = 500 lbs. (227 kgs.).

EPCRA Section 304 Reportable Quantity (RQ): EHS/CERCLA, 10 lbs. (4.54 kgs.).

EPCRA Section 313 Form R de minimus concentration reporting level: 0.1%.

CALIFORNIA'S PROPOSITION 65: Carcinogen.

EPA NAME: DIETHANOLAMINE

CAS: 111-42-2

SYNONYMS: N,N-BIS(2-HYDROXYETHYL)AMINE; ETHANOL, 2,2′-IMINOBIS-; 2,2′-IMINOBIS[ETHANOL]; 2,2′-IMINODIETHANOL

CLEAN AIR ACT: Hazardous Air Pollutants (Title I, Part A, Section 112).
EPCRA Section 304 Reportable Quantity (RQ): CERCLA, 100 lbs. (45.4 kgs.).
EPCRA Section 313 Form R de minimus concentration reporting level: 1.0%.

EPA NAME: DIETHATYL ETHYL
CAS: 38727-55-8
SYNONYMS: (CHLOROACETYL)-N-(2,6-DIETHYLPHENYL) GLYCINE ETHYL ESTER

EPCRA Section 313 Form R de minimus concentration reporting level: 1.0%.

EPA NAME: DIETHYLAMINE
CAS: 109-89-7
SYNONYMS: n-ETHYLETHANAMINE

CLEAN WATER ACT: Section 311 Hazardous Substances/RQ 100 lbs. (45.4 kgs.).
EPCRA Section 304 Reportable Quantity (RQ): CERCLA, 100 lbs. (45.4 kgs.).

EPA NAME: O-(2-(DIETHYLAMINO)-6-METHYL-4-PYRIMIDINYL)-O,O-DIMETHYL PHOSPHOROTHIOATE
[see PIRIMIPHOS METHYL]
CAS: 29232-93-7

EPA NAME: N,N-DIETHYLANILINE
CAS: 91-66-7
SYNONYMS: BENZENAMINE, N,N-DIETHYL-; DIETHYLPHENYLAMINE

EPCRA Section 304 Reportable Quantity (RQ): CERCLA, 1,000 lbs. (454 kgs.).

EPA NAME: DIETHYLARSINE
CAS: 692-42-2
SYNONYMS: ARSINE, DIETHYL-

EPA HAZARDOUS WASTE NUMBER (RCRA No.): P038.
RCRA Section 261 Hazardous Constituents.
EPCRA Section 304 Reportable Quantity (RQ): CERCLA, 1 lb. (0.454 kg.).

EPA NAME: DIETHYLCARBAMAZINE CITRATE
CAS: 1642-54-2
SYNONYMS: DIETHYLCARBAMAZINE ACID CITRATE; N,N-DIETHYL-4-METHYL-1-PIPERAZINE CARBOXAMIDE CITRATE

Removed from EHS list (FR Vol. 61, No. 89, page 20477).

EPA NAME: DIETHYL CHLOROPHOSPHATE
CAS: 814-49-3
SYNONYMS: CHLOROPHOSPHORIC ACID DIETHYL ESTER

EPCRA Section 302 Extremely Hazardous Substances: TPQ = 500 lbs. (227 kgs.).
EPCRA Section 304 Reportable Quantity (RQ): EHS, 500 lbs. (227 kgs.).

EPA NAME: DIETHYLDIISOCYANATOBENZENE
[see also DIISOCYANATES]
CAS: 134190-37-7
SYNONYMS: BENZENE, DIETHYLDIISOCYANATO-

EPCRA Section 313 Form R de minimus concentration reporting level: 1.0%. Form R Toxic Chemical Category Code: N120.

EPA NAME: DI(2-ETHYLHEXYL)PHTHALATE
CAS: 117-81-7
SYNONYMS: 1,2-BENZENEDICARBOXYLIC ACID, BIS(2-ETHYLHEXYL) ESTER; BIS(2-ETHYLHEXYL)PHTHALATE; BIS(2-ETHYLHEXYL)-1,2-BENZENEDICARBOXYLATE; DEHP; DI(2-ETHYLHEXYL)PHTHALATE; DI-sec-OCTYL PHTHALATE; DIOCTYL PHTHALATE; OCTYL PHTHALATE, DI-sec

CLEAN AIR ACT: Hazardous Air Pollutants (Title I, Part A, Section 112).
CLEAN WATER ACT: Section 307 Priority Pollutants.
EPA HAZARDOUS WASTE NUMBER (RCRA No.): U028.
RCRA Section 261 Hazardous Constituents.
RCRA Land Ban Waste.
RCRA Universal Treatment Standards: Wastewater (mg/L), 0.28; Nonwastewater (mg/kg), 28.
RCRA Ground Water Monitoring List. Suggested test method(s) (PQL µg/L): 8060(20); 8270(10).
SAFE DRINKING WATER ACT: MCL, 0.006 mg/L; MCLG, zero.
EPCRA Section 304 Reportable Quantity (RQ): CERCLA, 100 lbs. (45.4 kgs.).
EPCRA Section 313 Form R de minimus concentration reporting level: 0.1%.
CALIFORNIA'S PROPOSITION 65: Carcinogen.

EPA NAME: O,O-DIETHYL S-METHYL DITHIOPHOSPHATE
CAS: 3288-58-2
SYNONYMS: PHOSPHORODITHIOC ACID, O,O-DIETHYL S-METHYL ESTER

EPA HAZARDOUS WASTE NUMBER (RCRA No.): U087.
RCRA Section 261 Hazardous Constituents.
EPCRA Section 304 Reportable Quantity (RQ): CERCLA, 5,000 lbs. (2270 kgs.).

EPA NAME: DIETHYL-p-NITROPHENYL PHOSPHATE
CAS: 311-45-5
SYNONYMS: p-NITROPHENYL DIETHYLPHOSPHATE; PARAOXON

EPA HAZARDOUS WASTE NUMBER (RCRA No.): P041.
RCRA Section 261 Hazardous Constituents.

EPCRA Section 304 Reportable Quantity (RQ): CERCLA, 100 lbs. (45.4 kgs.).
MARINE POLLUTANT (49CFR, Subchapter 172.101, Appendix B).

EPA NAME: DIETHYL PHTHALATE
CAS: 84-66-2
SYNONYMS: 1,2-BENZENEDICARBOXYLIC ACID, DIETHYL ESTER; DIETHYL 1,2-BENZENEDICARBOXYLATE

CLEAN WATER ACT: Section 307 Priority Pollutants; Section 313 Priority Chemicals.
EPA HAZARDOUS WASTE NUMBER (RCRA No.): U088.
RCRA Section 261 Hazardous Constituents.
RCRA Land Ban Waste.
RCRA Universal Treatment Standards: Wastewater (mg/L), 0.20; Nonwastewater (mg/kg), 28.
RCRA Ground Water Monitoring List. Suggested test method(s) (PQL µg/L): 8060(5); 8270(10).
EPCRA Section 304 Reportable Quantity (RQ): CERCLA, 1,000 lbs. (454 kgs.).
EPCRA Section 313: Deleted from EPCRA/SARA 313 July 29, 1996 (FR Vol. 61, No. 146, p. 39356-39357).

EPA NAME: O,O-DIETHYL-o-PYRAZINYL PHOSPHOROTHIOATE
[see ZINOPHOS]
CAS: 297-97-2

EPA NAME: DIETHYLSTILBESTROL
CAS: 56-53-1
SYNONYMS: DES; α, α′-DIETHYL-(E)-4,4′D-STILBENEDIOL; PHENOL, 4,4′-(1,2-DIETHYL-1,2-ETHENEDIYL)BIS-, (E)-

EPA HAZARDOUS WASTE NUMBER (RCRA No.): U089.
RCRA Section 261 Hazardous Constituents.
EPCRA Section 304 Reportable Quantity (RQ): CERCLA, 1 lb. (0.454 kg.).
CALIFORNIA'S PROPOSITION 65: Carcinogen; reproductive toxin.

EPA NAME: DIETHYL SULFATE
CAS: 64-67-5
SYNONYMS: DIETHYL SULPHATE; SULFURIC ACID, DIETHYL ESTER

CLEAN AIR ACT: Hazardous Air Pollutants (Title I, Part A, Section 112).
EPCRA Section 304 Reportable Quantity (RQ): CERCLA, 10 lbs. (4.54 kgs.).
EPCRA Section 313 Form R de minimus concentration reporting level: 0.1%.
CALIFORNIA'S PROPOSITION 65: Carcinogen.

EPA NAME: DIFLUBENZURON
CAS: 35367-38-5

SYNONYMS: N-(((4-CHLOROPHENYL)AMINO)CARBONYL)-2,6-DIFLUOROBENZAMIDE; UREA, 1-(p-CHLOROPHENYL)-3-(2,6-DIFLUOROBENZOYL)-

EPCRA Section 313 Form R de minimus concentration reporting level: 1.0%.

EPA NAME: DIFLUOROETHANE

CAS: 75-37-6

SYNONYMS: 1,1-DIFLUROETHANE; ETHANE, 1,1-DIFLUORO-; FREON 152

CLEAN AIR ACT: Accidental Release Prevention/Flammable substances (Section 112[r], Table 3), TQ = 10,000 lbs. (4540 kgs.).

EPA NAME: DIGITOXIN

CAS: 71-63-6

SYNONYMS: DIGITALIN

EPCRA Section 302 Extremely Hazardous Substances: TPQ = 100/10,000 lbs. (45.4/4,540 kgs.).

EPCRA Section 304 Reportable Quantity (RQ): EHS, 100 lbs. (45.4 kgs.).

EPA NAME: 5,6-DIHYDRO-2-METHYL-N-PHENYL-1,4-OXATHIIN-3-CARBOXAMIDE

[see CARBOXIN]

CAS: 5234-68-4

EPA NAME: DIGLYCIDYL ETHER

CAS: 2238-07-5

SYNONYMS: BIS(2,3-EPOXYPROPYL)ETHER; DI(2,3-EPOXY) PROPYL ETHER; DGE; 2,2′-OXYBIS(METHYLENE)BISOXIRANE; OXIRANE, 2,2′-OXYBIS (METHYLENE) BIS-

EPCRA Section 302 Extremely Hazardous Substances: TPQ = 1,000 lbs. (454 kgs.).

EPCRA Section 304 Reportable Quantity (RQ): EHS, 1,000 lbs. (454 kgs.).

EPA NAME: DIGLYCIDYL RESORCINOL ETHER

CAS: 101-90-6

SYNONYMS: DGRE; 1,3-DIGLYCIDYLOXYBENZENE; OXIRANE,2,2′-(1,3-PHENYLENEBIS(OXYMETHYLENE))BIS-; RESORCINOL DIGLYCIDYL ETHER

EPCRA Section 313 Form R de minimus concentration reporting level: 0.1%.

CALIFORNIA'S PROPOSITION 65: Carcinogen.

EPA NAME: DIGOXIN

CAS: 20830-75-5

SYNONYMS: CHLOROFORMIC DIGITALIN

EPCRA Section 302 Extremely Hazardous Substances: TPQ=10/10,000 lbs. (0.454/4,540 kgs.).
EPCRA Section 304 Reportable Quantity (RQ): EHS, 10 lbs. (4.54 kgs.).

EPA NAME: 2,3,-DIHYDRO-5,6-DIMETHYL-1,4-DITHIIN-1,1,4,4-TETRAOXIDE
[see DIMETHIPIN]
CAS: 55290-64-7

EPA NAME: 5,6-DIHYDRO-2-METHYL-N-PHENYL-1,4-OXA-THIIN-3-CARBOXAMIDE
[see CARBOXIN]
CAS: 5234-68-4

EPA NAME: DIHYDROSAFROLE
CAS: 94-58-6
SYNONYMS: BENZENE, 1,2-(METHYLENEDIOXY)-4-PROPYL-; (1,2-(METHYLENEDIOXY)-4-PROPYL)BENZENE; 4-PROPYL-1,2-(METHYLENEDIOXY)BENZENE

EPA HAZARDOUS WASTE NUMBER (RCRA No.): U090.
RCRA Section 261 Hazardous Constituents.
EPCRA Section 304 Reportable Quantity (RQ): CERCLA, 10 lbs. (4.54 kgs.).
EPCRA Section 313 Form R de minimus concentration reporting level: 0.1%.
CALIFORNIA'S PROPOSITION 65: Carcinogen.

EPA NAME: DIISOCYANATES
EPCRA Section 313: This category includes only 20 chemicals. All are listed in this book with the notation "see also DIISOCYANATES." Form R de minimus concentration reporting level: 1.0%. Form R Toxic Chemical Category Code: N120.

EPA NAME: 4,4'-DIISOCYANATODIPHENYL ETHER
[see also DIISOCYANATES]
CAS: 4128-73-8
SYNONYMS: BENZENE, 1,1'-OXYBIS(4-ISOCYANATO)-; 1,1'-OXYBIS(4-ISOCYANATOBENZENE)

EPCRA Section 313 Form R de minimus concentration reporting level: 1.0%. Form R Toxic Chemical Category Code: N120.

EPA NAME: 2,4'-DIISOCYANATODIPHENYL SULFIDE
[see also DIISOCYANATES]
CAS: 75790-87-3
SYNONYMS: BENZENE, 1-ISOCYANATO-2-((4-ISOCYANATOPHENYL)THIO)-; o-((p-ISOCYANATOPHENYL)THIO)PHENYL ISOCYANATE

EPCRA Section 313 Form R de minimus concentration reporting level: 1.0%. Form R Toxic Chemical Category Code: N120.

EPA NAME: DIISOPROPYLFLUOROPHOSPHATE

CAS: 55-91-4

SYNONYMS: O,O-DIISOPROPYLFLUOROPHOSPHATE; DFP; ISOFLUORPHATE; PHOSPHOROFLUORIDIC ACID, BIS(1-METHYLETHYL)ESTER

EPA HAZARDOUS WASTE NUMBER (RCRA No.): P043.
RCRA Section 261 Hazardous Constituents.
EPCRA Section 302 Extremely Hazardous Substances: TPQ = 100 lbs. (45.4 kgs.).
EPCRA Section 304 Reportable Quantity (RQ): EHS/CERCLA, 100 lbs. (45.4 kgs.).

EPA NAME: DIMEFOX

CAS: 115-26-4

SYNONYMS: BIS(DIMETHYLAMIDO)FLUOROPHOSPHATE

EPCRA Section 302 Extremely Hazardous Substances: TPQ = 500 lbs. (227 kgs.).
EPCRA Section 304 Reportable Quantity (RQ): EHS, 500 lbs. (227 kgs.).

EPA NAME: 1,4,5,8-DIMETHANONAPHTHALENE, 1,2,3, 4,10,10-HEXACHLORO-1,4,4a,5,8,8a-HEXAHYDRO-(1α,4α,4aβ,5α,8α,8aβ)-

[see ALDRIN]

CAS: 309-00-2

EPA NAME: DIMETHIPIN

CAS: 55290-64-7

SYNONYMS: 2,3,-DIHYDRO-5,6-DIMETHYL-1,4-DITHIIN-1,1,4,4-TETRAOXIDE; P-DITHIANE, 2,3-DEHYDRO-2,3-DIMETHYL-, TETROXIDE; EPA PESTICIDE CHEMICAL CODE 118901

EPCRA Section 313 Form R de minimus concentration reporting level: 1.0%.

EPA NAME: DIMETHOATE

CAS: 60-51-5

SYNONYMS: ACETIC ACID, O,O-DIMETHYLDITHIOPHOSPHO-RYL-, N-MONOMETHYLAMIDE SALT; O,O-DIMETHYL METHYLCARBAMOYLMETHYL PHOSPHORODITHIOATE; EPA PESTICIDE CHEMICAL CODE 035001

EPA HAZARDOUS WASTE NUMBER (RCRA No.): P044.
RCRA Section 261 Hazardous Constituents.
RCRA Ground Water Monitoring List. Suggested test method(s) (PQL μg/L): 8270(10).
EPCRA Section 302 Extremely Hazardous Substances: TPQ = 500/10,000 lbs. (227/4,540 kgs.).
EPCRA Section 304 Reportable Quantity (RQ): EHS/CERCLA, 10 lbs. (4.54 kgs.).
EPCRA Section 313 Form R de minimus concentration reporting level: 1.0%.

MARINE POLLUTANT (49CFR, Subchapter 172.101, Appendix B).

EPA NAME: 3,3′-DIMETHOXYBENZIDINE
CAS: 119-90-4
SYNONYMS: BENZIDINE, 3,3′-DIMETHOXY-; 1,1′BIPHENYL]-4,4′-DIAMINE, 3,3′-DIMETHOXY-; o-DIANISIDINE; 3,3′-DIMETHOXY-4,4′-DIAMINODIPHENYL

CLEAN AIR ACT: Hazardous Air Pollutants (Title I, Part A, Section 112).
EPA HAZARDOUS WASTE NUMBER (RCRA No.): U091.
RCRA Section 261 Hazardous Constituents.
EPCRA Section 304 Reportable Quantity (RQ): CERCLA, 100 lbs. (45.4 kgs.).
EPCRA Section 313 Form R de minimus concentration reporting level: 0.1%.
CALIFORNIA'S PROPOSITION 65: Carcinogen.

EPA NAME: 3,3′-DIMETHOXYBENZIDINE DIHYDROCHLORIDE
CAS: 20325-40-0
SYNONYMS: BENZIDINE, 3,3′-DIMETHOXY-, DIHYDROCHLORIDE; o-DIANISIDINE DIHYDROCHLORIDE; 3,3′-DIMETHOXYBIPHENYL-4,4′-YLENEDIAMMONIUM DICHLORIDE

EPCRA Section 313 Form R de minimus concentration reporting level: 0.1%.
CALIFORNIA'S PROPOSITION 65: Carcinogen.

EPA NAME: 3,3′-DIMETHOXYBENZIDINE-4,4′-DIISOCYANATE
CAS: 91-93-0
[see also DIISOCYANATES]
SYNONYMS: 1,1′-BIPHENYL, 4,4′-DIISOCYANATO-3,3′-DIMETHOXY-

EPCRA Section 313 Form R de minimus concentration reporting level: 1.0%. Form R Toxic Chemical Category Code: N120.

EPA NAME: 3,3′-DIMETHOXYBENZIDINE HYDROCHLORIDE
CAS: 111984-09-9
SYNONYMS: (1,1′-BIPHENYL)-4,4′-DIAMINE, 3,3′-DIMETHOXY-, MONOHYDROCHLORIDE; o-DIANISIDINE HYDROCHLORIDE; 3,3′-DIMETHOXYBENZIDINE DIHYDROCHLORIDE

EPCRA Section 313 Form R de minimus concentration reporting level: 0.1%.
CALIFORNIA'S PROPOSITION 65: Carcinogen.

EPA NAME: DIMETHYLAMINE
CAS: 124-40-3
SYNONYMS: METHANAMINE, N-METHYL; N-METHYLMETHANAMINE

CLEAN AIR ACT: Accidental Release Prevention/Flammable substances (Section 112[r], Table 3), TQ = 10,000 lbs. (4540 kgs.).
CLEAN WATER ACT: Section 311 Hazardous Substances/RQ (same as CERCLA).
EPA HAZARDOUS WASTE NUMBER (RCRA No.): U092.
RCRA Section 261 Hazardous Constituents.
EPCRA Section 304 Reportable Quantity (RQ): CERCLA, 1,000 lbs. (454 kgs.).
EPCRA Section 313 Form R de minimus concentration reporting level: 1.0%.

EPA NAME: DIMETHYLAMINE DICAMBA
CAS: 2300-66-5
SYNONYMS: 3,6-DICHLORO-O-ANISIC ACID, COMPOUND WITH DIMETHYLAMINE (1:1); EPA PESTICIDE CHEMICAL CODE 029802
EPCRA Section 313 Form R de minimus concentration reporting level: 1.0%.

EPA NAME: 4-DIMETHYLAMINOAZOBENZENE
CAS: 60-11-7
SYNONYMS: BENZENAMINE, N,N-DIMETHYL-4-(PHENYLAZO)-; P-(DIMETHYLAMINO)AZOBENZENE; DIMETHYLAMINOAZOBENZENE; 4-(PHENYLAZO)-N,N-DIMETHYLANILINE
CLEAN AIR ACT: Hazardous Air Pollutants (Title I, Part A, Section 112).
EPA HAZARDOUS WASTE NUMBER (RCRA No.): U093.
RCRA Section 261 Hazardous Constituents.
RCRA Land Ban Waste.
RCRA Universal Treatment Standards: Wastewater (mg/L), 0.13; Nonwastewater (mg/kg), N/A.
RCRA Ground Water Monitoring List. Suggested test method(s) (PQL μg/L): 8270(10).
EPCRA Section 304 Reportable Quantity (RQ): CERCLA, 10 lbs. (4.54 kgs.).
EPCRA Section 313 Form R de minimus concentration reporting level: 0.1%.
CALIFORNIA'S PROPOSITION 65: Carcinogen.

EPA NAME: DIMETHYLAMINOAZOBENZENE
[see 4-DIMETHYLAMINOAZOBENZENE]
CAS: 60-11-7

EPA NAME: N,N-DIMETHYLANILINE
CAS: 121-69-7
SYNONYMS: ANILINE, N,N-DIMETHYL-; BENZENAMINE, N,N-DIMETHYL-; DIMETHYLANILINE; (DIMETHYLAMINO)BENZENE; N,N-DIMETHYLPHENYLAMINE
CLEAN AIR ACT: Hazardous Air Pollutants (Title I, Part A, Section 112).

EPCRA Section 304 Reportable Quantity (RQ): CERCLA, 100 lbs. (45.4 kgs.).
EPCRA Section 313 Form R de minimus concentration reporting level: 1.0%.

EPA NAME: 7,12-DIMETHYLBENZ(a)ANTHRACENE
[see also POLYCYCLIC AROMATIC COMPOUNDS]
CAS: 57-97-6
SYNONYMS: BENZ(a)ANTHRACENE, 9,10-DIMETHYL-; BENZ(a)ANTHRACENE, 7,12-DIMETHYL-; 7,12-DMBA; DMBA

CLEAN WATER ACT: Section 307 Toxic Pollutants as polynuclear aromatic hydrocarbons.
EPA HAZARDOUS WASTE NUMBER (RCRA No.): U094.
RCRA Section 261 Hazardous Constituents.
RCRA Ground Water Monitoring List. Suggested test method(s) (PQL μg/L): 8270(10).
EPCRA Section 304 Reportable Quantity (RQ): CERCLA, 1 lb. (0.454 kg.).
EPCRA Section 313 Form R de minimus concentration reporting level: 0.1%.
CALIFORNIA'S PROPOSITION 65: Carcinogen.

EPA NAME: 3,3′-DIMETHYLBENZIDINE
CAS: 119-93-7
SYNONYMS: [1,1′-BIPHENYL]-4,4′-DIAMINE, 3,3′-DIMETHYL-; 4,4′-DIAMINO-3,3′-DIMETHYLBIPHENYL; o-TOLIDINE; 3,3′-TOLIDINE; O,O′-TOLIDINE

CLEAN AIR ACT: Hazardous Air Pollutants (Title I, Part A, Section 112).
EPA HAZARDOUS WASTE NUMBER (RCRA No.): U095.
RCRA Section 261 Hazardous Constituents.
RCRA Ground Water Monitoring List. Suggested test method(s) (PQL μg/L): 8270(10).
EPCRA Section 304 Reportable Quantity (RQ): CERCLA, 10 lbs. (4.54 kgs.).
EPCRA Section 313 Form R de minimus concentration reporting level: 0.1%.
CALIFORNIA'S PROPOSITION 65: Carcinogen.

EPA NAME: 3,3′-DIMETHYLBENZIDINE DIHYDROCHLORIDE
CAS: 612-82-8
SYNONYMS: BENZIDINE, 3,3′-DIMETHYL-, DIHYDROCHLORIDE; o-TOLIDINE DIHYDROCHLORIDE

EPCRA Section 313 Form R de minimus concentration reporting level: 0.1%.
CALIFORNIA'S PROPOSITION 65: Carcinogen.

EPA NAME: 3,3-DIMETHYLBENZIDINE DIHYDROFLUORIDE
CAS: 41766-75-0
SYNONYMS: o-TOLIDINE DIHYDROFLUORIDE

EPCRA Section 313 Form R de minimus concentration reporting level: 0.1%.

EPA NAME: 2,2-DIMETHYL-1,3-BENZODIOXOL-4-OLMETHYLCARBAMATE

[see BENDIOCARB]

CAS: 22781-23-3

EPA NAME: DIMETHYLCARBAMOYL CHLORIDE

CAS: 79-44-7

SYNONYMS: CARBAMIC CHLORIDE, DIMETHYL-; DIMETHYLAMINOCARBONYL CHLORIDE; N,N-DIMETHYLCARBAMOYL CHLORIDE; DIMETHYLCARBAMOYL CHLORIDE; DIMETHYLCARBAMYL CHLORIDE

CLEAN AIR ACT: Hazardous Air Pollutants (Title I, Part A, Section 112).

EPA HAZARDOUS WASTE NUMBER (RCRA No.): U097.

RCRA Section 261 Hazardous Constituents.

EPCRA Section 304 Reportable Quantity (RQ): CERCLA, 1 lb. (0.454 kg.).

EPCRA Section 313 Form R de minimus concentration reporting level: 0.1%.

CALIFORNIA'S PROPOSITION 65: Carcinogen.

EPA NAME: DIMETHYL CHLOROTHIOPHOSPHATE

CAS: 2524-03-0

SYNONYMS: O,O-DIMETHYL CHLOROTHIONOPHOSPHATE; DIMETHYL PHOSPHOROCHLOROTHIOATE; DIMETHYL THIOPHOSPHORYL CHLORIDE; O,O-DIMETHYL THIOPHOSPHORYL CHLORIDE; PHOSPHONOTHIOIC ACID, CHLORO-, O,O-DIMETHYL ESTER

EPCRA Section 302 Extremely Hazardous Substances: TPQ = 500 lbs. (227 kgs.).

EPCRA Section 304 Reportable Quantity (RQ): EHS, 500 lbs. (227 kgs.).

EPCRA Section 313 Form R de minimus concentration reporting level: 1.0%.

EPA NAME: DIMETHYLDICHLOROSILANE

CAS: 75-78-5

SYNONYMS: DICHLORODIMETHYLSILANE; SILANE, DICHLORODIMETHYL-

CLEAN AIR ACT: Accidental Release Prevention/Flammable substances (Section 112[r], Table 3), TQ = 5,000 lbs. (2270 kgs.).

EPCRA Section 302 Extremely Hazardous Substances: TPQ = 500 lbs. (227 kgs.).

EPCRA Section 304 Reportable Quantity (RQ): EHS, 500 lbs. (227 kgs.).

EPCRA Section 313 Form R de minimus concentration reporting level: 1.0%.

EPA NAME: 3,3′-DIMETHYL-4,4′-DIPHENYLENE DIISOCYANATE

[see also DIISOCYANATES]

CAS: 91-97-4

SYNONYMS: 1,1′-BIPHENYL,4,4′-DIISOCYANATO-3,3′-DIMETHYL-; 4,4′-DIISOCYANATO-3,3′-DIMETHYL-1,1′-BIPHENYL

EPCRA Section 313 Form R de minimus concentration reporting level: 1.0%. Form R Toxic Chemical Category Code: N120.

EPA NAME: 3,3′-DIMETHYLDIPHENYLMETHANE-4,4′-DIISOCYANATE

[see also DIISOCYANATES]

CAS: 139-25-3

SYNONYMS: BENZENE, 1,1′-METHYLENEBIS(4-ISOCYANATO-3-METHYL-; 4,4′-DIISOCYANATO-3,3′-DIMETHYLDIPHENYLMETHANE

EPCRA Section 313 Form R de minimus concentration reporting level: 1.0%. Form R Toxic Chemical Category Code: N120.

EPA NAME: 2,4-D METHYL ESTER

CAS: 1928-38-7

SYNONYMS: ACETIC ACID, (2,4-DICHLOROPHENOXY)-, METHYL ESTER; 2,4-D ESTERS

CLEAN WATER ACT: Section 311 Hazardous Substances/RQ (same as CERCLA).

EPA HAZARDOUS WASTE NUMBER (RCRA No.): U240.

RCRA Section 261 Hazardous Constituents.

SAFE DRINKING WATER ACT: Regulated chemical (47 FR 9352) as 2,4-D.

EPCRA Section 304 Reportable Quantity (RQ): CERCLA, 100 lbs. (4.54 kgs.).

EPA NAME: N-(5-(1,1-DIMETHYLETHYL)-1,3,4-THIADIAZOL-2-YL)-N,N′-DIMETHYLUREA

[see TEBUTHIURON]

CAS: 34014-18-1

EPA NAME: DIMETHYLFORMAMIDE

CAS: 68-12-2

SYNONYMS: EPA PESTICIDE CHEMICAL CODE 366200; N,N-DIMETHYLFORMAMIDE; N,N-DIMETHYLMETHANAMIDE; FORMIC ACID, AMIDE, N,N-DIMETHYL-

CLEAN AIR ACT: Hazardous Air Pollutants (Title I, Part A, Section 112).

EPCRA Section 304 Reportable Quantity (RQ): CERCLA, 100 lbs. (45.4 kgs.).

EPCRA Section 313 Form R de minimus concentration reporting level: 1.0%.

EPA NAME: N,N-DIMETHYLFORMAMIDE

[see DIMETHYLFORMAMIDE]

CAS: 68-12-2

EPA NAME: 1,1-DIMETHYLHYDRAZINE

CAS: 57-14-7

SYNONYMS: DIMETHYLHYDRAZINE; HYDRAZINE, 1,1-DIMETHYL-; UNSYMMETRICAL DIMETHYLHYDRAZINE; asym-DIMETHYLHYDRAZINE; N,N-DIMETHYLHYDRAZINE

CLEAN AIR ACT: Hazardous Air Pollutants (Title I, Part A, Section 112); Accidental Release Prevention/Flammable substances (Section 112[r], Table 3), TQ = 15,000 lbs. (6810 kgs.).

EPA HAZARDOUS WASTE NUMBER (RCRA No.): U098.

RCRA Section 261 Hazardous Constituents.

EPCRA Section 302 Extremely Hazardous Substances: TPQ = 1,000 lbs. (454 kgs.).

EPCRA Section 304 Reportable Quantity (RQ): EHS/CERCLA, 10 lbs. (4.54 kgs.).

EPCRA Section 313 Form R de minimus concentration reporting level: 0.1%.

CALIFORNIA'S PROPOSITION 65: Carcinogen.

EPA NAME: DIMETHYLHYDRAZINE

[see 1,1-DIMETHYL HYDRAZINE]

CAS: 57-14-7

EPA NAME: O,O-DIMETHYL O-[3-METHYL-4-(METHYLTHIO)PHENYL] ESTER, PHOSPHOROTHIOIC ACID

[see FENTHION]

CAS: 55-38-9

EPA NAME: 2,2-DIMETHYL-3-(2-METHYL-1-PROPENYL)CYCLOPROPANECARBOXYLIC ACID (1,3,4,5,6,7-HEXAHYDRO-1,3-DIOXO-2H-ISOINDOL-2-YL)METHYL ESTER

[see TETRAMETHRIN]

CAS: 7696-12-0

EPA NAME: 2,2-DIMETHYL-3-(2-METHYL-1-PROPENYL)CYCLOPROPANECARBOXYLIC ACID (3-PHENOXYPHENYL)METHYL ESTER

[see PHENOTHRIN]

CAS: 26002-80-2

EPA NAME: 2,4-DIMETHYLPHENOL

CAS: 105-67-9

SYNONYMS: 4-HYDROXY-1,3-DIMETHYLBENZENE; PHENOL, 2,4-DIMETHYL-

CLEAN WATER ACT: Section 307 Toxic Pollutants; Section 307 Priority Pollutants.

EPA HAZARDOUS WASTE NUMBER (RCRA No.): U101.

RCRA Section 261 Hazardous Constituents.

RCRA Land Ban Waste.

RCRA Universal Treatment Standards: Wastewater (mg/L), 0.036; Nonwastewater (mg/kg), 1.4.

RCRA Ground Water Monitoring List. Suggested test method(s) (PQL μg/L): 8040(5); 8270(10).

EPCRA Section 304 Reportable Quantity (RQ): CERCLA, 100 lbs. (45.4 kgs.).

EPCRA Section 313 Form R de minimus concentration reporting level: 1.0%.

EPA NAME: 2,6-DIMETHYLPHENOL

CAS: 576-26-1

SYNONYMS: 1-HYDROXY-2,6-DIMETHYLBENZENE; PHENOL, 2,6-DIMETHYL-

EPCRA Section 313 Form R de minimus concentration reporting level: 1.0%.

EPA NAME: DIMETHYL-p-PHENYLENEDIAMINE

CAS: 99-98-9

SYNONYMS: N,N-DIMETHYL-p-PHENYLENEDIAMINE

EPCRA Section 302 Extremely Hazardous Substances: TPQ = 10/10,000 lbs. (4.5/4,540 kgs.).

EPCRA Section 304 Reportable Quantity (RQ): EHS, 10 lbs. (4.54 kgs.).

EPA NAME: DIMETHYL PHOSPHOROCHLOROTHIOATE

[see DIMETHYL CHLOROTHIOPHOSPHATE]

CAS: 2524-03-0

EPA NAME: DIMETHYL PHTHALATE

CAS: 131-11-3

SYNONYMS: 1,2-BENZENEDICARBOXYLIC ACID, DIMETHYL ESTER; DIMETHYL 1,2-BENZENEDICARBOXYLATE; DIMETHYL O-PHTHALATE; N,N-DIMETHYLPHTHALATE

CLEAN AIR ACT: Hazardous Air Pollutants (Title I, Part A, Section 112).

CLEAN WATER ACT: Section 307 Priority Pollutants; Section 313 Priority Chemicals; Toxic Pollutant (Section 401.15).

EPA HAZARDOUS WASTE NUMBER (RCRA No.): U102.

RCRA Section 261 Hazardous Constituents.

RCRA Land Ban Waste.

RCRA Universal Treatment Standards: Wastewater (mg/L), 0.047; Nonwastewater (mg/kg), 28.

RCRA Ground Water Monitoring List. Suggested test method(s) (PQL μg/L): 8060(5); 8270(10).

EPCRA Section 304 Reportable Quantity (RQ): CERCLA, 5,000 lbs. (2270 kgs.).

EPCRA Section 313 Form R de minimus concentration reporting level: 1.0%.

EPA NAME: 2,2-DIMETHYLPROPANE

CAS: 463-82-1

SYNONYMS: PROPANE, 2,2-DIMETHYL-

CLEAN AIR ACT: Accidental Release Prevention/Flammable substances (Section 112[r], Table 3), TQ = 10,000 lbs. (4540 kgs.).

EPA NAME: DIMETHYL SULFATE

CAS: 77-78-1

SYNONYMS: METHYL SULFATE; SULFURIC ACID, DIMETHYL ESTER

CLEAN AIR ACT: Hazardous Air Pollutants (Title I, Part A, Section 112).

EPA HAZARDOUS WASTE NUMBER (RCRA No.): U103.

RCRA Section 261 Hazardous Constituents.

EPCRA Section 302 Extremely Hazardous Substances: TPQ = 500 lbs. (227 kgs.).

EPCRA Section 304 Reportable Quantity (RQ): EHS/CERCLA, 100 lbs. (45.4 kgs.).

EPCRA Section 313 Form R de minimus concentration reporting level: 0.1%.

CALIFORNIA'S PROPOSITION 65: Carcinogen.

EPA NAME: O,O-DIMETHYL-O-(3,5,6-TRICHLORO-2-PYRIDYL)PHOSPHOROTHIOATE

[see CHLORPYRIFOS METHYL]

CAS: 5598-13-0

EPA NAME: DIMETILAN

CAS: 644-64-4

SYNONYMS: DIMETHYLCARBAMIC ACID-1-((DIMETHYLAMINO)CARBONYL)-5-METHYL-1H-PYRAZOL-3-YL ESTER; 5-METHYL-1H-PYRAZOL-3-YL DIMETHYLCARBAMATE

EPA HAZARDOUS WASTE NUMBER (RCRA No.): P191.

RCRA Section 261 Hazardous Constituents.

RCRA Land Ban Waste.

RCRA Universal Treatment Standards: Wastewater (mg/L), 0.056; Nonwastewater (mg/kg), 1.4.

EPCRA Section 302 Extremely Hazardous Substances: TPQ = 500/10,000 lbs. (227/4,540 kgs.).

EPCRA Section 304 Reportable Quantity (RQ): EHS, 1 lb. (0.454 kg.).

EPA NAME: DINITROBENZENE (MIXED ISOMERS)

CAS: 25154-54-5

SYNONYMS: BENZENE, DINITRO-

CLEAN WATER ACT: Section 311 Hazardous Substances/RQ (same as CERCLA).

RCRA Section 261 Hazardous Constituents, AS dinitrobenzene, n.o.s., waste number not listed.

EPCRA Section 304 Reportable Quantity (RQ): CERCLA, 100 lbs. (45.4 kgs.).

EPA NAME: m-DINITROBENZENE

CAS: 99-65-0

SYNONYMS: BENZENE, 1,3-DINITRO-; 1,3-DINITROBENZENE

CLEAN WATER ACT: Section 311 Hazardous Substances/RQ (same as CERCLA).

RCRA Ground Water Monitoring List. Suggested test method(s) (PQL µg/L): 8270(10).

EPCRA Section 304 Reportable Quantity (RQ): CERCLA, 100 lbs. (45.4 kgs.).

EPCRA Section 313 Form R de minimus concentration reporting level: 1.0%.

CALIFORNIA'S PROPOSITION 65: Reproductive toxin (male).

EPA NAME: o-DINITROBENZENE

CAS: 528-29-0

SYNONYMS: BENZENE, 1,2-DINITRO-; BENZENE, o-DINITRO-; 1,2-DINITROBENZENE; 1,2-DNB

CLEAN WATER ACT: Section 311 Hazardous Substances/RQ (same as CERCLA).

EPCRA Section 304 Reportable Quantity (RQ): CERCLA, 100 lbs. (45.4 kgs.).

EPCRA Section 313 Form R de minimus concentration reporting level: 1.0%.

CALIFORNIA'S PROPOSITION 65: Reproductive toxin (male).

EPA NAME: p-DINITROBENZENE

CAS: 100-25-4

SYNONYMS: BENZENE, 1,4-DINITRO-; BENZENE, p-DINITRO-

CLEAN WATER ACT: Section 311 Hazardous Substances/RQ (same as CERCLA).

RCRA Universal Treatment Standards: Wastewater (mg/L), 0.32; Non-wastewater (mg/kg), 2.3.

EPCRA Section 304 Reportable Quantity (RQ): CERCLA, 100 lbs. (45.4 kgs.).

EPCRA Section 313 Form R de minimus concentration reporting level: 1.0%.

CALIFORNIA'S PROPOSITION 65: Reproductive toxin (male).

EPA NAME: DINITROBUTYL PHENOL

CAS: 88-85-7

SYNONYMS: 2-sec-BUTYL-4,6-DINITROPHENOL; DINITROBUTYL PHENOL; 4,6-DINITRO-2-(1-METHYL-N-PROPYL)-PHENOL; DINOSEB; EPA PESTICIDE CHEMICAL CODE 037505; PHENOL, 2-(1-METHYLPROPYL)-4,6-DINITRO-

EPA HAZARDOUS WASTE NUMBER (RCRA No.): P020.

RCRA Section 261 Hazardous Constituents.

RCRA Land Ban Waste.

RCRA Universal Treatment Standards: Wastewater (mg/L), 0.066; Non-wastewater (mg/kg), 2.515.

RCRA Ground Water Monitoring List. Suggested test method(s) (PQL µg/L): 8150(1); 8270(10).

SAFE DRINKING WATER ACT: MCL, 0.007 mg/L; MCLG, 0.007 mg/L; Regulated chemical (47 FR 9352).

EPCRA Section 302 Extremely Hazardous Substances: TPQ = 100/10,000 lbs. (45.4/4,540 kgs.).

EPCRA Section 304 Reportable Quantity (RQ): EHS/CERCLA, 1,000 lbs. (454 kgs.).
EPCRA Section 313 Form R de minimus concentration reporting level: 1.0%.
MARINE POLLUTANT (49CFR, Subchapter 172.101, Appendix B).
CALIFORNIA'S PROPOSITION 65: Reproductive toxin (male).

EPA NAME: 4,6-DINITRO-o-CRESOL
CAS: 534-52-1
SYNONYMS: DINITROCRESOL; DINITRO-ortho-CRESOL; 4,6-DI-NITRO-2-METHYLPHENOL; DNOC; PHENOL, 2-METHYL-4,6-DINITRO-

CLEAN AIR ACT: Hazardous Air Pollutants (Title I, Part A, Section 112).
CLEAN WATER ACT: Section 313 Priority Chemicals.
EPA HAZARDOUS WASTE NUMBER (RCRA No.): P047.
RCRA Section 261 Hazardous Constituents.
RCRA Land Ban Waste.
RCRA Universal Treatment Standards: Wastewater (mg/L), 0.28; Non-wastewater (mg/kg), 160.
RCRA Ground Water Monitoring List. Suggested test method(s) (PQL µg/L): 8040(150); 8270(50).
EPCRA Section 302 Extremely Hazardous Substances: TPQ = 10/10,000 lbs. (4.54/4,540 kgs.).
EPCRA Section 304 Reportable Quantity (RQ): EHS/CERCLA, 10 lbs. (4.54 kgs.).
EPCRA Section 313 Form R de minimus concentration reporting level: 1.0%.
MARINE POLLUTANT (49CFR, Subchapter 172.101, Appendix B).

EPA NAME: DINITROCRESOL
[see 4,6-DINITRO-O-CRESOL]
CAS: 534-52-1

EPA NAME: 4,6-DINITRO-O-CRESOL AND SALTS
[see 4,6-DINITRO-O-CRESOL]
CAS: 534-52-1

EPA NAME: DINITROPHENOL
CAS: 25550-58-7
SYNONYMS: DINITROPHENOL, MIXED ISOMERS; PHENOL, DI-NITRO-

EPCRA Section 304 Reportable Quantity (RQ): CERCLA, 10 lbs. (4.54 kgs.).
MARINE POLLUTANT (49CFR, Subchapter 172.101, Appendix B).

EPA NAME: 2,4-DINITROPHENOL
CAS: 51-28-5
SYNONYMS: PHENOL, α-DINITRO-; PHENOL, 2,4-DINITRO-

CLEAN AIR ACT: Hazardous Air Pollutants (Title I, Part A, Section 112).

CLEAN WATER ACT: Section 311 Hazardous Substances/RQ (same as CERCLA); Section 313 Priority Chemicals.

EPA HAZARDOUS WASTE NUMBER (RCRA No.): P048.

RCRA Section 261 Hazardous Constituents.

RCRA Land Ban Waste.

RCRA Universal Treatment Standards: Wastewater (mg/L), 0.12; Non-wastewater (mg/kg), 160.

RCRA Ground Water Monitoring List. Suggested test method(s) (PQL μg/L): 8040(150); 8270(50).

SAFE DRINKING WATER ACT: Priority List (55 FR 1470).

EPCRA Section 304 Reportable Quantity (RQ): CERCLA, 10 lbs. (4.54 kgs.).

EPCRA Section 313 Form R de minimus concentration reporting level: 1.0%.

MARINE POLLUTANT (49CFR, Subchapter 172.101, Appendix B).

CALIFORNIA'S PROPOSITION 65: Carcinogen.

EPA NAME: 2,5-DINITROPHENOL

CAS: 329-71-5

SYNONYMS: γ-DINITROPHENOL; PHENOL, 2,5-DINITRO-

CLEAN WATER ACT: Section 311 Hazardous Substances/RQ (same as CERCLA).

EPCRA Section 304 Reportable Quantity (RQ): CERCLA, 10 lbs. (4.54 kgs.).

MARINE POLLUTANT (49CFR, Subchapter 172.101, Appendix B).

EPA NAME: 2,6-DINITROPHENOL

CAS: 573-56-8

SYNONYMS: β-DINITROPHENOL; PHENOL, 2,6-DINITRO-

CLEAN WATER ACT: Section 311 Hazardous Substances/RQ (same as CERCLA).

EPCRA Section 304 Reportable Quantity (RQ): CERCLA, 10 lbs. (4.54 kgs.).

MARINE POLLUTANT (49CFR, Subchapter 172.101, Appendix B).

CALIFORNIA'S PROPOSITION 65: Carcinogen.

EPA NAME: 2,4-DINITROTOLUENE

CAS: 121-14-2

SYNONYMS: BENZENE, 1-METHYL-2,4-DINITRO; 1-METHYL-2,4-DINITOBENZENE; TOLUENE, 2,4-DINITRO-

CLEAN AIR ACT: Hazardous Air Pollutants (Title I, Part A, Section 112).

CLEAN WATER ACT: Section 311 Hazardous Substances/RQ (same as CERCLA); Section 307 Priority Pollutants; Section 313 Priority Chemicals; Toxic Pollutant (Section 401.15) as dinitrotoluene.

EPA HAZARDOUS WASTE NUMBER (RCRA No.): U105; D030.

RCRA Section 261 Hazardous Constituents.

RCRA Toxicity Characteristic (Section 261.24), Maximum Concentration of Contaminants, regulatory level, 0.13 mg/L.

RCRA Land Ban Waste.

RCRA Universal Treatment Standards: Wastewater (mg/L), 0.32; Non-wastewater (mg/kg), 140.
RCRA Ground Water Monitoring List. Suggested test method(s) (PQL μg/L): 8090(0.2); 8270(10).
SAFE DRINKING WATER ACT: Priority List (55 FR 1470).
EPCRA Section 304 Reportable Quantity (RQ): CERCLA, 10 lbs. (4.54 kgs.).
EPCRA Section 313 Form R de minimus concentration reporting level: 1.0%.
CALIFORNIA'S PROPOSITION 65: Carcinogen.

EPA NAME: 2,6-DINITROTOLUENE
CAS: 606-20-2
SYNONYMS: BENZENE, 2-METHYL-1,3-DINITRO-; 1-METHYL-2,6-DINITOBENZENE; TOLUENE, 2,6-DINITRO-

CLEAN WATER ACT: Section 311 Hazardous Substances/RQ 10 lbs. (0.454 kgs.); Section 307 Priority Pollutants; Section 313 Priority Chemicals; Toxic Pollutant (Section 401.15) as dinitrotoluene.
EPA HAZARDOUS WASTE NUMBER (RCRA No.): U106.
RCRA Section 261 Hazardous Constituents.
RCRA Land Ban Waste.
RCRA Universal Treatment Standards: Wastewater (mg/L), 0.55; Non-wastewater (mg/kg), 28.
RCRA Ground Water Monitoring List. Suggested test method(s) (PQL μg/L): 8090(0.1); 8270(10).
SAFE DRINKING WATER ACT: Priority List (55 FR 1470).
EPCRA Section 313 Form R de minimus concentration reporting level: 1.0%.
EPCRA Section 304 Reportable Quantity (RQ): CERCLA, 100 lbs. (45.4 kgs.).
CALIFORNIA'S PROPOSITION 65: Carcinogen.

EPA NAME: 3,4-DINITROTOLUENE
CAS: 610-39-9
SYNONYMS: BENZENE, 4-METHYL-1,2-DINITRO-; 1-METHYL-3,4-DINITOBENZENE; TOLUENE, 3,4-DINITRO-

CLEAN WATER ACT: Section 311 Hazardous Substances/RQ (same as CERCLA); Toxic Pollutant (Section 401.15) as dinitrotoluene.
EPCRA Section 304 Reportable Quantity (RQ): CERCLA, 10 lbs. (4.54 kgs.).

EPA NAME: DINITROTOLUENE (MIXED ISOMERS)
CAS: 25321-14-6
SYNONYMS: BENZENE, METHYLDINITRO-; TOLUENE, DINI-TRO-

CLEAN WATER ACT: Section 311 Hazardous Substances/RQ (same as CERCLA); Toxic Pollutant (Section 401.15) as dinitrotoluene.
EPCRA Section 304 Reportable Quantity (RQ): CERCLA, 10 lbs. (4.54 kgs.).

EPCRA Section 313 Form R de minimus concentration reporting level: 1.0%.

EPA NAME: DINOCAP

CAS: 39300-45-3

SYNONYMS: 2-BUTENOIC ACID 2-(1-METHYLHEPTYL)-4,6-DINITROPHENYL ESTER; EPA PESTICIDE CHEMICAL CODE 036001; PHENOL, 2-(1-METHYLHEPTYL)-4,6-DINITRO-, CROTONATE (ESTER)

EPCRA Section 313 Form R de minimus concentration reporting level: 1.0%.

CALIFORNIA'S PROPOSITION 65: Reproductive toxin.

EPA NAME: DINOSEB

[see DINITROBUTYL PHENOL]

CAS: 88-85-7

EPA NAME: DINOTERB

CAS: 1420-07-1

SYNONYMS: 2,4-DINITRO-6-tert-BUTYLPHENOL

EPCRA Section 302 Extremely Hazardous Substances: TPQ = 500/10,000 lbs. (227/4,540 kgs.).

EPCRA Section 304 Reportable Quantity (RQ): EHS, 500 lbs. (227 kgs.).

EPA NAME: DI-n-OCTYL PHTHALATE

CAS: 117-84-0

SYNONYMS: 1,2-BENZENEDICARBOXYLIC ACID, DI-n-OCTYL ESTER; n-OCTYL PHTHALATE; PHTHALIC ACID, DIOCTYL ESTER

CLEAN WATER ACT: Section 307 Priority Pollutants; Section 313 Priority Chemicals.

EPA HAZARDOUS WASTE NUMBER (RCRA No.): U107.

RCRA Section 261 Hazardous Constituents.

RCRA Land Ban Waste.

RCRA Universal Treatment Standards: Wastewater (mg/L), 0.017; Nonwastewater (mg/kg), 28.

RCRA Ground Water Monitoring List. Suggested test method(s) (PQL µg/L): 8060(30); 8270(10).

EPCRA Section 304 Reportable Quantity (RQ): CERCLA, 5,000 lbs. (2270 kgs.).

EPA NAME: 1,4-DIOXANE

CAS: 123-91-1

SYNONYMS: 1,4-DIETHYLENE DIOXIDE; p-DIOXANE; 1,4-DIOXIN, TETRAHYDRO-

CLEAN AIR ACT: Hazardous Air Pollutants (Title I, Part A, Section 112).

EPA HAZARDOUS WASTE NUMBER (RCRA No.): U108.

RCRA Section 261 Hazardous Constituents.

RCRA Land Ban Waste.

RCRA Universal Treatment Standards: Wastewater (mg/L), 12.0; Non-wastewater (mg/kg), 170.
RCRA Ground Water Monitoring List. Suggested test method(s) (PQL μg/L): 8015(150).
EPCRA Section 304 Reportable Quantity (RQ): CERCLA, 100 lbs. (45.4 kgs.).
EPCRA Section 313 Form R de minimus concentration reporting level: 0.1%.
CALIFORNIA'S PROPOSITION 65: Carcinogen.

EPA NAME: DIOXATHION

CAS: 78-34-2
SYNONYMS: 2,3-DIOXANEDITHIOL S,S-BIS(O,O-DIETHYLPHOSPHORODITHIOATE); PHOSPHORODITHIOIC ACID-5-5′-1,4-DIOXANE-2,3-DIYL, O,O,O′,O′-TETRAETHYL ESTER

EPCRA Section 302 Extremely Hazardous Substances: TPQ = 500 lbs. (227 kgs.).
EPCRA Section 304 Reportable Quantity (RQ): EHS, 500 lbs. (227 kgs.).

EPA NAME: DIPHACIONE

CAS: 82-66-6
SYNONYMS: 2-(DIPHENYLACETYL)-1H-INDENE-1,3(2H)-DIONE

EPCRA Section 302 Extremely Hazardous Substances: TPQ = 10/10,000 lbs. (4.54/4,540 kgs.).
EPCRA Section 304 Reportable Quantity (RQ): EHS, 10 lbs. (4.54 kgs.).

EPA NAME: DIPHENAMID

CAS: 957-51-7
SYNONYMS: N,N-DIMETHYLDIPHENYLACETAMIDE; N,N-DIMETHYL-2,2-DIPHENYLACETAMIDE; DIPHENAMIDE; EPA PESTICIDE CHEMICAL CODE 036601

EPCRA Section 313 Form R de minimus concentration reporting level: 1.0%.

EPA NAME: DIPHENYLAMINE

CAS: 122-39-4
SYNONYMS: BENZENAMINE, N-PHENYL-; N,N-DIPHENYLAMINE; EPA PESTICIDE CHEMICAL CODE 038501; N-PHENYLBENZENAMINE

RCRA Section 261 Hazardous Constituents, waste number not listed.
RCRA Universal Treatment Standards: Wastewater (mg/L), 0.92; Non-wastewater (mg/kg), 13.
RCRA Ground Water Monitoring List. Suggested test method(s) (PQL μg/L): 8270(10).
EPCRA Section 313 Form R de minimus concentration reporting level: 1.0%.

EPA NAME: 1,2-DIPHENYLHYDRAZINE

CAS: 122-66-7

SYNONYMS: HYDRAZINE, 1,2-DIPHENYL-; HYDRAZOBENZENE; N,N'-DIPHENYLHYDRAZINE

CLEAN AIR ACT: Hazardous Air Pollutants (Title I, Part A, Section 112).
CLEAN WATER ACT: Section 307 Priority Pollutants; Section 313 Priority Chemicals; Section 307 Toxic Pollutants as diphenylhydrazine.
EPA HAZARDOUS WASTE NUMBER (RCRA No.): U109.
RCRA Section 261 Hazardous Constituents.
RCRA Land Ban Waste.
RCRA Universal Treatment Standards: Wastewater (mg/L), 0.087; Nonwastewater (mg/kg), N/A.
SAFE DRINKING WATER ACT: Priority List (55 FR 1470).
EPCRA Section 304 Reportable Quantity (RQ): CERCLA, 10 lbs. (4.54 kgs.).
EPCRA Section 313 Form R de minimus concentration reporting level: 0.1%.
CALIFORNIA'S PROPOSITION 65: Carcinogen.

EPA NAME: DIPHENYLHYDRAZINE
CAS: 55299-18-8
SYNONYMS: HYDRAZOBENZENE

CLEAN WATER ACT: Section 307 Toxic Pollutants as diphenylhydrazine.
EPA HAZARDOUS WASTE NUMBER (RCRA No.): U109.
RCRA Section 261 Hazardous Constituents.
EPCRA Section 304 Reportable Quantity (RQ): CERCLA, 10 lbs. (4.54 kgs.).
EPCRA Section 313 Form R de minimus concentration reporting level: 0.1%.

EPA NAME: DIPHOSPHORAMIDE, OCTAMETHYL-
CAS: 152-16-9
SYNONYMS: OCTAMETHYLPYROPHOSPHORAMIDE; SCHRADAN

EPA HAZARDOUS WASTE NUMBER (RCRA No.): P085.
RCRA Section 261 Hazardous Constituents.
EPCRA Section 302 Extremely Hazardous Substances: TPQ = 100 lbs. (45.4 kgs.).
EPCRA Section 304 Reportable Quantity (RQ): EHS/CERCLA, 100 lbs. (45.4 kgs.).

EPA NAME: DIPOTASSIUM ENDOTHALL
CAS: 2164-07-0
SYNONYMS: 7-OXABICYCLO(2.2.1)HEPTANE-2,3-DICARBOXYLIC ACID, DIPOTASSIUM SALT; ENDOTHALL DIPOTASSIUM SALT; EPA PESTICIDE CHEMICAL CODE 038904

EPCRA Section 313 Form R de minimus concentration reporting level: 1.0%.

EPA NAME: DIPROPYLAMINE
CAS: 142-84-7

SYNONYMS: 1-PROPANAMINE, n-PROPYL-; n-PROPYL-1-PROPANAMINE

EPA HAZARDOUS WASTE NUMBER (RCRA No.): U110.
RCRA Section 261 Hazardous Constituents.
EPCRA Section 304 Reportable Quantity (RQ): CERCLA, 5,000 lbs. (2270 kgs.).

EPA NAME: 4-(DIPROPYLAMINO)-3,5-DINITROBENZENE-SULFONAMIDE)
[see ORYZALIN]
CAS: 19044-88-3

EPA NAME: DIPROPYL ISOCINCHOMERONATE
CAS: 136-45-8
SYNONYMS: DIPROPYL 2,5-PYRIDINEDICARBOXYLATE; EPA PESTICIDE CHEMICAL CODE 047201; 2,5-PYRIDINEDICARBOXYLIC ACID, DIPROPYL ESTER

EPCRA Section 313 Form R de minimus concentration reporting level: 1.0%.

EPA NAME: DI-n-PROPYLNITROSAMINE
[see N-NITROSODI-N-PROPYLAMINE]
CAS: 621-64-7

EPA NAME: DIQUAT
CAS: 85-00-7
SYNONYMS: 9,10-DIHYDRO-8A,10A-DIAZONIAPHENANTHRENE(1,1′-ETHYLENE-2,2′-BIPYRIDYLIUM)DIBROMIDE; 5,6-DIHYDRO-DIPYRIDO(1,2A,2,1C)PYRAZINIUM DIBROMIDE; DIPYRIDO(1,2-A:2′,1′-C)PYRAZINEDIUM, 6,7-DIHYDRO-, DIBROMIDE

CLEAN WATER ACT: Section 311 Hazardous Substances/RQ (same as CERCLA).
SAFE DRINKING WATER ACT: MCL, 0.02 mg/L; MCLG, 0.02 mg/L; Regulated chemical (47 FR 9352).
EPCRA Section 304 Reportable Quantity (RQ): CERCLA, 1,000 lbs. (454 kgs.).

EPA NAME: DIQUAT
CAS: 2764-72-9
SYNONYMS: 6,7-DIHYDROPYRIDO (1,2-a:2′,1′-c)PYRAZINEDIIUM ION; DIPYRIDO(1,2-a:2′,1′-c)PYRAZINEDIIUM, 6,7-DIHYDRO-; DIQUAT, DIBROMIDE; REGLONE

CLEAN WATER ACT: Section 311 Hazardous Substances/RQ (same as CERCLA).
SAFE DRINKING WATER ACT: Regulated chemical (47 FR 9352) as diquat.
EPCRA Section 304 Reportable Quantity (RQ): CERCLA, 1,000 lbs. (454 kgs.).

EPA NAME: DISODIUM CYANODITHIOIMIDOCARBONATE

CAS: 138-93-2

SYNONYMS: CYANODITHIOIMIDOCARBONIC ACID DISODIUM SALT; EPA PESTICIDE CHEMICAL CODE 063301

EPCRA Section 313 Form R de minimus concentration reporting level: 1.0%.

EPA NAME: 2,4-D ISOOCTYL ESTER

CAS: 25168-26-7

SYNONYMS: ACETIC ACID(2,4-DICHLOROPHENOXY)-,ISOOCTYL ESTER; 2,4-DICHLOROPHENOXYACETIC ACID, ISOOCTYL ESTER

CLEAN WATER ACT: Section 311 Hazardous Substances/RQ (same as CERCLA).

EPA HAZARDOUS WASTE NUMBER (RCRA No.): U240.

RCRA Section 261 Hazardous Constituents.

SAFE DRINKING WATER ACT: Regulated chemical (47 FR 9352) as 2,4-D.

EPCRA Section 304 Reportable Quantity (RQ): CERCLA, 100 lbs. (45.4 kgs.).

EPA NAME: 2,4-D ISOPROPYL ESTER

CAS: 94-11-1

SYNONYMS: ACETIC ACID, (2,4-DICHLOROPHENOXY)-,ISOPROPYL ESTER; ACETIC ACID, (2,4-DICHLOROPHENOXY)-, 1-METHYLETHYL ESTER; 2,4-D ESTERS; 2,4-DIACETIC ACID, (CHLOROPHENOXY)-,1-METHYLETHYL ESTER; (2,4-DICHLOROPHENOXY)ACETIC ACID ESTER; 2,4-DICHLOROPHENOXYACETIC ACID, ISOPROPYL ESTER

CLEAN WATER ACT: Section 311 Hazardous Substances/RQ (same as CERCLA).

EPA HAZARDOUS WASTE NUMBER (RCRA No.): U240.

RCRA Section 261 Hazardous Constituents.

RCRA Land Ban Waste.

SAFE DRINKING WATER ACT: Regulated chemical (47 FR 9352) as 2,4-D.

EPCRA Section 304 Reportable Quantity (RQ): CERCLA, 100 lbs. (45.4 kgs.).

EPCRA Section 313 Form R de minimus concentration reporting level: 0.1%.

EPA NAME: DISULFOTON

CAS: 298-04-4

SYNONYMS: O,O-DIETHYL S-(2-ETHTHIOETHYL)PHOSPHORODITHIOATE; O,O-DIETHYL S-((2-ETHYLTHIO)ETHYL)ESTER; DI-SYSTON; PHOSPHORODITHIONIC ACID, S-2-(ETHYLTHIO)ETHYL-O,O-DIETHYLESTER

CLEAN WATER ACT: Section 311 Hazardous Substances/RQ (same as CERCLA).

EPA HAZARDOUS WASTE NUMBER (RCRA No.): P039.
RCRA Section 261 Hazardous Constituents.
RCRA Land Ban Waste.
RCRA Universal Treatment Standards: Wastewater (mg/L), 0.017; Nonwastewater (mg/kg), 6.2.
RCRA Ground Water Monitoring List. Suggested test method(s) (PQL µg/L): 8140(2).
EPCRA Section 302 Extremely Hazardous Substances: TPQ = 500 lbs. (227 kgs.).
EPCRA Section 304 Reportable Quantity (RQ): EHS/CERCLA, 1 lb. (0.454 kg.).
MARINE POLLUTANT (49CFR, Subchapter 172.101, Appendix B).

EPA NAME: DITHIAZANINE IODIDE
CAS: 514-73-8
SYNONYMS: 3-ETHYL-2-(5-(3-ETHYL-2-BENZOTHIAZOLINYLIDENE)-1,3-PENTADIENYL)BENZOTHIAZOLIUM IODIDE

EPCRA Section 302 Extremely Hazardous Substances: TPQ = 500/10,000 lbs. (227/4,540 kgs.).
EPCRA Section 304 Reportable Quantity (RQ): EHS, 500 lbs. (227 kgs.).

EPA NAME: DITHIOBIURET
CAS: 541-53-7
SYNONYMS: BIURET, 2,4-DITHIO-; 2,4-DITHIOBIURET; UREA, 2-THIO-1-(THIOCARBAMOYL)-

EPA HAZARDOUS WASTE NUMBER (RCRA No.): P049.
RCRA Section 261 Hazardous Constituents.
EPCRA Section 302 Extremely Hazardous Substances: TPQ = 100/10,000 lbs. (45.4/4,540 kgs.).
EPCRA Section 304 Reportable Quantity (RQ): EHS/CERCLA, 100 lbs. (45.4 kgs.).
EPCRA Section 313 Form R de minimus concentration reporting level: 1.0%.

EPA NAME: DIURON
CAS: 330-54-1
SYNONYMS: EPA PESTICIDE CHEMICAL CODE 035505; UREA, 3-(3,4-DICHLOROPHENYL)-1,1-DIMETHYL-; UREA, N′-(3,4-DICHLOROPHENYL)-N,N-DIMETHYL-

CLEAN WATER ACT: Section 311 Hazardous Substances/RQ (same as CERCLA).
EPCRA Section 304 Reportable Quantity (RQ): CERCLA, 100 lbs. (45.4 kgs.).
EPCRA Section 313 Form R de minimus concentration reporting level: 1.0%.

EPA NAME: DODECYLBENZENESULFONIC ACID
CAS: 27176-87-0
SYNONYMS: BENZENESULFONIC ACID, DODECYL-; BENZENE SULFONIC ACID, DODECYL ESTER

CLEAN WATER ACT: Section 311 Hazardous Substances/RQ (same as CERCLA).
EPCRA Section 304 Reportable Quantity (RQ): CERCLA, 1,000 lbs. (454 kgs.).

EPA NAME: DODECYLGUANIDINE MONOACETATE
[see DODINE]
CAS: 2439-10-3

EPA NAME: DODINE
CAS: 2439-10-3
SYNONYMS: DODECYLGUANIDINE MONOACETATE; 1-DODECYLGUANIDINIUM ACETATE; EPA PESTICIDE CHEMICAL CODE 044301

EPCRA Section 313 Form R de minimus concentration reporting level: 1.0%.

EPA NAME: 2,4-DP
CAS: 120-36-5
SYNONYMS: 2,4-DICHLOROPHENOXY-α-PROPIONIC ACID; 2,4-DICHLOROPHENOXYPROPIONIC ACID; EPA PESTICIDE CHEMICAL CODE 031401; PROPANOIC ACID, 2-(2,4-DICHLOROPHENOXY)-

EPCRA Section 313 Form R de minimus concentration reporting level: 0.1%.

EPA NAME: 2,4-D PROPYLENE GLYCOL BUTYL ETHER ESTER
CAS: 1320-18-9
SYNONYMS: ACETIC ACID, 2,4-DICHLOROPHENOXY-, BUTOXYPROPYL ESTER; ACETIC ACID, 2,4-DICHLOROPHENOXY-, BUTOXYMETHYLETHYL ESTER; 2,4-D ESTERS; 2,4-DICHLOROPHENOXY, 2-BUTOXYMETHYLETHYL ESTER; 2,4-D PROPYLENE GLYCOL BUTYL ETHER ESTER

CLEAN WATER ACT: Section 311 Hazardous Substances/RQ (same as CERCLA).
EPA HAZARDOUS WASTE NUMBER (RCRA No.): U240.
RCRA Section 261 Hazardous Constituents.
SAFE DRINKING WATER ACT: Regulated chemical (47 FR 9352) as 2,4-D.
EPCRA Section 304 Reportable Quantity (RQ): CERCLA, 100 lbs. (45.4 kgs.).
EPCRA Section 313 Form R de minimus concentration reporting level: 0.1%.

EPA NAME: 2,4-D PROPYL ESTER
CAS: 1928-61-6
SYNONYMS: ACETIC ACID, (2,4-DICHLOROPHENOXY)-, PROPYL ESTER; 2,4-D ESTERS; PROPYL 2,4-D; PROPYL 2,4-D ESTER

CLEAN WATER ACT: Section 311 Hazardous Substances/RQ (same as CERCLA).

EPA HAZARDOUS WASTE NUMBER (RCRA No.): U240.

RCRA Section 261 Hazardous Constituents.

SAFE DRINKING WATER ACT: Regulated chemical (47 FR 9352) as 2,4-D.

EPCRA Section 304 Reportable Quantity (RQ): CERCLA, 100 lbs. (45.4 kgs.).

EPA NAME: 2,4-D SODIUM SALT

CAS: 2702-72-9

SYNONYMS: 2,4-DICHLOROPHENOXYACETIC ACID, SODIUM SALT; SODIUM 2,4-DICHLOROPHENOXYACETATE; 2,4-D, SODIUM SALT; EPA PESTICIDE CHEMICAL CODE 030004

EPA HAZARDOUS WASTE NUMBER (RCRA No.): U240.

RCRA Section 261 Hazardous Constituents.

SAFE DRINKING WATER ACT: Regulated chemical (47 FR 9352) as 2,4-D.

EPCRA Section 304 Reportable Quantity (RQ): CERCLA, 100 lbs. (45.4 kgs.).

EPCRA Section 313 Form R de minimus concentration reporting level: 0.1%.

- E -

EPA NAME: EMETINE, DIHYDROCHLORIDE
CAS: 316-42-7
SYNONYMS: 1-EMETINE, DIHYDROCHLORIDE; EMETINE, HYDROCHLORIDE

EPCRA Section 302 Extremely Hazardous Substances: TPQ = 1/10,000 lbs. (0.454/4,540 kgs.).
EPCRA Section 304 Reportable Quantity (RQ): EHS, 1 lb. (0.454 kg.).

EPA NAME: ENDOSULFAN
CAS: 115-29-7
SYNONYMS: α,β-1,2,3,4,7,7-HEXACHLOROBICLO(2,2,1)HEPTEN-5,6-BIOXYMETHYLENESULFITE; 6,9-METHANO-2,4,3-BENZODIOXATHIEPIN, 6,7,8,9,10,10-HEXACHLORO-1,5,5A, 6,9,9A-HEXAHYDRO-, 3-OXIDE; THIODAN; 6,9-METHANO-2,4,3-BENZODIOXATHIEPIN, 6,7,8,9,10,10-HEXACHLORO-1,5,5a,6,9,9e-HEXAHYDRO-, 3-OXIDE

CLEAN WATER ACT: Section 311 Hazardous Substances/RQ (same as CERCLA); Toxic Pollutant (Section 401.15).
EPA HAZARDOUS WASTE NUMBER (RCRA No.): P050.
RCRA Section 261 Hazardous Constituents.
EPCRA Section 302 Extremely Hazardous Substances: TPQ = 10/10,000 lbs. (4.54/4,540 kgs.).
EPCRA Section 304 Reportable Quantity (RQ): EHS/CERCLA, 1 lb. (0.454 kg.).
MARINE POLLUTANT (49CFR, Subchapter 172.101, Appendix B): Severe pollutant.

EPA NAME: α-ENDOSULFAN
CAS: 959-98-8
SYNONYMS: ENDOSULFAN I; 6,9-METHANO-2,4,3-BENZODIOXATHIEPIN, 6,7,8,9,10,10-HEXACHLORO-1,5,5A,6,9,9A-HEXAHYDRO-, 3-OXIDE, (3α, 5Aβ,6α,9α,9Aβ)-; α-THIODAN; THIODAN

CLEAN WATER ACT: Section 307 Priority Pollutants; Toxic Pollutant (Section 401.15).
RCRA Land Ban Waste.
RCRA Universal Treatment Standards: Wastewater (mg/L), 0.023; Nonwastewater (mg/kg), 0.066.
RCRA Ground Water Monitoring List. Suggested test method(s) (PQL μg/L): 8080(0.1); 8250(10).
EPCRA Section 304 Reportable Quantity (RQ): CERCLA, 1 lb. (0.454 kg.).
MARINE POLLUTANT (49CFR, Subchapter 172.101, Appendix B). Severe pollutant; as endosulfan.

EPA NAME: β-ENDOSULFAN

CAS: 33213-65-9

SYNONYMS: ENDOSULFAN II; 6,9-METHANO-2,4,3-BENZODIOXATHIEPIN, 6,7,8,9,10,10-HEXACHLORO-1,5,5A,6,9,9A-HEXAHYDRO-, 3-OXIDE, 3α, 5Aα,6β,9β,9Aα)-; β-THIODAN

CLEAN WATER ACT: Section 307 Priority Pollutants; Toxic Pollutant (Section 401.15).

RCRA Land Ban Waste.

RCRA Universal Treatment Standards: Wastewater (mg/L), 0.029; Nonwastewater (mg/kg), 0.13.

RCRA Ground Water Monitoring List. Suggested test method(s) (PQL μg/L): 8080(0.05).

EPCRA Section 304 Reportable Quantity (RQ): CERCLA, 1 lb. (0.454 kg.).

MARINE POLLUTANT (49CFR, Subchapter 172.101, Appendix B): Severe pollutant; as endosulfan.

EPA NAME: ENDOSULFAN and METABOLITES

CLEAN WATER ACT: Toxic Pollutant (Section 401.15).

EPA NAME: ENDOSULFAN SULFATE

CAS: 1031-07-8

SYNONYMS: 1,4,5,6,7,7-HEXACHLORO-5-NORBORNENE-2,3-DIMETHANOL, CYCLIC SULFITE; 6,9-METHANO-2,4,3-BENZODIOXATHIEPIN, 6,7,8,9,10,10-HEXACHLORO-1,5,5A,6,9,9A-HEXAHYDRO-, 3-DIOXIDE

CLEAN WATER ACT: Section 307 Priority Pollutants; Toxic Pollutant (Section 401.15).

RCRA Land Ban Waste.

RCRA Universal Treatment Standards: Wastewater (mg/L), 0.029; Nonwastewater (mg/kg), 0.13.

RCRA Ground Water Monitoring List. Suggested test method(s) (PQL μg/L): 8080(0.5); 8270(10).

EPCRA Section 304 Reportable Quantity (RQ): CERCLA, 1 lb. (0.454 kg.).

MARINE POLLUTANT (49CFR, Subchapter 172.101, Appendix B): Severe pollutant; as endosulfan.

EPA NAME: ENDOTHALL

CAS: 145-73-3

SYNONYMS: 3,6-ENDOOXOHEXAHYDROPHTHALIC ACID; 7-OXABICYCLO(2,2,1)HEPTANE-2,3-DICARBOXYLIC ACID

EPA HAZARDOUS WASTE NUMBER (RCRA No.): P088.

RCRA Section 261 Hazardous Constituents.

SAFE DRINKING WATER ACT: MCL, 0.1 mg/L; MCLG, 0.1 mg/L; Regulated chemical (47 FR 9352).

EPCRA Section 304 Reportable Quantity (RQ): CERCLA, 1,000 lbs. (454 kgs.).

EPA NAME: ENDOTHION

CAS: 2778-04-3

SYNONYMS: O,O-DIMETHYL-S-(5-METHOXY-4-OXO-4H-PYRAN-2-YL)PHOSPHOROTHIOATE

EPCRA Section 302 Extremely Hazardous Substances: TPQ= 500/10,000 lb. (227/4,540 kg.).

EPCRA Section 304 Reportable Quantity (RQ): EHS, 500 lbs. (227 kgs.).

EPA NAME: ENDRIN

CAS: 72-20-8

SYNONYMS: 2,7:3,6-DIMETHANONAPHTH(2, 3-B)OXIRENE, 3,4,5,6,9,9-HEXACHLORO-1A,2,2A,3,6,6A,7,7A-OCTAHYDRO-, (Aα,2β,2Aβ,2Aβ,3 α,6α,6Aβ,7β,7Aα)-; 1,2,3,4,10,10-HEXACHLORO-6,7-EPOXY-1,4,4A,5,6,7,8,8A-OCTAHYDRO-1,4-endo-endo-1,4,5,8-DIMETHANONAPHTHALENE

CLEAN WATER ACT: Section 311 Hazardous Substances/RQ (same as CERCLA); Section 307 Priority Pollutants.

EPA HAZARDOUS WASTE NUMBER (RCRA No.): P051; D012.

RCRA Toxicity Characteristic (Section 261.24), Maximum Concentration of Contaminants, regulatory level, 0.02 mg/L.

RCRA Section 261 Hazardous Constituents.

RCRA Land Ban Waste.

RCRA Universal Treatment Standards: Wastewater (mg/L), 0.0028; Nonwastewater (mg/kg), 0.13.

RCRA Ground Water Monitoring List. Suggested test method(s) (PQL μg/L): 8080(0.1); 8250(10).

SAFE DRINKING WATER ACT: MCL, 0.002 mg/L; MCLG, 0.002 mg/L; Regulated chemical (47 FR 9352).

EPCRA Section 302 Extremely Hazardous Substances: TPQ= 500/10,000 lbs. (227/4,540 kgs.).

EPCRA Section 304 Reportable Quantity (RQ): EHS/CERCLA, 1 lb. (0.454 kg.).

MARINE POLLUTANT (49CFR, Subchapter 172.101, Appendix B).

EPA NAME: ENDRIN ALDEHYDE

CAS: 7421-93-4

SYNONYMS: 1,2,4-METHENOCYCLOPENTA(CD)PENTALENE-5-CARBOXALDEHYDE, 2,2A,3,3,4,7-HEXACHLORODECAHYDRO-,(1α,2β,2Aβ,4β,4Aβ,5β,6Aβ,6Bβ,7R*)-

CLEAN WATER ACT: Section 307 Priority Pollutants.

RCRA Land Ban Waste.

RCRA Universal Treatment Standards: Wastewater (mg/L), 0.025; Nonwastewater (mg/kg), 0.13.

RCRA Ground Water Monitoring List. Suggested test method(s) (PQL μg/L): 8080(0.2); 8250(10).

EPCRA Section 304 Reportable Quantity (RQ): CERCLA, 1 lb. (0.454 kg.).

EPA NAME: EPICHLOROHYDRIN

CAS: 106-89-8

SYNONYMS: 3-CHLORO-1,2-EPOXYPROPANE; 1-CHLORO-2,3-EPOXYPROPANE; OXIRANE, (CHLOROMETHYL)-; PROPANE, 1-CHLORO-2,3-EPOXY-

CLEAN AIR ACT: Hazardous Air Pollutants (Title I, Part A, Section 112); Accidental Release Prevention/Flammable substances (Section 112[r], Table 3), TQ=20,000 lbs. (9080 kgs.).

CLEAN WATER ACT: Section 311 Hazardous Substances/RQ (same as CERCLA); Section 313 Priority Chemicals.

EPA HAZARDOUS WASTE NUMBER (RCRA No.): U041.

RCRA Section 261 Hazardous Constituents.

SAFE DRINKING WATER ACT: Regulated chemical (47 FR 9352).

EPCRA Section 302 Extremely Hazardous Substances: TPQ=1,000 lbs. (454 kgs.).

EPCRA Section 304 Reportable Quantity (RQ): EHS/CERCLA, 100 lbs. (45.4 kgs.).

EPCRA Section 313 Form R de minimus concentration reporting level: 0.1%.

EPA NAME: EPINEPHRINE

CAS: 51-43-4

SYNONYMS: 1,2-BENZENEDIOL-4-3,4-(1-DIHYDROXY-2-[METHYLAMINO]ETHYL)-

EPA HAZARDOUS WASTE NUMBER (RCRA No.): P042.

RCRA Section 261 Hazardous Constituents.

EPCRA Section 304 Reportable Quantity (RQ): CERCLA, 1,000 lbs. (454 kgs.).

EPA NAME: EPN

CAS: 2104-64-5

SYNONYMS: O-ETHYL-O-(4-NITROPHENYL)-BENZENETHIONOPHOSPHONATE; O-ETHYL-O-p-NITROPHENYL PHENYLPHOSPHONOTHIOATE; O-ETHYL-O-(4-NITROPHENYL PHENYL)PHENYLPHOSPHONOTHIOATE; PHOSPHONOTHIOIC ACID, PHENYL-, O-ETHYL, O-(4-NITROPHENYL)ESTER

EPCRA Section 302 Extremely Hazardous Substances: TPQ= 100/10,000 lb. (45.4/4,540 kgs.).

EPCRA Section 304 Reportable Quantity (RQ): EHS, 100 lbs. (45.4 kgs.).

EPA NAME: EPTC

[see ETHYL DIPROPYLTHIOCARBAMATE]

CAS: 759-94-4

EPA NAME: ERGOCALCIFEROL

CAS: 50-14-6

SYNONYMS: 1,2-ETHYLIDENE DICHLORIDE; 9,10,SECOERGOSTA-5,7,10(19),22-TETRAEN-3-β-OL

EPCRA Section 302 Extremely Hazardous Substances: TPQ= 1,000/10,000 lbs. (454/4,540 kgs.).

EPCRA Section 304 Reportable Quantity (RQ): EHS, 1,000 lbs. (454 kgs.).

EPA NAME: ERGOTAMINE TARTRATE
CAS: 379-79-3
SYNONYMS: ERGOTAMINE BITARTRATE

EPCRA Section 302 Extremely Hazardous Substances: TPQ = 500/10,000 lbs. (227/4,540 kgs.).
EPCRA Section 304 Reportable Quantity (RQ): EHS, 500 lbs. (227 kgs.).

EPA NAME: ETHANAMINE
CAS: 75-04-7
SYNONYMS: ETHYLAMINE; MONOETHYLAMINE

CLEAN AIR ACT: Accidental Release Prevention/Flammable substances (Section 112[r], Table 3), TQ = 10,000 lbs. (4540 kgs.).
CLEAN WATER ACT: Section 311 Hazardous Substances/RQ (same as CERCLA).
EPCRA Section 304 Reportable Quantity (RQ): CERCLA, 100 lbs. (45.4 kgs.).

EPA NAME: ETHANE
CAS: 74-84-0
SYNONYMS: ETHYL HYDRIDE; METHYLMETHANE

CLEAN AIR ACT: Accidental Release Prevention/Flammable substances (Section 112[r], Table 3), TQ = 10,000 lbs. (4540 kgs.).

EPA NAME: ETHANE, CHLORO-
[see CHLOROETHANE]
CAS: 75-00-3

EPA NAME: 1,2-ETHANEDIAMINE
[see ETHYLENEDIAMINE]
CAS: 107-15-3

EPA NAME: ETHANE, 1,1-DIFLUORO-
[see DIFLUOROETHANE]
CAS: 75-37-6

EPA NAME: ETHANEDINITRILE
[see CYANOGEN]
CAS: 460-19-5

EPA NAME: ETHANE, 1,1′-OXYBIS-
[see ETHYL ETHER]
CAS: 60-29-7

EPA NAME: ETHANEPEROXOIC ACID
[see PERACETIC ACID]
CAS: 79-21-0

EPA NAME: ETHANESULFONYL CHLORIDE, 2-CHLORO-
CAS: 1622-32-8

SYNONYMS: 2-CHLOROETHANESULFONYL CHLORIDE

EPCRA Section 302 Extremely Hazardous Substances: TPQ = 500 lbs. (227 kgs.).
EPCRA Section 304 Reportable Quantity (RQ): EHS, 500 lbs. (227 kgs.).

EPA NAME: ETHANE, 1,1,1,2-TETRACHLORO-
[see 1,1,1,2-TETRACHLOROETHANE]

EPA NAME: ETHANE, 1,1′-THIOBIS[2-CHLORO-
[see MUSTARD GAS]
CAS: 505-60-2

EPA NAME: ETHANETHIOL
[see ETHYL MERCAPTAN]
CAS: 75-08-1

EPA NAME: ETHANE, 1,1,2-TRICHLORO-1,2,2,-TRIFLUORO-
[see FREON 113]
CAS: 76-13-1

EPA NAME: ETHANIMIDOTHIOIC ACID, N-[[METHYLAMINO]CARBONYL]OXY]-, METHYL ESTER
[see METHOMYL]
CAS: 16752-77-5

EPA NAME: ETHANOL, 1,2-DICHLORO-, ACETATE
CAS: 10140-87-1
SYNONYMS: 1,2-DICHLOROETHYL ACETATE

EPCRA Section 302 Extremely Hazardous Substances: TPQ = 1,000 lbs. (454 kgs.).
EPCRA Section 304 Reportable Quantity (RQ): EHS, 1,000 lbs. (454 kgs.).

EPA NAME: ETHANOL, 2-ETHOXY-
[see 2-ETHOXYETHANOL]
CAS: 110-80-5

EPA NAME: ETHENE
[see ETHYLENE]
CAS: 74-85-1

EPA NAME: ETHENE, BROMOTRIFLUORO-
[see BROMOTRIFLUORETHYLENE]
CAS: 598-73-2

EPA NAME: ETHENE, CHLORO-
[see VINYL CHLORIDE]
CAS: 75-01-4

EPA NAME: ETHENE, CHLOROTRIFLUORO-
[see TRIFLUOROCHLOROETHYLENE]
CAS: 79-38-9

EPA NAME: ETHENE, 1,1 DICHLORO
[see VINYLIDENE CHLORIDE]
CAS: 75-35-4

EPA NAME: ETHENE, 1,1-DIFLUORO-
[see VINYLIDENE FLUORIDE]
CAS: 75-38-7

EPA NAME: ETHENE, ETHOXY-
[see VINYL ETHYL ETHER]
CAS: 109-92-2

EPA NAME: ETHENE, FLUORO-
[see VINYL FLUORIDE]
CAS: 75-02-5

EPA NAME: ETHENE, METHOXY-
[see VINYL METHYL ETHER]
CAS: 107-25-5

EPA NAME: ETHENE, TETRAFLUORO-
[see TETRAFLUOROETHYLENE]
CAS: 116-14-3

EPA NAME: ETHION
CAS: 563-12-2
SYNONYMS: NIALATE; PHOSPHORODITHIOIC ACID, O,O-DIETHYL ESTER, S,S-DIESTER with METHANEDITHIOL; PHOSPHORODITHIOIC ACID, S,S′-METHYLENE O,O,O′,O′-TETRAETHYL ESTER

CLEAN WATER ACT: Section 311 Hazardous Substances/RQ (same as CERCLA).
EPCRA Section 302 Extremely Hazardous Substances: TPQ = 1,000 lbs. (454 kgs.).
EPCRA Section 304 Reportable Quantity (RQ): EHS/CERCLA, 10 lbs. (4.54 kgs.).

EPA NAME: ETHOPROP
CAS: 13194-48-4
SYNONYMS: EPA PESTICIDE CHEMICAL CODE 041101; ETHOPROPHOS; O-ETHYL S,S-DIPROPYL DITHIOPHOSPHATE; O-ETHYL S,S-DIPROPYL PHOSPHORODITHIOATE; PHOSPHORODITHIOIC ACID, O-ETHYL S,S-DIPROPYL ESTER

EPCRA Section 302 Extremely Hazardous Substances: TPQ = 1,000 lbs. (454 kgs.).
EPCRA Section 304 Reportable Quantity (RQ): EHS, 1,000 lbs. (454 kgs.).
EPCRA Section 313 Form R de minimus concentration reporting level: 1.0%.
MARINE POLLUTANT (49CFR, Subchapter 172.101, Appendix B).

EPA NAME: ETHOPROPHOS
[see ETHOPROP]
CAS: 13194-48-4

EPA NAME: 2-ETHOXYETHANOL
[see also GLYCOL ETHERS]
CAS: 110-80-5
SYNONYMS: EGEE; ETHANOL, 2-ETHOXY-; 2-ETHOXYACETYLENE; ETHYLENE GLYCOL ETHYL ETHER; ETHYLENE GLYCOL MONOETHYL ETHER

EPA HAZARDOUS WASTE NUMBER (RCRA No.): U359.
RCRA Section 261 Hazardous Constituents.
EPCRA Section 304 Reportable Quantity (RQ): CERCLA, 1,000 lbs. (454 kgs.).
EPCRA Section 313 Form R de minimus concentration reporting level: 1.0%.
CALIFORNIA'S PROPOSITION 65: Reproductive toxin (male).

EPA NAME: 2-[1-(ETHOXYIMINO) BUTYL]-5-[2-(ETHYLTHIO)PROPYL]-3-HYDROXYL-2-CYCLOHEXEN-1-ONE
[see SETHOXYDIM]
CAS: 74051-80-5

EPA NAME: 2-[[ETHOXYL[(1-METHYLETHYL)AMINO] PHOSPHINOTHIOYL]OXY] BENZOIC ACID 1-METHYLETHYL ESTER
[see ISOFENPHOS]
CAS: 25311-71-1

EPA NAME: ETHYL ACETATE
CAS: 141-78-6
SYNONYMS: ACETIC ACID ETHYL ESTER; ACETIC ETHER

EPA HAZARDOUS WASTE NUMBER (RCRA No.): U112.
RCRA Section 261 Hazardous Constituents.
RCRA Land Ban Waste.
RCRA Universal Treatment Standards: Wastewater (mg/L), 0.34; Nonwastewater (mg/kg), 33.
EPCRA Section 304 Reportable Quantity (RQ): CERCLA, 5,000 lbs. (2270 kgs.).

EPA NAME: ETHYL ACETYLENE
[see 1-BUTYNE]
CAS: 107-00-6

EPA NAME: ETHYL ACRYLATE
CAS: 140-88-5
SYNONYMS: ACRYLIC ACID, ETHYL ESTER; ETHOXY CARBONYL ETHYLENE; 2-PROPENOIC ACID, ETHYL ESTER

CLEAN AIR ACT: Hazardous Air Pollutants (Title I, Part A, Section 112).
EPA HAZARDOUS WASTE NUMBER (RCRA No.): U113.

RCRA Section 261 Hazardous Constituents.
EPCRA Section 304 Reportable Quantity (RQ): CERCLA, 1,000 lbs. (454 kgs.).
EPCRA Section 313 Form R de minimus concentration reporting level: 0.1%.
MARINE POLLUTANT (49CFR, Subchapter 172.101, Appendix B).
CALIFORNIA'S PROPOSITION 65: Carcinogen.

EPA NAME: 3-[(ETHYLAMINO)METHOXYPHOSPHINOTHIOYL]OXY]-2-BUTENOIC ACID, 1-METHYLETHYL ESTER
[see PROPETAMPHOS]
CAS: 31218-83-4

EPA NAME: ETHYLBENZENE
CAS: 100-41-4
SYNONYMS: BENZENE, ETHYL-; PHENYLETHANE

CLEAN AIR ACT: Hazardous Air Pollutants (Title I, Part A, Section 112).
CLEAN WATER ACT: Section 311 Hazardous Substances/RQ (same as CERCLA); Section 307 Priority Pollutants; Section 313 Priority Chemicals; Section 307 Toxic Pollutants.
RCRA Universal Treatment Standards: Wastewater (mg/L), 0.057; Nonwastewater (mg/kg), 10.
RCRA Ground Water Monitoring List. Suggested test method(s) (PQL µg/L): 8020(2); 8240(5).
SAFE DRINKING WATER ACT: MCL, 0.7 mg/L; MCLG, 0.7 mg/L.
EPCRA Section 304 Reportable Quantity (RQ): CERCLA, 1,000 lbs. (454 kgs.).
EPCRA Section 313 Form R de minimus concentration reporting level: 1.0%.

EPA NAME: ETHYLBIS(2-CHLOROETHYL)AMINE
CAS: 538-07-8
SYNONYMS: 2,2′-DICHLOROTRIETHYLAMINE

EPCRA Section 302 Extremely Hazardous Substances: TPQ = 500 lbs. (227 kgs.).
EPCRA Section 304 Reportable Quantity (RQ): EHS, 500 lbs. (227 kgs.).

EPA NAME: ETHYLCARBAMATE
[see URETHANE]
CAS: 51-79-6

EPA NAME: ETHYL CHLORIDE
[see CHLOROETHANE]
CAS: 75-00-3

EPA NAME: ETHYL CHLOROFORMATE
CAS: 541-41-3
SYNONYMS: CARBONOCHLORIDIC ACID, ETHYL ESTER; CHLOROCARBONIC ACID ETHYL ESTER

EPCRA Section 313 Form R de minimus concentration reporting level: 1.0%.

EPA NAME: ETHYL-2-[[[(4-CHLORO-6-METHOXYPYRIMIDIN-2-YL)-CARBONYL]-AMINO]SULFONYL]BENZOATE

[see CHLORIMURON ETHYL]

CAS: 90982-32-4

EPA NAME: ETHYL CYANIDE

[see PROPIONITRILE]

CAS: 107-12-0

EPA NAME: ETHYL DIPROPYLTHIOCARBAMATE

CAS: 759-94-4

SYNONYMS: CARBAMIC ACID, DIPROPYLTHIO-, S-ETHYL ESTER; EPA PESTICIDE CHEMICAL CODE 041401; EPTC; ETHYL N,N-DIPROPYLTHIOLCARBAMATE

EPA HAZARDOUS WASTE NUMBER (RCRA No.): U390.
RCRA Section 261 Hazardous Constituents.
RCRA Universal Treatment Standards: Wastewater (mg/L), 0.003; Nonwastewater (mg/kg), 1.4.
EPCRA Section 304 Reportable Quantity (RQ): CERCLA, 1 lb. (0.454 kgs.).
EPCRA Section 313 Form R de minimus concentration reporting level: 1.0%.

EPA NAME: ETHYLENE

CAS: 74-85-1

SYNONYMS: ACETENE; BICARBURRETTED HYDROGEN; ELAYL; ETHENE; OLEFIANT GAS

CLEAN AIR ACT: Accidental Release Prevention/Flammable substances (Section 112[r], Table 3), TQ = 10,000 lbs. (4540 kgs.).
EPCRA Section 313 Form R de minimus concentration reporting level: 1.0%.

EPA NAME: ETHYLENEBISDITHIOCARBAMIC ACID

CAS: 111-54-6

SYNONYMS: CARBAMODITHIOIC ACID, 1,2-ETHANEDIYLBIS, SALTS AND ESTERS; EBDCs; 1,2-ETHANEDIYLBISCARBAMODITHIOIC ACID; ETHYLENEBIS(DITHIOCARBAMIC ACID)

EPA HAZARDOUS WASTE NUMBER (RCRA No.): U114.
RCRA Section 261 Hazardous Constituents.
EPCRA Section 304 Reportable Quantity (RQ): CERCLA, 5,000 lbs. (2270 kgs.).
EPCRA Section 313 Form R de minimus concentration reporting level: 1.0%. Form R Toxic Chemical Category Code: N171.

EPA NAME: ETHYLENEBISDITHIOCARBAMIC ACID, SALTS and ESTERS (EBDCs)

EPCRA Section 313: Includes any unique chemical substance that contains EDBC or an EDBC salt as part of that chemical's infrastructure. Form R de minimus concentration reporting level: 1.0%. Form R Toxic Chemical Category Code: N171.

EPA NAME: ETHYLENEDIAMINE

CAS: 107-15-3

SYNONYMS: 1,2-DIAMINOETHANE, ANHYDROUS; DIMETHYLENEDIAMINE; 1,2-ETHANEDIAMINE

CLEAN AIR ACT: Accidental Release Prevention/Flammable substances (Section 112[r], Table 3), TQ=20,000 lbs. (9080 kgs.).

CLEAN WATER ACT: Section 311 Hazardous Substances/RQ (same as CERCLA).

EPCRA Section 302 Extremely Hazardous Substances: TPQ = 10,000 lbs. (4,540 kgs.).

EPCRA Section 304 Reportable Quantity (RQ): EHS/CERCLA, 5,000 lbs. (2270 kgs.).

EPA NAME: ETHYLENEDIAMINE-TETRAACETIC ACID (EDTA)

CAS: 60-00-4

SYNONYMS: ACETIC ACID (ETHYLENEDINITRILO)TETRA-; ETHYLENEDIAMINETETRAACETIC ACID; GLYCINE, N,N'-1,2-ETHANEDIYLBIS(N-(CARBOXYMETHYL)-9CI)

CLEAN WATER ACT: Section 311 Hazardous Substances/RQ (same as CERCLA).

EPCRA Section 304 Reportable Quantity (RQ): CERCLA, 5,000 lbs. (2270 kgs.).

EPA NAME: ETHYLENE DIBROMIDE

[see 1,2-DIBROMOETHANE]

CAS: 106-93-4

EPA NAME: ETHYLENE DICHLORIDE

[see 1,2-DICHLOROETHANE]

CAS: 107-06-2

EPA NAME: ETHYLENE FLUOROHYDRIN

CAS: 371-62-0

SYNONYMS: 2-FLUROETHANOL

EPCRA Section 302 Extremely Hazardous Substances: TPQ = 10 lbs. (4.54 kgs.).

EPCRA Section 304 Reportable Quantity (RQ): EHS, 10 lbs. (4.54 kgs.).

EPA NAME: ETHYLENE GLYCOL

CAS: 107-21-1

SYNONYMS: 1,2-DIHYDROXYETHANE; 1,2-ETHANEDIOL

CLEAN AIR ACT: Hazardous Air Pollutants (Title I, Part A, Section 112).

EPCRA Section 304 Reportable Quantity (RQ): CERCLA, 5,000 lbs. (2270 kgs.).
EPCRA Section 313 Form R de minimus concentration reporting level: 1.0%.

EPA NAME: ETHYLENEIMINE
CAS: 151-56-4
SYNONYMS: AMINOETHYLENE; AZACYCLOPROPANE; AZIRANE; AZIRIDINE; AZIRINE; 1H-AZIRINE,DIHYDRO-; DIHYDROAZIRINE; DIHYDRO-1-AZIRINE; DIMETHYLENEIMINE; DIMETHYLENIMINE; ETHYLIMINE

CLEAN AIR ACT: Hazardous Air Pollutants (Title I, Part A, Section 112); Section 112[r], Accidental Release Prevention/Flammable substances (40CFR/68.130; 59 FR 4497), TQ = 10,000 lbs (4550 kgs). Reportable Quantity (RQ): CERCLA, 1 lb. (0.454 kg.).
EPA HAZARDOUS WASTE NUMBER (RCRA No.): P054.
RCRA Section 261 Hazardous Constituents.
EPCRA Section 302, Extremely Hazardous Substances: TPQ = 500 lbs. (228 kgs.).
EPCRA Section 304 Reportable Quantity (RQ): CERCLA, 1 lb. (0.454 kg.).
EPCRA Section 313 Form R de minimus concentration reporting level: 0.1%.
CALIFORNIA'S PROPOSITION 65: Carcinogen.

EPA NAME: ETHYLENE OXIDE
CAS: 75-21-8
SYNONYMS: ETO; OXACYCLOPROPANE; OXIRANE

CLEAN AIR ACT: Hazardous Air Pollutants (Title I, Part A, Section 112); Accidental Release Prevention/Flammable substances (Section 112[r], Table 3), TQ = 10,000 lbs. (4540 kgs.).
EPA HAZARDOUS WASTE NUMBER (RCRA No.): U115.
RCRA Section 261 Hazardous Constituents.
RCRA Land Ban Waste.
RCRA Universal Treatment Standards: Wastewater (mg/L), 0.12; Nonwastewater (mg/kg), N/A.
EPCRA Section 302 Extremely Hazardous Substances: TPQ = 1,000 lbs. (454 kgs.).
EPCRA Section 304 Reportable Quantity (RQ): EHS/CERCLA, 10 lbs. (4.54 kgs.).
EPCRA Section 313 Form R de minimus concentration reporting level: 0.1%.
CALIFORNIA'S PROPOSITION 65: Carcinogen; reproductive toxin (female).

EPA NAME: ETHYLENE THIOUREA
CAS: 96-45-7
SYNONYMS: 1,3-ETHYLENETHIOUREA; 2-IMIDAZOLIDINETHIONE

CLEAN AIR ACT: Hazardous Air Pollutants (Title I, Part A, Section 112).

EPA HAZARDOUS WASTE NUMBER (RCRA No.): U116.
RCRA Section 261 Hazardous Constituents.
SAFE DRINKING WATER ACT: Priority List (55 FR 1470).
EPCRA Section 304 Reportable Quantity (RQ): CERCLA, 10 lbs. (4.54 kgs.).
EPCRA Section 313 Form R de minimus concentration reporting level: 0.1%.
CALIFORNIA'S PROPOSITION 65: Carcinogen; reproductive toxin.

EPA NAME: ETHYL ETHER
CAS: 60-29-7
SYNONYMS: ETHANE, 1,1'-OXYBIS-

CLEAN AIR ACT: Accidental Release Prevention/Flammable substances (Section 112[r], Table 3), TQ = 10,000 lbs. (4540 kgs.).
EPA HAZARDOUS WASTE NUMBER (RCRA No.): U117.
RCRA Section 261 Hazardous Constituents.
RCRA Land Ban Waste.
EPCRA Section 304 Reportable Quantity (RQ): CERCLA, 100 lbs. (45.4 kgs.).
RCRA Universal Treatment Standards: Wastewater (mg/L), 0.12; Non-wastewater (mg/kg), 160.

EPA NAME: ETHYLIDENE DICHLORIDE
CAS: 75-34-3
SYNONYMS: 1,1-DICHLOROETHANE; ETHANE, 1,1-DICHLORO-; ETHYLIDENE DICHLORIDE; DICHLOROMETHYLETHANE

CLEAN AIR ACT: Hazardous Air Pollutants (Title I, Part A, Section 112).
CLEAN WATER ACT: Section 307 Priority Pollutants.
EPA HAZARDOUS WASTE NUMBER (RCRA No.): U076.
RCRA Section 261 Hazardous Constituents.
RCRA Land Ban Waste.
RCRA Universal Treatment Standards: Wastewater (mg/L), 0.059; Non-wastewater (mg/kg), 6.0.
RCRA Ground Water Monitoring List. Suggested test method(s) (PQL μg/L): 8010(1); 8240(5).
SAFE DRINKING WATER ACT: Priority List (55 FR 1470).
EPCRA Section 304 Reportable Quantity (RQ): CERCLA, 1,000 lbs. (454 kgs.).
EPCRA Section 313 Form R de minimus concentration reporting level: 1.0%.
MARINE POLLUTANT (49CFR, Subchapter 172.101, Appendix B).
CALIFORNIA'S PROPOSITION 65: Carcinogen.

EPA NAME: ETHYL MERCAPTAN
CAS: 75-08-1
SYNONYMS: ETHANETHIOL; 1-ETHANETHIOL; MERCAPTO-ETHANE

CLEAN AIR ACT: Accidental Release Prevention/Flammable substances (Section 112[r], Table 3), TQ = 10,000 lbs. (4540 kgs.).

EPA NAME: ETHYL METHACRYLATE
CAS: 97-63-2
SYNONYMS: 1-2-METHACRYLIC ACID, ETHYL ESTER; 2-METHYLE-2-PROPENOIC ACID, ETHYL ESTER; 2-PROPENOIC ACID, 1-METHYL-, ETHYL ESTER

EPA HAZARDOUS WASTE NUMBER (RCRA No.): U118.
RCRA Section 261 Hazardous Constituents.
RCRA Land Ban Waste.
RCRA Universal Treatment Standards: Wastewater (mg/L), 0.14; Nonwastewater (mg/kg), 160.
RCRA Ground Water Monitoring List. Suggested test method(s) (PQL μg/L): 8015(10); 8240(5).
EPCRA Section 304 Reportable Quantity (RQ): CERCLA, 1,000 lbs. (454 kgs.).

EPA NAME: ETHYL METHANESULFONATE
CAS: 62-50-0
SYNONYMS: METHANESULPHONIC ACID, ETHYL ESTER

EPA HAZARDOUS WASTE NUMBER (RCRA No.): U119.
RCRA Section 261 Hazardous Constituents.
RCRA Ground Water Monitoring List. Suggested test method(s) (PQL μg/L): 8270(10).
EPCRA Section 304 Reportable Quantity (RQ): CERCLA, 1 lb. (0.454 kg.).
CALIFORNIA'S PROPOSITION 65: Carcinogen.

EPA NAME: N-ETHYL-N'-(1-METHYLETHYL)-6-(METHYLTHIOL)-1,3,5,-TRIAZINE-2,4-DIAMINE
[see AMETRYN]
CAS: 834-12-8

EPA NAME: O-ETHYL O-[4-(METHYLTHIO)PHENYL]PHOSPHORODITHIOIC ACID S-PROPYL ESTER
[see SULPROFOS]
CAS: 35400-43-2

EPA NAME: ETHYL NITRITE
CAS: 109-95-5
SYNONYMS: NITROUS ACID ETHYL ESTER; NITROSYL ETHOXIDE; NITROUS ETHYL ETHER

CLEAN AIR ACT: Accidental Release Prevention/Flammable substances (Section 112[r], Table 3), TQ = 10,000 lbs. (4540 kgs.).

EPA NAME: N-(1-ETHYLPROPYL)-3,4-DIMETHYL-2,6-DINITROBENZENAMINE)
[see PENDIMETHALIN]
CAS: 40487-42-1

EPA NAME: S-(2-(ETHYLSULFINYL)ETHYL) O,O-DIMETHYL ESTER PHOSPHOROTHIOIC ACID
[see OXYDEMETON METHYL]

CAS: 301-12-2

EPA NAME: ETHYLTHIOCYANATE
[see also CYANIDE COMPOUNDS]
CAS: 542-90-5
SYNONYMS: THIOCYANATOETHANE

EPCRA Section 302 Extremely Hazardous Substances: TPQ = 10,000 lbs. (4,540 kgs.).
EPCRA Section 304 Reportable Quantity (RQ): EHS, 10,000 lbs. (4,540 kgs.).

EPA NAME: ETHYNE
[see ACETYLENE]
CAS: 74-86-2

- F -

EPA NAME: FAMPHUR

CAS: 52-85-7

SYNONYMS: O-(4-((DIMETHYLAMINO)SULFONYL)PHENYL) O,O-DIMETHYL PHOSPHOROTHIOATE; EPA PESTICIDE CHEMICAL CODE 059901; PHOSPHOROTHIOIC ACID, O-(4-((DIMETHYLAMINO)SULFONYL)PHENYL) O,O-DIMETHYL ESTER

EPA HAZARDOUS WASTE NUMBER (RCRA No.): P097.

RCRA Section 261 Hazardous Constituents.

RCRA Land Ban Waste.

RCRA Universal Treatment Standards: Wastewater (mg/L), 0.017; Non-wastewater (mg/kg), 15.

RCRA Ground Water Monitoring List. Suggested test method(s) (PQL µg/L): 8270(10).

EPCRA Section 304 Reportable Quantity (RQ): CERCLA, 1,000 lbs. (454 kgs.).

EPCRA Section 313 Form R de minimus concentration reporting level: 1.0%.

EPA NAME: FENAMIPHOS

CAS: 22224-92-6

SYNONYMS: ETHYL 3-METHYL-4-(METHYLTHIO)PHENYL(1-METHYLETHYL)PHOSPHORAMIDATE; ETHYL 4-(METHYLTHIO)-m-TOLYLISOPROPYLPHOSPHORAMIDATE; FENAMINPHOS(DOT); 1-(METHYLETHYL)-ETHYL 3-METHYL-4-(METHYLTHIO)PHENYLPHOSPHORAMIDATE; PHOSPHORAMIDIC ACID, (1-METHYLETHYL)-, ETHYL(3-METHYL-4-(METHYLTHIO)PHENYL)ESTER

EPCRA Section 302 Extremely Hazardous Substances: TPQ = 10/10,000 lbs. (4.54/4,540 kgs.).

EPCRA Section 304 Reportable Quantity (RQ): EHS, 10 lbs. (45.4 kgs.).

MARINE POLLUTANT (49CFR, Subchapter 172.101, Appendix B).

EPA NAME: FENARIMOL

CAS: 60168-88-9

SYNONYMS: α-(2-CHLOROPHENYL)-α-(4-CHLOROPHENYL)-5-PYRIMIDINEMETHANOL; (2-CHLOROPHENYL)-α-(4-CHLOROPHENYL)-5-PYRIMIDINEMETHANOL; EPA PESTICIDE CHEMICAL CODE 206600

EPCRA Section 313 Form R de minimus concentration reporting level: 1.0%.

EPA NAME: FENBUTATIN OXIDE

CAS: 13356-08-6

SYNONYMS: BIS(TRIS(2-METHYL-2-PHENYLPROPYL)TIN)OXIDE; DI(TRI-(2,2-DIMETHYL-2-PHENYLETHYL)TIN)OXIDE; EPA PESTICIDE CHEMICAL CODE 104601; HEXAKIS(2-METHYL-2-PHENYLPROPYL)DISTANNOXANE

EPCRA Section 313 Form R de minimus concentration reporting level: 1.0%.

EPA NAME: FENITROTHION

CAS: 122-14-5

SYNONYMS: O,O-DIMETHYL-O-(3-METHYL-4-NITROPHENYL)-PHOSPHOROTHIOATE

Removed from EHS list (FR Vol. 61, No. 89, page 20477).

EPA NAME: FENOXAPROP ETHYL-

CAS: 66441-23-4

SYNONYMS: 2-(4-((6-CHLORO-2-BENZOXAZOLYLEN)OXY)PHENOXY)PROPANOIC ACID, ETHYL ESTER; EPA PESTICIDE CHEMICAL CODE 128701; ETHYL-2-((4-(6-CHLORO-2-BENZOXAZOLYLOXY))-PHENOXY)PROPIONATE; PROPIONIC ACID, 2-(4-((6-CHLORO-2-BENZOXAZOLYL)OXY)PHENOXY)-,ETHYL ESTER, (+-)-

EPCRA Section 313 Form R de minimus concentration reporting level: 1.0%.

EPA NAME: FENOXYCARB

CAS: 72490-01-8

SYNONYMS: EPA PESTICIDE CHEMICAL CODE 125301; ETHYL(2-(4-PHENOXYPHENOXY)ETHYL)CARBAMATE; N-(2-(P-PHENOXY PHENOXY)ETHYL)CARBAMIC ACID; 2-(4-PHENOXYPHENOXY)ETHYL-CARBAMIC ACID ETHYL ESTER

EPCRA Section 313 Form R de minimus concentration reporting level: 1.0%.

EPA NAME: FENPROPATHRIN

CAS: 39515-41-8

SYNONYMS: CYANO-3-PHENOXY BENZYL-2,2,3,3-TETRAMETHYLCYCLOPROPANECARBOXYLATE; EPA PESTICIDE CHEMICAL CODE 127901; 2,2,3,3-TETRAMETHYLCYCLOPROPANE CARBOXYLIC ACID CYANO(3-PHENOXYPHENYL)METHYL ESTER

EPCRA Section 313 Form R de minimus concentration reporting level: 1.0%.

MARINE POLLUTANT (49CFR, Subchapter 172.101, Appendix B): Severe pollutant.

EPA NAME: FENSULFOTHION

CAS: 115-90-2

SYNONYMS: O,O-DIETHYLO(P-(METHYLSULFINYL)PHENYL)PHOSPHOROTHIOATE; DMSP; ENTPHOSPHOROTHIOATE; PHOSPHOROTHIOIC ACID, O,O-DIETHYLO-(P-(METHYLSULFINYL)PHENYL) ESTER

EPCRA Section 302 Extremely Hazardous Substances: TPQ = 500 lbs. (227 kgs.).

EPCRA Section 304 Reportable Quantity (RQ): EHS, 500 lbs. (227 kgs.).

MARINE POLLUTANT (49CFR, Subchapter 172.101, Appendix B).

EPA NAME: FENTHION

CAS: 55-38-9

SYNONYMS: BAYTEX; O,O-DIMETHYL O-[3-METHYL-4-(METHYLTHIO)PHENYL] ESTER, PHOSPHOROTHIOIC ACID; MERCAPTOPHOS; EPA PESTICIDE CHEMICAL CODE 053301; 4-METHYLMERCAPTO-3-METHYLPHENYLDIMETHYLTHIOPHOSPHATE; PHOSPHOROTHIOIC ACID, O,O-DIMETHYL O-(3-METHYL-4-(METHYLTHIO)PHENYL) ESTER

EPCRA Section 313 Form R de minimus concentration reporting level: 1.0%.

MARINE POLLUTANT (49CFR, Subchapter 172.101, Appendix B): Severe pollutant.

EPA NAME: FENVALERATE

CAS: 51630-58-1

SYNONYMS: BENZENEACETIC ACID, 4-CHLORO-α-(1-METHYLETHYL)-, CYANO(3-PHENOXYPHENYL)METHYL ESTER; 4-CHLORO-α-(1-METHYLETHYL)BENZENEACETIC ACID CYANO(3-PHENOXYPHENYL)METHYL ESTER; EPA PESTICIDE CHEMICAL CODE 109301

EPCRA Section 313 Form R de minimus concentration reporting level: 1.0%.

MARINE POLLUTANT (49CFR, Subchapter 172.101, Appendix B): Severe pollutant.

EPA NAME: FERBAM

CAS: 14484-64-1

SYNONYMS: EPA PESTICIDE CHEMICAL CODE 034801; IRON, TRIS(DIMETHYLCARBAMODITHIOATO-S,S')-, (OC-6-11)-; IRON, TRIS(DIMETHYLCARBAMODITHIOATO-S,S'-); TRIS(DIMETHYLCARBAMODITHIOATO-S,S')IRON

EPA HAZARDOUS WASTE NUMBER (RCRA No.): U396.

RCRA Section 261 Hazardous Constituents.

EPCRA Section 313 Form R de minimus concentration reporting level: 1.0%.

EPA NAME: FERRIC AMMONIUM CITRATE

CAS: 1185-57-5

SYNONYMS: AMMONIUM FERRIC CITRATE

CLEAN WATER ACT: Section 311 Hazardous Substances/RQ (same as CERCLA).

EPCRA Section 304 Reportable Quantity (RQ): CERCLA, 1,000 lbs. (454 kgs.).

EPA NAME: FERRIC AMMONIUM OXALATE

CAS: 2944-67-4

SYNONYMS: AMMONIUM FERRIC OXALATE TRIHYDRATE; ETHANEDIOIC ACID, AMMONIUM IRON(3+) SALT; ETHANEDIOIC ACID, AMMONIUM IRON(III) SALT; OXALIC ACID, AMMONIUM IRON(III) SALT (3:3:1)

CLEAN WATER ACT: Section 311 Hazardous Substances/RQ (same as CERCLA).

EPCRA Section 304 Reportable Quantity (RQ): CERCLA, 1,000 lbs. (454 kgs.).

EPA NAME: FERRIC AMMONIUM OXALATE

CAS: 55488-87-4

SYNONYMS: ETHANEDIOIC ACID, AMMONIUM SALT; FERRATE(3-), TRIS(OXALATO)-,TRIAMMONIUM

CLEAN WATER ACT: Section 311 Hazardous Substances/RQ (same as CERCLA).

EPCRA Section 304 Reportable Quantity (RQ): CERCLA, 1,000 lbs. (454 kgs.).

EPA NAME: FERRIC CHLORIDE

CAS: 7705-08-0

SYNONYMS: IRON CHLORIDE; IRON(III) CHLORIDE; IRON TRICHLORIDE

CLEAN WATER ACT: Section 311 Hazardous Substances/RQ (same as CERCLA).

EPCRA Section 304 Reportable Quantity (RQ): CERCLA, 1,000 lbs. (454 kgs.).

EPA NAME: FERRIC FLUORIDE

CAS: 7783-50-8

SYNONYMS: IRON FLUORIDE

CLEAN WATER ACT: Section 311 Hazardous Substances/RQ (same as CERCLA).

EPCRA Section 304 Reportable Quantity (RQ): CERCLA, 100 lbs. (45.4 kgs.).

EPA NAME: FERRIC NITRATE

CAS: 10421-48-4

SYNONYMS: IRON(III)NITRATE; IRON TRINITRATE

CLEAN WATER ACT: Section 311 Hazardous Substances/RQ (same as CERCLA).

EPCRA Section 304 Reportable Quantity (RQ): CERCLA, 1,000 lbs. (454 kgs.).
MARINE POLLUTANT (49CFR, Subchapter 172.101, Appendix B).

EPA NAME: FERRIC SULFATE
CAS: 10028-22-5
SYNONYMS: IRON(III) SULFATE

CLEAN WATER ACT: Section 311 Hazardous Substances/RQ (same as CERCLA).
EPCRA Section 304 Reportable Quantity (RQ): CERCLA, 1,000 lbs. (454 kgs.).

EPA NAME: FERROUS AMMONIUM SULFATE
CAS: 10045-89-3
SYNONYMS: AMMONIUM IRON SULFATE

CLEAN WATER ACT: Section 311 Hazardous Substances/RQ (same as CERCLA).
EPCRA Section 304 Reportable Quantity (RQ): CERCLA, 1,000 lbs. (454 kgs.).

EPA NAME: FERROUS CHLORIDE
CAS: 7758-94-3
SYNONYMS: IRON(II) CHLORIDE (1:2)

CLEAN WATER ACT: Section 311 Hazardous Substances/RQ (same as CERCLA).
EPCRA Section 304 Reportable Quantity (RQ): CERCLA, 100 lbs. (45.4 kgs.).

EPA NAME: FERROUS SULFATE
CAS: 7720-78-7
SYNONYMS: GREEN VITRIOL; IRON(II) SULFATE; SULFURIC ACID, IRON(II) SALT (1:1)

CLEAN WATER ACT: Section 311 Hazardous Substances/RQ (same as CERCLA).
EPCRA Section 304 Reportable Quantity (RQ): CERCLA, 1,000 lbs. (454 kgs.).

EPA NAME: FERROUS SULFATE
CAS: 7782-63-0
SYNONYMS: IRON VITRIOL; IRON PROTOSULFATE; IRON(II) SULFATE(1:1) HEPTAHYDRATE

CLEAN WATER ACT: Section 311 Hazardous Substances/RQ (same as CERCLA).
EPCRA Section 304 Reportable Quantity (RQ): CERCLA, 1,000 lbs. (454 kgs.).

EPA NAME: FINE MINERAL FIBERS

CLEAN AIR ACT: Hazardous Air Pollutants (Title I, Part A, Section 112). Includes mineral fiber emissions from facilities manufacturing or processing glass, rock, or slag fibers (or mineral derived fibers) of average diameter 1 micrometer or less.

EPA NAME: FLUAZIFOP-BUTYL

CAS: 69806-50-4

SYNONYMS: BUTYL(RS)-2-(4-((5-(TRIFLUOROMETHYL)-2-PYRIDINYL)OXY)PHENOXY)PROPANOATE; EPA PESTICIDE CHEMICAL CODE 122805; PROPANOIC ACID, 2-(4-((5-(TRIFLUOROMETHYL)-2-PYRIDINYL)OXY)PHENOXY)-,BUTYL ESTER; 2-[4-[[5-(TRIFLUOROMETHYL)-2-PYRIDINYL]OXY]-PHENOXY]PROPANOIC ACID, BUTYL ESTER

EPCRA Section 313 Form R de minimus concentration reporting level: 1.0%.

EPA NAME: FLUENETIL

CAS: 4301-50-2

SYNONYMS: 4-BIPHENYLACETIC ACID, 2-FLUOROETHYL ESTER

EPCRA Section 302 Extremely Hazardous Substances: TPQ = 100/10,000 lbs. (45.4/4,540 kgs.).

EPCRA Section 304 Reportable Quantity (RQ): EHS, 100 lbs. (45.4 kgs.).

EPA NAME: FLUOMETURON

CAS: 2164-17-2

SYNONYMS: N,N-DIMETHYL-N'-[3-(TRIFLUOROMETHYL)PHENYL]UREA; UREA, N,N-DIMETHYL-N'-[3-(TRIFLUOROMETHYL)PHENYL]-

EPCRA Section 313 Form R de minimus concentration reporting level: 1.0%.

EPA NAME: FLUORANTHENE

CAS: 206-44-0

SYNONYMS: BENZO(jk)FLUORENE; 1,2-(1,8-NAPHTHYLENE)BENZENE

CLEAN WATER ACT: Section 307 Toxic Pollutants; Section 307 Priority Pollutants.

EPA HAZARDOUS WASTE NUMBER (RCRA No.): U120.

RCRA Section 261 Hazardous Constituents.

RCRA Land Ban Waste.

RCRA Universal Treatment Standards: Wastewater (mg/L), 0.068; Nonwastewater (mg/kg), 3.4.

RCRA Ground Water Monitoring List. Suggested test method(s) (PQL µg/L): 8100(200); 8270(10).

EPCRA Section 304 Reportable Quantity (RQ): CERCLA, 100 lbs. (45.4 kgs.).

EPA NAME: FLUORENE
[see also POLYNUCLEAR AROMATIC HYDROCARBONS]
CAS: 86-73-7
SYNONYMS: 2,2'-METHYLENEBIPHENYL; 9H-FLUORENE

CLEAN WATER ACT: Section 307 Priority Pollutants.
RCRA Land Ban Waste.
RCRA Universal Treatment Standards: Wastewater (mg/L), 0.059; Non-wastewater (mg/kg), 3.4.
RCRA Ground Water Monitoring List. Suggested test method(s) (PQL µg/L): 8100(200); 8270(10).
EPCRA Section 304 Reportable Quantity (RQ): CERCLA, 5,000 lbs. (2270 kgs.).

EPA NAME: FLUORIDE
CAS: 16984-48-8
SAFE DRINKING WATER ACT: Regulated chemical (47 FR 9352); MCL, 4.0 mg/L; MCLG, 4.0 mg/L; SMCL, 2.0 mg/L.

EPA NAME: FLUORINE
CAS: 7782-41-4
SYNONYMS: FLUORINE-19

CLEAN AIR ACT: Accidental Release Prevention/Flammable substances (Section 112[r], Table 3), TQ = 1,000 lbs. (454 kgs.).
EPA HAZARDOUS WASTE NUMBER (RCRA No.): P056.
RCRA Section 261 Hazardous Constituents.
EPCRA Section 302 Extremely Hazardous Substances: TPQ = 500 lbs. (227 kgs.).
EPCRA Section 304 Reportable Quantity (RQ): EHS/CERCLA, 10 lbs. (4.54 kgs.).
EPCRA Section 313 Form R de minimus concentration reporting level: 1.0%.

EPA NAME: FLUOROACETAMIDE
CAS: 640-19-7
SYNONYMS: ACETAMIDE, 2-FLUORO; 2-FLUOROACETAMIDE

EPA HAZARDOUS WASTE NUMBER (RCRA No.): P057.
RCRA Section 261 Hazardous Constituents.
EPCRA Section 302 Extremely Hazardous Substances: TPQ = 100/10,000 lbs. (45.4/4,540 kgs.).
EPCRA Section 304 Reportable Quantity (RQ): EHS/CERCLA, 100 lbs. (45.4 kgs.).

EPA NAME: FLUOROACETIC ACID
CAS: 144-49-0
SYNONYMS: 2-FLUOROACETIC ACID

EPCRA Section 302 Extremely Hazardous Substances: TPQ = 10/10,000 lbs. (4.54/4,540 kgs.).
EPCRA Section 304 Reportable Quantity (RQ): EHS, 10 lbs. (4.54 kgs.).

EPA NAME: FLUOROACETIC ACID, SODIUM SALT
[see SODIUM FLUORACETATE]
CAS: 62-74-8

EPA NAME: FLUOROACETYL CHLORIDE
CAS: 359-06-8
SYNONYMS: ACETYL CHLORIDE, FLUORO-

EPCRA Section 302 Extremely Hazardous Substances: TPQ = 10 lbs. (4.54 kgs.).
EPCRA Section 304 Reportable Quantity (RQ): EHS, 10 lbs. (4.54 kgs.).

EPA NAME: FLUOROURACIL
CAS: 51-21-8
SYNONYMS: 2,4-DIOXO-5-FLUOROPYRIMIDINE; 5-FLUOROURACIL; URACIL, 5-FLUORO-

EPCRA Section 302 Extremely Hazardous Substances: TPQ = 500/10,000 lbs. (227/4,540 kgs.).
EPCRA Section 304 Reportable Quantity (RQ): EHS, 500 lbs. (227 kgs.).
EPCRA Section 313 Form R de minimus concentration reporting level: 1.0%.
CALIFORNIA'S PROPOSITION 65: Reproductive toxin.

EPA NAME: 5-FLUOROURACIL
[see FLUOROURACIL]
CAS: 51-21-8

EPA NAME: FLUVALINATE
CAS: 69409-94-5
SYNONYMS: N-[2-CHLORO-4-(TRIFLUOROMETHYL)PHENYL]-DL-VALINE(+)-CYANO(3-PHENOXYLPHENYL)METHYL-ESTER; EPA PESTICIDE CHEMICAL CODE 109301

EPCRA Section 313 Form R de minimus concentration reporting level: 1.0%.

EPA NAME: FOLPET
CAS: 133-07-3
SYNONYMS: EPA PESTICIDE CHEMICAL CODE 081601; 1H-ISOINDOLE-1,3(2H)-DIONE, 2-((TRICHLOROMETHYL) THIO)-; (TRICHLORMETHYLTHIO)PHTHALIMIDE

EPCRA Section 313 Form R de minimus concentration reporting level: 1.0%.
CALIFORNIA'S PROPOSITION 65: Carcinogen.

EPA NAME: FOMESAFEN
CAS: 72178-02-0
SYNONYMS: BENZAMIDE, 5-(2-CHLORO-4-(TRIFLUOROMETHYL)PHENOXY)-N-(METHYLSULFONYL)-2-NITRO-; 5-(2-CHLORO-4-(TRIFLUOROMETHYL)PHENOXY)-N-METHYLSULFONYL)-2-NITROBENZAMIDE

SAFE DRINKING WATER ACT: Priority List (55 FR 1470).

EPCRA Section 313 Form R de minimus concentration reporting level: 1.0%.

EPA NAME: FONOFOS

CAS: 944-22-9

SYNONYMS: DIFONATE; DYFONATE; PHOSPHONODITHIOIC ACID, ETHYL-O-ETHYL S-PHENYL ESTER

EPCRA Section 302 Extremely Hazardous Substances: TPQ = 500 lbs. (227 kgs.).
EPCRA Section 304 Reportable Quantity (RQ): EHS, 500 lbs. (227 kgs.).
MARINE POLLUTANT (49CFR, Subchapter 172.101, Appendix B) as fonofos.

EPA NAME: FORMALDEHYDE

CAS: 50-00-0

SYNONYMS: FORMALDEHYDE (solution); FORMALIN

CLEAN AIR ACT: Hazardous Air Pollutants (Title I, Part A, Section 112); Accidental Release Prevention/Flammable substances (Section 112[r], Table 3), TQ = 15,000 lbs. (6810 kgs.).
CLEAN WATER ACT: Section 311 Hazardous Substances/RQ (same as CERCLA); Section 313 Priority Chemicals.
EPA HAZARDOUS WASTE NUMBER (RCRA No.): U122.
RCRA Section 261 Hazardous Constituents.
EPCRA Section 302 Extremely Hazardous Substances: TPQ = 500 lbs. (227 kgs.).
EPCRA Section 304 Reportable Quantity (RQ): EHS/CERCLA, 100 lbs. (45.4 kgs.).
EPCRA Section 313 Form R de minimus concentration reporting level: 0.1%.
CALIFORNIA'S PROPOSITION 65: Carcinogen as formaldehyde (gas).

EPA NAME: FORMALDEHYDE (solution)

[see FORMALDEHYDE]

CAS: 50-00-0

EPA NAME: FORMALDEHYDE CYANOHYDRIN

CAS: 107-16-4

SYNONYMS: GYYCOLIC NITRILE; GLYCOLONITRILE; HYDROXYACETONITRILE

EPCRA Section 302 Extremely Hazardous Substances: TPQ = 1,000 lbs. (454 kgs.).
EPCRA Section 304 Reportable Quantity (RQ): EHS, 1 lb. (0.454 kg.).

EPA NAME: FORMETANATE HYDROCHLORIDE

CAS: 23422-53-9

SYNONYMS: m-(((DI-METHYLAMINO)METHYLENE)AMINO) PHENYLCARBAMATE, HYDROCHLORIDE; FORMETANATE

EPA HAZARDOUS WASTE NUMBER (RCRA No.): P198.
RCRA Section 261 Hazardous Constituents.
RCRA Land Ban Waste.

RCRA Universal Treatment Standards: Wastewater (mg/L), 0.056; Nonwastewater (mg/kg), 1.4.
EPCRA Section 302 Extremely Hazardous Substances: TPQ = 500/10,000 lbs. (227/4,540 kgs.).
EPCRA Section 304 Reportable Quantity (RQ): EHS, 1 lb. (0.454 kg.).
MARINE POLLUTANT (49CFR, Subchapter 172.101, Appendix B), severe pollutant as formetanate.

EPA NAME: FORMIC ACID
CAS: 64-18-6
SYNONYMS: FORMYLIC ACID; HYDROGEN CARBOXYLIC ACID

CLEAN WATER ACT: Section 311 Hazardous Substances/RQ (same as CERCLA).
EPCRA Section 304 Reportable Quantity (RQ): CERCLA, 5,000 lbs. (2270 kgs.).
EPA HAZARDOUS WASTE NUMBER (RCRA No.): U123.
RCRA Section 261 Hazardous Constituents.
EPCRA Section 313 Form R de minimus concentration reporting level: 1.0%.

EPA NAME: FORMIC ACID, METHYL ESTER
[see METHYL FORMATE]
CAS: 107-31-3

EPA NAME: FORMOTHION
CAS: 2540-82-1
SYNONYMS: O,O-DIMETHYL-S-(N-FORMYL-N-METHYLCARBAMOYLMETHYL)PHOSPHORODITHIOATE

EPCRA Section 302 Extremely Hazardous Substances: TPQ = 100 lbs. (45.4 kgs.).
EPCRA Section 304 Reportable Quantity (RQ): EHS, 100 lbs. (45.4 kgs.).

EPA NAME: FORMPARANATE
CAS: 17702-57-7
SYNONYMS: N,N-DIMETHYL-N'(2-METHYL-4(((METHYLAMINO) CARBONYL)OXY) PHENYL) METHANIMIDAMIDE

EPA HAZARDOUS WASTE NUMBER (RCRA No.): P197.
RCRA Section 261 Hazardous Constituents.
RCRA Land Ban Waste.
RCRA Universal Treatment Standards: Wastewater (mg/L), 0.056; Nonwastewater (mg/kg), 1.4.
EPCRA Section 302 Extremely Hazardous Substances: TPQ = 100/10,000 lbs. (45.4/4,540 kgs.).
EPCRA Section 304 Reportable Quantity (RQ): EHS, 1 lb. (0.454 kg.).

EPA NAME: FOSTHIETAN
CAS: 21548-32-3
SYNONYMS: (DIETHOXYPHOSPHINYLIMINO)-1,3-DITHIETANE

EPCRA Section 302 Extremely Hazardous Substances: TPQ = 500 lbs. (227 kgs.).
EPCRA Section 304 Reportable Quantity (RQ): EHS, 500 lbs. (227 kgs.).

EPA NAME: FREON 113
CAS: 76-13-1
SYNONYMS: CFC 113; CHLORINATED FLUOROCARBON 113; ETHANE, 1,1,2-TRICHLORO-1,2,2,-TRIFLUORO-; 1,1,2-TRICHLORO-1,2,2-TRIFLUOROETHANE; 1,1,2-TRICHLOROTRIFLUOROETHANE

RCRA Land Ban Waste.
CLEAN AIR ACT: Stratospheric ozone protection (Title VI, Subpart A, Appendix A), Class I, Ozone Depletion Potential = 0.8.
RCRA Universal Treatment Standards: Wastewater (mg/L), 0.057; Nonwastewater (mg/kg), 30.
EPCRA Section 313 Form R de minimus concentration reporting level: 1.0%.

EPA NAME: FUBERIDAZOLE
CAS: 3878-19-1
SYNONYMS: 2-(2-FURANYL)-1H-BENZIMIDAZOLE

EPCRA Section 302 Extremely Hazardous Substances: TPQ = 100/10,000 lbs. (45.4/4,540 kgs.).
EPCRA Section 304 Reportable Quantity (RQ): EHS, 100 lbs. (45.4 kgs.).

EPA NAME: FUMARIC ACID
CAS: 110-17-8
SYNONYMS: 2-BUTENEDIOIC ACID (E)-; trans-BUTENEDIOIC ACID; (E)-BUTENEDIOIC ACID

CLEAN WATER ACT: Section 311 Hazardous Substances/RQ (same as CERCLA).
EPCRA Section 304 Reportable Quantity (RQ): CERCLA, 5,000 lbs. (2270 kgs.).

EPA NAME: FURAN
CAS: 110-00-9
SYNONYMS: OXACYCLOPENTADIENE

CLEAN AIR ACT: Accidental Release Prevention/Flammable substances (Section 112[r], Table 3), TQ = 5,000 lbs. (2270 kgs.).
EPA HAZARDOUS WASTE NUMBER (RCRA No.): U124.
RCRA Section 261 Hazardous Constituents.
EPCRA Section 302 Extremely Hazardous Substances: TPQ = 500 lbs. (227 kgs.).
EPCRA Section 304 Reportable Quantity (RQ): EHS/CERCLA, 100 lbs. (45.4 kgs.).
CALIFORNIA'S PROPOSITION 65: Carcinogen.

EPA NAME: FURAN, TETRAHYDRO-
CAS: 109-99-9

SYNONYMS: CYCLOTETRAMETHYLENE OXIDE; TETRAHY-DROFURAN

EPA HAZARDOUS WASTE NUMBER (RCRA No.): U213.
RCRA Section 261 Hazardous Constituents.
EPCRA Section 304 Reportable Quantity (RQ): CERCLA, 1,000 lbs. (454 kgs.).

EPA NAME: FURFURAL
CAS: 98-01-1
SYNONYMS: 2-FURALDEHYDE; 2-FURANACARBOXALDE-HYDE; FURFURALDEHYDE

CLEAN WATER ACT: Section 311 Hazardous Substances/RQ (same as CERCLA).
EPA HAZARDOUS WASTE NUMBER (RCRA No.): U125.
RCRA Section 261 Hazardous Constituents.
RCRA Land Ban Waste.
EPCRA Section 304 Reportable Quantity (RQ): CERCLA, 5,000 lbs. (2270 kgs.).

- G -

EPA NAME: GALLIUM TRICHLORIDE
CAS: 13450-90-3
SYNONYMS: GALLIUM(3+) CHLORIDE; GALLIUM(III) CHLORIDE

EPCRA Section 302 Extremely Hazardous Substances: TPQ = 500/10,000 lbs. (227/4,540 kgs.).
EPCRA Section 304 Reportable Quantity (RQ): EHS, 500 lbs. (227 kgs.).

EPA NAME: D-GLUCOSE, 2-DEOXY-2-[[(METHYLNITROSOAMINO)CARBONYL]AMINO]-
CAS: 18883-66-4
SYNONYMS: N-d-GLUCOSYL-(2)-N′-NITROSOMETHYLUREA; STREPTOZOCIN

EPA HAZARDOUS WASTE NUMBER (RCRA No.): U206.
RCRA Section 261 Hazardous Constituents.
RCRA Land Ban Waste.
EPCRA Section 304 Reportable Quantity (RQ): CERCLA, 1 lb. (0.454 kg.).

EPA NAME: GLYCIDYLALDEHYDE
CAS: 765-34-4
SYNONYMS: 2,3-EPOXY-1-PROPANAL

EPA HAZARDOUS WASTE NUMBER (RCRA No.): U126.
RCRA Section 261 Hazardous Constituents.
RCRA Land Ban Waste.
EPCRA Section 304 Reportable Quantity (RQ): CERCLA, 10 lbs. (4.54 kgs.).
CALIFORNIA'S PROPOSITION 65: Carcinogen.

EPA NAME: GLYCOL ETHERS
CLEAN AIR ACT: Hazardous Air Pollutants (Title I, Part A, Section 112) includes mono- and diethers of ethylene glycol, diethyl glycol, and triethylene glycol $R-(OCH_2CH_2)_n-OR'$ where n = 1,2, or 3; R = alkyl or aryl groups; R′ = R, H, or groups that when removed, yield glycol ethers with the structure: $R-(OCH_2CH)_n-OH$. Polymers are excluded from the glycol category.
EPCRA Section 313: Certain glycol ethers are covered. $R-(OCH_2CH_2)_n-OR'$; where n = 1,2 or 3; R = alkyl C7 or less; or R = phenyl or alkyl substituted phenyl; R′ + H, or alkyl C7 or less; or OR′ consisting of carboxylic ester, sulfate, phosphate, nitrate, or sulfonate. Form R de minimus concentration reporting level: 1.0%. Form R Toxic Chemical Category Code: N230.

EPA NAME: GUANIDINE, N-METHYL-N′-NITRO-N-NITROSO-
CAS: 70-25-7

SYNONYMS: 1-METHYL-3-NITRO-1-NITROSOGUANIDINE; N-METHYL-N'-NITRO-N-NITROSOGUANIDINE; N'-NITRO-N-NITROSO-N-METHYLGUANIDINE

EPA HAZARDOUS WASTE NUMBER (RCRA No.): U163.
RCRA Section 261 Hazardous Constituents.
RCRA Land Ban Waste.
EPCRA Section 304 Reportable Quantity (RQ): CERCLA, 10 lbs. (4.54 kgs.).

EPA NAME: GUTHION
[see AZINPHOS-METHYL]
CAS: 86-50-0

- H -

EPA NAME: HALOETHERS
CLEAN WATER ACT: Section 307 Toxic Pollutants.

EPA NAME: HALOMETHANES
CLEAN WATER ACT: Section 307 Toxic Pollutants.

EPA NAME: HALON 1211
[see BROMOCHLORODIFLUOROMETHANE]
CAS: 353-59-3

EPA NAME: HALON 1301
[see BROMOTRIFLUROMETHANE]
CAS: 75-63-8

EPA NAME: HALON 2402
[see DIBROMOTETRAFLUOROETHANE]
CAS: 124-73-2

EPA NAME: HCFC-121
[see 1,1,2,2-TETRACHLORO-1-FLUOROETHANE]
CAS: 354-14-3

EPA NAME: HCFC-123a
[see 1,2-DICHLORO-1,1,2-TRIFLUOROETHANE]
CAS: 354-23-4

EPA NAME: HCFC-123b
[see 1,1-DICHLORO-1,2,2-TRIFLUOROETHANE]
CAS: 812-04-4

EPA NAME: HCFC-124
[see 2-CHLORO-1,1,1,2-TETRAFLUOROETHANE]
CAS: 2837-89-0

EPA NAME: HCFC-124a
[see 1-CHLORO-1,1,2,2-TETRAFLUOROETHANE]
CAS: 354-25-6

EPA NAME: HCFC-132b
[see 1,2-DICHLORO-1,1-DIFLUOROETHANE]
CAS: 1649-08-7

EPA NAME: HCFC-133a
[see 2-CHLORO-1,1,1-TRIFLUOROETHANE]
CAS: 75-88-7

EPA NAME: HCFC-141b
[see 1,1-DICHLORO-1-FLUOROETHANE]
CAS: 1717-00-6

EPA NAME: HCFC-142b
[see 1-CHLORO-1,1-DIFLUOROETHANE]
CAS: 75-68-3

EPA NAME: HCFC-21
[see DICHLOROFLUOROMETHANE]
CAS: 75-43-4

EPA NAME: HCFC-22
[see CHLORODIFLUOROMETHANE]
CAS: 75-45-6

EPA NAME: HCFC-225aa
[see 2,2-DICHLORO-1,1,1,3,3-PENTAFLUOROPROPANE]
CAS: 128903-21-9

EPA NAME: HCFC-225ba
[see 2,3-DICHLORO-1,1,1,2,3-PENTAFLUOROPROPANE]
CAS: 422-48-0

EPA NAME: HCFC-225bb
[see 1,2-DICHLORO-1,1,2,3,3-PENTAFLUOROPROPANE]
CAS: 422-44-6

EPA NAME: HCFC 225ca
[see 3,3-DICHLORO-1,1,1,2,2-PENTAFLUOROPROPANE]
CAS: 422-56-0

EPA NAME: HCFC-225cb
[see 1,3-DICHLORO-1,1,2,2,3-PENTAFLUOROPROPANE]
CAS: 507-55-1

EPA NAME: HCFC-225cc
[see 1,1-DICHLORO-1,2,2,3,3-PENTAFLUOROPROPANE]
CAS: 13474-88-9

EPA NAME: HCFC-225da
[see 1,2-DICHLORO-1,1,3,3,3-PENTAFLUOROPROPANE]
CAS: 431-86-7

EPA NAME: HCFC-225ea
[see 1,3-DICHLORO-1,1,2,3,3-PENTAFLUOROPROPANE]
CAS: 136013-79-1

EPA NAME: HCFC-225eb
[see 1,1-DICHLORO-1,2,3,3,3-PENTAFLUOROPROPANE]
CAS: 111512-56-2

EPA NAME: HCFC 253fb
[see 3-CHLORO-1,1,1-TRIFLUOROPROPANE]
CAS: 460-35-5

EPA NAME: HEPTACHLOR
CAS: 76-44-8

SYNONYMS: 3-CHLOROCHLORDENE; 4,7-METHANOINDENE, 1,4,5,6,7,8,8-HEPTACHLORO-3A,4,7,7A-TETRAHYDRO-; 1,4,5,6,7,8,8-HEPTACHLORO-3A,4,7,7A-TETRAHYDRO-4,7-METHANO-1H-INDENE

CLEAN AIR ACT: Hazardous Air Pollutants (Title I, Part A, Section 112).
CLEAN WATER ACT: Section 311 Hazardous Substances/RQ (same as CERCLA); Section 307 Priority Pollutants; Section 313 Priority Chemicals; Section 307 Toxic Pollutants.
EPA HAZARDOUS WASTE NUMBER (RCRA No.): P059; D031.
RCRA Section 261 Hazardous Constituents.
RCRA Toxicity Characteristic (Section 261.24), Maximum Concentration of Contaminants, regulatory level, 0.008 mg/L.
RCRA Land Ban Waste.
RCRA Universal Treatment Standards: Wastewater (mg/L), 0.0012; Nonwastewater (mg/kg), 0.066.
RCRA Ground Water Monitoring List. Suggested test method(s) (PQL μg/L): 8080(0.05); 8270(10).
SAFE DRINKING WATER ACT: MCL, 0.0004 mg/L; MCLG, zero.
EPCRA Section 304 Reportable Quantity (RQ): CERCLA, 1 lb. (0.454 kg.).
EPCRA Section 313 Form R de minimus concentration reporting level: 1.0%.
MARINE POLLUTANT (49CFR, Subchapter 172.101, Appendix B): Severe pollutant.
CALIFORNIA'S PROPOSITION 65: Carcinogen.

EPA NAME: HEPTACHLOR AND METABOLITES

[see also HEPTACHLOR and HEXACHLOROCYCLOHEXANE]
CLEAN WATER ACT: Section 307 Toxic Pollutants.
MARINE POLLUTANT (49CFR, Subchapter 172.101, Appendix B): Severe pollutant, as heptachlor.
CALIFORNIA'S PROPOSITION 65: Carcinogen as heptachlor.

EPA NAME: HEPTACHLOR EPOXIDE

[see also HEXACHLOROCYCLOHEXANE]
CAS: 1024-57-3
SYNONYMS: BHC-HEXACHLOROCYCLOHEXANE

CLEAN WATER ACT: Section 307 Priority Pollutants; Section 307 Toxic Pollutants as hexachlorocyclohexane.
EPA HAZARDOUS WASTE NUMBER (RCRA No.): D031.
RCRA Section 261 Hazardous Constituents, waste number not listed.
RCRA Toxicity Characteristic (Section 261.24), Maximum Concentration of Contaminants, regulatory level, 0.008 mg/L.
RCRA Land Ban Waste.
RCRA Universal Treatment Standards: Wastewater (mg/L), 0.016; Nonwastewater (mg/kg), 0.066.
RCRA Ground Water Monitoring List. Suggested test method(s) (PQL μg/L): 8080(1); 8270(10).
SAFE DRINKING WATER ACT: MCL, 0.0002 mg/L; MCLG, zero.

EPCRA Section 304 Reportable Quantity (RQ): CERCLA, 1 lb. (0.454 kg.).
CALIFORNIA'S PROPOSITION 65: Carcinogen.

EPA NAME: 1,4,5,6,7,8,8-HEPTACHLORO-3A,4,7,7A-TETRAHYDRO-4,7-METHANO-1H-INDENE

[see HEPTACHLOR]
CAS: 76-44-8

EPA NAME: HEXACHLOROBENZENE

[see also CHLORINATED BENZENES]
CAS: 118-74-1
SYNONYMS: BENZENE, HEXACHLORO-; PERCHLOROBENZENE

CLEAN AIR ACT: Hazardous Air Pollutants (Title I, Part A, Section 112).
CLEAN WATER ACT: Section 313 Priority Chemicals.
EPA HAZARDOUS WASTE NUMBER (RCRA No.): U127; D032.
RCRA Section 261 Hazardous Constituents.
RCRA Toxicity Characteristic (Section 261.24), Maximum Concentration of Contaminants, regulatory level, 0.13 mg/L.
RCRA Land Ban Waste.
RCRA Universal Treatment Standards: Wastewater (mg/L), 0.055; Nonwastewater (mg/kg), 10.
RCRA Ground Water Monitoring List. Suggested test method(s) (PQL µg/L): 8120(0.05); 8270(10).
SAFE DRINKING WATER ACT: MCL, 0.001 mg/L; MCLG, zero.
EPCRA Section 304 Reportable Quantity (RQ): CERCLA, 10 lbs. (4.54 kgs.).
EPCRA Section 313 Form R de minimus concentration reporting level: 0.1%.
CALIFORNIA'S PROPOSITION 65: Carcinogen; reproductive toxin.

EPA NAME: HEXACHLOROBUTADIENE

[see HEXACHLORO-1,3-BUTADIENE]
CAS: 87-68-3

EPA NAME: HEXACHLORO-1,3-BUTADIENE

CAS: 87-68-3
SYNONYMS: 1,3-BUTADIENE, 1,1,2,3,4,4-HEXACHLORO-; BUTADIENE, HEXACHLORO-; HEXACHLOROBUTADIENE; PERCHLOROBUTADIENE

CLEAN AIR ACT: Hazardous Air Pollutants (Title I, Part A, Section 112).
CLEAN WATER ACT: Section 313 Priority Chemicals; Section 307 Toxic Pollutants.
EPA HAZARDOUS WASTE NUMBER (RCRA No.): U128; D033.
RCRA Section 261 Hazardous Constituents.
RCRA Toxicity Characteristic (Section 261.24), Maximum Concentration of Contaminants, regulatory level, 0.5 mg/L.
RCRA Land Ban Waste.
RCRA Universal Treatment Standards: Wastewater (mg/L), 0.055; Nonwastewater (mg/kg), 5.6.

RCRA Ground Water Monitoring List. Suggested test method(s) (PQL µg/L): 8120(5); 8270(10).
SAFE DRINKING WATER ACT: Priority List (55 FR 1470).
EPCRA Section 304 Reportable Quantity (RQ): CERCLA, 1 lb. (0.454 kg.).
EPCRA Section 313 Form R de minimus concentration reporting level: 1.0%.
MARINE POLLUTANT (49CFR, Subchapter 172.101, Appendix B).

EPA NAME: HEXACHLOROCYCLOHEXANE (ALL ISOMERS)
CAS: 608-73-1
SYNONYMS: BHC; BENZENE HEXACHLORIDE; CYCLOHEXANE, 1,2,3,4,5,6-HEXACHLORO-; 1,2,3,4,5,6-HEXACHLOROCYCLOHEXANE

CLEAN WATER ACT: Section 307 Toxic Pollutants.
MARINE POLLUTANT (49CFR, Subchapter 172.101, Appendix B): Severe pollutant.
CALIFORNIA'S PROPOSITION 65: Carcinogen.

EPA NAME: α-HEXACHLOROCYCLOHEXANE
[see also HEXACHLOROCYCLOHEXANE (ALL ISOMERS)]
CAS: 319-84-6
SYNONYMS: α-BHC; α-BHC-α; CYCLOHEXANE 1,2,3,4,5,6-HEXACHLORO-(1α,2α,3β,4α,5β,6β)-; α-HCH; α-HEXACHLOROCYCLOHEXANE; α-HEXACHLOROCYCLOHEXANE; HEXACHLORCYCLOHEXANE, ALPHA ISOMER; HEXACHLOROCYCLOHEXANE, α-

CLEAN WATER ACT: Section 307 Priority Pollutants; Section 307 Toxic Pollutants as hexachlorocyclohexane.
RCRA Land Ban Waste.
RCRA Universal Treatment Standards: Wastewater (mg/L), 0.00014; Nonwastewater (mg/kg), 0.066.
RCRA Ground Water Monitoring List. Suggested test method(s) (PQL µg/L): 8080(0.05); 8250 (10).
EPCRA Section 304 Reportable Quantity (RQ): CERCLA, 10 lbs. (4.54 kgs.).
EPCRA Section 313 Form R de minimus concentration reporting level: 1.0
CALIFORNIA'S PROPOSITION 65: Carcinogen, as hexachlorocyclohexane.

EPA NAME: β-HEXACHLOROCYCLOHEXANE
[see also HEXACHLOROCYCLOHEXANE (ALL ISOMERS)]
CAS: 319-85-7
SYNONYMS: β-BHC; β-BHC-β; β-BENZENEHEXACHLORIDE; CYCLOHEXANE 1,2,3,4,5,6-HEXACHLORO-(1α,2β,3α,4β,5α,6β)-; β-HCH; HEXACHLOROCYCLOHEXANE, BETA ISOMER; β-HEXACHLOROCYCLOHEXANE

CLEAN WATER ACT: Section 307 Priority Pollutants; Section 307 Toxic Pollutants as hexachlorocyclohexane.

RCRA Land Ban Waste.
RCRA Universal Treatment Standards: Wastewater (mg/L), 0.00014; Nonwastewater (mg/kg), 0.066.
RCRA Ground Water Monitoring List. Suggested test method(s) (PQL μg/L): 8080(0.05); 8250 (40).
EPCRA Section 304 Reportable Quantity (RQ): CERCLA, 1 lb. (0.454 kg.).
CALIFORNIA'S PROPOSITION 65: Carcinogen, as hexachlorocyclohexane.

EPA NAME: δ-HEXACHLOROCYCLOHEXANE
[see also HEXACHLOROCYCLOHEXANE (ALL ISOMERS)]
CAS: 319-86-8
SYNONYMS: δ-BENZENEHEXACHLORIDE; δ-BHC; δ-BHC; CYCLOHEXANE 1,2,3,4,5,6-HEXACHLORO-,(1α,2α,3α,4β,5β,6β)-; δ-BHC (PCB-POLYCHLORINATED BIPHENYLS); δ-BHC-δ; δ-HCH; δ-HEXACHLOROCYCLOHEXANE; δ-1,2,3,4,5,6-HEXACHLOROCYCLOHEXANE-δ; δ-LINDANE; HEXACHLOROCYCLOHEXANE, DELTA ISOMER

CLEAN WATER ACT: Section 307 Priority Pollutants; Section 307 Toxic Pollutants as hexachlorocyclohexane.
RCRA Land Ban Waste.
RCRA Universal Treatment Standards: Wastewater (mg/L), 0.023; Nonwastewater (mg/kg), 0.066.
RCRA Ground Water Monitoring List. Suggested test method(s) (PQL μg/L): 8080(0.1); 8250 (30).
EPCRA Section 304 Reportable Quantity (RQ): CERCLA, 1 lb. (0.454 kg.).
CALIFORNIA'S PROPOSITION 65: Carcinogen, as hexachlorocyclohexane.

EPA NAME: HEXACHLOROCYCLOHEXANE (γ isomer)
[see LINDANE]
CAS: 58-89-9

EPA NAME: HEXACHLOROCYCLOPENTADIENE
CAS: 77-47-4
SYNONYMS: 1,3-CYCLOPENTADIENE, 1,2,3,4,5,5-HEXACHLORO-; PERCHLOROCYCLOPENTADIENE

CLEAN AIR ACT: Hazardous Air Pollutants (Title I, Part A, Section 112).
CLEAN WATER ACT: Section 311 Hazardous Substances/RQ (same as CERCLA); Section 313 Priority Chemicals; Section 307 Toxic Pollutants.
EPA HAZARDOUS WASTE NUMBER (RCRA No.): U130.
RCRA Section 261 Hazardous Constituents.
RCRA Land Ban Waste.
RCRA Universal Treatment Standards: Wastewater (mg/L), 0.057; Nonwastewater (mg/kg), 2.4.
RCRA Ground Water Monitoring List. Suggested test method(s) (PQL μg/L): 8120(5); 8270(10).

SAFE DRINKING WATER ACT: MCL, 0.05 mg/L; MCLG, 0.05 mg/L.
EPCRA Section 302 Extremely Hazardous Substances: TPQ = 100 lbs. (45.4 kgs.).
EPCRA Section 304 Reportable Quantity (RQ): EHS/CERCLA, 10 lbs. (4.54 kgs.).
EPCRA Section 313 Form R de minimus concentration reporting level: 1.0%.

EPA NAME: HEXACHLOROETHANE
[see also CHLORINATED ETHANES]
CAS: 67-72-1
SYNONYMS: ETHANE, HEXACHLORO-; 1,1,1,2,2,2-HEXACHLOROETHANE

CLEAN AIR ACT: Hazardous Air Pollutants (Title I, Part A, Section 112).
CLEAN WATER ACT: Section 313 Priority Chemicals; Toxic Pollutant (Section 401.15) as chlorinated ethanes.
EPA HAZARDOUS WASTE NUMBER (RCRA No.): U131; D034.
RCRA Section 261 Hazardous Constituents.
RCRA Toxicity Characteristic (Section 261.24), Maximum Concentration of Contaminants, regulatory level, 3.0 mg/L.
RCRA Land Ban Waste.
RCRA Universal Treatment Standards: Wastewater (mg/L), 0.055; Nonwastewater (mg/kg), 30.
RCRA Ground Water Monitoring List. Suggested test method(s) (PQL µg/L): 8120(0.5); 8270(10).
SAFE DRINKING WATER ACT: Priority List (55 FR 1470).
EPCRA Section 302 Extremely Hazardous Substances: TPQ = 100 lbs. (45.4 kgs.).
EPCRA Section 304 Reportable Quantity (RQ): CERCLA, 100 lbs. (45.4 kgs.).
EPCRA Section 313 Form R de minimus concentration reporting level: 1.0%.
CALIFORNIA'S PROPOSITION 65: Carcinogen.

EPA NAME: HEXACHLORONAPHTHALENE
[see also CHLORINATED NAPHTHALENES]
CAS: 1335-87-1
SYNONYMS: NAPHTHALENE, HEXACHLORO-

EPCRA Section 313 Form R de minimus concentration reporting level: 1.0%.

EPA NAME: HEXACHLOROPHENE
CAS: 70-30-4
SYNONYMS: BIS(3,5,6-TRICHLORO-2-HYDROXYPHENYL)METHANE; HCP

EPCRA Section 304 Reportable Quantity (RQ): CERCLA, 100 lbs. (45.4 kgs.).
EPA HAZARDOUS WASTE NUMBER (RCRA No.): U132.
RCRA Ground Water Monitoring List. Suggested test method(s) (PQL µg/L): 8270(10).

EPCRA Section 304 Reportable Quantity (RQ): CERCLA, 100 lbs. (45.4 kgs.).
EPCRA Section 313 Form R de minimus concentration reporting level: 1.0%.

EPA NAME: HEXACHLOROPROPENE
CAS: 1888-71-7
SYNONYMS: HEXACHLOROPROPYLENE; 1-PROPENE, 1,1,2,3,3-HEXACHLORO-

EPA HAZARDOUS WASTE NUMBER (RCRA No.): U243.
RCRA Land Ban Waste.
RCRA Universal Treatment Standards: Wastewater (mg/L), 0.035; Non-wastewater (mg/kg), 30.
RCRA Ground Water Monitoring List. Suggested test method(s) (PQL µg/L): 8270(10).
EPCRA Section 304 Reportable Quantity (RQ): CERCLA, 1,000 lbs. (454 kgs.).

EPA NAME: HEXAETHYL TETRAPHOSPHATE
CAS: 757-58-4
SYNONYMS: ETHYLTETRAPHOSPHATE; TETRAPHOSPHORIC ACID, HEXAETHYL ESTER

EPA HAZARDOUS WASTE NUMBER (RCRA No.): P062.
RCRA Section 261 Hazardous Constituents.
EPCRA Section 304 Reportable Quantity (RQ): CERCLA, 100 lbs. (45.4 kgs.).
MARINE POLLUTANT (49CFR, Subchapter 172.101, Appendix B), solid or liquid.

EPA NAME: HEXAKIS(2-METHYL-2-PHENYLPROPYL) DISTANNOXANE
[see FENBUTATIN OXIDE]
CAS: 13356-08-6

EPA NAME: HEXAMETHYLENE-1,6-DIISOCYANATE
[see also DIISOCYANATES]
CAS: 822-06-0
SYNONYMS: 1,6-HEXAMETHYLENE DIISOCYANATE; HEXAMETHYLENE DIISOCYANATE; HEXAMETHYLENE DIISOCYANATE, 1,6-; HEXANE, 1,6-DIISOCYANATO-

CLEAN AIR ACT: Hazardous Air Pollutants (Title I, Part A, Section 112).
EPCRA Section 304 Reportable Quantity (RQ): CERCLA, 100 lbs. (45.4 kgs.).
EPCRA Section 313 Form R de minimus concentration reporting level: 1.0%. Form R Toxic Chemical Category Code: N120.

EPA NAME: HEXAMETHYLENEDIAMINE, N,N′-DIBUTYL-
CAS: 4835-11-4
SYNONYMS: N,N′-DIBUTYLHEXAMETHYLENEDIAMINE; 1,6-N,N′-DIBUTYLHEXANEDIAMINE

EPCRA Section 302 Extremely Hazardous Substances: TPQ = 500 lbs. (227 kgs.).

EPCRA Section 304 Reportable Quantity (RQ): EHS, 500 lbs. (227 kgs.).

EPA NAME: HEXAMETHYLPHOSPHORAMIDE

CAS: 680-31-9

SYNONYMS: HEMPA; HEXAMETHYLORTHOPHOSPHORIC TRIAMIDE; HEXAMETHYLPHOSPHORIC ACID TRIAMIDE; HMPT

CLEAN AIR ACT: Hazardous Air Pollutants (Title I, Part A, Section 112).

EPCRA Section 304 Reportable Quantity (RQ): CERCLA, 1 lb. (0.454 kg.).

EPCRA Section 313 Form R de minimus concentration reporting level: 0.1%.

CALIFORNIA'S PROPOSITION 65: Carcinogen; reproductive toxin.

EPA NAME: n-HEXANE

CAS: 110-54-3

SYNONYMS: HEXANE; HEXYL HYDRIDE

CLEAN AIR ACT: Hazardous Air Pollutants (Title I, Part A, Section 112).

EPCRA Section 304 Reportable Quantity (RQ): CERCLA, 5,000 lbs. (2,270 kgs.).

EPCRA Section 313 Form R de minimus concentration reporting level: 1.0%.

EPA NAME: HEXANE

[see n-HEXANE]

CAS: 110-54-3

EPA NAME: HEXAZINONE

CAS: 51235-04-2

SYNONYMS: 3-CYCLOHEXYL-6-(DIMETHYLAMINO)-1-METHYL-1,3,5-TRIAZINE-2,4(1H,3H)-DIONE; EPA PESTICIDE CHEMICAL CODE 107201

EPCRA Section 313 Form R de minimus concentration reporting level: 1.0%.

EPA NAME: HYDRAMETHYLNON

CAS: 67485-29-4

SYNONYMS: EPA PESTICIDE CHEMICAL CODE 118401; 2(1H)-PYRIMIDINONE, TETRAHYDRO-5,5-DIMETHYL-,(3-(4-(TRIFLUOROMETHYL)PHENYL)-1-(2-(4-(TRIFLUOROMETHYL)PHENYL)ETHENYL)-2-PROPENYLIDENE)HYDRAZONE; TETRAHYDRO-5,5-DIMETHYL-2(1H)-PYRIMIDINONE[3-[4-(TRIFLUOROMETHYL)PHENYL]-1-[2-[4-(TRIFLUOROMETHYL)PHENYL]ETHENYL]-2-PRO PENYLIDENE] HYDRAZONE

EPCRA Section 313 Form R de minimus concentration reporting level: 1.0%.

EPA NAME: HYDRAZINE
CAS: 302-01-2
SYNONYMS: DIAMINE

CLEAN AIR ACT: Hazardous Air Pollutants (Title I, Part A, Section 112); Accidental Release Prevention/Flammable substances (Section 112[r], Table 3), TQ = 15,000 lbs. (6810 kgs.).
EPA HAZARDOUS WASTE NUMBER (RCRA No.): U133.
RCRA Section 261 Hazardous Constituents.
EPCRA Section 302 Extremely Hazardous Substances: TPQ = 1,000 lbs. (454 kgs.).
EPCRA Section 304 Reportable Quantity (RQ): EHS/CERCLA, 1 lb. (0.454 kg.).
EPCRA Section 313 Form R de minimus concentration reporting level: 0.1%.
CALIFORNIA'S PROPOSITION 65: Carcinogen.

EPA NAME: HYDRAZINE, 1,2-DIETHYL-
CAS: 1615-80-1
SYNONYMS: 1,2-DIETHYLHYDRAZINE

EPA HAZARDOUS WASTE NUMBER (RCRA No.): U086.
RCRA Section 261 Hazardous Constituents.
EPCRA Section 304 Reportable Quantity (RQ): CERCLA, 10 lbs. (4.54 kgs.).
CALIFORNIA'S PROPOSITION 65: Carcinogen.

EPA NAME: HYDRAZINE, 1,1-DIMETHYL-
[see DIMETHYLHYDRAZINE]
CAS: 57-14-7

EPA NAME: HYDRAZINE, 1,2-DIMETHYL-
CAS: 540-73-8
SYNONYMS: 1,2-DIMETHYLHYDRAZINE; N,N'-DIMETHYLHYDRAZINE

EPA HAZARDOUS WASTE NUMBER (RCRA No.): U099.
RCRA Section 261 Hazardous Constituents.
EPCRA Section 304 Reportable Quantity (RQ): CERCLA, 1 lb. (0.454 kg.).

EPA NAME: HYDRAZINE, 1,2-DIPHENYL-
[see 1,2-DIPHENYLHYDRAZINE]
CAS: 122-66-7

EPA NAME: HYDRAZINE, METHYL-
[see METHYL HYDRAZINE]
CAS: 60-34-4

EPA NAME: HYDRAZINE SULFATE
CAS: 10034-93-2
SYNONYMS: HYDRAZINE HYDROGEN SULFATE; HYDRAZINE MONOSULFATE

EPCRA Section 313 Form R de minimus concentration reporting level: 0.1%.
CALIFORNIA'S PROPOSITION 65: Carcinogen.

EPA NAME: HYDRAZOBENZENE
[see 1,2-DIPHENYLHYDRAZINE]
CAS: 122-66-7

EPA NAME: HYDROCHLORIC ACID
[see also HYDROCHLORIC ACID conc. 30% or greater)
CAS: 7647-01-0
SYNONYMS: HYDROGEN CHLORIDE (ANHYDROUS); HYDROGEN CHLORIDE (GAS ONLY); MURIATIC ACID

CLEAN AIR ACT: Hazardous Air Pollutants (Title I, Part A, Section 112); (anhydrous or gas) Accidental Release Prevention/Flammable substances (Section 112[r], Table 3), TQ = 5,000 lbs. (2270 kgs.).
CLEAN WATER ACT: Section 311 Hazardous Substances/RQ (same as CERCLA); Section 313 Priority Chemicals.
EPCRA Section 304 Reportable Quantity (RQ): CERCLA, 5,000 lbs. (2270 kgs.).
EPCRA Section 313 Form R de minimus concentration reporting level: 1.0%. Note: Non-aerosol forms of hydrochloric acid have been deleted from EPCRA/SARA 313 reporting, 7/29/96 (FR vol. 61, No. 146, pp. 39356-39357).

EPA NAME: HYDROCHLORIC ACID (conc. 30% or greater)
[see also HYDROCHLORIC ACID]
CAS: 7647-01-0
CLEAN AIR ACT: Hazardous Air Pollutants (Title I, Part A, Section 112); (conc. 30% or greater) Accidental Release Prevention/Flammable substances (Section 112[r], Table 3), TQ = 15,000 lbs. (6810 kgs.).
CLEAN WATER ACT: Section 311 Hazardous Substances/RQ (same as CERCLA); Section 313 Priority Chemicals.
EPCRA Section 304 Reportable Quantity (RQ): CERCLA, 5,000 lbs. (2270 kgs.).
EPCRA Section 313 Form R de minimus concentration reporting level: 1.0%. Note: Non-aerosol forms of hydrochloric acid have been deleted from EPCRA/SARA 313 reporting, 7/29/96 (FR vol. 61, No. 146, pp. 39356-39357).

EPA NAME: HYDROCYANIC ACID
[see HYDROGEN CYANIDE]
CAS: 74-90-8

EPA NAME: HYDROFLUORIC ACID
[see HYDROGEN FLUORIDE]
CAS: 7664-39-3

EPA NAME: HYDROFLUORIC ACID (conc. 50% or greater, or anhydrous)
[see also HYDROFLUORIC ACID]

CAS: 7664-39-3

SYNONYMS: HYDROGEN FLUORIDE (ANHYDROUS); FLUORHYDRIC ACID; FLUORIC ACID

CLEAN AIR ACT: Hazardous Air Pollutants (Title I, Part A, Section 112); (conc. 50% or greater, or anhydrous) Accidental Release Prevention/Flammable substances (Section 112[r], Table 3), TQ = 1,000 lbs. (454 kgs.).

CLEAN WATER ACT: Section 311 Hazardous Substances/RQ (same as CERCLA); Section 313 Priority Chemicals.

EPA HAZARDOUS WASTE NUMBER (RCRA No.): U134.

EPCRA Section 302 Extremely Hazardous Substances: TPQ = 100 lbs. (45.4 kgs.).

EPCRA Section 304 Reportable Quantity (RQ): CERCLA, 100 lbs. (45.4 kgs.).

EPCRA Section 313 Form R de minimus concentration reporting level: 1.0%.

EPA NAME: HYDROGEN

CAS: 1333-74-0

SYNONYMS: HYDROGEN, COMPRESSED

CLEAN AIR ACT: Accidental Release Prevention/Flammable substances (Section 112[r], Table 3), TQ = 10,000 lbs. (4540 kgs.).

EPA NAME: HYDROGEN CHLORIDE (anhydrous)

[see HYDROCHLORIC ACID]

CAS: 7647-01-0

EPA NAME: HYDROGEN CHLORIDE (gas only)

[see HYDROCHLORIC ACID]

CAS: 7647-01-0

EPCRA Section 302 Extremely Hazardous Substances: TPQ = 500 lbs. (227 kgs.).

EPCRA Section 304 Reportable Quantity (RQ): EHS, 5,000 lbs. (2,270 kgs.).

EPA NAME: HYDROGEN CYANIDE

CAS: 74-90-8

SYNONYMS: CARBON HYDRIDE NITRIDE (CHN); CYANHYDRIC ACID; HYDROCYANIC ACID

CLEAN AIR ACT: Accidental Release Prevention/Flammable substances (Section 112[r], Table 3), TQ = 2,500 lbs. (1135 kgs.).

CLEAN WATER ACT: Section 311 Hazardous Substances/RQ (same as CERCLA); Section 313 Priority Chemicals.

EPA HAZARDOUS WASTE NUMBER (RCRA No.): P063.

EPCRA Section 302 Extremely Hazardous Substances: TPQ = 100 lbs. (45.4 kgs.).

EPCRA Section 304 Reportable Quantity (RQ): EHS/CERCLA, 10 lbs. (4.54 kgs.).

EPCRA Section 313 Form R de minimus concentration reporting level: 1.0%.

MARINE POLLUTANT (49CFR, Subchapter 172.101, Appendix B).

EPA NAME: HYDROGEN FLUORIDE
[see also HYDROFLUORIC ACID (conc. 50% or greater, or anhydrous)]
CAS: 7664-39-3
SYNONYMS: HYDROFLUORIC ACID; HYDROFLUORIC ACID (conc. 50% or greater); HYDROGEN FLUORIDE (anhydrous); FLUORHYDRIC ACID; FLUORIC ACID

CLEAN AIR ACT: Hazardous Air Pollutants (Title I, Part A, Section 112).
CLEAN WATER ACT: Section 311 Hazardous Substances/RQ (same as CERCLA); Section 313 Priority Chemicals.
EPA HAZARDOUS WASTE NUMBER (RCRA No.): U134.
EPCRA Section 302 Extremely Hazardous Substances: TPQ = 100 lbs. (45.4 kgs.).
EPCRA Section 304 Reportable Quantity (RQ): EHS/CERCLA, 100 lbs. (45.4 kgs.).
EPCRA Section 313 Form R de minimus concentration reporting level: 1.0%.

EPA NAME: HYDROGEN FLUORIDE (anhydrous)
[see HYDROGEN FLUORIDE]

EPA NAME: HYDROGEN PEROXIDE (conc. > 52%)
CAS: 7722-84-1
SYNONYMS: HYDROGEN DIOXIDE

EPCRA Section 302 Extremely Hazardous Substances: TPQ = 1,000 lbs. (454 kgs.).
EPCRA Section 304 Reportable Quantity (RQ): EHS, 1,000 lbs. (454 kgs.).

EPA NAME: HYDROGEN SELENIDE
CAS: 7783-07-5
SYNONYMS: SELENIUM HYDRIDE

CLEAN AIR ACT: Accidental Release Prevention/Flammable substances (Section 112[r], Table 3), TQ = 500 lbs. (227 kgs.).
EPCRA Section 302 Extremely Hazardous Substances: TPQ = 10 lbs. (4.54 kgs.).
EPCRA Section 304 Reportable Quantity (RQ): EHS, 10 lbs. (4.54 kgs.).

EPA NAME: HYDROGEN SULFIDE
CAS: 7783-06-4
SYNONYMS: HYDROSULFURIC ACID; SULFUR HYDRIDE

CLEAN AIR ACT: Accidental Release Prevention/Flammable substances (Section 112[r], Table 3), TQ = 10,000 lbs. (4540 kgs.).
CLEAN WATER ACT: Section 311 Hazardous Substances/RQ (same as CERCLA).
EPA HAZARDOUS WASTE NUMBER (RCRA No.): U135.
RCRA Section 261 Hazardous Constituents.

EPCRA Section 302 Extremely Hazardous Substances: TPQ = 500 lbs. (227 kgs.).

EPCRA Section 304 Reportable Quantity (RQ): EHS/CERCLA, 100 lbs. (45.4 kgs.).

EPCRA Section 313: Subject to an administrative stay until fuurther notice. See 59 FR 43048, August 22, 1994.

EPA NAME: HYDROPEROXIDE, 1-METHYL-1-PHENYL-ETHYL-

[see CUMENE HYDROPEROXIDE]

CAS: 80-15-9

EPA NAME: HYDROQUINONE

CAS: 123-31-9

SYNONYMS: 1,4-BENZENEDIOL; 1,4-DIHYDROXYBENZENE; p-DIHYDROXYBENZENE; p-HYDROQUINONE

CLEAN AIR ACT: Hazardous Air Pollutants (Title I, Part A, Section 112).

EPCRA Section 302 Extremely Hazardous Substances: TPQ = 500/10,000 lbs. (227/4,540 kgs.).

EPCRA Section 304 Reportable Quantity (RQ): EHS/CERCLA, 100 lbs. (45.4 kgs.).

EPCRA Section 313 Form R de minimus concentration reporting level: 1.0%.

EPA NAME: IMAZALIL
CAS: 35554-44-0
SYNONYMS: ALLYL-1-(2,4-DICHLOROPHENYL)-2-IMIDAZOL-1-YLETHYL ETHER; 1-[2-(2,4-DICHLOROPHENYL)-2-(2-PROPENYLOXY)ETHYL]-1H-IMIDAZOLE; EPA PESTICIDE CHEMICAL CODE 111901

EPCRA Section 313 Form R de minimus concentration reporting level: 1.0%.

EPA NAME: INDENO[1,2,3-cd]PYRENE
[see also POLYCYCLIC AROMATIC COMPOUNDS]
CAS: 193-39-5
SYNONYMS: o-PHENYLENEPYRENE

CLEAN WATER ACT: Section 307 Priority Pollutants; Section 307 Toxic Pollutants as polynuclear aromatic hydrocarbons.
EPA HAZARDOUS WASTE NUMBER (RCRA No.): U137.
RCRA Section 261 Hazardous Constituents.
RCRA Land Ban Waste.
RCRA Universal Treatment Standards: Wastewater (mg/L), 0.0055; Nonwastewater (mg/kg), 3.4.
RCRA Ground Water Monitoring List. Suggested test method(s) (PQL µg/L): 8100(200); 8270(10).
EPCRA Section 313 Form R de minimus concentration reporting level: 0.1%. Form R Toxic Chemical Category Code: N590.
CALIFORNIA'S PROPOSITION 65: Carcinogen.

EPA NAME: 3-IODO-2-PROPYNYL BUTYLCARBAMATE
CAS: 55406-53-6
SYNONYMS: CARBAMIC ACID, BUTYL-, 3-IODO-2-PROPYNYL ESTER; EPA PESTICIDE CHEMICAL CODE 107801

EPA HAZARDOUS WASTE NUMBER (RCRA No.): U375.
RCRA Land Ban Waste.
RCRA Universal Treatment Standards: Wastewater (mg/L), 0.056; Nonwastewater (mg/kg), 1.4.
EPCRA Section 304 Reportable Quantity (RQ): CERCLA, 1 lb. (0.454 kg.).
EPCRA Section 313 Form R de minimus concentration reporting level: 1.0%.

EPA NAME: IRON
CAS: 7439-89-6
SYNONYMS: IRON, ELEMENTAL

SAFE DRINKING WATER ACT: SMCL, 0.3 mg/L.

EPA NAME: IRON CARBONYL ($Fe(CO)_5$), (TB-5-11)-
[see IRON PENTACARBONYL]

CAS: 13463-40-6

EPA NAME: IRONPENTACARBONYL

CAS: 13463-40-6

SYNONYMS: IRON CARBONYL ($Fe(CO)_5$), (TB-5-11)-; PENTACARBONYLIRON

CLEAN AIR ACT: Accidental Release Prevention/Flammable substances (Section 112[r], Table 3), TQ = 2,500 lbs. (1,135 kgs.).

EPCRA Section 302 Extremely Hazardous Substances: TPQ = 100 lbs. (45.4 kgs.).

EPCRA Section 304 Reportable Quantity (RQ): EHS, 100 lbs. (45.4 kgs.).

EPCRA Section 313 Form R de minimus concentration reporting level: 1.0%.

EPA NAME: ISOBENZAN

CAS: 297-78-9

SYNONYMS: 1,3,4,5,6,8,8-OCTOCHLORO-1,3,3a,4,7,7a-HEXAHYDRO-4,7-METHANOISOBENZOFURAN

EPCRA Section 302 Extremely Hazardous Substances: TPQ = 100/10,000 lbs. (45.4/4,540 kgs.).

EPCRA Section 304 Reportable Quantity (RQ): EHS, 100 lbs. (45.4 kgs.).

EPA NAME: ISOBUTANE

CAS: 75-28-5

SYNONYMS: 1,1-DIMETHYLETHANE; 2-METHYLPROPANE; PROPANE, 2-METHYL

CLEAN AIR ACT: Accidental Release Prevention/Flammable substances (Section 112[r], Table 3), TQ = 10,000 lbs. (4,540 kgs.).

EPA NAME: ISOBUTYL ALCOHOL

CAS: 78-83-1

SYNONYMS: 1-HYDROXYMETHYLPROPANE; ISOBUTANOL; 2-METHYL-1-PROPANOL; 2-METHYLPROPAN-1-OL; 1-PROPANOL, 2-METHYL-

EPA HAZARDOUS WASTE NUMBER (RCRA No.): U140.

RCRA Section 261 Hazardous Constituents.

RCRA Land Ban Waste.

RCRA Universal Treatment Standards: Wastewater (mg/L), 5.6; Nonwastewater (mg/kg), 170.

RCRA Ground Water Monitoring List. Suggested test method(s) (PQL μg/L): 8015(50).

EPCRA Section 304 Reportable Quantity (RQ): CERCLA, 5,000 lbs. (2,270 kgs.).

EPA NAME: ISOBUTYRALDEHYDE

CAS: 78-84-2

SYNONYMS: ISOBUTANAL; PROPANAL, 2-METHYL-

EPCRA Section 313 Form R de minimus concentration reporting level: 1.0%.

EPA NAME: ISOBUTYRONITRILE

CAS: 78-82-0

SYNONYMS: 2-METHYLPROPIONITRILE; PROPANENITRILE, 2-METHYL-

CLEAN AIR ACT: Accidental Release Prevention/Flammable substances (Section 112[r], Table 3), TQ=20,000 lbs. (9,080 kgs.).

EPCRA Section 302 Extremely Hazardous Substances: TPQ=1,000 lbs. (454 kgs.).

EPCRA Section 304 Reportable Quantity (RQ): EHS, 1,000 lbs. (454 kgs.).

EPA NAME: ISOCYANIC ACID, 3,4-DICHLOROPHENYL ESTER

CAS: 102-36-3

SYNONYMS: 1,2-DICHLORO-4-PHENYL ISOCYANATE; DICHLOROPHENYL ISOCYANATE

EPCRA Section 302 Extremely Hazardous Substances: TPQ= 500/10,000 lbs. (227/4,540 kgs.).

EPCRA Section 304 Reportable Quantity (RQ): EHS, 500 lbs. (227 kgs.).

EPA NAME: ISODRIN

CAS: 465-73-6

SYNONYMS: 1,2,3,4,10,10-HEXACHLORO-1,4,4A,5,8,8A-HEXAHYDRO-1,4:5,8-ENDO-ENDO-DIMETHANONAPHTHALENE

EPA HAZARDOUS WASTE NUMBER (RCRA No.): P060.

RCRA Section 261 Hazardous Constituents.

RCRA Land Ban Waste.

RCRA Universal Treatment Standards: Wastewater (mg/L), 0.021; Nonwastewater (mg/kg), 0.066.

RCRA Ground Water Monitoring List. Suggested test method(s) (PQL µg/L): 8270(10).

EPCRA Section 302 Extremely Hazardous Substances: TPQ= 100/10,000 lbs. (45.4/4,540 kgs.).

EPCRA Section 304 Reportable Quantity (RQ): EHS, 1 lb. (0.454 kg.).

EPCRA Section 313 Form R de minimus concentration reporting level: 1.0%.

EPA NAME: ISOFENPHOS

CAS: 25311-71-1

SYNONYMS: BENZOIC ACID, 2-((ETHOXY((1-METHYLETHYL)AMINO)PHOSPHINOTHIOYL)OXY), 1-METHYLETHYL ESTER; EPA PESTICIDE CHEMICAL CODE 109401; 2-[[ETHOXYL[(1-METHYLETHYL)AMINO]PHOSPHINOTHIOYL]OXY] BENZOIC ACID 1-METHYLETHYL ESTER

EPCRA Section 313 Form R de minimus concentration reporting level: 1.0%.

MARINE POLLUTANT (49CFR, Subchapter 172.101, Appendix B).

EPA NAME: ISOFLUORPHATE
[see DIISOPROPYLFLUOROPHOSPHATE]
CAS: 55-91-4

EPA NAME: 1H-ISOINDOLE-1,3(2H)-DIONE,3a,4,7,7a-TETRAHYDRO-2-[(TRICHLOROMETHYL)THIO]-
[see CAPTAN]
CAS: 133-06-2

EPA NAME: ISOLAN
[see ISOPROPYLMETHYLPYRAZOYL DIMETHYLCARBAMATE]
CAS: 119-38-0

EPA NAME: ISOPENTANE
CAS: 78-78-4
SYNONYMS: BUTANE, 2-METHYL-; ETHYL DIMETHYL METHANE; ISOAMYL HYDRIDE; 2-METHYLBUTANE

CLEAN AIR ACT: Accidental Release Prevention/Flammable substances (Section 112[r], Table 3), TQ = 10,000 lbs. (4,540 kgs.).

EPA NAME: ISOPHORONE
CAS: 78-59-1
SYNONYMS: 2-CYCLOHEXEN-1-ONE,3,5,5-TRIMETHYL-; 3,5,5-TRIMETHYL-2-CYCLOHEXENE-1-ONE

CLEAN AIR ACT: Hazardous Air Pollutants (Title I, Part A, Section 112).
CLEAN WATER ACT: Section 307 Toxic Pollutants; Section 307 Priority Pollutants.
RCRA Ground Water Monitoring List. Suggested test method(s) (PQL µg/L): 8090 (60); 8270 (10).
SAFE DRINKING WATER ACT: Priority List (55 FR 1470).
EPCRA Section 304 Reportable Quantity (RQ): CERCLA, 5,000 lbs. (2,270 kgs.).

EPA NAME: ISOPHORONE DIISOCYANATE
[see also DIISOCYANTES]
CAS: 4098-71-9
SYNONYMS: CYCLOHEXANE, 5-ISOCYANATO-1-(ISOCYANATOMETHYL)-1,3,3-TRIMETHYL-; ISOCYANIC ACID, METHYLENE(3,5,5-TRIMETHYL-3,1-CYCLOHEXYLENE) ESTER

EPCRA Section 302 Extremely Hazardous Substances: TPQ = 100 lbs. (45.4 kgs.).
EPCRA Section 304 Reportable Quantity (RQ): EHS, 100 lbs. (45.4 kgs.).
EPCRA Section 313 Form R de minimus concentration reporting level: 1.0%. Form R Toxic Chemical Category Code: N120.

EPA NAME: ISOPRENE
CAS: 78-79-5
SYNONYMS: 1,3-BUTADIENE, 2-METHYL; β-METHYLBIVINYL; 2-METHYL-1,3-BUTADIENE; 3-METHYL-1,3-BUTADIENE

CLEAN AIR ACT: Hazardous Air Pollutants (Title I, Part A, Section 112); Accidental Release Prevention/Flammable substances (Section 112[r], Table 3), TQ = 10,000 lbs. (4,540 kgs.).
CLEAN WATER ACT: Section 311 Hazardous Substances/RQ (same as CERCLA); Section 307 Priority Pollutants; Toxic Pollutant (Section 401.15).
RCRA Ground Water Monitoring List. Suggested test method(s) (PQL μg/L): 8090 (60); 8270 (10).
EPCRA Section 304 Reportable Quantity (RQ): CERCLA, 100 lbs. (45.4 kgs.).

EPA NAME: ISOPROPANOLAMINE DODECYLBENZENE SULFONATE
CAS: 42504-46-1
SYNONYMS: BENZENE SULFONIC ACID, DODECYL-, compd. with 1-AMINO-2-PROPANOL (1:1)

CLEAN WATER ACT: Section 311 Hazardous Substances/RQ (same as CERCLA).
EPCRA Section 304 Reportable Quantity (RQ): CERCLA, 1,000 lbs. (454 kgs.).

EPA NAME: ISOPROPYL ALCOHOL (mfg.-strong acid process)
CAS: 67-63-0
SYNONYMS: DIMETHYLCARBINOL; ISOPROPANOL; 2-PROPANOL

EPCRA Section 313 (reportable if being manufactured by the strong process only; no supplier notification.) Form R de minimus concentration reporting level: 1.0%.

EPA NAME: ISOPROPYLAMINE
CAS: 75-31-0
SYNONYMS: 1-METHYLETHYLAMINE; 2-PROPANAMINE; 2-PROPYLAMINE;

CLEAN AIR ACT: Accidental Release Prevention/Flammable substances (Section 112[r], Table 3), TQ = 10,000 lbs. (4,540 kgs.).

EPA NAME: ISOPROPYL CHLORIDE
CAS: 75-29-6
SYNONYMS: 2-CHLOROPROPANE; PROPANE, 2-CHLORO

CLEAN AIR ACT: Accidental Release Prevention/Flammable substances (Section 112[r], Table 3), TQ = 10,000 lbs. (4,540 kgs.).
MARINE POLLUTANT (49CFR, Subchapter 172.101, Appendix B).

EPA NAME: ISOPROPYL CHLOROFORMATE
CAS: 108-23-6
SYNONYMS: CARBONOCHLORIDIC ACID, 1-METHYLETHYL ESTER; CHLOROFORMIC ACID ISOPROPYL ESTER; ISOPROPYL CHLOROCARBONATE

CLEAN AIR ACT: Accidental Release Prevention/Flammable substances (Section 112[r], Table 3), TQ = 15,000 lbs. (6,810 kgs.).

EPCRA Section 302 Extremely Hazardous Substances: TPQ = 1,000 lbs. (454 kgs.).

EPCRA Section 304 Reportable Quantity (RQ): EHS, 1,000 lbs. (454 kgs.).

EPA NAME: 4,4′-ISOPROPYLIDENEDIPHENOL

CAS: 80-05-7

SYNONYMS: p,p′-BISPHENOL A; BISPHENOL A; PHENOL, 4,4′-ISOPROPYLIDENEDI-; PHENOL, 4,4′-(1-METHYLETHYLIDENE)BIS-

EPCRA Section 313 Form R de minimus concentration reporting level: 1.0%.

EPA NAME: ISOPROPYLMETHYLPYRAZOYL DIMETHYLCARBAMATE

CAS: 119-38-0

SYNONYMS: ISOLAN; (1-ISOPROPYL-3-METHYL-1H-PYRAZOL-5-YL)-N,N-DIMETHYL CARBAMATE

EPA HAZARDOUS WASTE NUMBER (RCRA No.): P192.

RCRA Section 261 Hazardous Constituents.

RCRA Land Ban Waste.

RCRA Universal Treatment Standards: Wastewater (mg/L), 0.056; Nonwastewater (mg/kg), 1.4.

EPCRA Section 302 Extremely Hazardous Substances: TPQ = 500 lbs. (227 kgs.).

EPCRA Section 304 Reportable Quantity (RQ): EHS, 1 lb. (0.454 kg.).

EPA NAME: ISOSAFROLE

CAS: 120-58-1

SYNONYMS: BENZENE, 1,2-(METHYLENEDIOXY)-4-PROPENYL-; 1,2-(METHYLENEDIOXY)-4-PROPENYLBENZENE; 3,4-(METHYLENEDIOXY)-1-PROPENYLBENZENE; 5-(1-PROPENYL)-1,3-BENZODIOXOLE

EPA HAZARDOUS WASTE NUMBER (RCRA No.): U141.

RCRA Land Ban Waste.

RCRA Universal Treatment Standards: Wastewater (mg/L), 0.081; Nonwastewater (mg/kg), 2.6.

RCRA Ground Water Monitoring List. Suggested test method(s) (PQL μg/L): 8270(10).

EPCRA Section 304 Reportable Quantity (RQ): CERCLA, 100 lbs. (45.4 kgs.).

EPCRA Section 313 Form R de minimus concentration reporting level: 1.0%.

CALIFORNIA'S PROPOSITION 65: Carcinogen.

EPA NAME: ISOTHIOCYANATOMETHANE

[see METHYL ISOTHIOCYANATE]

CAS: 556-61-6

- K -

EPA NAME: KEPONE

CAS: 143-50-0

SYNONYMS: 1,3,4-METHENO-2H-CYCLOBUTA(cd)PENTALEN-2-ONE,1,1a,3,3a,4,5,5a,5b,6-DECACHLORO-OCTAHYDRO-

CLEAN WATER ACT: Section 311 Hazardous Substances/RQ (same as CERCLA).

EPA HAZARDOUS WASTE NUMBER (RCRA No.): U142.

RCRA Section 261 Hazardous Constituents.

RCRA Land Ban Waste.

RCRA Universal Treatment Standards: Wastewater (mg/L), 0.0011; Nonwastewater (mg/kg), 0.13.

RCRA Ground Water Monitoring List. Suggested test method(s) (PQL µg/L): 8270(10).

EPCRA Section 304 Reportable Quantity (RQ): CERCLA, 1 lb. (0.454 kg.).

CALIFORNIA'S PROPOSITION 65: Reproductive toxin.

- L -

EPA NAME: LACTOFEN

CAS: 77501-63-4

SYNONYMS: BENZOIC ACID, 5-(2-CHLORO-4-(TRIFLUORO-METHYL)PHENOXY)-2-NITRO-2-ETHOXY-1-METHYL-2-OX-OETHYL ESTER

SAFE DRINKING WATER ACT: Priority List (55 FR 1470).

EPCRA Section 313 Form R de minimus concentration reporting level: 1.0%.

CALIFORNIA'S PROPOSITION 65: Carcinogen.

EPA NAME: LACTONITRILE

CAS: 78-97-7

SYNONYMS: 2-HYDROXYPROPIONITRILE

EPCRA Section 302 Extremely Hazardous Substances: TPQ = 1,000 lbs. (454 kgs.).

EPCRA Section 304 Reportable Quantity (RQ): EHS, 1,000 lbs. (454 kgs.).

EPA NAME: LASIOCARPINE

CAS: 303-34-4

SYNONYMS: 2-BUTENOIC ACID, 2-METHYL-, 7-((2,3-DIHY-DROXY-2-(1-METHOXYETHYL)-3-METHYL-1-OXOBUTOX-Y)METHYL)-2,3,5,7A-TETRAHYDRO-1H-PYRROLIZIN-1-YL ESTER,(IS(1α(Z),7(2S*,3R*),7Aα)-

EPA HAZARDOUS WASTE NUMBER (RCRA No.): U143.

RCRA Section 261 Hazardous Constituents.

EPCRA Section 304 Reportable Quantity (RQ): CERCLA, 10 lbs. (4.54 kgs.).

CALIFORNIA'S PROPOSITION 65: Carcinogen.

EPA NAME: LEAD

[see also LEAD COMPOUNDS]

CAS: 7439-92-1

SYNONYMS: INORGANIC LEAD; LEAD ELEMENTAL

CLEAN WATER ACT: Section 307 Priority Pollutants; Section 313. Priority Chemicals; Section 307 Toxic Pollutants as lead and compounds.

EPA HAZARDOUS WASTE NUMBER (RCRA No.): D008.

RCRA Toxicity Characteristic (Section 261.24), Maximum Concentration of Contaminants, regulatory level, 5.0 mg/L.

RCRA Section 261 Hazardous Constituents, as lead compounds, n.o.s., waste number not listed.

RCRA Universal Treatment Standards: Wastewater (mg/L), 0.69; Non-wastewater (mg/L), 0.37 TCLP.

RCRA Ground Water Monitoring List. Suggested test method(s) (PQL μg/L): 6010(40); 7420(1,000); 7421(10).

SAFE DRINKING WATER ACT: MCL, zero; MCLG, zero; Regulated chemical (47 FR 9352).
EPCRA Section 304 Reportable Quantity (RQ): CERCLA, 10 lbs. (4.54 kgs.).
EPCRA Section 313 Form R de minimus concentration reporting level: 0.1%.
CALIFORNIA'S PROPOSITION 65: Carcinogen; Reproductive toxin (male, female).

EPA NAME: LEAD ACETATE
[see also LEAD and LEAD COMPOUNDS]
CAS: 301-04-2
SYNONYMS: ACETIC ACID, LEAD(II) SALT; LEAD DIACETATE; LEAD(II) ACETATE; LEAD(2+) ACETATE

CLEAN WATER ACT: Section 311 Hazardous Substances/RQ (same as CERCLA); Section 313 Priority Chemicals; Section 307 Toxic Pollutants as lead and compounds.
EPA HAZARDOUS WASTE NUMBER (RCRA No.): U144.
RCRA Section 261 Hazardous Constituents.
EPCRA Section 304 Reportable Quantity (RQ): CERCLA, 10 lbs. (4.54 kgs.).
EPCRA Section 313 Form R de minimus concentration reporting level: 1.0%. Form R Toxic Chemical Category Code: N420.
MARINE POLLUTANT (49CFR, Subchapter 172.101, Appendix B).
CALIFORNIA'S PROPOSITION 65: Carcinogen.

EPA NAME: LEAD ARSENATE
[see also LEAD, LEAD COMPOUNDS, ARSENIC COMPOUNDS]
CAS: 7645-25-2
SYNONYMS: ARSENIC ACID, LEAD SALT

CLEAN WATER ACT: Section 311 Hazardous Substances/RQ (same as CERCLA); Section 313 Priority Chemicals; Section 307 Toxic Pollutants as lead and compounds.
RCRA Section 261 Hazardous Constituents, waste number not listed, as lead compounds, n.o.s.
EPCRA Section 304 Reportable Quantity (RQ): CERCLA, 1 lb. (0.454 kg.).
EPCRA Section 313 Form R de minimus concentration reporting level: 0.1%. Form R Toxic Chemical Category Code: N420.
MARINE POLLUTANT (49CFR, Subchapter 172.101, Appendix B).
CALIFORNIA'S PROPOSITION 65: Carcinogen.

EPA NAME: LEAD ARSENATE
[see also LEAD, LEAD COMPOUNDS, ARSENIC COMPOUNDS]
CAS: 7784-40-9
SYNONYMS: ARSENIC ACID, LEAD(II) SALT (1:1)

CLEAN WATER ACT: Section 311 Hazardous Substances/RQ (same as CERCLA); Section 313 Priority Chemicals; Section 307 Toxic Pollutants as lead and compounds.

RCRA Section 261 Hazardous Constituents, waste number not listed, as lead compounds, n.o.s.

EPCRA Section 304 Reportable Quantity (RQ): CERCLA, 1 lb. (0.454 kg.).

EPCRA Section 313 Form R de minimus concentration reporting level: 0.1%. Form R Toxic Chemical Category Code: N420.

MARINE POLLUTANT (49CFR, Subchapter 172.101, Appendix B).

CALIFORNIA'S PROPOSITION 65: Carcinogen.

EPA NAME: LEAD ARSENATE

[see also LEAD, LEAD COMPOUNDS, ARSENIC COMPOUNDS]

CAS: 10102-48-4

SYNONYMS: ARSENIC ACID, LEAD(IV) SALT (3:2); LEAD(IV) ARSENATE; LEAD(4+) ARSENATE

CLEAN WATER ACT: Section 311 Hazardous Substances/RQ (same as CERCLA); Section 313 Priority Chemicals; Section 307 Toxic Pollutants as lead and compounds.

RCRA Section 261 Hazardous Constituents, waste number not listed, as lead compounds, n.o.s.

EPCRA Section 304 Reportable Quantity (RQ): CERCLA, 1 lb. (0.454 kg.).

EPCRA Section 313 Form R de minimus concentration reporting level: 0.1%. Form R Toxic Chemical Category Code: N420.

MARINE POLLUTANT (49CFR, Subchapter 172.101, Appendix B).

CALIFORNIA'S PROPOSITION 65: Carcinogen.

EPA NAME: LEAD CHLORIDE

[see also LEAD and LEAD COMPOUNDS]

CAS: 7758-95-4

SYNONYMS: LEAD(II) CHLORIDE; LEAD(2+) CHLORIDE; LEAD DICHLORIDE

CLEAN WATER ACT: Section 311 Hazardous Substances/RQ (same as CERCLA); Section 313 Priority Chemicals; Section 307 Toxic Pollutants as lead and compounds.

RCRA Section 261 Hazardous Constituents, waste number not listed, as lead compounds, n.o.s.

EPCRA Section 304 Reportable Quantity (RQ): CERCLA, 10 lbs. (4.54 kgs.).

EPCRA Section 313 Form R de minimus concentration reporting level: 0.1%. Form R Toxic Chemical Category Code: N420.

MARINE POLLUTANT (49CFR, Subchapter 172.101, Appendix B).

CALIFORNIA'S PROPOSITION 65: Carcinogen.

EPA NAME: LEAD COMPOUNDS

CLEAN WATER ACT: Section 307 Toxic Pollutants as lead and compounds.

RCRA Section 261 Hazardous Constituents, waste number not listed, as lead compounds, n.o.s.

EPCRA Section 313: Includes any unique chemical substance that contains lead as part of that chemical's infrastructure. Form R de minimus concentration reporting level: inorganic compounds 0.1%.; organic compounds 1.0%. Form R Toxic Chemical Category Code: N420.
MARINE POLLUTANT (49CFR, Subchapter 172.101, Appendix B) as lead compounds, soluble, n.o.s.
CALIFORNIA'S PROPOSITION 65: Carcinogen.

EPA NAME: LEAD FLUOBORATE
[see also LEAD and LEAD COMPOUNDS]
CAS: 13814-96-5
SYNONYMS: BORATE (1-), TETRAFLUORO-, LEAD (2+) (2:1); LEAD BORON FLUORIDE; TETRAFLUORO BORATE(1-), LEAD (2+)

CLEAN WATER ACT: Section 311 Hazardous Substances/RQ (same as CERCLA); Section 313 Priority Chemicals; Section 307 Toxic Pollutants as lead and compounds.
RCRA Section 261 Hazardous Constituents, waste number not listed, as lead compounds, n.o.s.
EPCRA Section 304 Reportable Quantity (RQ): CERCLA, 10 lbs. (4.54 kgs.).
EPCRA Section 313 Form R de minimus concentration reporting level: 0.1%. Form R Toxic Chemical Category Code: N420.
MARINE POLLUTANT (49CFR, Subchapter 172.101, Appendix B).
CALIFORNIA'S PROPOSITION 65: Carcinogen.

EPA NAME: LEAD FLUORIDE
[see also LEAD and LEAD COMPOUNDS]
CAS: 7783-46-2
SYNONYMS: LEAD DIFLUORIDE; LEAD(II) FLUORIDE; LEAD(2+) FLUORIDE

CLEAN WATER ACT: Section 311 Hazardous Substances/RQ (same as CERCLA); Section 313 Priority Chemicals; Section 307 Toxic Pollutants as lead and compounds.
RCRA Section 261 Hazardous Constituents, waste number not listed, as lead compounds, n.o.s.
EPCRA Section 304 Reportable Quantity (RQ): CERCLA, 10 lbs. (4.54 kgs.).
EPCRA Section 313 Form R de minimus concentration reporting level: 0.1%. Form R Toxic Chemical Category Code: N420.
MARINE POLLUTANT (49CFR, Subchapter 172.101, Appendix B).
CALIFORNIA'S PROPOSITION 65: Carcinogen.

EPA NAME: LEAD IODIDE
[see also LEAD and LEAD COMPOUNDS]
CAS: 10101-63-0
SYNONYMS: LEAD(II) IODIDE

CLEAN WATER ACT: Section 311 Hazardous Substances/RQ (same as CERCLA); Section 313 Priority Chemicals; Section 307 Toxic Pollutants as lead and compounds.

RCRA Section 261 Hazardous Constituents, waste number not listed, as lead compounds, n.o.s.
EPCRA Section 304 Reportable Quantity (RQ): CERCLA, 10 lbs. (4.54 kgs.).
EPCRA Section 313 Form R de minimus concentration reporting level: 0.1%. Form R Toxic Chemical Category Code: N420.
MARINE POLLUTANT (49CFR, Subchapter 172.101, Appendix B).
CALIFORNIA'S PROPOSITION 65: Carcinogen.

EPA NAME: LEAD NITRATE
[see also LEAD and LEAD COMPOUNDS]
CAS: 10099-74-8
SYNONYMS: LEAD(2+) NITRATE; LEAD(II) NITRATE; NITRIC ACID, LEAD(II) SALT

CLEAN WATER ACT: Section 311 Hazardous Substances/RQ (same as CERCLA); Section 313 Priority Chemicals; Section 307 Toxic Pollutants as lead and compounds.
RCRA Section 261 Hazardous Constituents, waste number not listed, as lead compounds, n.o.s.
EPCRA Section 304 Reportable Quantity (RQ): CERCLA, 10 lbs. (4.54 kgs.).
EPCRA Section 313 Form R de minimus concentration reporting level: 0.1% (lead); 1.0% (nitrate compounds). Water-dissociable nitrate; reportable only when in aqueous solution. Molecular weight 331.21. Form R Toxic Chemical Category Code: N420 (lead); N511 (nitrate).
MARINE POLLUTANT (49CFR, Subchapter 172.101, Appendix B).
CALIFORNIA'S PROPOSITION 65: Carcinogen.

EPA NAME: LEAD PHOSPHATE
[see also LEAD and LEAD COMPOUNDS]
CAS: 7446-27-7
SYNONYMS: LEAD(II) PHOSPHATE; PHOSPHORIC ACID, LEAD(II) SALT (2:3)

CLEAN WATER ACT: Section 307 Toxic Pollutants as lead and compounds.
EPA HAZARDOUS WASTE NUMBER (RCRA No.): U145.
RCRA Section 261 Hazardous Constituents.
EPCRA Section 304 Reportable Quantity (RQ): CERCLA, 10 lbs. (4.54 kgs.).
EPCRA Section 313 Form R de minimus concentration reporting level: 0.1%. Form R Toxic Chemical Category Code: N420.
MARINE POLLUTANT (49CFR, Subchapter 172.101, Appendix B).
CALIFORNIA'S PROPOSITION 65: Carcinogen.

EPA NAME: LEAD STEARATE
[see also LEAD and LEAD COMPOUNDS]
CAS: 1072-35-1
SYNONYMS: OCTADECANOIC ACID, LEAD(II) SALT; STEARIC ACID, LEAD(II) SALT

CLEAN WATER ACT: Section 311 Hazardous Substances/RQ (same as CERCLA); Section 313 Priority Chemicals; Section 307 Toxic Pollutants as lead and compounds.
RCRA Section 261 Hazardous Constituents, waste number not listed, as lead compounds, n.o.s.
EPCRA Section 304 Reportable Quantity (RQ): CERCLA, 10 lbs. (4.54 kgs.).
EPCRA Section 313 Form R de minimus concentration reporting level: 1.0%. Form R Toxic Chemical Category Code: N420.
MARINE POLLUTANT (49CFR, Subchapter 172.101, Appendix B).
CALIFORNIA'S PROPOSITION 65: Carcinogen.

EPA NAME: LEAD STEARATE
[see also LEAD and LEAD COMPOUNDS]
CAS: 7428-48-0
SYNONYMS: OCTADECANOIC ACID, LEAD SALT; STEARIC ACID, LEAD SALT

CLEAN WATER ACT: Section 311 Hazardous Substances/RQ (same as CERCLA); Section 313 Priority Chemicals; Section 307 Toxic Pollutants as lead and compounds.
RCRA Section 261 Hazardous Constituents, waste number not listed, as lead compounds, n.o.s.
EPCRA Section 304 Reportable Quantity (RQ): CERCLA, 10 lbs. (4.54 kgs.).
EPCRA Section 313 Form R de minimus concentration reporting level: 1.0%. Form R Toxic Chemical Category Code: N420.
MARINE POLLUTANT (49CFR, Subchapter 172.101, Appendix B).
CALIFORNIA'S PROPOSITION 65: Carcinogen.

EPA NAME: LEAD STEARATE
[see also LEAD and LEAD COMPOUNDS]
CAS: 52652-59-2
SYNONYMS: LEAD STERATE, DIBASIC; OCTADECANOIC ACID, LEAD SALT, DIBASIC; STEARIC ACID, LEAD SALT, DIBASIC

CLEAN WATER ACT: Section 311 Hazardous Substances/RQ (same as CERCLA); Section 313 Priority Chemicals; Section 307 Toxic Pollutants as lead and compounds.
RCRA Section 261 Hazardous Constituents, waste number not listed, as lead compounds, n.o.s.
EPCRA Section 304 Reportable Quantity (RQ): CERCLA, 10 lbs. (4.54 kgs.).
EPCRA Section 313 Form R de minimus concentration reporting level: 1.0%. Form R Toxic Chemical Category Code: N420.
MARINE POLLUTANT (49CFR, Subchapter 172.101, Appendix B).
CALIFORNIA'S PROPOSITION 65: Carcinogen.

EPA NAME: LEAD STEARATE
[see also LEAD and LEAD COMPOUNDS]
CAS: 56189-09-4
SYNONYMS: LEAD, BIS(OCTADECANOATO)DIOXODI-

CLEAN WATER ACT: Section 311 Hazardous Substances/RQ (same as CERCLA); Section 313 Priority Chemicals; Section 307 Toxic Pollutants as lead and compounds.
RCRA Section 261 Hazardous Constituents, waste number not listed, as lead compounds, n.o.s.
EPCRA Section 304 Reportable Quantity (RQ): CERCLA, 10 lbs. (4.54 kgs.).
EPCRA Section 313 Form R de minimus concentration reporting level: 1.0%. Form R Toxic Chemical Category Code: N420.
MARINE POLLUTANT (49CFR, Subchapter 172.101, Appendix B).
CALIFORNIA'S PROPOSITION 65: Carcinogen.

EPA NAME: LEAD SUBACETATE
[see also LEAD and LEAD COMPOUNDS]
CAS: 1335-32-6
SYNONYMS: BIS(ACETATO)TETRAHYDROXYTRILEAD; LEAD ACETATE, BASIC; LEAD, BIS(ACETATO)TETRA-HYDROXYTRI-

CLEAN WATER ACT: Section 307 Toxic Pollutants as lead and compounds.
EPA HAZARDOUS WASTE NUMBER (RCRA No.): U146.
RCRA Section 261 Hazardous Constituents.
EPCRA Section 304 Reportable Quantity (RQ): CERCLA, 10 lbs. (4.54 kgs.).
EPCRA Section 313 Form R de minimus concentration reporting level: 1.0%. Form R Toxic Chemical Category Code: N420.
MARINE POLLUTANT (49CFR, Subchapter 172.101, Appendix B).
CALIFORNIA'S PROPOSITION 65: Carcinogen.

EPA NAME: LEAD SULFATE
[see also LEAD and LEAD COMPOUNDS]
CAS: 7446-14-2
SYNONYMS: LEAD(II) SULFATE(1:1); SULFURIC ACID, LEAD(II) SALT(1:1)

CLEAN WATER ACT: Section 311 Hazardous Substances/RQ (same as CERCLA); Section 313 Priority Chemicals; Section 307 Toxic Pollutants as lead and compounds.
RCRA Section 261 Hazardous Constituents, waste number not listed, as lead compounds, n.o.s.
RCRA Land Ban Waste.
EPCRA Section 304 Reportable Quantity (RQ): CERCLA, 10 lbs. (4.54 kgs.).
EPCRA Section 313 Form R de minimus concentration reporting level: 0.1%. Form R Toxic Chemical Category Code: N420.
MARINE POLLUTANT (49CFR, Subchapter 172.101, Appendix B).
CALIFORNIA'S PROPOSITION 65: Carcinogen.

EPA NAME: LEAD SULFATE
[see also LEAD and LEAD COMPOUNDS]
CAS: 15739-80-7

SYNONYMS: LEAD(II) SULFATE(1:1); SULFURIC ACID, LEAD(II) SALT(1:1)

CLEAN WATER ACT: Section 307 Toxic Pollutants as lead and compounds.
RCRA Section 261 Hazardous Constituents, waste number not listed, as lead compounds, n.o.s.
RCRA Land Ban Waste.
RCRA Section 261 Hazardous Constituents.
EPCRA Section 304 Reportable Quantity (RQ): CERCLA, 10 lbs. (4.54 kgs.).
EPCRA Section 313 Form R de minimus concentration reporting level: 0.1%. Form R Toxic Chemical Category Code: N420.
MARINE POLLUTANT (49CFR, Subchapter 172.101, Appendix B).
CALIFORNIA'S PROPOSITION 65: Carcinogen.

EPA NAME: LEAD SULFIDE
[see also LEAD and LEAD COMPOUNDS]
CAS: 1314-87-0
SYNONYMS: LEAD MONOSULFIDE

CLEAN WATER ACT: Section 311 Hazardous Substances/RQ (same as CERCLA); Section 313 Priority Chemicals; Section 307 Toxic Pollutants as lead and compounds.
RCRA Section 261 Hazardous Constituents, waste number not listed, as lead compounds, n.o.s.
EPCRA Section 304 Reportable Quantity (RQ): CERCLA, 10 lbs. (4.54 kgs.).
EPCRA Section 313 Form R de minimus concentration reporting level: 0.1%. Form R Toxic Chemical Category Code: N420.
MARINE POLLUTANT (49CFR, Subchapter 172.101, Appendix B).
CALIFORNIA'S PROPOSITION 65: Carcinogen.

EPA NAME: LEAD THIOCYANATE
[see also LEAD, LEAD COMPOUNDS, and CYANIDE COMPOUNDS]
CAS: 592-87-0
SYNONYMS: LEAD SULFOCYANATE; LEAD(II) THIOCYANATE

CLEAN AIR ACT: Hazardous Air Pollutants (Title I, Part A, Section 112) as cyanide compounds.
CLEAN WATER ACT: Section 311 Hazardous Substances/RQ (same as CERCLA); Section 313 Priority Chemicals; Section 307 Toxic Pollutants as lead and compounds.
RCRA Section 261 Hazardous Constituents, waste number not listed, as lead compounds, n.o.s.
EPCRA Section 304 Reportable Quantity (RQ): CERCLA, 10 lbs. (4.54 kgs.).
EPCRA Section 313 Form R de minimus concentration reporting level: 0.1%. Form R Toxic Chemical Category Code: N420.
MARINE POLLUTANT (49CFR, Subchapter 172.101, Appendix B).
CALIFORNIA'S PROPOSITION 65: Carcinogen.

EPA NAME: LEPTOPHOS

CAS: 21609-90-5

SYNONYMS: O-(4-BROMO-2,5-DICHLOROPHENYL)O-METHYL PHENYLPHOSPHONOTHIOATE; O-(2,5-DICHLORO-4-BROMOPHENYL) O-METHYL PHENYLTHIOPHOSPHONATE; PHENYLPHOSPHONOTHIOIC ACID O-(4-BROMO-2,5-BROMO-2,5-DICHLOROPHENYL)O-METHYL ESTER

EPCRA Section 302 Extremely Hazardous Substances: TPQ = 500/10,000 lbs. (227/4,540 kgs.).

EPCRA Section 304 Reportable Quantity (RQ): EHS, 500 lbs. (227 kgs.).

EPCRA Section 313 Form R de minimus concentration reporting level: inorganic compounds 0.1%; organic compounds 1.0%. Form R Toxic Chemical Category Code: N420.

MARINE POLLUTANT (49CFR, Subchapter 172.101, Appendix B).

EPA NAME: LEWISITE

[see also ARENIC and ARSENIC COMPOUNDS]

CAS: 541-25-3

SYNONYMS: (2-CHLOROETHENYL)ARSONOUS DICHLORIDE; β-CHLOROVINYLBICHLOROARSINE

CLEAN AIR ACT: Hazardous Air Pollutants (Title I, Part A, Section 112); List of high risk pollutants (Section 63.74), as arsenic compounds.

CLEAN WATER ACT: Toxic Pollutant (Section 401.15) as arsenic and compounds.

RCRA Section 261 Hazardous Constituents, as arsenic compounds, n.o.s., waste number not listed.

EPCRA Section 302 Extremely Hazardous Substances: TPQ = 10 lbs. (4.54 kgs.).

EPCRA Section 304 Reportable Quantity (RQ): EHS, 10 lbs. (4.54 kgs.).

EPCRA Section 313 Form R de minimus concentration reporting level: 1.0%. as organic arsenic compound Form R, Toxic Chemical Category Code: N020.

MARINE POLLUTANT (49CFR, Subchapter 172.101, Appendix B) as arsenates, liquid, n.o.s.; arsenates, solid, n.o.s.; arsenical pesticides liquid, toxic, flammable, n.o.s.

EPA NAME: LINDANE

[see also HEXACHLOROCYCLOHEXANE]

CAS: 58-89-9

SYNONYMS: 2,5-CYCLOHEXANE,1,2,3,4,5,6-HEXACHLORO-, (1α,2 α,3β,4α,5α,6β)-; CYCLOHEXANE, 1,2,3,4,5,6-HEXACHLORO-, (1α,2α,3β,4α, 5α,6β); γ-BHC; γ-HCH; γ-HEXACHLOROCYCLOHEXANE; HEXACHLOROCYCLOHEXANE, γ; γ-LINDANE; HEXACHLOROCYCLOHEXANE (γ ISOMER)

CLEAN AIR ACT: Hazardous Air Pollutants (Title I, Part A, Section 112).

CLEAN WATER ACT: Section 311 Hazardous Substances/RQ (same as CERCLA); Section 307 Priority Pollutants; Section 313 Priority Chemicals; Section 307 Toxic Pollutants as hexachlorocyclohexane.

EPA HAZARDOUS WASTE NUMBER (RCRA No.): U129; D013.

RCRA Toxicity Characteristic (Section 261.24), Maximum Concentration of Contaminants, regulatory level, 0.4 mg/L.
RCRA Section 261 Hazardous Constituents.
RCRA Land Ban Waste.
SAFE DRINKING WATER ACT: MCL, 0.0002 mg/L; MCLG, 0.0002 mg/L; Regulated chemical (47 FR 9352).
RCRA Ground Water Monitoring List. Suggested test method(s) (PQL µg/L): 8080(0.05).
RCRA Universal Treatment Standards: Wastewater (mg/L), 0.0017; Nonwastewater (mg/kg), 0.066.
EPCRA Section 302 Extremely Hazardous Substances: TPQ = 1,000/10,000 lbs. (454/4,540 kgs.).
EPCRA Section 304 Reportable Quantity (RQ): EHS/CERCLA, 1 lb. (0.454 kg.).
EPCRA Section 313 Form R de minimus concentration reporting level: 0.1%.
MARINE POLLUTANT (49CFR, Subchapter 172.101, Appendix B).
CALIFORNIA'S PROPOSITION 65: Carcinogen.

EPA NAME: LINURON

CAS: 330-55-2
SYNONYMS: EPA PESTICIDE CHEMICAL CODE 035506; N'-(3,4-DICHLOROPHENYL)-N-METHOXY-N-METHYLUREA; 1-METHOXY-1-METHYL-3-(3,4-DICHLOROPHENYL)UREA

EPCRA Section 313 Form R de minimus concentration reporting level: 1.0%.

EPA NAME: LITHIUM CARBONATE

CAS: 554-13-2
SYNONYMS: CARBONIC ACID, DILITHIUM SALT

EPCRA Section 313 Form R de minimus concentration reporting level: 1.0%.
CALIFORNIA'S PROPOSITION 65: Reproductive toxin.

EPA NAME: LITHIUM CHROMATE

[see also CHROMIUM COMPOUNDS]
CAS: 14307-35-8
SYNONYMS: CHROMIC ACID, DILITHIUM SALT; CHROMIUM LITHIUM OXIDE; DILITHIUM CHROMATE

CLEAN AIR ACT: Hazardous Air Pollutants (Title I, Part A, Section 112) as chromium compounds.
CLEAN WATER ACT: Section 311 Hazardous Substances/RQ (same as CERCLA); Toxic Pollutant (Section 401.15); Section 313 Priority Chemicals.
RCRA Section 261 Hazardous Constituents, as chromium compounds, waste number not listed.
RCRA Land Ban Waste.
EPCRA Section 304 Reportable Quantity (RQ): CERCLA, 10 lbs. (4.54 kgs.).

EPCRA Section 313: See chromium compounds. Form R Toxic Chemical Category Code: N090.

EPA NAME: LITHIUM HYDRIDE
CAS: 7580-67-8
SYNONYMS: LITHIUM HYDRIDE (LiH)

EPCRA Section 302 Extremely Hazardous Substances: TPQ = 100 lbs. (45.4 kgs.).
EPCRA Section 304 Reportable Quantity (RQ): EHS, 100 lbs. (45.4 kgs.).

- M -

EPA NAME: MALATHION
CAS: 121-75-5
SYNONYMS: S-(1,2-BIS(CARBETHOXY)ETHYL) O,O-DIMETHYL-DITHIOPHOSPHATE; DIETHYL MERCAPTOSUCCINATE, S-ESTER WITH O,O-DIMETHYL PHOSPHORODITHIOATE; EPA PESTICIDE CHEMICAL CODE 057701

CLEAN WATER ACT: Section 311 Hazardous Substances/RQ (same as CERCLA).
EPCRA Section 304 Reportable Quantity (RQ): CERCLA, 100 lbs. (45.4 kgs.).
EPCRA Section 313 Form R de minimus concentration reporting level: 1.0%.
MARINE POLLUTANT (49CFR, Subchapter 172.101, Appendix B).

EPA NAME: MALEIC ACID
CAS: 110-16-7
SYNONYMS: BUTENEDIOIC ACID, (Z)-; cis-BUTENEDIOIC ACID, (Z); cis-1,2-ETHYLENEDICARBOXYLIC ACID, TOXILIC ACID

CLEAN WATER ACT: Section 311 Hazardous Substances/RQ (same as CERCLA).
EPCRA Section 304 Reportable Quantity (RQ): CERCLA, 5,000 lbs. (2270 kgs.).

EPA NAME: MALEIC ANHYDRIDE
CAS: 108-31-6
SYNONYMS: DIHYDRO-2,5-DIOXOFURAN; 2,5-FURANDIONE

CLEAN AIR ACT: Hazardous Air Pollutants (Title I, Part A, Section 112).
CLEAN WATER ACT: Section 311 Hazardous Substances/RQ (same as CERCLA); Section 313 Priority Chemicals.
EPA HAZARDOUS WASTE NUMBER (RCRA No.): U147.
RCRA Section 261 Hazardous Constituents.
EPCRA Section 304 Reportable Quantity (RQ): CERCLA, 5,000 lbs. (2270 kgs.).
EPCRA Section 313 Form R de minimus concentration reporting level: 1.0%.

EPA NAME: MALEIC HYDRAZIDE
CAS: 123-33-1
SYNONYMS: 1,2-DIHYDROPYRIDAZINE-3,6-DIONE; 1,2-DIHYDRO-3,6-PYRIDAZINEDIONE; 3,6-PYRIDAZINEDIONE, 1,2-DIHYDRO-

EPCRA Section 304 Reportable Quantity (RQ): CERCLA, 5,000 lbs. (2270 kgs.).

EPA NAME: MALONONITRILE
CAS: 109-77-3

SYNONYMS: PROPANEDINITRILE

EPA HAZARDOUS WASTE NUMBER (RCRA No.): U149.
RCRA Section 261 Hazardous Constituents.
EPCRA Section 302 Extremely Hazardous Substances: TPQ = 500/10,000 lbs. (227/4,540 kgs.).
EPCRA Section 304 Reportable Quantity (RQ): EHS/CERCLA, 1,000 lbs. (454 kgs.).
EPCRA Section 313 Form R de minimus concentration reporting level: 1.0%.

EPA NAME: MANEB
[see also MANGANESE COMPOUNDS and ETHYLENEBISDITHIOCARBAMIC ACID, SALTS, and ESTERS (EBDCs)]
CAS: 12427-38-2
SYNONYMS: CARBAMODITHIOIC ACID, 1,2-ETHANEDIYLBIS-, MANGANESE SALT; 1,2-ETHANEDIYLBIS(CARBAMODITHIOATO)(2-)-MANGANESE; 1,2-ETHANEDIYLBISCARBAMODITHIOIC ACID, MANGANESE COMPLEX

EPCRA Section 313 Form R de minimus concentration reporting level: 1.0%. Form R Toxic Chemical Category Code: N171 (EBDCs); N450 (manganese).
MARINE POLLUTANT (49CFR, Subchapter 172.101, Appendix B).
CALIFORNIA'S PROPOSITION 65: Carcinogen.

EPA NAME: MANGANESE
[see also MANGANESE COMPOUNDS]
CAS: 7439-96-5
SYNONYMS: MANGANESE ELEMENTAL

SAFE DRINKING WATER ACT: SMCL, 0.05 mg/L.
EPCRA Section 313 Form R de minimus concentration reporting level: 1.0%. Form R Toxic Chemical Category Code: N450.

EPA NAME: MANGANESE DIMETHYLDITHIOCARBAMATE
[see also MANGANESE and MANGANESE COMPOUNDS]
CAS: 15339-36-3
SYNONYMS: MANGANESE, BIS(DIMETHYLCARBAMODITHIOATO-S,S')-

CLEAN AIR ACT: Hazardous Air Pollutants (Title I, Part A, Section 112) as manganese compounds.
EPA HAZARDOUS WASTE NUMBER (RCRA No.): P196.
RCRA Section 261 Hazardous Constituents.
EPCRA Section 313 (as manganese compound) Form R de minimus concentration reporting level: 1.0%. Form R Toxic Chemical Category Code: N450.

EPA NAME: MANGANESE COMPOUNDS
CLEAN AIR ACT: Hazardous Air Pollutants (Title I, Part A, Section 112).

EPCRA Section 313: Includes any unique chemical substance that contains manganese as part of that chemical's infrastructure. Form R de minimus concentration reporting level: 1.0%. Form R Toxic Chemical Category Code: N450.

EPA NAME: MANGANESE TRICARBONYL METHYLCYCLOPENTADIENYL
[see also MANGANESE and MANGANESE COMPOUNDS]
CAS: 12108-13-3
SYNONYMS: MANGANESE, (METHYLCYCLOPENTADIENYL)TRICARBONYL-; METHYLCYCLOPENTADIENYL MANGANESE TRICARBONYL; 2-METHYLCYCLOPENTADIENYL MANGANESETRICARBONYL

CLEAN AIR ACT: Hazardous Air Pollutants (Title I, Part A, Section 112), as manganese compounds.
EPCRA Section 302 Extremely Hazardous Substances: TPQ = 100 lbs. (45.4 kgs.
EPCRA Section 304 Reportable Quantity (RQ): EHS, 100 lbs. (45.4 kgs.).
EPCRA Section 313 (as manganese compound) Form R de minimus concentration reporting level: 1.0%. Form R Toxic Chemical Category Code: N450.

EPA NAME: MBOCA
[see 4,4′-METHYLENEBIS(2-CHLOROANILINE)]
CAS: 101-14-4

EPA NAME: MBT
[see 2-MERCAPTOBENZOTHIAZOLE]
CAS: 149-30-4

EPA NAME: MCPA
[see METHOXONE]
CAS: 94-74-6

EPA NAME: MDI
[see METHYLBIS(PHENYLISOCYANATE)]
CAS: 101-68-8

EPA NAME: MECHLORETHAMINE
[see NITROGEN MUSTARD]
CAS: 51-75-2

EPA NAME: MECOPROP
CAS: 93-65-2
SYNONYMS: 2-(2-METHYL-4-CHLOROPHENOXY)PROPANOIC ACID; PROPIONIC ACID, 2-(4-CHLORO-2-METHYLPHENOXY)

EPCRA Section 313 Form R de minimus concentration reporting level: 0.1%.

EPA NAME: MELPHALAN

CAS: 148-82-3

SYNONYMS: 4-(BIS(2-CHLOROETHYL)AMINO)-L-PHENYLALANINE; L-PHENYLALANINE, 4(BIS(2-CHLOROETHYL) AMINO)-

EPA HAZARDOUS WASTE NUMBER (RCRA No.): U150.

RCRA Section 261 Hazardous Constituents.

EPCRA Section 304 Reportable Quantity (RQ): CERCLA, 1 lb. (0.454 kg.).

CALIFORNIA'S PROPOSITION 65: Carcinogen; reproductive toxin.

EPA NAME: MEPHOSFOLAN

CAS: 950-10-7

SYNONYMS: DIETHYL(4-METHYL-1,3-DITHIOLAN-2-YLIDENE)PHOSPHOROAMIDATE

EPCRA Section 302 Extremely Hazardous Substances: TPQ = 500 lbs. (227 kgs.).

EPCRA Section 304 Reportable Quantity (RQ): EHS, 500 lbs. (227 kgs.).

MARINE POLLUTANT (49CFR, Subchapter 172.101, Appendix B).

EPA NAME: 2-MERCAPTOBENZOTHIAZOLE

CAS: 149-30-4

SYNONYMS: 2-BENZOTHIAZOLETHIOL; MBT; CAPTOBENZOTHIAZOL; MERCAPTOBENZOTHIAZOLE

EPCRA Section 313 Form R de minimus concentration reporting level: 1.0%.

EPA NAME: MERCAPTODIMETHUR

[see METHIOCARB]

CAS: 2032-65-7

EPA NAME: MERCURIC ACETATE

[see also MERCURY and MERCURY COMPOUNDS]

CAS: 1600-27-7

SYNONYMS: MERCURY ACETATE; MERCURY(II) ACETATE

CLEAN WATER ACT: Section 307 Toxic Pollutants as mercury and compounds.

RCRA Section 261 Hazardous Constituents, waste number not listed, as mercury compounds, n.o.s.

EPCRA Section 302 Extremely Hazardous Substances: TPQ = 500/10,000 lbs. (227/4,540 kgs.).

EPCRA Section 304 Reportable Quantity (RQ): EHS, 500 lbs. (227 kgs.).

EPCRA Section 313 (as mercury compound) Form R de minimus concentration reporting level: 1.0%. Form R Toxic Chemical Category Code: N458.

MARINE POLLUTANT (49CFR, Subchapter 172.101, Appendix B): Severe pollutant.

CALIFORNIA'S PROPOSITION 65: Reproductive toxin.

EPA NAME: MERCURIC CHLORIDE
[see also MERCURY and MERCURY COMPOUNDS]
CAS: 7487-94-7
SYNONYMS: MERCURY(II) CHLORIDE

CLEAN WATER ACT: Section 307 Toxic Pollutants as mercury and compounds.
RCRA Section 261 Hazardous Constituents, waste number not listed, as mercury compounds, n.o.s.
EPCRA Section 302 Extremely Hazardous Substances: TPQ = 500/10,000 lbs. (227/4,540 kgs.).
EPCRA Section 304 Reportable Quantity (RQ): EHS, 500 lbs. (227 kgs.).
EPCRA Section 313 (as mercury compound) Form R de minimus concentration reporting level: 1.0%. Form R Toxic Chemical Category Code: N458.
MARINE POLLUTANT (49CFR, Subchapter 172.101, Appendix B): Severe pollutant.
CALIFORNIA'S PROPOSITION 65: Reproductive toxin.

EPA NAME: MERCURIC CYANIDE
[see also MERCURY, MERCURY COMPOUNDS, CYANIDE, and CYANIDE COMPOUNDS]
CAS: 592-04-1
SYNONYMS: MERCURY(II) CYANIDE

CLEAN WATER ACT: Section 311 Hazardous Substances/RQ (same as CERCLA); Section 307 Toxic Pollutants as mercury and compounds; Section 313 Priority Chemicals.
RCRA Section 261 Hazardous Constituents, waste number not listed, as mercury compounds, n.o.s.
EPCRA Section 304 Reportable Quantity (RQ): CERCLA, 1 lb. (0.454 kg.).
EPCRA Section 313 (as mercury compound) Form R de minimus concentration reporting level: 1.0%. Form R Toxic Chemical Category Code: N458.
MARINE POLLUTANT (49CFR, Subchapter 172.101, Appendix B): Severe pollutant.
CALIFORNIA'S PROPOSITION 65: Reproductive toxin.

EPA NAME: MERCURIC NITRATE
[see also MERCURY and MERCURY COMPOUNDS]
CAS: 10045-94-0
SYNONYMS: MERCURY(II) NITRATE (1:2); NITRIC ACID, MERCURY(II) SALT

CLEAN WATER ACT: Section 311 Hazardous Substances/RQ (same as CERCLA); Section 307 Toxic Pollutants as mercury and compounds; Section 313 Priority Chemicals.
RCRA Section 261 Hazardous Constituents, waste number not listed, as mercury compounds, n.o.s.
EPCRA Section 304 Reportable Quantity (RQ): CERCLA, 10 lbs. (4.54 kgs.).

EPCRA Section 313 (as mercury compound) Form R de minimus concentration reporting level: 1.0%. Form R Toxic Chemical Category Code: N458.

MARINE POLLUTANT (49CFR, Subchapter 172.101, Appendix B): Severe pollutant.

CALIFORNIA'S PROPOSITION 65: Reproductive toxin.

EPA NAME: MERCURIC OXIDE

[see also MERCURY and MERCURY COMPOUNDS]

CAS: 21908-53-2

SYNONYMS: MERCURY OXIDE

CLEAN WATER ACT: Section 307 Toxic Pollutants as mercury and compounds.

RCRA Section 261 Hazardous Constituents, waste number not listed, as mercury compounds, n.o.s.

EPCRA Section 302 Extremely Hazardous Substances: TPQ = 500/10,000 lbs. (227/4,540 kgs.).

EPCRA Section 304 Reportable Quantity (RQ): EHS, 500 lbs. (227 kgs.).

EPCRA Section 313 (as mercury compound) Form R de minimus concentration reporting level: 1.0%. Form R Toxic Chemical Category Code: N458.

MARINE POLLUTANT (49CFR, Subchapter 172.101, Appendix B): Severe pollutant.

CALIFORNIA'S PROPOSITION 65: Reproductive toxin.

EPA NAME: MERCURIC SULFATE

[see also MERCURY and MERCURY COMPOUNDS]

CAS: 7783-35-9

SYNONYMS: MERCURY(II) SULFATE (1:1); SULFURIC ACID, MERCURY(II) SALT (1:1)

CLEAN WATER ACT: Section 311 Hazardous Substances/RQ (same as CERCLA); Section 307 Toxic Pollutants as mercury and compounds; Section 313 Priority Chemicals.

RCRA Section 261 Hazardous Constituents, waste number not listed, as mercury compounds, n.o.s.

EPCRA Section 304 Reportable Quantity (RQ): CERCLA, 10 lbs. (4.54 kgs.).

EPCRA Section 313 (as mercury compound) Form R de minimus concentration reporting level: 1.0%. Form R Toxic Chemical Category Code: N458.

MARINE POLLUTANT (49CFR, Subchapter 172.101, Appendix B): Severe pollutant as mercuric sulphate.

CALIFORNIA'S PROPOSITION 65: Reproductive toxin.

EPA NAME: MERCURIC THIOCYANATE

[see also MERCURY and MERCURY COMPOUNDS]

CAS: 592-85-8

SYNONYMS: BIS(THYOCYANATO)- MERCURY; MERCURIC SULFOCYANATE

CLEAN WATER ACT: Section 311 Hazardous Substances/RQ (same as CERCLA); Section 307 Toxic Pollutants as mercury and compounds; Section 313 Priority Chemicals.
RCRA Section 261 Hazardous Constituents, waste number not listed, as mercury compounds, n.o.s.
EPCRA Section 304 Reportable Quantity (RQ): CERCLA, 10 lbs. (4.54 kgs.).
EPCRA Section 313 (as mercury compound) Form R de minimus concentration reporting level: 1.0%. Form R Toxic Chemical Category Code: N458.
MARINE POLLUTANT (49CFR, Subchapter 172.101, Appendix B): Severe pollutant.
CALIFORNIA'S PROPOSITION 65: Reproductive toxin.

EPA NAME: MERCUROUS NITRATE
[see also MERCURY and MERCURY COMPOUNDS]
CAS: 7782-86-7
SYNONYMS: MERCUROUS NITRATE MONOHYDRATE

CLEAN WATER ACT: Section 311 Hazardous Substances/RQ (same as CERCLA); Section 307 Toxic Pollutants as mercury and compounds; Section 313 Priority Chemicals.
RCRA Section 261 Hazardous Constituents, waste number not listed, as mercury compounds, n.o.s.
EPCRA Section 304 Reportable Quantity (RQ): CERCLA, 10 lbs. (4.54 kgs.).
EPCRA Section 313 (as mercury compound) Form R de minimus concentration reporting level: 1.0%. Form R Toxic Chemical Category Code: N458.
MARINE POLLUTANT (49CFR, Subchapter 172.101, Appendix B): Severe pollutant.
CALIFORNIA'S PROPOSITION 65: Reproductive toxin.

EPA NAME: MERCUROUS NITRATE
[see also MERCURY and MERCURY COMPOUNDS]
CAS: 10415-75-5
SYNONYMS: MERCURY(I) NITRATE (1:1); NITRIC ACID, MERCURY(I) SALT

CLEAN WATER ACT: Section 311 Hazardous Substances/RQ (same as CERCLA); Section 307 Toxic Pollutants as mercury and compounds.
RCRA Section 261 Hazardous Constituents, waste number not listed, as mercury compounds, n.o.s.
EPCRA Section 304 Reportable Quantity (RQ): CERCLA, 10 lbs. (4.54 kgs.).
EPCRA Section 313 (as mercury compound) Form R de minimus concentration reporting level: 1.0%. Form R Toxic Chemical Category Code: N458.
MARINE POLLUTANT (49CFR, Subchapter 172.101, Appendix B): Severe pollutant.
CALIFORNIA'S PROPOSITION 65: Reproductive toxin.

EPA NAME: MERCURY
[see also MERCURY COMPOUNDS]
CAS: 7439-97-6
SYNONYMS: METALLIC MERCURY

CLEAN WATER ACT: Section 307 Toxic Pollutants as mercury and compounds; Section 313 Priority Chemicals.
EPA HAZARDOUS WASTE NUMBER (RCRA No.): U151.
RCRA Section 261 Hazardous Constituents.
RCRA Toxicity Characteristic (Section 261.24), Maximum Concentration of Contaminants, regulatory level, 0.2 mg/L.
RCRA Land Ban Waste.
RCRA Universal Treatment Standards: Wastewater (mg/L), 0.15; Nonwastewater (mg/L), 0.25 TCLP; Wastewater from retort, N/A; Nonwastewater from retort (mg/L), 0.20 TCLP.
RCRA Ground Water Monitoring List. Suggested test method(s) (PQL μg/L): 7420(2) as mercury (total).
SAFE DRINKING WATER ACT: MCL, 0.002 mg/L; MCLG, 0.002 mg/L.
EPCRA Section 304 Reportable Quantity (RQ): CERCLA, 1 lb. (0.454 kg.).
EPCRA Section 313 Form R de minimus concentration reporting level: 1.0%. Form R Toxic Chemical Category Code: N458.
CALIFORNIA'S PROPOSITION 65: Reproductive toxin.

EPA NAME: MERCURY COMPOUNDS
CLEAN AIR ACT: Hazardous Air Pollutants (Title I, Part A, Section 112).
CLEAN WATER ACT: Section 307 Toxic Pollutants as mercury and compounds.
RCRA Section 261 Hazardous Constituents, waste number not listed, as mercury compounds, n.o.s.
EPCRA Section 313: Includes any unique chemical substance that contains mercury as part of that chemical's infrastructure. Form R de minimus concentration reporting level: 1.0%. Form R Toxic Chemical Category Code: N458.
MARINE POLLUTANT (49CFR, Subchapter 172.101, Appendix B): Severe pollutant as mercury based pesticides, liquid, flammable, toxic, n.o.s.; mercury based pesticides, liquid, toxic, n.o.s.; mercury based pesticides, solid, toxic, n.o.s.; mercury compounds, liquid, n.o.s. mercury compounds, solid, n.o.s.; mercury(I) (mercurous) compounds (pesticides); mercury(II) (mercuric) compounds (pesticides).
CALIFORNIA'S PROPOSITION 65: Reproductive toxin.

EPA NAME: MERCURY FULMINATE
[see also MERCURY and MERCURY COMPOUNDS]
CAS: 628-86-4
SYNONYMS: FULMINIC ACID, MERCURY(2+) SALT; FULMINATE OF MERCURY

CLEAN WATER ACT: Section 307 Toxic Pollutants as mercury and compounds.
EPA HAZARDOUS WASTE NUMBER (RCRA No.): P065.
RCRA Section 261 Hazardous Constituents.

EPCRA Section 304 Reportable Quantity (RQ): CERCLA, 10 lbs. (4.54 kgs.).
EPCRA Section 313 (as mercury compound) Form R de minimus concentration reporting level: 1.0%. Form R Toxic Chemical Category Code: N458.
MARINE POLLUTANT (49CFR, Subchapter 172.101, Appendix B): Severe pollutant.
CALIFORNIA'S PROPOSITION 65: Reproductive toxin.

EPA NAME: MERPHOS
CAS: 150-50-5
SYNONYMS: EPA PESTICIDE CHEMICAL CODE 074901; PHOSPHOROTRITHIOUS ACID, TRIBUTYL ESTER; S,S',S-TRIBUTYL PHOSPHOROTRITHIOITE; S,S,S-TRIBUTYL PHOSPHOROTRITHIOITE

EPCRA Section 313 Form R de minimus concentration reporting level: 1.0%.

EPA NAME: METHACROLEIN DIACETATE
CAS: 10476-95-6
SYNONYMS: ACETIC ACID-2-METHYL-2-PROPENE-1,1-DIOL DIESTER; 2-METHYL-2-PROPENE-1,1'-DIOL DIACETATE

EPCRA Section 302 Extremely Hazardous Substances: TPQ = 1,000 lbs. (454 kgs.).
EPCRA Section 304 Reportable Quantity (RQ): EHS, 1,000 lbs. (454 kgs.).

EPA NAME: METHACRYLIC ANHYDRIDE
CAS: 760-93-0
SYNONYMS: METHACRYLIC ACID ANHYDRIDE; METHACRYLOYL ANHYDRIDE; 2-METHYL-2-PROPENOIC ACID ANHYDRIDE (9CI)

EPCRA Section 302 Extremely Hazardous Substances: TPQ = 500 lbs. (227 kgs.).
EPCRA Section 304 Reportable Quantity (RQ): EHS, 500 lbs. (227 kgs.).

EPA NAME: METHACRYLONITRILE
CAS: 126-98-7
SYNONYMS: α-METHACRYLONITRILE; α-METHYLACRYLONITRILE; 2-METHYLACRYLONITRILE; 2-METHYL-2-PROPENENITRILE; 2-PROPENENITRILE, 2-METHYL-

CLEAN AIR ACT: Accidental Release Prevention/Flammable substances (Section 112[r], Table 3), TQ = 10,000 lbs. (4540 kgs.).
EPA HAZARDOUS WASTE NUMBER (RCRA No.): U152.
RCRA Section 261 Hazardous Constituents.
RCRA Land Ban Waste.
RCRA Universal Treatment Standards: Wastewater (mg/L), 0.24; Nonwastewater (mg/kg), 84.

RCRA Ground Water Monitoring List. Suggested test method(s) (PQL µg/L): 8015(5); 8240(5).
EPCRA Section 302 Extremely Hazardous Substances: TPQ = 500 lbs. (227 kgs.).
EPCRA Section 304 Reportable Quantity (RQ): EHS/CERCLA, 1,000 lbs. (454 kgs.).
EPCRA Section 313 Form R de minimus concentration reporting level: 1.0%.

EPA NAME: METHACRYLOYL CHLORIDE
CAS: 920-46-7
SYNONYMS: METHACRYLIC CHLORIDE; 2-METHYL-2-PROPENOYL CHLORIDE

EPCRA Section 302 Extremely Hazardous Substances: TPQ = 100 lbs. (45.4 kgs.).
EPCRA Section 304 Reportable Quantity (RQ): EHS, 100 lbs. (45.4 kgs.).

EPA NAME: METHACRYLOYLOXYETHYL ISOCYANATE
CAS: 30674-80-7
SYNONYMS: 2-ISOCYANOTOETHYLMETHACRYLATE; β-ISOCYANOTOETHYLMETHACRYLATE

EPCRA Section 302 Extremely Hazardous Substances: TPQ = 100 lbs. (45.4 kgs.).
EPCRA Section 304 Reportable Quantity (RQ): EHS, 1 lb. (0.454 kg.).

EPA NAME: METHAMIDOPHOS
CAS: 10265-92-6
SYNONYMS: O,S-DIMETHYLPHOSPHORAMIDOTHIOATE; PHOSPHORAMIDOTHIOIC ACID, O,S-DIMETHYL ESTER

EPCRA Section 302 Extremely Hazardous Substances: TPQ = 100/10,000 lbs. (45.4/4,540 kgs.).
EPCRA Section 304 Reportable Quantity (RQ): EHS, 100 lbs. (45.4 kgs.).

EPA NAME: METHAM SODIUM
CAS: 137-42-8
SYNONYMS: CARBAMIC ACID, METHYLDITHIO-, MONOSODIUM SALT; N-METHYLAMINOMETHANETHIONOTHIOLIC ACID SODIUM SALT; SODIUM METHYLDITHIOCARBAMATE

EPA HAZARDOUS WASTE NUMBER (RCRA No.): U384.
RCRA Section 261 Hazardous Constituents.
EPCRA Section 304 Reportable Quantity (RQ): CERCLA, 1 lb. (0.454 kg.).
EPCRA Section 313 Form R de minimus concentration reporting level: 1.0%.
MARINE POLLUTANT (49CFR, Subchapter 172.101, Appendix B).

EPA NAME: METHANAMINE
CAS: 74-89-5
SYNONYMS: AMINOMETHANE; MONOMETHYLAMINE

CLEAN AIR ACT: Accidental Release Prevention/Flammable substances (Section 112[r], Table 3), TQ=10,000 lbs. (4540 kgs.).
CLEAN WATER ACT: Section 311 Hazardous Substances/RQ (same as CERCLA).
EPCRA Section 304 Reportable Quantity (RQ): CERCLA, 100 lbs. (45.4 kgs.).

EPA NAME: METHAMINE, N,N-DIMETHYL-
[see TRIMETHYLAMINE]
CAS: 75-50-3

EPA NAME: METHANAMINE, N-METHYL
[see DIMETHYLAMINE]
CAS: 124-40-3

EPA NAME: METHANAMINE, N-METHYL-N-NITROSO-
[see N-NITROSODIMETHYLAMINE]
CAS: 62-75-9

EPA NAME: METHANE
CAS: 74-82-8
SYNONYMS: METHYL HYDRIDE; NATURAL GAS

CLEAN AIR ACT: Accidental Release Prevention/Flammable substances (Section 112[r], Table 3), TQ=10,000 lbs. (4540 kgs.).

EPA NAME: METHANE, CHLORO-
[see CHLOROMETHANE]
CAS: 74-87-3

EPA NAME: METHANE, CHLOROMETHOXY-
[see CHLOROMETHYL METHYL ETHER]
CAS: 107-30-2

EPA NAME: METHANE, ISOCYANATO-
[see METHYL ISOCYANATE]
CAS: 624-83-9

EPA NAME: METHANE OXYBIS-
[see METHYL ETHER]
CAS: 115-10-6

EPA NAME: METHANE, OXYBIS[CHLORO]
[see BIS(CHLOROMETHYL)ETHER]
CAS: 542-88-1

EPA NAME: METHANESULFENYL CHLORIDE, TRICHLORO-
[see PERCHLOROMETHYL MERCAPTAN]
CAS: 594-42-3

EPA NAME: METHANESULFONYL FLUORIDE
CAS: 558-25-8
SYNONYMS: METHANESULPHONYL FLUORIDE

EPCRA Section 302 Extremely Hazardous Substances: TPQ = 1,000 lbs. (454 kgs.).
EPCRA Section 304 Reportable Quantity (RQ): EHS, 1,000 lbs. (454 kgs.).

EPA NAME: METHANE, TETRANITRO-
[see TETRANITROMETHANE]
CAS: 509-14-8

EPA NAME: METHANETHIOL
[see METHYL MERCAPTAN]

EPA NAME: METHANE, TRICHLORO-
[see CHLOROFORM]
CAS: 67-66-3

EPA NAME: 4,7-METHANOINDAN, 1,2,3,4,5,6,7,8,8-OCTA-CHLORO-2,3,3a,4,7,7a-HEXAHYDRO-
[see CHLORDANE]
CAS: 57-74-9

EPA NAME: METHANOL
CAS: 67-56-1
SYNONYMS: METHYL ALCOHOL

CLEAN AIR ACT: Hazardous Air Pollutants (Title I, Part A, Section 112); Accidental Release Prevention/Flammable substances (Section 112[r], Table 3), TQ = 5,000 lbs. (2270 kgs.).
EPA HAZARDOUS WASTE NUMBER (RCRA No.): U154.
RCRA Section 261 Hazardous Constituents.
RCRA Land Ban Waste.
RCRA Universal Treatment Standards: Wastewater (mg/L), 5.6; Non-wastewater (mg/L/TCLP), 0.75.
EPCRA Section 304 Reportable Quantity (RQ): CERCLA, 5,000 lbs. (2270 kgs.).
EPCRA Section 313 Form R de minimus concentration reporting level: 1.0%.

EPA NAME: METHAPYRILENE
CAS: 91-80-5
SYNONYMS: N,N-DIMETHYL-N′-2-PYRIDINYL-N′-(2-THIENYL-METHYL)-1,2-ETHANEDIAMIDE; 1,2-ETHANEDIAMINE, N,N-DIETHYL N′-PYRIDINYL-N′(2-THIENYLMETHYL)-

EPA HAZARDOUS WASTE NUMBER (RCRA No.): U155.
RCRA Section 261 Hazardous Constituents.
RCRA Land Ban Waste.
RCRA Universal Treatment Standards: Wastewater (mg/L), 0.081; Non-wastewater (mg/kg), 1.5.
RCRA Ground Water Monitoring List. Suggested test method(s) (PQL µg/L): 8270(10).
EPCRA Section 304 Reportable Quantity (RQ): CERCLA, 5,000 lbs. (2270 kgs.).

EPA NAME: METHAZOLE

CAS: 20354-26-1

SYNONYMS: 2-(3,4-DICHLOROPHENYL)-4-METHYL-1,2,4-OXADIAZOLIDINE-3,5-DIONE; EPA PESTICIDE CHEMICAL CODE 106001; 1,2,4-OXADIAZOLIDINE-3,5-DIONE, 2-(3,4-DICHLOROPHENYL)-4-METHYL-

EPCRA Section 313 Form R de minimus concentration reporting level: 1.0%.

EPA NAME: METHIDATHION

CAS: 950-37-8

SYNONYMS: O,O-DIMETHYL)-S-(2-METHOXY-1,3,4-THIADIAZOLE-5(4H)-ONYL-(4)-METHYL)-PHOSPHORODITHIOATE; O,O-DIMETHYL PHOSPHORODITHIOATE, S-ESTER with 4-(MERCAPTOMETHYL)-2-METHOXY-2-OXO-1,3,4-THIADIAZOLIN-5-ONE; PHOSPHORODITHIOIC ACID, S-((5-METHOXY-2-OXO-1,3,4-THIADIAZOL-3(2H)-YL)METHYL) O,O-DIMETHYL ESTER S-((5-METHOXY-2-OXO-1,3,4-THIADIAZOL-3(2H)-YL)METHYL)-O,O-DIMETHYL PHOSPHORDITHIOATE

EPCRA Section 302 Extremely Hazardous Substances: TPQ = 500/10,000 lbs. (227/4,540 kgs.).

EPCRA Section 304 Reportable Quantity (RQ): EHS, 500 lbs. (227 kgs.).

MARINE POLLUTANT (49CFR, Subchapter 172.101, Appendix B).

EPA NAME: METHIOCARB

CAS: 2032-65-7

SYNONYMS: 3,5-DIMETHYL-4-(METHYLTHIO)PHENOL METHYLCARBAMATE; EPA PESTICIDE CHEMICAL CODE 100501; MESUROL; MERCAPTODIMETHUR; 4-(METHYLTHIO)-3,5-XYLYL-N-METHYLCARBAMATE; PHENOL, 3,5-DIMETHYL-4-(METHYLTHIO)-, METHYLCARBAMATE

CLEAN WATER ACT: Section 311 Hazardous Substances/RQ (same as CERCLA).

EPA HAZARDOUS WASTE NUMBER (RCRA No.): P199.

RCRA Section 261 Hazardous Constituents.

RCRA Land Ban Waste.

RCRA Universal Treatment Standards: Wastewater (mg/L), 0.056; Nonwastewater (mg/kg), 1.4.

EPCRA Section 302 Extremely Hazardous Substances: TPQ = 500/10,000 lbs. (227/4,540 kgs.).

EPCRA Section 304 Reportable Quantity (RQ): EHS/CERCLA, 10 lbs. (4.54 kgs.).

EPCRA Section 313 Form R de minimus concentration reporting level: 1.0%.

MARINE POLLUTANT (49CFR, Subchapter 172.101, Appendix B) as mercaptodimethur.

EPA NAME: METHOMYL

CAS: 16752-77-5

SYNONYMS: ACETIMIDIC ACID, THIO-N-(METHYLCARBAMOYL)OXY-,METHYL ESTER; ETHANIMIDOTHIC ACID, N-[[METHYLAMINO]CARBONYL]OXY]-, METHYL ESTER; LANNATE; METHYL-N-(METHYL(CARBAMOYL)OXY) THIOACETIMIDATE; s-METHYL N-((METHYLCARBAMOYL)OXY)THIOACETIMIDATE

EPA HAZARDOUS WASTE NUMBER (RCRA No.): P066.
RCRA Section 261 Hazardous Constituents.
RCRA Land Ban Waste.
RCRA Universal Treatment Standards: Wastewater (mg/L), 0.028; Nonwastewater (mg/kg), 0.14.
SAFE DRINKING WATER ACT: Priority List (55 FR 1470).
EPCRA Section 302 Extremely Hazardous Substances: TPQ = 500/10,000 lbs. (227/4,540 kgs.).
EPCRA Section 304 Reportable Quantity (RQ): EHS/CERCLA, 100 lbs. (45.4 kgs.).
MARINE POLLUTANT (49CFR, Subchapter 172.101, Appendix B).

EPA NAME: METHOXONE
CAS: 94-74-6
SYNONYMS: ACETIC ACID (4-CHLORO-2-METHYLPHENOXY)-; (4-CHLORO-2-METHYLPHENOXY) ACETIC ACID; (4-CHLORO-o-TOLOXY)ACETIC ACID; EPA PESTICIDE CHEMICAL CODE 030501; MCPA

EPCRA Section 313 Form R de minimus concentration reporting level: 0.1%.

EPA NAME: METHOXONE, SODIUM SALT
CAS: 3653-48-3
SYNONYMS: ACETIC ACID, (4-CHLORO-2-METHYLPHENOXY)-, SODIUM SALT; ACETIC ACID, ((4-CHLORO-o-TOLYL)OXY)-SODIUM SALT; (4-CHLORO-2-METHYLPHENOXY) ACETATE SODIUM SALT; EPA PESTICIDE CHEMICAL CODE 030502

EPCRA Section 313 Form R de minimus concentration reporting level: 1.0%.

EPA NAME: METHOXYCHLOR
[see also CHLORINATED ETHANES]
CAS: 72-43-5
SYNONYMS: BENZENE, 1,1′-(2,2,2-TRICHLOROETHYLIDENE)BIS[4-METHOXY-]; p,p′-DIMETHOXYDIPHENYLTRICHLOROETHANE; METHOXY-DDT; 2,2-BIS(p-METHOXYPHENYL)-1,1,1-TRICHLOROETHANE; 1,1,1-TRICHLORO-2,2-BIS(p-METHOXYPHENYL)ETHANE

CLEAN AIR ACT: Hazardous Air Pollutants (Title I, Part A, Section 112).
CLEAN WATER ACT: Section 311 Hazardous Substances/RQ (same as CERCLA); Section 313 Priority Chemicals; Section 313 Priority Chemicals
EPA HAZARDOUS WASTE NUMBER (RCRA No.): U247.

RCRA Toxicity Characteristic (Section 261.24), Maximum Concentration of Contaminants, regulatory level, 10.0 mg/L.
RCRA Section 261 Hazardous Constituents.
RCRA Land Ban Waste.
RCRA Universal Treatment Standards: Wastewater (mg/L), 0.25; Nonwastewater (mg/kg), 0.18.
RCRA Ground Water Monitoring List. Suggested test method(s) (PQL μg/L): 8080(2); 8270(10).
SAFE DRINKING WATER ACT: MCL, 0.04 mg/L; MCLG, 0.04 mg/L; Regulated chemical (47 FR 9352); Priority List (55 FR 1470).
EPCRA Section 304 Reportable Quantity (RQ): CERCLA, 1 lb. (0.45 kg.).
EPCRA Section 313 Form R de minimus concentration reporting level: 1.0%.

EPA NAME: 2-METHOXYETHANOL
CAS: 109-86-4
SYNONYMS: EGME; EGMME; ETHANOL, 2-METHOXY-; ETHYLENE GLYCOL MONOMETHYL ETHER; GLYCOL METHYL ETHER; METHYL CELLOSOLVE; 2-METHOXYETHANOL

EPCRA Section 313 Form R de minimus concentration reporting level: 1.0%.
CALIFORNIA'S PROPOSITION 65: Reproductive toxin (male).

EPA NAME: METHOXYETHYLMERCURIC ACETATE
[see also MERCURY COMPOUNDS]
CAS: 151-38-2
SYNONYMS: ACETATO(2-METHOXYETHYL)MERCURY

CLEAN WATER ACT: Section 307 Toxic Pollutants as mercury and compounds.
EPCRA Section 302 Extremely Hazardous Substances: TPQ = 500/10,000 lbs. (227/4,540 kgs.).
EPCRA Section 304 Reportable Quantity (RQ): EHS, 500 lbs. (227 kgs.).
MARINE POLLUTANT (49CFR, Subchapter 172.101, Appendix B).
CALIFORNIA'S PROPOSITION 65: Reproductive toxin.

EPA NAME: 2-(((((4-METHOXY-6-METHYL-1,3,5-TRIAZIN-2-YL)-METHYLAMINO)CARBONYL)AMINO)SULFONYL)-, METHYL ESTER
[see TRIBENURON METHYL-]
CAS: 101200-48-0

EPA NAME: METHYL ACRYLATE
CAS: 96-33-3
SYNONYMS: ACRYLIC ACID METHYL ESTER; METHYL-2-PROPENOATE; 2-PROPENOIC ACID, METHYL ESTER

EPA HAZARDOUS WASTE NUMBER (RCRA No.): U328.
RCRA Section 261 Hazardous Constituents.
EPCRA Section 304 Reportable Quantity (RQ): CERCLA, 100 lbs. (45.4 kgs.).

EPCRA Section 313 Form R de minimus concentration reporting level: 1.0%.

EPA NAME: METHYL BROMIDE
[see BROMOMETHANE]
CAS: 74-83-9
SYNONYMS: BROMOMETHANE; HALON 1001; METHANE, BROMO-

EPA NAME: 2-METHYL-1-BUTENE
CAS: 563-46-2
SYNONYMS: 1-BUTENE, 2-METHYL

CLEAN AIR ACT: Accidental Release Prevention/Flammable substances (Section 112[r], Table 3), TQ = 10,000 lbs. (4540 kgs.).

EPA NAME: 3-METHYL-1-BUTENE
CAS: 563-45-1
SYNONYMS: 1-BUTENE, 3-METHYL; ISOPENTENE

CLEAN AIR ACT: Accidental Release Prevention/Flammable substances (Section 112[r], Table 3), TQ = 10,000 lbs. (4540 kgs.).

EPA NAME: METHYL tert-BUTYL ETHER
CAS: 1634-04-4
SYNONYMS: tert-BUTOXYMETHANE; T-BUTYL METHYL ETHER; 1,1-DIMETHYLETHYL METHYL ETHER; METHYL-T-BUTYL ETHER; PROPANE, 2-METHOXY-2-METHYL-

CLEAN AIR ACT: Hazardous Air Pollutants (Title I, Part A, Section 112).
SAFE DRINKING WATER ACT: Priority List (55 FR 1470).
EPCRA Section 304 Reportable Quantity (RQ): CERCLA, 1,000 lbs. (454 kgs.).
EPCRA Section 313 Form R de minimus concentration reporting level: 1.0%.

EPA NAME: METHYL CHLORIDE
[see CHLOROMETHANE]
CAS: 74-87-3

EPA NAME: METHYL 2-CHLOROACRYLATE
CAS: 80-63-7
SYNONYMS: 2-CHLOROACRYLIC ACID, METHYL ESTER; 2-CHLORO-2-PROPENOIC ACID METHYL ESTER (9CI)

EPCRA Section 302 Extremely Hazardous Substances: TPQ = 500 lbs. (227 kgs.).
EPCRA Section 304 Reportable Quantity (RQ): EHS, 500 lbs. (227 kgs.).

EPA NAME: METHYL CHLOROCARBONATE
CAS: 79-22-1
SYNONYMS: CARBONOCHLORIDIC ACID, METHYL ESTER; METHYL CARBONOCHLORIDATE; METHYL CHLOROFORMATE

CLEAN AIR ACT: Accidental Release Prevention/Flammable substances (Section 112[r], Table 3), TQ = 5,000 lbs. (2270 kgs.).
EPA HAZARDOUS WASTE NUMBER (RCRA No.): U156.
RCRA Section 261 Hazardous Constituents.
EPCRA Section 302 Extremely Hazardous Substances: TPQ = 500 lbs. (227 kgs.).
EPCRA Section 304 Reportable Quantity (RQ): EHS/CERCLA, 1,000 lbs. (454 kgs.).
EPCRA Section 313 Form R de minimus concentration reporting level: 1.0%.

EPA NAME: METHYL CHLOROFORM
[see 1,1,1-TRICHLOROETHANE]
CAS: 71-55-6

EPA NAME: METHYL CHLOROFORMATE
[see METHYL CHLOROCARBONATE]
CAS: 79-22-1

EPA NAME: 3-METHYLCHLOANTHRENE
CAS: 56-49-5
SYNONYMS: 1,2-DIHYDRO-3-METHYL-BENZ(j)ACEANTHRYLENE; 3-MCA; METHYLCHOLANTHRENE; 20-METHYLCHOLANTHRENE

EPA HAZARDOUS WASTE NUMBER (RCRA No.): U157.
RCRA Section 261 Hazardous Constituents.
RCRA Land Ban Waste.
RCRA Universal Treatment Standards: Wastewater (mg/L), 0.0055; Nonwastewater (mg/kg), 15.
RCRA Ground Water Monitoring List. Suggested test method(s) (PQL µg/L): 8270(10).
EPCRA Section 304 Reportable Quantity (RQ): CERCLA, 10 lbs. (4.54 kgs.).
CALIFORNIA'S PROPOSITION 65: Carcinogen.

EPA NAME: 5-METHYLCHRYSENE
[see also POLYCYCLIC AROMATIC COMPOUNDS (PACs)]
CAS: 3697-24-3
SYNONYMS: CHRYSENE, 5-METHYL-

EPCRA Section 313 Form R de minimus concentration reporting level: 0.1%.
CALIFORNIA'S PROPOSITION 65: Carcinogen.

EPA NAME: 4-METHYLDIPHENYLMETHANE-3,4-DIISOCYANATE
[see also DIISOCYANATES]
CAS: 75790-84-0
SYNONYMS: BENZENE, 1-ISOCYANATO-2-((4-ISOCYANATOPHENYL)THIO)-; BENZENE, 2-ISOCYANATO-4((4-ISOCYANATOPHENYL)METHYL)-1-METHYL-3,4′-DIISOCYANATO-4-METHYL DIPHENYLMETHANE

EPCRA Section 313 Form R de minimus concentration reporting level: 1.0%.

EPA NAME: 6-METHYL-1,3-DITHIOLO[4,5-B]QUINOXALIN-2-ONE

[see CHINOMETHIONAT]

CAS: 2439-01-2

EPA NAME: 4,4′-METHYLENEBIS(2-CHLOROANILINE)

CAS: 101-14-4

SYNONYMS: BENZENEAMINE, 4-4′-METHYLENEBIS(2-CHLORO)-; 2,2′-DICHLORO-4,4′-METHYLENEDIANILINE; 3,3′-DICHLORO-4,4′-DIAMINODIPHENYLMETHANE; MBOCA; 4,4′-METHYLENEBIS(2-CHLOROANILINE); METHYLENEBIS(3-CHLORO-4-AMINOBENZENE); 4,4′-METHYLENEBIS (2-CHLORO-BENZENEAMINE)

CLEAN AIR ACT: Hazardous Air Pollutants (Title I, Part A, Section 112).

EPA HAZARDOUS WASTE NUMBER (RCRA No.): U158.

RCRA Section 261 Hazardous Constituents.

RCRA Land Ban Waste.

RCRA Universal Treatment Standards: Wastewater (mg/L), 0.50; Nonwastewater (mg/kg), 30.

EPCRA Section 304 Reportable Quantity (RQ): CERCLA, 10 lbs. (4.54 kgs.).

EPCRA Section 313 Form R de minimus concentration reporting level: 0.1%.

CALIFORNIA'S PROPOSITION 65: Carcinogen.

EPA NAME: 2,2′-METHYLENEBIS(4-CHLOROPHENOL)

[see DICHLOROPHENE]

CAS: 97-23-4

EPA NAME: 4,4′-METHYLENEBIS (N,N-DIMETHYL) BENZENAMINE

CAS: 101-61-1

SYNONYMS: BENZENAMINE, 4,4′-METHYLENEBIS(N,N-DIMETHYL)-; 4,4′-TETRAMETHYLDIAMINO-DIPHENYLMETHANE

EPCRA Section 313 Form R de minimus concentration reporting level: 0.1%.

CALIFORNIA'S PROPOSITION 65: Carcinogen.

EPA NAME: 1,1-METHYLENE BIS(4-ISOCYANATOCYCLOHEXANE)

[see also DIISOCYANATES]

CAS: 5124-30-1

SYNONYMS: CYCLOHEXANE, 1,1′-METHYLENEBIS(4-ISOCYANATO-; DICYCLOHEXYLMETHANE-4,4′-DIISOCYANATE; METHYLENE BIS(4-CYCLOHEXYLISOCYANATE)

EPCRA Section 313 Form R de minimus concentration reporting level: 1.0%.

EPA NAME: METHYLBIS(PHENYLISOCYANATE)
[see also DIISOCYANATES]
CAS: 101-68-8
SYNONYMS: BENZENE, 1,1′-METHYLENEBIS(4-ISOCYANATO-; 4,4′-DIPHENYLMETHANE DIISOCYANATE; METHYLENEDIPHENYL DIISOCYANATE; METHYLENEDIPHENYL DIISOCYANATE (4,4′)-; 4,4′-METHYLENEDI(PHENYLDIISOCYANATE); MDI

CLEAN AIR ACT: Hazardous Air Pollutants (Title I, Part A, Section 112).
EPCRA Section 304 Reportable Quantity (RQ): CERCLA, 5,000 lbs. (2,270 kgs.).
EPCRA Section 313 Form R de minimus concentration reporting level: 1.0%.

EPA NAME: METHYLENE BROMIDE
CAS: 74-95-3
SYNONYMS: DIBROMOMETHANE; METHANE, DIBROMO-; METHYLENE DIBROMIDE

EPA HAZARDOUS WASTE NUMBER (RCRA No.): U068.
RCRA Section 261 Hazardous Constituents.
RCRA Land Ban Waste.
RCRA Universal Treatment Standards: Wastewater (mg/L), 0.11; Nonwastewater (mg/kg), 15.
RCRA Ground Water Monitoring List. Suggested test method(s) (PQL μg/L): 8010(15); 8240(5).
EPCRA Section 304 Reportable Quantity (RQ): CERCLA, 1,000 lbs. (454 kgs.).
EPCRA Section 313 Form R de minimus concentration reporting level: 1.0%.

EPA NAME: METHYLENE CHLORIDE
[see DICHLOROMETHANE]
CAS: 75-09-2

EPA NAME: 4,4′-METHYLENEDIANILINE
CAS: 101-77-9
SYNONYMS: BENZENAMINE, 4,4′-METHYLENEBIS-; 4,4′-DIAMINODIPHENYLMETHANE; 4,4′-METHYLENEBIS(ANILINE); METHYLENEDIANILINE

CLEAN AIR ACT: Hazardous Air Pollutants (Title I, Part A, Section 112).
EPCRA Section 304 Reportable Quantity (RQ): CERCLA, 10 lbs. (4.54 kgs.).
EPCRA Section 313 Form R de minimus concentration reporting level: 0.1%.

EPA NAME: METHYL ETHER
CAS: 115-10-6

SYNONYMS: DIMETHYL ETHER; METHANE, OXYBIS-; OXYBISMETHANE

CLEAN AIR ACT: Accidental Release Prevention/Flammable substances (Section 112[r], Table 3), TQ=10,000 lbs. (4540 kgs.).

EPA NAME: METHYL ETHYL KETONE

CAS: 78-93-3

SYNONYMS: 2-BUTANONE; BUTAN-2-ONE; ETHYL METHYL KETONE; MEK; METHYL ACETONE

CLEAN AIR ACT: Hazardous Air Pollutants (Title I, Part A, Section 112).
EPA HAZARDOUS WASTE NUMBER (RCRA No.): U159.
RCRA Section 261 Hazardous Constituents.
RCRA Land Ban Waste.
RCRA Toxicity Characteristic (Section 261.24), Maximum Concentration of Contaminants, regulatory level, 200.0 mg/L.
RCRA Universal Treatment Standards: Wastewater (mg/L), 0.28; Nonwastewater (mg/kg), 36.
RCRA Ground Water Monitoring List. Suggested test method(s) (PQL µg/L): 8015(10); 8240(100).
SAFE DRINKING WATER ACT: Priority List (55 FR 1470).
EPCRA Section 304 Reportable Quantity (RQ): CERCLA, 5,000 lbs. (2270 kgs.).
EPCRA Section 313 Form R de minimus concentration reporting level: 1.0%.

EPA NAME: METHYL ETHYL KETONE PEROXIDE

CAS: 1338-23-4

SYNONYMS: 2-BUTANONE, PEROXIDE

EPA HAZARDOUS WASTE NUMBER (RCRA No.): U160.
RCRA Section 261 Hazardous Constituents.
EPCRA Section 304 Reportable Quantity (RQ): CERCLA, 10 lbs. (4.54 kgs.).

EPA NAME: METHYL FORMATE

CAS: 107-31-3

SYNONYMS: FORMIC ACID, METHYL ESTER; METHYL METHANOATE

CLEAN AIR ACT: Accidental Release Prevention/Flammable substances (Section 112[r], Table 3), TQ=10,000 lbs. (4540 kgs.).

EPA NAME: METHYL HYDRAZINE

CAS: 60-34-4

SYNONYMS: HYDRAZINE, METHYL-; 1-METHYL-HYDRAZINE; N-METHYL HYDRAZINE; MONOMETHYLHYDRAZINE

CLEAN AIR ACT: Hazardous Air Pollutants (Title I, Part A, Section 112); Accidental Release Prevention/Flammable substances (Section 112[r], Table 3), TQ=15,000 lbs. (6810 kgs.).
EPA HAZARDOUS WASTE NUMBER (RCRA No.): P068.
RCRA Section 261 Hazardous Constituents.

EPCRA Section 302 Extremely Hazardous Substances: TPQ = 500 lbs. (227 kgs.).
EPCRA Section 304 Reportable Quantity (RQ): EHS/CERCLA, 10 lbs. (4.54 kgs.).
EPCRA Section 313 Form R de minimus concentration reporting level: 1.0%.
CALIFORNIA'S PROPOSITION 65: Carcinogen as methylhydrazine and its salts.

EPA NAME: METHYL IODIDE
CAS: 74-88-4
SYNONYMS: IODOMETHANE; METHANE, IODO

CLEAN AIR ACT: Hazardous Air Pollutants (Title I, Part A, Section 112).
EPA HAZARDOUS WASTE NUMBER (RCRA No.): U138.
RCRA Section 261 Hazardous Constituents.
RCRA Universal Treatment Standards: Wastewater (mg/L), 0.19; Non-wastewater (mg/kg), 65.
RCRA Ground Water Monitoring List. Suggested test method(s) (PQL μg/L): 8010(40); 8240(5).
EPCRA Section 304 Reportable Quantity (RQ): CERCLA, 100 lbs. (45.4 kgs.).
EPCRA Section 313 Form R de minimus concentration reporting level: 1.0%.
CALIFORNIA'S PROPOSITION 65: Carcinogen.

EPA NAME: METHYL ISOBUTYL KETONE
CAS: 108-10-1
SYNONYMS: HEXONE; ISOBUTYL METHYL KETONE; 2-METHYL-4-PENTANONE; 4-METHYL-2-PENTANONE; 4-METHYL-PENTAN-2-ONE

CLEAN AIR ACT: Hazardous Air Pollutants (Title I, Part A, Section 112).
EPA HAZARDOUS WASTE NUMBER (RCRA No.): U161.
RCRA Section 261 Hazardous Constituents.
RCRA Land Ban Waste.
RCRA Universal Treatment Standards: Wastewater (mg/L), 0.14; Non-wastewater (mg/kg), 33.
SAFE DRINKING WATER ACT: Priority List (55 FR 1470).
EPCRA Section 304 Reportable Quantity (RQ): CERCLA, 5,000 lbs. (2270 kgs.).
EPCRA Section 313 Form R de minimus concentration reporting level: 1.0%.

EPA NAME: METHYL ISOCYANATE
CAS: 624-83-9
SYNONYMS: METHYL CARBONIMIDE; METHANE, ISOCYANATO-

CLEAN AIR ACT: Hazardous Air Pollutants (Title I, Part A, Section 112); Accidental Release Prevention/Flammable substances (Section 112[r], Table 3), TQ = 10,000 lbs. (4540 kgs.).
EPA HAZARDOUS WASTE NUMBER (RCRA No.): P064.

RCRA Section 261 Hazardous Constituents.

EPCRA Section 302 Extremely Hazardous Substances: TPQ = 500 lbs. (227 kgs.).

EPCRA Section 304 Reportable Quantity (RQ): EHS/CERCLA, 10 lbs. (4.54 kgs.).

EPCRA Section 313 Form R de minimus concentration reporting level: 1.0%.

EPA NAME: METHYL ISOTHIOCYANATE

CAS: 556-61-6

SYNONYMS: EPA PESTICIDE CHEMICAL CODE 068103; ISOTHIOCYANATOMETHANE; METHANE, ISOTHIOCYANATO-

EPCRA Section 302 Extremely Hazardous Substances: TPQ = 500 lbs. (227 kgs.).

EPCRA Section 304 Reportable Quantity (RQ): EHS, 1 lb. (0.454 kg.).

EPCRA Section 313 Form R de minimus concentration reporting level: 1.0%.

EPA NAME: 2-METHYLLACTONITRILE

CAS: 75-86-5

SYNONYMS: ACETONE CYANOHYDRIN; 2-METHYLACTONITRILE; PROPANENITRILE,2-HYDROXY-2-METHYL-

CLEAN WATER ACT: Section 311 Hazardous Substances/RQ (same as CERCLA); 313 Priority Pollutant (57FR 41331).

EPA HAZARDOUS WASTE NUMBER (RCRA No.): P069.

RCRA Section 261 Hazardous Constituents.

EPCRA Section 302, Extremely Hazardous Substances: TPQ = 1,000 lbs. (454 kgs.).

EPCRA Section 304 Reportable Quantity (RQ): EHS/CERCLA, 10 lbs. (4.54 kgs.).

EPCRA Section 313 Form R de minimus concentration reporting level: 1.0%.

MARINE POLLUTANT (49CFR, Subchapter 172.101, Appendix B).

EPA NAME: METHYL MERCAPTAN

CAS: 74-93-1

SYNONYMS: MERCAPTOMETHANE; METHANETHIOL; 1-METHANETHIOL; THIOMETHANOL

CLEAN AIR ACT: Accidental Release Prevention/Flammable substances (Section 112[r], Table 3), TQ = 10,000 lbs. (4540 kgs.).

CLEAN WATER ACT: Section 311 Hazardous Substances/RQ (same as CERCLA).

EPA HAZARDOUS WASTE NUMBER (RCRA No.): U153.

RCRA Section 261 Hazardous Constituents.

EPCRA Section 302 Extremely Hazardous Substances: TPQ = 500 lbs. (227 kgs.).

EPCRA Section 304 Reportable Quantity (RQ): EHS/CERCLA, 100 lbs. (45.4 kgs.).

EPCRA Section 313 Form R de minimus concentration reporting level: 1.0%. Note: Subject to an administrative stay under EPCRA Section 313. Not reportable until stay is lifted. See 8/22/94 (59 FR 43048).
MARINE POLLUTANT (49CFR, Subchapter 172.101, Appendix B).

EPA NAME: METHYLMERCURIC DICYANAMIDE
[see also MERCURY COMPOUNDS]
CAS: 502-39-6
SYNONYMS: METHYLHYDRAZINE; METHYLMERCURIC CYANOGUANIDINE; METHYLMERCURY DICYANANDIMIDE

CLEAN WATER ACT: Section 307 Toxic Pollutants as mercury and compounds.
EPCRA Section 302 Extremely Hazardous Substances: TPQ = 500/10,000 lbs. (227/4,540 kgs.).
EPCRA Section 304 Reportable Quantity (RQ): EHS, 500 lbs. (227 kgs.).
CALIFORNIA'S PROPOSITION 65: Reproductive toxin.

EPA NAME: METHYL METHACRYLATE
CAS: 80-62-6
SYNONYMS: METHACRYLIC ACID, METHYL ESTER; METHYL-2-METHYLPROPENOATE; 2-PROPENOIC ACID, 2-METHYL-, METHYL ESTER

CLEAN AIR ACT: Hazardous Air Pollutants (Title I, Part A, Section 112).
CLEAN WATER ACT: Section 311 Hazardous Substances/RQ (same as CERCLA); Section 313 Priority Chemicals.
EPA HAZARDOUS WASTE NUMBER (RCRA No.): U162.
RCRA Section 261 Hazardous Constituents.
RCRA Land Ban Waste.
RCRA Universal Treatment Standards: Wastewater (mg/L), 0.14; Nonwastewater (mg/kg), 160.
RCRA Ground Water Monitoring List. Suggested test method(s) (PQL µg/L): 8015(2); 8240(5).
EPCRA Section 304 Reportable Quantity (RQ): CERCLA, 1,000 lbs. (454 kgs.).
EPCRA Section 313 Form R de minimus concentration reporting level: 1.0%.
MARINE POLLUTANT (49CFR, Subchapter 172.101, Appendix B).

EPA NAME: METHYLMETHANESULFONATE
CAS: 66-27-3
SYNONYMS: METHANESULFONIC ACID, METHYL ESTER

RCRA Land Ban Waste.
RCRA Universal Treatment Standards: Wastewater (mg/L), 0.018; Nonwastewater (mg/kg), N/A.
RCRA Ground Water Monitoring List. Suggested test method(s) (PQL µg/L): 8270(10).
CALIFORNIA'S PROPOSITION 65: Carcinogen.

EPA NAME: N-METHYLOLACRYLAMIDE
CAS: 924-42-5
SYNONYMS: N-(HYDROXYMETHYL)ACRYLAMIDE; METHYLOLACRYLAMIDE; 2-PROPENAMIDE, N-(HYDROXYMETHYL)-

EPCRA Section 313 Form R de minimus concentration reporting level: 1.0%.

EPA NAME: 2-METHYLNAPHTHALENE
CAS: 91-57-6
SYNONYMS: NAPHTHALENE, 2-METHYL-

RCRA Ground Water Monitoring List. Suggested test method(s) (PQL μg/L): 8270(10).
MARINE POLLUTANT (49CFR, Subchapter 172.101, Appendix B).

EPA NAME: METHYL PARATHION
CAS: 298-00-0
SYNONYMS: EPA PESTICIDE CHEMICAL CODE 053501; PARATHION-METHYL; PHOSPHOROTHIOIC ACID, O,O-DIMETHYL O-(4-NITROPHENYL) ESTER; O,O-DIMETHYL O-(p-NITROPHENYL) PHOSPHOROTHIOATE

CLEAN WATER ACT: Section 311 Hazardous Substances/RQ (same as CERCLA).
EPA HAZARDOUS WASTE NUMBER (RCRA No.): P071.
RCRA Section 261 Hazardous Constituents.
RCRA Land Ban Waste.
RCRA Universal Treatment Standards: Wastewater (mg/L), 0.014; Nonwastewater (mg/kg), 4.6.
RCRA Ground Water Monitoring List. Suggested test method(s) (PQL μg/L): 8140(0.5); 8270(10).
EPCRA Section 302 Extremely Hazardous Substances: TPQ = 100/10,000 lbs. (45.4/4,540 kgs.).
EPCRA Section 304 Reportable Quantity (RQ): EHS/CERCLA, 100 lbs. (45.4 kgs.).
EPCRA Section 313 Form R de minimus concentration reporting level: 1.0%.

EPA NAME: METHYL PHENKAPTON
CAS: 3735-23-7
SYNONYMS: O,O-DIMETHYL S-(2,5-DICHLOROPHENYLTHIO)METHYL PHOSPHORODITHIOATE

EPCRA Section 302 Extremely Hazardous Substances: TPQ = 500 lbs. (227 kgs.).
EPCRA Section 304 Reportable Quantity (RQ): EHS, 500 lbs. (227 kgs.).

EPA NAME: METHYL PHOSPHONIC DICHLORIDE
CAS: 676-97-1
EPCRA Section 302 Extremely Hazardous Substances: TPQ = 100 lbs. (45.4 kgs.).

EPCRA Section 304 Reportable Quantity (RQ): EHS, 100 lbs. (45.4 kgs.).

EPA NAME: 2-METHYLPROPENE
CAS: 115-11-7
SYNONYMS: γ-BUYLENE; ISOBUTENE; ISOBUTYLENE; 1-PROPENE, 2-METHYL

CLEAN AIR ACT: Accidental Release Prevention/Flammable substances (Section 112[r], Table 3), TQ = 10,000 lbs. (4540 kgs.).

EPA NAME: 2-METHYLPYRIDINE
CAS: 109-06-8
SYNONYMS: 2-METHYLPYRIDINE; 2-PICOLINE; o-PICOLINE; PYRIDINE, 2-METHYL-

EPA HAZARDOUS WASTE NUMBER (RCRA No.): U191.
RCRA Section 261 Hazardous Constituents.
EPCRA Section 304 Reportable Quantity (RQ): CERCLA, 5,000 lbs. (2270 kgs.).
EPCRA Section 313 Form R de minimus concentration reporting level: 1.0%.

EPA NAME: N-METHYL-2-PYRROLIDONE
CAS: 872-50-4
SYNONYMS: N-METHYL-2-PYRROLIDINONE; 1-METHYL-2-PYRROLIDONE; N-METHYLPYRROLIDINONE; 2-PYRROLIDINONE, 1-METHYL-

EPCRA Section 313 Form R de minimus concentration reporting level: 1.0%.

EPA NAME: METHYL t-BUTYL ETHER
[see METHYL tert-BUTYL ETHER]
CAS: 1634-04-4

EPA NAME: METHYL THIOCYANATE
CAS: 556-64-9
SYNONYMS: METHYL SULFOCYANATE; METHYLTHIOKYANAT; THIOCYANIC ACID, METHYL ESTER

CLEAN AIR ACT: Accidental Release Prevention/Flammable substances (Section 112[r], Table 3), TQ = 20,000 lbs. (9080 kgs.).
EPCRA Section 302 Extremely Hazardous Substances: TPQ = 10,000 lbs. (4,540 kgs.).
EPCRA Section 304 Reportable Quantity (RQ): EHS, 10,000 lbs. (4,540 kgs.).

EPA NAME: METHYLTHIOURACIL
CAS: 56-04-2
SYNONYMS: 4(1H)-PYRIMIDIONE, 2,3-DIHYDRO-6-METHYL-2-THIOXO-; 6-METHYL-2-THIOURACIL; 2-THIO-6-METHYL-1,3-PYRIMIDIN-4-ONE

EPA HAZARDOUS WASTE NUMBER (RCRA No.): U164.
RCRA Section 261 Hazardous Constituents.

EPCRA Section 304 Reportable Quantity (RQ): CERCLA, 10 lbs. (4.54 kgs.).
CALIFORNIA'S PROPOSITION 65: Carcinogen.

EPA NAME: METHYLTRICHLOROSILANE
CAS: 75-79-6
SYNONYMS: METHYLCHLOROSILANE; SILANE, METHYLTRICHLORO-; SILANE, TRICHLOROMETHYL-

CLEAN AIR ACT: Accidental Release Prevention/Flammable substances (Section 112[r], Table 3), TQ = 5,000 lbs. (2270 kgs.).
EPCRA Section 302 Extremely Hazardous Substances: TPQ = 500 lbs. (227 kgs.).
EPCRA Section 304 Reportable Quantity (RQ): EHS, 500 lbs. (227 kgs.).
EPCRA Section 313 Form R de minimus concentration reporting level: 1.0%.

EPA NAME: METHYL VINYL KETONE
CAS: 78-94-4
SYNONYMS: METHYLENE ACETONE; VINYL METHYL KETONE

EPCRA Section 302 Extremely Hazardous Substances: TPQ = 10 lbs. (4.54 kgs.).
EPCRA Section 304 Reportable Quantity (RQ): EHS, 10 lbs. (4.54 kgs.).

EPA NAME: METIRAM
[see also ETHYLENEBISDITHIOCARBAMIC ACID]
CAS: 9006-42-2
SYNONYMS: EPA PESTICIDE CHEMICAL CODE 014601; ZINC AMMONIATE ETHYLENEBIS(DITHIOCARBAMATE)-POLY(ETHYLENETHIURAM DISULFIDE); ZINC METIRAM

EPCRA Section 313 Form R de minimus concentration reporting level: 1.0%. Form R Toxic Chemical Category Code: N171.
CALIFORNIA'S PROPOSITION 65: Carcinogen.

EPA NAME: METOLCARB
CAS: 1129-41-5
SYNONYMS: CARBAMIC ACID, METHYL-, 3-METHYLPHENYL ESTER

EPA HAZARDOUS WASTE NUMBER (RCRA No.): P190.
RCRA Section 261 Hazardous Constituents.
RCRA Land Ban Waste.
RCRA Universal Treatment Standards: Wastewater (mg/L), 0.056; Nonwastewater (mg/kg), 1.4.
EPCRA Section 302 Extremely Hazardous Substances: TPQ = 100/10,000 lbs. (45.4/4,540 kgs.).
EPCRA Section 304 Reportable Quantity (RQ): EHS, 1 lb. (0.454 kg.).

EPA NAME: METRIBUZIN
CAS: 21087-64-9

SYNONYMS: 4-AMINO-6-tert-BUTYL-3-(METHYLTHIO)-1,2,4-TRIAZIN-5-ONE; as-TRIAZIN-5(4H)-ONE,4-AMINO-6-tert-BUTYL-3-(METHYLTHIO)-; 1,2,4-TRIAZIN-5(4H)-ONE,4-AMINO-6-(1,1-DIMETHYLETHYL)-3-(METHYLTHIO)-; SENCOR

EPCRA Section 313 Form R de minimus concentration reporting level: 1.0%.

SAFE DRINKING WATER ACT: Priority List (55 FR 1470).

EPA NAME: MEVINPHOS

CAS: 7786-34-7

SYNONYMS: 2-BUTENOIC ACID, 3-((DIMETHOXYPHOSPHINYL)OXY)-, METHYL ESTER; α-2-CARBOMETHOXY-1-METHYLVINYL DIMETHYL PHOSPHATE; 2-CARBOMETHOXY-1-METHYLVINYL DIMETHYL PHOSPHATE; EPA PESTICIDE CHEMICAL CODE 015801; PHOSDRIN

CLEAN WATER ACT: Section 311 Hazardous Substances/RQ (same as CERCLA).

EPCRA Section 302 Extremely Hazardous Substances: TPQ = 500 lbs. (227 kgs.).

EPCRA Section 304 Reportable Quantity (RQ): EHS/CERCLA, 10 lbs. (4.54 kgs.).

EPCRA Section 313 Form R de minimus concentration reporting level: 1.0%.

EPA NAME: MEXACARBATE

CAS: 315-18-4

SYNONYMS: CARBAMATE,4-DIMETHYLAMINO-3,5-XYLYL N-METHYL-; PHENOL, 4-(DIMETHYLAMINO)-3,5-DIMETHYL-METHYLCARBAMATE (ESTER); ZACTRAN; ZECTANE

CLEAN WATER ACT: Section 311 Hazardous Substances/RQ (same as CERCLA).

EPA HAZARDOUS WASTE NUMBER (RCRA No.): P128.

RCRA Section 261 Hazardous Constituents.

RCRA Land Ban Waste.

RCRA Universal Treatment Standards: Wastewater (mg/L), 0.056; Nonwastewater (mg/kg), 1.4.

EPCRA Section 302 Extremely Hazardous Substances: TPQ = 500/10,000 lbs. (227/4,540 kgs.).

EPCRA Section 304 Reportable Quantity (RQ): EHS/CERCLA, 1,000 lbs. (454 kgs.).

MARINE POLLUTANT (49CFR, Subchapter 172.101, Appendix B).

EPA NAME: MICHLER'S KETONE

CAS: 90-94-8

SYNONYMS: BENZOPHENONE, 4,4′-BIS(DIMETHYLAMINO)-; METHANONE, BIS[4-(DIMETHYLAMINO)PHENYL]-

EPCRA Section 313 Form R de minimus concentration reporting level: 0.1%.

CALIFORNIA'S PROPOSITION 65: Carcinogen.

EPA NAME: MITOMYCIN C
CAS: 50-07-7
SYNONYMS: AMETYCIN; AZIRINOL(2′,3′:3,4)PYRROLO(1,2-A)INDOLE-4,7-DIONE,6-AMINO-8[[(AMINOCARBONYL)O-XY]METHYL]-1,1A,2,8,8A,8B-HEXAHYDRO-8A-METHOXY-5-METHYL-, [1AS-(Aα,8β,8Aα,8Bα)]-; 7-AMINO-9-α-METHOXY-MITOSANE

EPA HAZARDOUS WASTE NUMBER (RCRA No.): U010.
RCRA Section 261 Hazardous Constituents.
EPCRA Section 302 Extremely Hazardous Substances: TPQ = 500/10,000 lbs. (227/4,540 kgs.).
EPCRA Section 304 Reportable Quantity (RQ): EHS/CERCLA, 10 lbs. (4.54 kgs.).
CALIFORNIA'S PROPOSITION 65: Carcinogen.

EPA NAME: MOLINATE
CAS: 2212-67-1
SYNONYMS: 1H-AZEPINE-1-CARBOTHIOIC ACID, HEXAHY-DRO-S-ETHYL ESTER; EPA PESTICIDE CHEMICAL CODE 041402; S-ETHYL AZEPANE-1-CARBOTHIOATE; ORDRAM

EPA HAZARDOUS WASTE NUMBER (RCRA No.): U365.
RCRA Section 261 Hazardous Constituents.
RCRA Land Ban Waste.
RCRA Universal Treatment Standards: Wastewater (mg/L), 0.003; Non-wastewater (mg/kg), 1.4.
RCRA Section 261 Hazardous Constituents.
EPCRA Section 304 Reportable Quantity (RQ): CERCLA, 1 lb. (0.454 kg.).
EPCRA Section 313 Form R de minimus concentration reporting level: 1.0%.

EPA NAME: MOLYBDENUM TRIOXIDE
CAS: 1313-27-5
SYNONYMS: MOLYBDENUM(VI) OXIDE; MOLYBDIC ACID AN-HYDRIDE

EPCRA Section 313 Form R de minimus concentration reporting level: 1.0%.
SAFE DRINKING WATER ACT: Priority List (55 FR 1470) as molybde-num.

EPA NAME: MONOCHLOROBENZENE
[see CHLOROBENZENE]
CAS: 108-90-7

EPA NAME: MONOCHLOROPENTAFLUOROETHANE
CAS: 76-15-3
SYNONYMS: CFC-115; 1-CHLORO-1,1,2,2,2-PENTAFLUORO-METHANE; CHLOROPENTAFLUOROETHANE; FREON 115; PENTAFLUOROMONOCHLOROETHANE

CLEAN AIR ACT: Stratospheric ozone protection (Title VI, Subpart A, Appendix A), Class I, Ozone Depletion Potential = 0.6
EPCRA Section 313 Form R de minimus concentration reporting level: 1.0%.

EPA NAME: MONOCROPTOPHOS
CAS: 6923-22-4
SYNONYMS: AZODRIN; 3-(DIMETHOXYPHOSPHINYLOXY)N-METHYL-cis-CROTONAMIDE; PHOSPHORIC ACID, DIMETHYL ESTER, ESTER WITH cis-3-HYDROXY-N-METHYLCROTONAMIDE; PHOSPHORIC ACID, DIMETHYL 1-METHYL-3-(METHYLAMINO)-3-OXO-1-PROPENYL ESTER (E)-

EPCRA Section 302 Extremely Hazardous Substances: TPQ = 10/10,000 lbs. (4.54/4,540 kgs.).
EPCRA Section 304 Reportable Quantity (RQ): EHS, 10 lbs. (4.54 kgs.).
MARINE POLLUTANT (49CFR, Subchapter 172.101, Appendix B).

EPA NAME: MONOETHYLAMINE
[see ETHANAMINE]
CAS: 75-04-7

EPA NAME: MONOMETHYLAMINE
[see METHANAMINE]
CAS: 74-89-5

EPA NAME: MONURON
CAS: 150-68-5
SYNONYMS: 3-para-CHLOROPHENYL-1,1-DIMETHYLUREA; N-(p-CHLOROPHENYL)-N′,N′-DIMETHYLUREA; EPA PESTICIDE CHEMICAL CODE 035501; UREA, N′-(4-CHLOROPHENYL)-N,N-DIMETHYL-

EPCRA Section 313 Form R de minimus concentration reporting level: 1.0%.

EPA NAME: MUSCIMOL
CAS: 2763-96-4
SYNONYMS: 5-(AMINOMETHYL)-3-ISOXAZOLOL; AGARIN; 5-AMINOMETHYL-3-HYDROXYISOXAZOLE; 5-(AMINOMETHYL)-3-(2H)ISOXAZOLONE; 5-HYDROXY-5-AMINOMETHYLISOXAZOLE

EPA HAZARDOUS WASTE NUMBER (RCRA No.): P007.
RCRA Section 261 Hazardous Constituents.
EPCRA Section 302, Extremely Hazardous Substances, TPQ = 500/10,000 lbs (227/4,550 kgs.).
EPCRA Section 304 Reportable Quantity (RQ): EHS/CERCLA, 1,000 lbs. (455 kgs.).
CALIFORNIA'S PROPOSITION 65: Carcinogen.

EPA NAME: MUSTARD GAS
[see also CHLORINATED ETHANES]

CAS: 505-60-2

SYNONYMS: 2,2′-DICHLORODIETHYL SULFIDE; ETHANE, 1,1′-THIOBIS[2-CHLORO-; 1,1′-THIOBIS(2-CHLOROETHANE)

RCRA Section 261 Hazardous Constituents, waste number not listed.

EPCRA Section 302 Extremely Hazardous Substances: TPQ = 500 lbs. (227 kgs.).

EPCRA Section 304 Reportable Quantity (RQ): EHS, 500 lbs. (227 kgs.).

EPCRA Section 313 Form R de minimus concentration reporting level: 0.1%.

CALIFORNIA'S PROPOSITION 65: Carcinogen.

EPA NAME: MYCLOBUTANIL

CAS: 88671-89-0

SYNONYMS: α-BUTYL-α-(4-CHLOROPHENYL)-1H-1,2,4-TRIAZOLE-1-PROPANENITRILE; 2(4-CHLOROPHENYL)-2-(1H-1,2,4-TRIAZOLE-1-YLMETHYL)HEXANENITRILE; 2-p-CHLOROPHENYL-2-(1H-1,2,4-TRIAZOLE-1-YLMETHYL) HEXANENITRILE; EPA PESTICIDE CHEMICAL CODE 128857; 1H-1,2,4-TRIAZOLE-1-PROPNENITRILE,α-BUTYL-α-(4-CHLOROPHENYL)

EPCRA Section 313 Form R de minimus concentration reporting level: 1.0%.

- N -

EPA NAME: NABAM
[See also ETHYLENEBISDITHIOCARBAMIC ACID]
CAS: 142-59-6
SYNONYMS: EPA PESTICIDE CHEMICAL CODE 014503; 1,2-ETHANEDIYLBIS CARBAMODITHIOIC ACID DISODIUM SALT; ETHYLENE BIS(DITHIOCARBAMATE), DISODIUM SALT

EPCRA Section 313 Form R de minimus concentration reporting level: 1.0%. Form R Toxic Chemical Category Code: N171.
MARINE POLLUTANT (49CFR, Subchapter 172.101, Appendix B).

EPA NAME: NALED
CAS: 300-76-5
SYNONYMS: DIBROM; DIMETHYL 1,2-DIBROMO-2,2-DICHLOROETHYL PHOSPHATE; O,O-DIMETHYL-O-(1,2-DIBROMO-2,2-DICHLOROETHYL)PHOSPHATE; EPA PESTICIDE CHEMICAL CODE 034401; PHOSPHORIC ACID, 1,2-DIBROMO-2,2-DICHLOROETHYL DIMETHYL ESTER

CLEAN WATER ACT: Section 311 Hazardous Substances/RQ (same as CERCLA).
EPCRA Section 304 Reportable Quantity (RQ): CERCLA, 10 lbs. (4.54 kgs.).
EPCRA Section 313 Form R de minimus concentration reporting level: 1.0%.
MARINE POLLUTANT (49CFR, Subchapter 172.101, Appendix B).

EPA NAME: NAPHTHALENE
[see also POLYNUCLEAR AROMATIC HYDROCARBONS]
CAS: 91-20-3
SYNONYMS: TAR CAMPHOR

CLEAN AIR ACT: Hazardous Air Pollutants (Title I, Part A, Section 112).
CLEAN WATER ACT: Section 311 Hazardous Substances/RQ (same as CERCLA); Section 307 Toxic Pollutants.
EPA HAZARDOUS WASTE NUMBER (RCRA No.): U165.
RCRA Section 261 Hazardous Constituents.
RCRA Land Ban Waste.
RCRA Universal Treatment Standards: Wastewater (mg/L), 0.059; Nonwastewater (mg/kg), 5.6.
RCRA Ground Water Monitoring List. Suggested test method(s) (PQL µg/L): 8100(200); 8270(10).
SAFE DRINKING WATER ACT: Regulated chemical (47 FR 9352).
EPCRA Section 304 Reportable Quantity (RQ): CERCLA, 100 lbs. (45.4 kgs.).
EPCRA Section 313 Form R de minimus concentration reporting level: 1.0%.
MARINE POLLUTANT (49CFR, Subchapter 172.101, Appendix B).

EPA NAME: 1,5-NAPHTHALENE DIISOCYANATE
[see also DIISOCYANATES]
CAS: 3173-72-6
SYNONYMS: ISOCYANIC ACID, 1,5-NAPHTHYLENE ESTER; NAPHTHALENE, 1,5-DIISOCYANATO-

EPCRA Section 313 Form R de minimus concentration reporting level: 1.0%.

EPA NAME: 1-NAPHTHOL N-METHYLCARBAMATE
[see CARBARYL]
CAS: 63-25-2

EPA NAME: NAPHTHENIC ACID
CAS: 1338-24-5
SYNONYMS: SUNAPTIC ACID B; SUNAPTIC ACID C

CLEAN WATER ACT: Section 311 Hazardous Substances/RQ (same as CERCLA).
EPCRA Section 304 Reportable Quantity (RQ): CERCLA, 100 lbs. (45.4 kgs.).
MARINE POLLUTANT (49CFR, Subchapter 172.101, Appendix B).

EPA NAME: 1,4-NAPHTHOQUINONE
CAS: 130-15-4
SYNONYMS: 1,4-NAPHTHALENEDIONE; α-NAPHTHOQUINONE

EPA HAZARDOUS WASTE NUMBER (RCRA No.): U166.
RCRA Section 261 Hazardous Constituents.
RCRA Ground Water Monitoring List. Suggested test method(s) (PQL μg/L): 8270(10).
EPCRA Section 304 Reportable Quantity (RQ): CERCLA, 5,000 lbs. (2270 kgs.).

EPA NAME: α-NAPHTHYLAMINE
CAS: 134-32-7
SYNONYMS: NAPHTHYLAMINE; 1-NAPHTHALENAMINE

EPA HAZARDOUS WASTE NUMBER (RCRA No.): U167.
RCRA Section 261 Hazardous Constituents.
RCRA Ground Water Monitoring List. Suggested test method(s) (PQL μg/L): 8270(10).
EPCRA Section 304 Reportable Quantity (RQ): CERCLA, 100 lbs. (45.4 kgs.).
EPCRA Section 313 Form R de minimus concentration reporting level: 0.1%.
CALIFORNIA'S PROPOSITION 65: Carcinogen.

EPA NAME: β-NAPHTHYLAMINE
CAS: 91-59-8
SYNONYMS: 2-NAPHTHYLAMINE; 2-NAPHTHALENAMINE

EPA HAZARDOUS WASTE NUMBER (RCRA No.): U168.
RCRA Section 261 Hazardous Constituents.

RCRA Land Ban Waste.
RCRA Universal Treatment Standards: Wastewater (mg/L), 0.52; Non-wastewater (mg/kg), N/A.
RCRA Ground Water Monitoring List. Suggested test method(s) (PQL µg/L): 8270(10).
EPCRA Section 304 Reportable Quantity (RQ): CERCLA, 10 lbs. (4.54 kgs.).
EPCRA Section 313 Form R de minimus concentration reporting level: 0.1%.
CALIFORNIA'S PROPOSITION 65: Carcinogen.

EPA NAME: α-NAPHTHYLTHIOUREA
[see ANTU]
CAS: 86-88-4

EPA NAME: NICKEL
[see also NICKEL COMPOUNDS]
CAS: 7440-02-0
SYNONYMS: METALLIC NICKEL; NICKEL, ELEMENTAL; NICKEL, METAL

CLEAN AIR ACT: Hazardous Air Pollutants (Title I, Part A, Section 112) as nickel compounds.
CLEAN WATER ACT: Section 307 Toxic Pollutants as nickel and compounds; Section 313 Priority Chemicals.
RCRA Section 261 Hazardous Constituents, waste number not listed.
RCRA Land Ban Waste.
RCRA Universal Treatment Standards: Wastewater (mg/L), 3.98; Non-wastewater (mg/L), 5.0 TCLP.
RCRA Ground Water Monitoring List. Suggested test method(s) (PQL µg/L): (total) 6010(50); 7520(400).
SAFE DRINKING WATER ACT: Regulated chemical (47 FR 9352).
EPCRA Section 304 Reportable Quantity (RQ): CERCLA, 100 lbs. (45.4 kgs.).
EPCRA Section 313 Form R de minimus concentration reporting level: 0.1%. Form R Toxic Chemical Category Code: N495.
CALIFORNIA'S PROPOSITION 65: Carcinogen as nickel and certain nickel compounds.

EPA NAME: NICKEL AMMONIUM SULFATE
[see also NICKEL and NICKEL COMPOUNDS]
CAS: 15699-18-0
SYNONYMS: NICKEL(II) AMMONIUM SULFATE; SULFURIC ACID, AMMONIUM NICKEL(II) SALT (2:2:1)

CLEAN AIR ACT: Hazardous Air Pollutants (Title I, Part A, Section 112) as nickel compounds.
CLEAN WATER ACT: Section 311 Hazardous Substances/RQ (same as CERCLA); Section 307 Toxic Pollutants as nickel and compounds; Section 313 Priority Chemicals.
RCRA Section 261 Hazardous Constituents, waste number not listed, as nickel compounds, n.o.s.

EPCRA Section 304 Reportable Quantity (RQ): CERCLA, 100 lbs. (45.4 kgs.).
EPCRA Section 313 Form R de minimus concentration reporting level: 0.1%. Form R Toxic Chemical Category Code: N495.

EPA NAME: NICKEL CARBONYL
[see also NICKEL and NICKEL COMPOUNDS]
CAS: 13463-39-3
SYNONYMS: NICKEL TETRACARBONYL; TETRACARBONYL-NICKEL

CLEAN AIR ACT: Hazardous Air Pollutants (Title I, Part A, Section 112) as nickel compounds; Accidental Release Prevention/Flammable substances (Section 112[r], Table 3), TQ = 1,000 lbs. (454.0 kgs.).
CLEAN WATER ACT: Section 307 Toxic Pollutants as nickel and compounds.
EPA HAZARDOUS WASTE NUMBER (RCRA No.): P073.
RCRA Section 261 Hazardous Constituents.
EPCRA Section 302 Extremely Hazardous Substances: TPQ = 1 lb. (0454 kgs.).
EPCRA Section 304 Reportable Quantity (RQ): EHS/CERCLA, 10 lbs. (4.54 kgs.).
EPCRA Section 313 Form R de minimus concentration reporting level: 0.1%. Form R Toxic Chemical Category Code: N495.
MARINE POLLUTANT (49CFR, Subchapter 172.101, Appendix B).
CALIFORNIA'S PROPOSITION 65: Carcinogen.

EPA NAME: NICKEL CHLORIDE
[see also NICKEL and NICKEL COMPOUNDS]
CAS: 7718-54-9; 37211-05-5
SYNONYMS: NICKEL(II) CHLORIDE; NICKEL(II) CHLORIDE (1:2)

CLEAN AIR ACT: Hazardous Air Pollutants (Title I, Part A, Section 112) as nickel compounds.
CLEAN WATER ACT: Section 311 Hazardous Substances/RQ (same as CERCLA); Section 307 Toxic Pollutants as nickel and compounds; Section 313 Priority Chemicals.
RCRA Section 261 Hazardous Constituents, waste number not listed, as nickel compounds, n.o.s.
EPCRA Section 304 Reportable Quantity (RQ): CERCLA, 100 lbs. (45.4 kgs.).
EPCRA Section 313 Form R de minimus concentration reporting level: 0.1%. Form R Toxic Chemical Category Code: N495.

EPA NAME: NICKEL COMPOUNDS
CLEAN AIR ACT: Hazardous Air Pollutants (Title I, Part A, Section 112).
CLEAN WATER ACT: Section 307 Toxic Pollutants as nickel and compounds.
RCRA Section 261 Hazardous Constituents, waste number not listed, as nickel compounds, n.o.s.

EPCRA Section 313: Includes any unique chemical substance that contains nickel as part of that chemical's infrastructure. Form R de minimus concentration reporting level: 0.1%. Form R Toxic Chemical Category Code: N495.
CALIFORNIA'S PROPOSITION 65: Carcinogen, as nickel and certain nickel compounds; nickel refinery dust from pyrometallurgical process.

EPA NAME: NICKEL CYANIDE
[see also NICKEL, NICKEL COMPOUNDS, CYANIDE, and CYANIDE COMPOUNDS]
CAS: 557-19-7
SYNONYMS: NICKEL(II)CYANIDE

CLEAN AIR ACT: Hazardous Air Pollutants (Title I, Part A, Section 112) as nickel compounds.
CLEAN WATER ACT: Section 307 Toxic Pollutants as nickel and compounds.
EPA HAZARDOUS WASTE NUMBER (RCRA No.): P074.
RCRA Section 261 Hazardous Constituents.
EPCRA Section 304 Reportable Quantity (RQ): CERCLA, 10 lbs. (4.54 kgs.).
EPCRA Section 313 Form R de minimus concentration reporting level: 0.1%. Form R Toxic Chemical Category Code: N495 (nickel).
MARINE POLLUTANT (49CFR, Subchapter 172.101, Appendix B).

EPA NAME: NICKEL HYDROXIDE
[see also NICKEL and NICKEL COMPOUNDS]
CAS: 12054-48-7
SYNONYMS: NICKEL(II) HYDROXIDE

CLEAN AIR ACT: Hazardous Air Pollutants (Title I, Part A, Section 112) as nickel compounds.
CLEAN WATER ACT: Section 311 Hazardous Substances/RQ (same as CERCLA); Section 307 Toxic Pollutants as nickel and compounds; Section 313 Priority Chemicals.
RCRA Section 261 Hazardous Constituents, waste number not listed, as nickel compounds, n.o.s.
EPCRA Section 304 Reportable Quantity (RQ): CERCLA, 10 lbs. (4.54 kgs.).
EPCRA Section 313 Form R de minimus concentration reporting level: 0.1%. Form R Toxic Chemical Category Code: N495.

EPA NAME: NICKEL NITRATE
[see also NICKEL and NICKEL COMPOUNDS]
CAS: 14216-75-2
SYNONYMS: NITRIC ACID, NICKEL SALT

CLEAN AIR ACT: Hazardous Air Pollutants (Title I, Part A, Section 112) as nickel compounds.
CLEAN WATER ACT: Section 311 Hazardous Substances/RQ (same as CERCLA); Section 307 Toxic Pollutants as nickel and compounds; Section 313 Priority Chemicals.

RCRA Section 261 Hazardous Constituents, waste number not listed, as nickel compounds, n.o.s.

EPCRA Section 304 Reportable Quantity (RQ): CERCLA, 100 lbs. (45.4 kgs.).

EPCRA Section 313 Form R de minimus concentration reporting level: 0.1%. Form R Toxic Chemical Category Code: N495.

EPA NAME: NICKEL NITRATE

[see also NICKEL and NICKEL COMPOUNDS]

CAS: 13138-45-9

SYNONYMS: NICKEL(II) NITRATE; NITRIC ACID, NICKEL(2+) SALT

CLEAN AIR ACT: Hazardous Air Pollutants (Title I, Part A, Section 112) as nickel compounds.

CLEAN WATER ACT: Section 307 Toxic Pollutants as nickel and compounds.

RCRA Section 261 Hazardous Constituents, waste number not listed, as nickel compounds, n.o.s.

EPCRA Section 313 Form R de minimus concentration reporting level: 0.1%. Form R Toxic Chemical Category Code: N495.

EPA NAME: NICKEL SULFATE

[see also NICKEL and NICKEL COMPOUNDS]

CAS: 7786-81-4

SYNONYMS: NICKEL(II) SULFATE; NICKEL(2+) SULFATE(1:1); SULFURIC ACID, NICKEL(II) SALT

CLEAN AIR ACT: Hazardous Air Pollutants (Title I, Part A, Section 112) as nickel compounds.

CLEAN WATER ACT: Section 311 Hazardous Substances/RQ (same as CERCLA); Section 307 Toxic Pollutants as nickel and compounds; Section 313 Priority Chemicals.

RCRA Section 261 Hazardous Constituents, waste number not listed, as nickel compounds, n.o.s.

EPCRA Section 304 Reportable Quantity (RQ): CERCLA, 100 lbs. (45.4 kgs.).

EPCRA Section 313 Form R de minimus concentration reporting level: 0.1%. Form R Toxic Chemical Category Code: N495.

EPA NAME: NICOTINE

CAS: 54-11-5

SYNONYMS: PYRIDINE, 3-(1-METHYL-2-PYRROLIDINYL)-, (S)-; 3-(1-METHYL-2-PYRROLIDYL) PYRIDINE

EPA HAZARDOUS WASTE NUMBER (RCRA No.): P075.

RCRA Section 261 Hazardous Constituents.

EPCRA Section 302 Extremely Hazardous Substances: TPQ = 100 lbs. (45.4 kgs.).

EPCRA Section 304 Reportable Quantity (RQ): EHS/CERCLA, 100 lbs. (45.4 kgs.).

EPCRA Section 313 Form R de minimus concentration reporting level: 1.0%. Form R Toxic Chemical Category Code: N503.

EPA NAME: NICOTINE AND SALTS
[see also NICOTINE]
RCRA Section 261 Hazardous Constituents.
EPCRA Section 313: Includes any unique chemical substance that contains nicotine or a nicotine salt as part of that chemical's infrastructure. Form R de minimus concentration reporting level: 1.0%. Form R Toxic Chemical Category Code: N503.

EPA NAME: NICOTINE SULFATE
CAS: 65-30-5
SYNONYMS: (S)-3-(1-METHYL-2-PYRROLIDINYL)PYRIDINE SULFATE (2:1)
EPCRA Section 302 Extremely Hazardous Substances: TPQ = 100/10,000 lbs. (45.4/4,540 kgs.).
EPCRA Section 304 Reportable Quantity (RQ): EHS, 1 lb. (0.454 kg.).
EPCRA Section 313 Form R de minimus concentration reporting level: 1.0% Form R Toxic Chemical Category Code: N503.

EPA NAME: NITRAPYRIN
CAS: 1929-82-4
SYNONYMS: 4-CHLORO-6-(TRICHLOROMETHYL)PYRIDINE; EPA PESTICIDE CHEMICAL CODE 069203; PYRIDINE, 2-CHLORO-6(TRICHLOROMETHYL)-
EPCRA Section 313 Form R de minimus concentration reporting level: 1.0%.

EPA NAME: NITRATE COMPOUNDS (water dissociable)
EPCRA Section 313: Reportable only when in aqueous solution. Form R de minimus concentration reporting level: 1.0%. Form R Toxic Chemical Category Code: N511.
MARINE POLLUTANT (49CFR, Subchapter 172.101, Appendix B), as nitrates, inorganic, n.o.s.
EPCRA Section 313 Form R de minimus concentration reporting level: 1.0% (water dissociable; reportable only when in aqueous solution). Form R Toxic Chemical Category Code: N511.

EPA NAME: NITRIC ACID
[see also NITRIC ACID (conc. 80% or greater)]
CAS: 7697-37-2
CLEAN WATER ACT: Section 311 Hazardous Substances/RQ (same as CERCLA); Section 313 Priority Chemicals.
EPCRA Section 302 Extremely Hazardous Substances: TPQ = 1,000 lbs. (454 kgs.).
EPCRA Section 304 Reportable Quantity (RQ): EHS/CERCLA, 1,000 lbs. (454 kgs.).
EPCRA Section 313 Form R de minimus concentration reporting level: 1.0%.

EPA NAME: NITRIC ACID (conc. 80% or greater)
[see also NITRIC ACID]
CAS: 7697-37-2

CLEAN AIR ACT: Accidental Release Prevention/Flammable substances (Section 112[r], Table 3), TQ = 15,000 lbs. (1362 kgs.).
EPCRA Section 302 Extremely Hazardous Substances: TPQ = 1,000 lbs. (454 kgs.).
EPCRA Section 304 Reportable Quantity (RQ): EHS/CERCLA, 1,000 lbs. (454 kgs.).
EPCRA Section 313 Form R de minimus concentration reporting level: 1.0%.

EPA NAME: NITRIC OXIDE
CAS: 10102-43-9
SYNONYMS: NITROGEN MONOXIDE

CLEAN AIR ACT: Accidental Release Prevention/Flammable substances (Section 112[r], Table 3), TQ = 10,000 lbs. (4540 kgs.).
EPA HAZARDOUS WASTE NUMBER (RCRA No.): P076.
RCRA Section 261 Hazardous Constituents.
EPCRA Section 302 Extremely Hazardous Substances: TPQ = 100 lbs. (45.4 kgs.).
EPCRA Section 304 Reportable Quantity (RQ): EHS/CERCLA, 10 lbs. (4.54 kgs.).

EPA NAME: NITRILOTRIACETIC ACID
CAS: 139-13-9
SYNONYMS: AMINOTRIETHANOIC ACID; GLYCINE, N,N-BIS(CARBOXYMETHYL)-; α,α′,α″-TRIMETHYLAMINETRI-CARBOXYLIC ACID

EPCRA Section 313 Form R de minimus concentration reporting level: 0.1%.
CALIFORNIA'S PROPOSITION 65: Carcinogen.

EPA NAME: o-NITROANILINE
CAS: 88-74-4
SYNONYMS: BENZENAMINE, 2-NITRO-; 2-NITROANILINE; ortho-NITROANILINE; o-NITROPHENYLAMINE

RCRA Universal Treatment Standards: Wastewater (mg/L), 0.27; Non-wastewater (mg/kg), 14.

EPA NAME: p-NITROANILINE
CAS: 100-01-6
SYNONYMS: BENZENAMINE, 4-NITRO-; 4-NITROANILINE; para-NITROANILINE; p-NITROPHENYLAMINE

EPA HAZARDOUS WASTE NUMBER (RCRA No.): P077.
RCRA Section 261 Hazardous Constituents.
RCRA Land Ban Waste.
RCRA Universal Treatment Standards: Wastewater (mg/L), 0.028; Non-wastewater (mg/kg), 28.
EPCRA Section 304 Reportable Quantity (RQ): CERCLA, 5,000 lbs. (2270 kgs.).

EPCRA Section 313 Form R de minimus concentration reporting level: 1.0%.

EPA NAME: 5-NITRO-o-ANISIDINE
CAS: 99-59-2
SYNONYMS: 2-AMINO-1-METHOXY-4-NITROBENZENE; BENZENAMINE, 2-METHOXY-5-NITRO-; 5-NITRO-2-METHOXYANILINE; 5-NITRO-ortho-ANISIDINE

EPCRA Section 313 Form R de minimus concentration reporting level: 1.0%.
CALIFORNIA'S PROPOSITION 65: Carcinogen.

EPA NAME: NITROBENZENE
CAS: 98-95-3
SYNONYMS: BENZENE, NITRO-

CLEAN AIR ACT: Hazardous Air Pollutants (Title I, Part A, Section 112).
CLEAN WATER ACT: Section 311 Hazardous Substances/RQ (same as CERCLA); Section 307 Toxic Pollutants; Section 313 Priority Chemicals.
EPA HAZARDOUS WASTE NUMBER (RCRA No.): U169.
RCRA Section 261 Hazardous Constituents.
RCRA Land Ban Waste.
RCRA Universal Treatment Standards: Wastewater (mg/L), 0.068; Nonwastewater (mg/kg), 14.
RCRA Ground Water Monitoring List. Suggested test method(s) (PQL μg/L): 8090(40); 8270(10).
SAFE DRINKING WATER ACT: Regulated chemical (47 FR 9352).
EPCRA Section 302 Extremely Hazardous Substances: TPQ = 10,000 lbs. (4,540 kgs.).
EPCRA Section 304 Reportable Quantity (RQ): EHS/CERCLA, 1,000 lbs. (454 kgs.).
EPCRA Section 313 Form R de minimus concentration reporting level: 1.0%.

EPA NAME: 4-NITROBIPHENYL
CAS: 92-93-3
SYNONYMS: 1,1′-BIPHENYL, 4-NITRO-; BIPHENYL, 4-NITRO-

CLEAN AIR ACT: Hazardous Air Pollutants (Title I, Part A, Section 112).
EPCRA Section 304 Reportable Quantity (RQ): CERCLA, 10 lbs. (4.54 kgs.).
EPCRA Section 313 Form R de minimus concentration reporting level: 0.1%.
CALIFORNIA'S PROPOSITION 65: Carcinogen.

EPA NAME: NITROCYCLOHEXANE
CAS: 1122-60-7
SYNONYMS: CYCLOHEXANE, NITRO-

EPCRA Section 302 Extremely Hazardous Substances: TPQ = 500 lbs. (227 kgs.).

EPCRA Section 304 Reportable Quantity (RQ): EHS, 500 lbs. (227 kgs.).

EPA NAME: NITROFEN

CAS: 1836-75-5

SYNONYMS: BENZENE,2,4-DICHLORO-1-(4-NITROPHENOXY)-; BENZENAMINE, 4-ETHOXY-N-(5-NITRO-2-FURANYL)METHYLENE-; 2,4-DICHLORO-4′-NITRODIPHENYL ETHER; 2,4-DICHLORO-1-(4-NITROPHENOXY) BENZENE; 4-(2,4-DICHLOROPHENOXY)NITROBENZENE; 2,4-DICHLOROPHENYL 4-NITROPHENYL ETHER; 2,4-DICHLOROPHENYL p-NITROPHENYL ETHER; ETHER,2,4-DICHLOROPHENYL p-NITROPHENYL; 4′-NITRO-2,4-DICHLORODIPHENYL ETHER; 4-NITRO-2′,4′-DICHLORODIPHENYL ETHER

EPCRA Section 313 Form R de minimus concentration reporting level: 0.1%.

MARINE POLLUTANT (49CFR, Subchapter 172.101, Appendix B).

CALIFORNIA'S PROPOSITION 65: Carcinogen as nitrofen (technical grade).

EPA NAME: NITROGEN DIOXIDE

CAS: 10102-44-0

SYNONYMS: NITROGEN OXIDE (NO_2)

CLEAN WATER ACT: Section 311 Hazardous Substances/RQ (same as CERCLA).

EPA HAZARDOUS WASTE NUMBER (RCRA No.): P078.

RCRA Section 261 Hazardous Constituents.

EPCRA Section 302 Extremely Hazardous Substances: TPQ = 100 lbs. (45.4 kgs.).

EPCRA Section 304 Reportable Quantity (RQ): EHS/CERCLA, 10 lbs. (4.54 kgs.).

EPA NAME: NITROGEN DIOXIDE

CAS: 10544-72-6

SYNONYMS: DINITROGEN DIOXIDE; NITROGEN OXIDE (N_2O_4); DINITROGEN TETROXIDE

EPCRA Section 304 Reportable Quantity (RQ): CERCLA, 10 lbs. (4.54 kgs.).

EPA NAME: NITROGEN MUSTARD

CAS: 51-75-2

SYNONYMS: 2-CHLORO-N-(2-CHLOROETHYL)-N-METHYLETHANAMINE; ETHANAMINE, 2-CHLORO-N-(2-CHLOROETHYL)-N-METHYL-; MECHLORETHAMINE; N-METHYLBIS(2-CHLOROETHYL)AMINE; METHYLBIS(2-CHLOROETHYL)AMINE; N,N-BIS(2-CHLOROETHYL)-METHYLAMINE

RCRA Section 261 Hazardous Constituents, waste number not listed.

EPCRA Section 302 Extremely Hazardous Substances: TPQ = 10 lbs. (4.54 kgs.).
EPCRA Section 304 Reportable Quantity (RQ): EHS, 10 lbs. (4.54 kgs.).
EPCRA Section 313 Form R de minimus concentration reporting level: 0.1%.
CALIFORNIA'S PROPOSITION 65: Carcinogen; reproductive toxin.

EPA NAME: NITROGEN OXIDE (NO)
[see NITRIC OXIDE]
CAS: 10102-43-9

EPA NAME: NITROGLYCERIN
CAS: 55-63-0
SYNONYMS: GLYCEROL TRINITRATE; 1,2,3-PROPANETROL, TRINITRATE

EPA HAZARDOUS WASTE NUMBER (RCRA No.): P081.
RCRA Section 261 Hazardous Constituents.
EPCRA Section 304 Reportable Quantity (RQ): CERCLA, 10 lbs. (4.54 kgs.).
EPCRA Section 313 Form R de minimus concentration reporting level: 1.0%.

EPA NAME: NITROPHENOL (MIXED ISOMERS)
CAS: 25154-55-6
SYNONYMS: NITROPHENOLS

CLEAN WATER ACT: Section 311 Hazardous Substances/RQ (same as CERCLA); Section 307 Toxic Pollutants as nitrophenols.
EPCRA Section 304 Reportable Quantity (RQ): CERCLA, 100 lbs. (45.4 kgs.).

EPA NAME: m-NITROPHENOL
CAS: 554-84-7
SYNONYMS: 3-NITROPHENOL; PHENOL, 3-NITRO-

CLEAN WATER ACT: Section 307 Toxic Pollutants as nitrophenols.
EPCRA Section 304 Reportable Quantity (RQ): CERCLA, 100 lbs. (45.4 kgs.).

EPA NAME: p-NITROPHENOL
CAS: 100-02-7
SYNONYMS: DEGRADATION PRODUCT OF PARATHION; 4-HYDROXYNITROBENZENE; 4-NITROPHENOL; PHENOL, 4-NITRO

CLEAN AIR ACT: Hazardous Air Pollutants (Title I, Part A, Section 112).
CLEAN WATER ACT: Section 307 Toxic Pollutants as nitrophenols; Section 313 Priority Chemicals.
EPA HAZARDOUS WASTE NUMBER (RCRA No.): U170.
RCRA Section 261 Hazardous Constituents.
RCRA Land Ban Waste.
RCRA Universal Treatment Standards: Wastewater (mg/L), 0.12; Nonwastewater (mg/kg), 29.

RCRA Ground Water Monitoring List. Suggested test method(s) (PQL µg/L): 8040(10); 8270(50).
EPCRA Section 304 Reportable Quantity (RQ): CERCLA, 100 lbs. (45.4 kgs.).
EPCRA Section 313 Form R de minimus concentration reporting level: 1.0%.

EPA NAME: 2-NITROPHENOL
CAS: 88-75-5
SYNONYMS: o-NITROPHENOL; PHENOL, 2-NITRO-; PHENOL, o-NITRO-

CLEAN WATER ACT: Section 307 Toxic Pollutants as nitrophenols; Section 313 Priority Chemicals.
RCRA Land Ban Waste.
RCRA Universal Treatment Standards: Wastewater (mg/L), 0.028; Nonwastewater (mg/kg), 13.
RCRA Ground Water Monitoring List. Suggested test method(s) (PQL µg/L): 8040(5); 8270(10).
EPCRA Section 304 Reportable Quantity (RQ): CERCLA, 100 lbs. (45.4 kgs.).
EPCRA Section 313 Form R de minimus concentration reporting level: 1.0%.

EPA NAME: 3-NITROPHENOL
[see m-NITROPHENOL]
CAS: 554-84-7

EPA NAME: 4-NITROPHENOL
[see p-NITROPHENOL]
CAS: 100-02-7

EPA NAME: NITROPHENOLS
CLEAN WATER ACT: Section 307 Toxic Pollutants as nitrophenols.
EPCRA Section 304 Reportable Quantity (RQ): CERCLA, 100 lbs. (45.4 kgs.).

EPA NAME: 2-NITROPROPANE
CAS: 79-46-9
SYNONYMS: β-NITROPROPANE; PROPANE, 2-NITRO

CLEAN AIR ACT: Hazardous Air Pollutants (Title I, Part A, Section 112).
EPA HAZARDOUS WASTE NUMBER (RCRA No.): U171.
RCRA Section 261 Hazardous Constituents.
EPCRA Section 304 Reportable Quantity (RQ): CERCLA, 10 lbs. (4.54 kgs.).
EPCRA Section 313 Form R de minimus concentration reporting level: 0.1%.
CALIFORNIA'S PROPOSITION 65: Carcinogen.

EPA NAME: 1-NITROPYRENE
[see also POLYCYCLIC AROMATIC COMPOUNDS]
CAS: 5522-43-0

SYNONYMS: PYRENE, 1-NITRO-

EPCRA Section 313 Form R de minimus concentration reporting level: 1.0%.

CALIFORNIA'S PROPOSITION 65: Carcinogen.

EPA NAME: NITROSAMINES

CAS: 35576-91-1

CLEAN WATER ACT: Section 307 Toxic Pollutants as nitrosamines.

RCRA Section 261 Hazardous Constituents, waste number not listed.

EPA NAME: N-NITROSODI-n-BUTYLAMINE

CAS: 924-16-3

SYNONYMS: 1-BUTANAMINE, N-BUTYL-N-NITROSO-; N,N-DI-n-BUTYLNITROSAMINE; N-NITROSODIBUTYLAMINE

EPA HAZARDOUS WASTE NUMBER (RCRA No.): U172.

RCRA Section 261 Hazardous Constituents.

RCRA Land Ban Waste.

RCRA Universal Treatment Standards: Wastewater (mg/L), 0.40; Non-wastewater (mg/kg), 17.

RCRA Ground Water Monitoring List. Suggested test method(s) (PQL μg/L): 8270(10).

EPCRA Section 304 Reportable Quantity (RQ): CERCLA, 10 lbs. (4.54 kgs.).

EPCRA Section 313 Form R de minimus concentration reporting level: 0.1%.

CALIFORNIA'S PROPOSITION 65: Carcinogen.

EPA NAME: N-NITROSODIETHANOLAMINE

CAS: 1116-54-7

SYNONYMS: 2,2′-DIHYDROXY-N-NITROSODIETHYLAMINE; ETHANOL, 2,2′-(NITROSOIMINO)BIS-; N-NITROSOAMINO-DIETHANOL

EPA HAZARDOUS WASTE NUMBER (RCRA No.): U173.

RCRA Section 261 Hazardous Constituents.

EPCRA Section 304 Reportable Quantity (RQ): CERCLA, 1 lb. (0.454 kg.).

CALIFORNIA'S PROPOSITION 65: Carcinogen.

EPA NAME: N-NITROSODIETHYLAMINE

CAS: 55-18-5

SYNONYMS: DIETHYLNITROSAMIDE; N,N-DIETHYLNITRO-SOAMINE

EPA HAZARDOUS WASTE NUMBER (RCRA No.): U174.

RCRA Section 261 Hazardous Constituents.

RCRA Land Ban Waste.

RCRA Universal Treatment Standards: Wastewater (mg/L), 0.40; Non-wastewater (mg/kg), 28.

RCRA Ground Water Monitoring List. Suggested test method(s) (PQL μg/L): 8270(10).

EPCRA Section 304 Reportable Quantity (RQ): CERCLA, 1 lb. (0.454 kg.).
EPCRA Section 313 Form R de minimus concentration reporting level: 0.1%.
CALIFORNIA'S PROPOSITION 65: Carcinogen.

EPA NAME: N-NITROSODIMETHYLAMINE
CAS: 62-75-9
SYNONYMS: DIMETHYLNITROSAMINE; METHANAMINE, N-METHYL-N-NITROSO-; N-METHYL-N-NITROSOMETHANAMINE; NITROSODIMETHYLAMINE

CLEAN AIR ACT: Hazardous Air Pollutants (Title I, Part A, Section 112).
CLEAN WATER ACT: Section 313 Priority Chemicals.
EPA HAZARDOUS WASTE NUMBER (RCRA No.): P082.
RCRA Section 261 Hazardous Constituents.
RCRA Land Ban Waste.
RCRA Universal Treatment Standards: Wastewater (mg/L), 0.40; Nonwastewater (mg/kg), 2.3.
RCRA Ground Water Monitoring List. Suggested test method(s) (PQL µg/L): 8270(10).
EPCRA Section 302 Extremely Hazardous Substances: TPQ = 1,000 lbs. (454 kgs.).
EPCRA Section 304 Reportable Quantity (RQ): EHS/CERCLA, 10 lbs. (4.54 kgs.).
EPCRA Section 313 Form R de minimus concentration reporting level: 0.1%.
CALIFORNIA'S PROPOSITION 65: Carcinogen.

EPA NAME: NITROSODIMETHYLAMINE
[see N-NITROSODIMETHYLAMINE]
CAS: 62-75-9

EPA NAME: N-NITROSODIPHENYLAMINE
CAS: 86-30-6
SYNONYMS: BENZENAMINE, N-NITROSO-N-PHENYL-; DIPHENYLAMINE, N-NITROSO-; DIPHENYLNITROSAMINE

CLEAN WATER ACT: Section 313 Priority Chemicals.
RCRA Land Ban Waste.
RCRA Universal Treatment Standards: Wastewater (mg/L), 0.92; Nonwastewater (mg/kg), 13.
RCRA Ground Water Monitoring List. Suggested test method(s) (PQL µg/L): 8270(10).
EPCRA Section 304 Reportable Quantity (RQ): CERCLA, 100 lbs. (45.4 kgs.).
EPCRA Section 313 Form R de minimus concentration reporting level: 1.0%.
CALIFORNIA'S PROPOSITION 65: Carcinogen.

EPA NAME: p-NITROSODIPHENYLAMINE
CAS: 156-10-5

SYNONYMS: BENZENAMINE, 4-NITROSO-N-PHENYL-; 4-NITROSODIPHENYLAMINE; para-NITROSOPHENYLANILINE

EPCRA Section 313 Form R de minimus concentration reporting level: 1.0%.

CALIFORNIA'S PROPOSITION 65: Carcinogen.

EPA NAME: N-NITROSODI-n-PROPYLAMINE

CAS: 621-64-7

SYNONYMS: DIPROPYLAMINE, N-NITROSO-; DI-n-PROPYLNITROSAMINE; DIPROPYLNITROSAMINE; NITROSODIPROPYLAMINE; N-NITROSODIPROPYLAMINE; 1-PROPANAMINE, N-NITROSO-n-PROPYL-

CLEAN WATER ACT: Section 313 Priority Chemicals.
EPA HAZARDOUS WASTE NUMBER (RCRA No.): U111.
RCRA Section 261 Hazardous Constituents.
RCRA Land Ban Waste.
RCRA Universal Treatment Standards: Wastewater (mg/L), 0.40; Nonwastewater (mg/kg), 14.
RCRA Ground Water Monitoring List. Suggested test method(s) (PQL µg/L): 8270(10).
EPCRA Section 304 Reportable Quantity (RQ): CERCLA, 10 lbs. (4.54 kgs.).
EPCRA Section 313 Form R de minimus concentration reporting level: 0.1%.
CALIFORNIA'S PROPOSITION 65: Carcinogen.

EPA NAME: N-NITROSO-N-ETHYLUREA

CAS: 759-73-9

SYNONYMS: 1-ETHYL-1-NITROSOUREA; N-ETHYL-N-NITROSOUREA; UREA, N-ETHYL-N-NITROSO-

EPA HAZARDOUS WASTE NUMBER (RCRA No.): U176.
RCRA Section 261 Hazardous Constituents.
EPCRA Section 304 Reportable Quantity (RQ): CERCLA, 1 lb. (0.454 kg.).
EPCRA Section 313 Form R de minimus concentration reporting level: 0.1%.
CALIFORNIA'S PROPOSITION 65: Carcinogen.

EPA NAME: N-NITROSO-N-METHYLUREA

CAS: 684-93-5

SYNONYMS: N-METHYL-N-NITROSOUREA; 1-NITROSO-1-METHYLUREA; N-NITROSO-N-METHYLUREA CARBAMIDE; UREA, N-METHYL-N-NITROSO-; UREA, 1-METHYL-1-NITROSO-

CLEAN AIR ACT: Hazardous Air Pollutants (Title I, Part A, Section 112).
EPA HAZARDOUS WASTE NUMBER (RCRA No.): U177.
RCRA Section 261 Hazardous Constituents.
EPCRA Section 304 Reportable Quantity (RQ): CERCLA, 1 lb. (0.454 kg.).

EPCRA Section 313 Form R de minimus concentration reporting level: 0.1%.

EPA NAME: N-NITROSO-N-METHYLURETHANE

CAS: 615-53-2

SYNONYMS: CARBAMIC ACID, METHYLNITROSO-, ETHYL ESTER; METHYLNITROSOURETHANE; N-METHYL-N-NITROSOURETHANE; NITROSOMETHYLURETHANE; N-NITROSO-N-METHYLURETHANE

EPA HAZARDOUS WASTE NUMBER (RCRA No.): U178.

RCRA Section 261 Hazardous Constituents.

EPCRA Section 304 Reportable Quantity (RQ): CERCLA, 1 lb. (0.454 kg.).

CALIFORNIA'S PROPOSITION 65: Carcinogen.

EPA NAME: N-NITROSOMETHYLVINYLAMINE

CAS: 4549-40-0

SYNONYMS: ETHENAMINE, N-METHYL-N-NITROSO-; ETHYLENE, N-METHYL-N-NITROSO-

EPA HAZARDOUS WASTE NUMBER (RCRA No.): P084.

RCRA Section 261 Hazardous Constituents.

EPCRA Section 304 Reportable Quantity (RQ): CERCLA, 10 lbs. (4.54 kgs.).

EPCRA Section 313 Form R de minimus concentration reporting level: 0.1%.

CALIFORNIA'S PROPOSITION 65: Carcinogen.

EPA NAME: N-NITROSOMORPHOLINE

CAS: 59-89-2

SYNONYMS: MORPHOLINE, 4-NITROSO-; 4-NITROSOMORPHOLINE

CLEAN AIR ACT: Hazardous Air Pollutants (Title I, Part A, Section 112).

RCRA Land Ban Waste.

RCRA Universal Treatment Standards: Wastewater (mg/L), 0.40; Nonwastewater (mg/kg), 2.3.

RCRA Ground Water Monitoring List. Suggested test method(s) (PQL μg/L): 8270(10).

EPCRA Section 304 Reportable Quantity (RQ): CERCLA, 1 lb. (0.454 kg.).

EPCRA Section 313 Form R de minimus concentration reporting level: 0.1%.

CALIFORNIA'S PROPOSITION 65: Carcinogen.

EPA NAME: N-NITROSONORNICOTINE

CAS: 16543-55-8

SYNONYMS: NICOTINE, 1′-DEMETHYL-1′-NITROSO-; N′-NITROSONORNICOTINE; PYRIDINE, 3-(1-NITROSO-2-PYRROLIDINYL)-, (S)-

EPCRA Section 313 Form R de minimus concentration reporting level: 0.1%.
CALIFORNIA'S PROPOSITION 65: Carcinogen.

EPA NAME: n-NITROSOPIPERIDINE
CAS: 100-75-4
SYNONYMS: 1-NITROSOPIPERIDINE; PIPERIDINE, 1-NITROSO

EPA HAZARDOUS WASTE NUMBER (RCRA No.): U179.
RCRA Section 261 Hazardous Constituents.
RCRA Land Ban Waste.
RCRA Universal Treatment Standards: Wastewater (mg/L), 0.013; Non-wastewater (mg/kg), 35.
RCRA Ground Water Monitoring List. Suggested test method(s) (PQL µg/L): 8270(10).
EPCRA Section 304 Reportable Quantity (RQ): CERCLA, 10 lbs. (4.54 kgs.).
EPCRA Section 313 Form R de minimus concentration reporting level: 0.1%.
CALIFORNIA'S PROPOSITION 65: Carcinogen.

EPA NAME: N-NITROSOPYRROLIDINE
CAS: 930-55-2
SYNONYMS: 1-NITROSOPYRROLIDINE; PYRROLIDINE, 1-NITROSO-

EPA HAZARDOUS WASTE NUMBER (RCRA No.): U180.
RCRA Section 261 Hazardous Constituents.
RCRA Land Ban Waste.
RCRA Universal Treatment Standards: Wastewater (mg/L), 0.013; Non-wastewater (mg/kg), 35.
RCRA Ground Water Monitoring List. Suggested test method(s) (PQL µg/L): 8270(10).
EPCRA Section 304 Reportable Quantity (RQ): CERCLA, 1 lb. (0.454 kg.).
CALIFORNIA'S PROPOSITION 65: Carcinogen.

EPA NAME: NITROTOLUENE
CAS: 1321-12-6
SYNONYMS: BENZENE, METHYLNITRO-; NITROTOLUENE (all isomers); NITROTOLUENE (mixed isomers)

CLEAN WATER ACT: Section 311 Hazardous Substances/RQ (same as CERCLA).
EPCRA Section 304 Reportable Quantity (RQ): CERCLA, 1,000 lbs. (454 kgs.).

EPA NAME: m-NITROTOLUENE
CAS: 99-08-1
SYNONYMS: BENZENE, 1-METHYL-3-NITRO-; m-METHYLNITROBENZENE; 3-NITROTOLUENE; METANITROTOLUENE

EPCRA Section 304 Reportable Quantity (RQ): CERCLA, 1,000 lbs. (454 kgs.).

EPA NAME: o-NITROTOLUENE

CAS: 88-72-2

SYNONYMS: BENZENE, 1-METHYL-2-NITRO-; o-METHYLNITROBENZENE; 2-NITROTOLUENE; ORTHONITROTOLUENE

EPCRA Section 304 Reportable Quantity (RQ): CERCLA, 1,000 lbs. (454 kgs.).

EPA NAME: p-NITROTOLUENE

CAS: 99-99-0

SYNONYMS: BENZENE, 1-METHYL-4-NITRO-; 4-METHYLNITROBENZENE; p-METHYLNITROBENZENE; 4-NITROTOLUENE; PARANITROTOLUENE

EPCRA Section 304 Reportable Quantity (RQ): CERCLA, 1,000 lbs. (454 kgs.).

EPA NAME: 5-NITRO-o-TOLUENE

CAS: 99-55-8

SYNONYMS: BENZENAMINE, 2-METHYL-5-NITRO-; 4-NITRO-2-AMINOTOLUENE; 3-NITRO-6-METHYLANILINE; 5-NITRO-2-METHYLANILINE; 5-NITRO-2-TOLUIDINE; 5-NITRO-ortho-TOLUIDINE

EPA HAZARDOUS WASTE NUMBER (RCRA No.): U181.

RCRA Section 261 Hazardous Constituents.

RCRA Universal Treatment Standards: Wastewater (mg/L), 0.32; Nonwastewater (mg/kg), 28.

RCRA Ground Water Monitoring List. Suggested test method(s) (PQL µg/L): 8270(10).

EPCRA Section 304 Reportable Quantity (RQ): CERCLA, 100 lbs. (45.4 kgs.).

EPCRA Section 313 Form R de minimus concentration reporting level: 1.0%.

EPA NAME: NITROUS ACID ETHYL ESTER

[see ETHYL NITRITE]

CAS: 109-95-5

EPA NAME: NORBORMIDE

CAS: 991-42-4

SYNONYMS: 5(α-HYDROXY-α-2-PYRIDYLBENZYL)-7-(α-2-PYRIDYLBENZYLIDENE)-5-NORBORENE-2,3-DICARBOXIDE

EPCRA Section 302 Extremely Hazardous Substances: TPQ = 100/10,000 lbs. (45.4/4,540 kgs.).

EPCRA Section 304 Reportable Quantity (RQ): EHS, 100 lbs. (45.4 kgs.).

EPA NAME: NORFLURAZON

CAS: 27314-13-2

SYNONYMS: 4-CHLORO-5-(METHYLAMINO)-2-[3-(TRIFLUORO-METHYL)PHENYL]-3(2H)-PYRIDAZINONE; 4-CHLORO-5-(METHYLAMINO)-2-(α, α, α-TRIFLUORO-m-TOLYL)-3(2H)-PYRIDAZINONE; EPA PESTICIDE CHEMICAL CODE 105801

EPCRA Section 313 Form R de minimus concentration reporting level: 1.0%.

- O -

EPA NAME: OCTACHLORONAPHTHALENE
[see also CHLORINATED NAPHTHALENES]
CAS: 2234-13-1
SYNONYMS: NAPHTHALENE, OCTACHLORO-; 1,2,3,4,5,6,7,8-OCTACHLORONAPHTHALENE

EPCRA Section 313 Form R de minimus concentration reporting level: 1.0%.

EPA NAME: OCTANOIC ACID,2,6-DIBROMO-4-CYANOPHENYL ESTER
[see BROMOXYNIL OCTANOATE]
CAS: 1689-99-2

EPA NAME: OLEUM (FUMING SULFURIC ACID)
CAS: 8014-95-7
SYNONYMS: DISULPHURIC ACID; PYROSULFURIC ACID; SULFURIC ACID (FUMING); SULFURIC ACID, MIXTURE WITH SULFUR TRIOXIDE

CLEAN AIR ACT: Accidental Release Prevention/Flammable substances (Section 112[r], Table 3), TQ = 10,000 lbs. (4540 kgs.).
EPCRA Section 304 Reportable Quantity (RQ): CERCLA, 1,000 lbs. (454 kgs.).

EPA NAME: ORGANORHODIUM COMPLEX (PMN-82-147)
CAS: N/A
EPCRA Section 302 Extremely Hazardous Substances: TPQ = 10/10,000 lbs. (4.54/4,540 kgs.).
EPCRA Section 304 Reportable Quantity (RQ): EHS, 10 lbs. (4.54 kgs.).

EPA NAME: ORYZALIN
CAS: 19044-88-3
SYNONYMS: BENZENESULFONAMIDE, 4-(DIPROPYLAMINO)-3,5-DINITRO-; 4-(DIPROPYLAMINO)-3,5-DINITROBENZENESULFONAMIDE); EPA PESTICIDE CHEMICAL CODE 104201

EPCRA Section 313 Form R de minimus concentration reporting level: 1.0%.

EPA NAME: OSMIUM OXIDE (OsO_4) (T-4)-
[see OSMIUM TETROXIDE]
CAS: 20816-12-0

EPA NAME: OSMIUM TETROXIDE
[OSMIUM OXIDE (OsO_4) (T-4)-]
CAS: 20816-12-0
SYNONYMS: MILAS' REAGENT; NAMED REAGENTS AND SOLUTIONS, MILAS'; OSMIUM(IV) OXIDE; OSMIUM OXIDE (OsO_4); OSMIUM OXIDE (OsO_4), (T-4)-; RTECS NO. RN1140000

EPA HAZARDOUS WASTE NUMBER (RCRA No.): P087.
RCRA Section 261 Hazardous Constituents.
EPCRA Section 304 Reportable Quantity (RQ): CERCLA, 1,000 lbs. (454 kgs.).
EPCRA Section 313 Form R de minimus concentration reporting level: 1.0%.

EPA NAME: OUABAIN
CAS: 630-60-4
SYNONYMS: OUABAINE; PUROSTROPHAN

EPCRA Section 302 Extremely Hazardous Substances: TPQ = 100/10,000 lbs. (45.4/4,540 kgs.).
EPCRA Section 304 Reportable Quantity (RQ): EHS, 100 lbs. (45.4 kgs.).

EPA NAME: 7-OXABICYCLO(2.2.1)HEPTANE-2,3-DICARBOXYLIC ACID, DIPOTASSIUM SALT
[see DIPOTASSIUM ENDOTHALL]
CAS: 2164-07-0

EPA NAME: OXAMYL
CAS: 23135-22-0
SYNONYMS: 2-(DIMETHYLAMINO)-N(((METHYLAMINO)CARBONYL)OXY)2-OXOETHANIMIDOTHIOIC ACID METHYL ESTER; VYDATE; VYDATE OXAMYL INSECTICIDE/NEMATOCIDE

EPA HAZARDOUS WASTE NUMBER (RCRA No.): P194.
RCRA Section 261 Hazardous Constituents.
RCRA Land Ban Waste.
RCRA Universal Treatment Standards: Wastewater (mg/L), 0.056; Nonwastewater (mg/kg), 0.28.
EPCRA Section 302 Extremely Hazardous Substances: TPQ = 100/10,000 lbs. (45.4/4,540 kgs.).
EPCRA Section 304 Reportable Quantity (RQ): EHS, 500 lbs. (227 kgs.).

EPA NAME: OXETANE, 3,3-BIS(CHLOROMETHYL)-
CAS: 78-71-7
SYNONYMS: 3,3-BIS(CHLOROMETHYL)OXETANE; 3,3-DICHLOROMETHYLOXYCYCLOBUTANE

EPCRA Section 302 Extremely Hazardous Substances: TPQ = 500 lbs. (227 kgs.).
EPCRA Section 304 Reportable Quantity (RQ): EHS, 1 lb. (0.454 kg.).

EPA NAME: OXIRANE
[see ETHYLENE OXIDE]
CAS: 75-21-8

EPA NAME: OXIRANE, (CHLOROMETHYL)-
[see EPICHLOROHYDRIN]
CAS: 106-89-8

EPA NAME: OXIRANE, METHYL-
[see PROPYLENE OXIDE]
CAS: 75-56-9

EPA NAME: OXYDEMETON METHYL
CAS: 301-12-2
SYNONYMS: DEMETON-S METHYL SULFOXIDE; S-(2-(ETHYLSULFINYL)ETHYL) O,O-DIMETHYL ESTER PHOSPHOROTHIOIC ACID

EPCRA Section 313 Form R de minimus concentration reporting level: 1.0%.

EPA NAME: OXYDIAZON
CAS: 19666-30-9
SYNONYMS: 3-[2,4-DICHLORO-5-(1-METHYLETHOXY)PHENYL]-5-(1,1-DIMETHYLETHYL)-1,3,4-OXADI AZOL-2(3H)-ONE; 3-(2,4-DICHLORO-5-ISOPROPYLOXY-PHENYL)-DELTA(SUP4)-5-(TERT-BUTYL)-1,3,4-OXADIA ZOLINE-2-ONE; EPA PESTICIDE CHEMICAL CODE 109001; OXADIAZON; 1,3,4-OXAZOL-2(3 H)-ONE, 3-(2,4-DICHLORO-5-(1-METHYLETHOXY)PHENYL)-5-(1,1-DIMETHYLETHYL)-

EPCRA Section 313 Form R de minimus concentration reporting level: 1.0%.

CALIFORNIA'S PROPOSITION 65: Carcinogen.

EPA NAME: OXYDISULFOTON
CAS: 2497-07-6
SYNONYMS: BAY 23323; O,O-DIEYHYL-S-((ETHYLSULFINYL)ETHYL)PHOSPHORODITHIOATE

EPCRA Section 302 Extremely Hazardous Substances: TPQ = 500 lbs. (227 kgs.).
EPCRA Section 304 Reportable Quantity (RQ): EHS, 500 lbs. (227 kgs.).

EPA NAME: OXYFLUORFEN
CAS: 42874-03-3
SYNONYMS: BENZENE, 2-CHLORO-1-(3-ETHOXY-4-NITROPHENOXY)-4-(TRIFLUOROMETHYL)-; EPA PESTICIDE CHEMICAL CODE 111601

EPCRA Section 313 Form R de minimus concentration reporting level: 1.0%.

EPA NAME: OZONE
CAS: 10028-15-6
SYNONYMS: OXYGEN mol (O_3); TRIATOMIC OXYGEN

EPCRA Section 302 Extremely Hazardous Substances: TPQ = 100 lbs. (45.4 kgs.).
EPCRA Section 304 Reportable Quantity (RQ): EHS, 100 lbs. (45.4 kgs.).
EPCRA Section 313 Form R de minimus concentration reporting level: 1.0%.

- P -

EPA NAME: PARAFORMALDEHYDE
CAS: 30525-89-4
SYNONYMS: TRIOXYMETHYLENE

CLEAN WATER ACT: Section 311 Hazardous Substances/RQ (same as CERCLA).
EPCRA Section 304 Reportable Quantity (RQ): CERCLA, 1,000 lbs.(454 kgs.).

EPA NAME: PARALDEHYDE
CAS: 123-63-7
SYNONYMS: 2,4,6-TRIMETHYL-1,3,5-TRIOXANE; 1,3,5-TRIOXANE,2,4,6-TRIMETHYL

EPA HAZARDOUS WASTE NUMBER (RCRA No.): U182.
RCRA Section 261 Hazardous Constituents.
EPCRA Section 304 Reportable Quantity (RQ): CERCLA, 1,000 lbs. (454 kgs.).
EPCRA Section 313 Form R de minimus concentration reporting level: 1.0%.

EPA NAME: PARAQUAT DICHLORIDE
CAS: 1910-42-5
SYNONYMS: BIPYRIDINIUM, 1,1′-DIMETHYL-4,4′-, DICHLORIDE; 4,4′-BIPYRIDINIUM, 1,1′-DIMETHYL-, DICHLORIDE; EPA PESTICIDE CHEMICAL CODE 061601; GRAMOXONE

EPCRA Section 302 Extremely Hazardous Substances: TPQ = 100/10,000 lbs. (45.4/4,540 kgs.).
EPCRA Section 304 Reportable Quantity (RQ): EHS, 10 lbs. (45.4 kgs.).
EPCRA Section 313 Form R de minimus concentration reporting level: 1.0%.

EPA NAME: PARAQUAT METHOSULFATE
CAS: 2074-50-2
SYNONYMS: 4,4′-BIPYRIDYNIUM, 1,1-DIMETHYLSULFATE, BIS (METHYL SULFATE); PARAQUAT BIS(METHYL SULFATE); PARAQUAT DIMETHYL SULFATE

EPCRA Section 302 Extremely Hazardous Substances: TPQ = 100/10,000 lbs. (45.4/4,540 kgs.).
EPCRA Section 304 Reportable Quantity (RQ): EHS, 10 lbs. (4.54 kgs.).

EPA NAME: PARATHION
CAS: 56-38-2
SYNONYMS: O,O-DIETHYL O-(p-NITROPHENYL) PHOSPHOROTHIOATE; PHOSPHOROTHIOIC ACID, O,O-DIETHYL-O-(4-NITROPHENYL) ESTER

CLEAN WATER ACT: Section 311 Hazardous Substances/RQ (same as CERCLA); Section 313 Priority Chemicals.

EPA HAZARDOUS WASTE NUMBER (RCRA No.): P089.

RCRA Section 261 Hazardous Constituents.

RCRA Land Ban Waste.

RCRA Universal Treatment Standards: Wastewater (mg/L), 0.014; Nonwastewater (mg/kg), 4.6.

RCRA Ground Water Monitoring List. Suggested test method(s) (PQL μg/L): 8270(10).

SAFE DRINKING WATER ACT: Priority List (55 FR 1470) as parathion degradation.

EPCRA Section 302 Extremely Hazardous Substances: TPQ = 100 lbs. (45.4 kgs.).

EPCRA Section 304 Reportable Quantity (RQ): EHS/CERCLA, 10 lbs. (4.54 kgs.).

EPCRA Section 313 Form R de minimus concentration reporting level: 1.0%.

MARINE POLLUTANT (49CFR, Subchapter 172.101, Appendix B): Severe pollutant.

EPA NAME: PARATHION-METHYL

[see METHYL PARATHION]

CAS: 298-00-0

EPA NAME: PARIS GREEN

[see CUPRIC ACETOARSENITE]

CAS: 12002-03-8

EPA NAME: PCBs

[see POLYCHLORINATED BIPHENYLS]

CAS: 1336-36-3

EPA NAME: PCNB

[see QUINTOZENE]

CAS: 82-68-8

EPA NAME: PCP

[see PENTACHLOROPHENOL]

CAS: 87-86-5

EPA NAME: PEBULATE

CAS: 1114-71-2

SYNONYMS: BUTYLETHYLCARBAMOTHIOIC ACID S-PROPYL ESTER; CARBAMOTHIOIC ACID, BUTYLETHYL-, S-PROPYL ESTER; PROPYL-ETHYLBUTYLTHIOCARBAMATE

EPA HAZARDOUS WASTE NUMBER (RCRA No.): U391.

RCRA Section 261 Hazardous Constituents.

RCRA Land Ban Waste.

RCRA Universal Treatment Standards: Wastewater (mg/L), 0.003; Nonwastewater (mg/kg), 1.4.

EPCRA Section 304 Reportable Quantity (RQ): CERCLA, 1 lb. (0.454 kg.).
EPCRA Section 313 Form R de minimus concentration reporting level: 1.0%.

EPA NAME: PENDIMETHALIN N-(1-ETHYLPROPYL)-3,4-DIMETHYL-2,6-DINITROBENZENAMINE
CAS: 40487-42-1
SYNONYMS: ANILINE, 3,4-DIMETHYL-2,6-DINITRO-N-(1-ETHYLPROPYL)-; BENZENAMINE, 3,4-DIMETHYL-2,6-DINITRO-N-(1-ETHYLPROPYL)-; EPA PESTICIDE CHEMICAL CODE 108501

EPCRA Section 313 Form R de minimus concentration reporting level: 1.0%.

EPA NAME: PENTABORANE
CAS: 19624-22-7
SYNONYMS: DIHYDROPENTABORANE(9); PENTABORANE(9)

EPCRA Section 302 Extremely Hazardous Substances: TPQ = 500 lbs. (227 kgs.).
EPCRA Section 304 Reportable Quantity (RQ): EHS, 500 lbs. (227 kgs.).

EPA NAME: PENTACHLOROBENZENE
[see also CHLORINATED BENZENES]
CAS: 608-93-5
SYNONYMS: BENZENE, PENTACHLORO-

EPA HAZARDOUS WASTE NUMBER (RCRA No.): U183.
RCRA Section 261 Hazardous Constituents.
RCRA Land Ban Waste.
RCRA Universal Treatment Standards: Wastewater (mg/L), 0.055; Nonwastewater (mg/kg), 10.
RCRA Ground Water Monitoring List. Suggested test method(s) (PQL µg/L): 8270(10).
EPCRA Section 304 Reportable Quantity (RQ): CERCLA, 10 lbs. (4.54 kgs.).

EPA NAME: PENTACHLOROETHANE
[see also CHLORINATED ETHANES]
CAS: 76-01-7
SYNONYMS: ETHANE PENTACHLORIDE; ETHANE, PENTACHLORO-

EPA HAZARDOUS WASTE NUMBER (RCRA No.): U184.
RCRA Section 261 Hazardous Constituents.
RCRA Land Ban Waste.
RCRA Universal Treatment Standards: Wastewater (mg/L), 0.055; Nonwastewater (mg/kg), 6.0.
RCRA Ground Water Monitoring List. Suggested test method(s) (PQL µg/L): 8240(5); 8270(10).

EPCRA Section 304 Reportable Quantity (RQ): CERCLA, 10 lbs. (4.54 kgs.).
EPCRA Section 313 Form R de minimus concentration reporting level: 1.0%.
MARINE POLLUTANT (49CFR, Subchapter 172.101, Appendix B).

EPA NAME: PENTACHLORONITROBENZENE
[see QUINTOZENE]
CAS: 82-68-8

EPA NAME: PENTACHLOROPHENOL
CAS: 87-86-5
SYNONYMS: PCP; PHENOL, PENTACHLORO-

CLEAN AIR ACT: Hazardous Air Pollutants (Title I, Part A, Section 112).
CLEAN WATER ACT: Section 311 Hazardous Substances/RQ (same as CERCLA); Section 307 Toxic Pollutants; Section 307 Priority Pollutants; Section 313 Priority Chemicals.
EPA HAZARDOUS WASTE NUMBER (RCRA No.): D037; F027.
RCRA Section 261 Hazardous Constituents.
RCRA Toxicity Characteristic (Section 261.24), Maximum Concentration of Contaminants, regulatory level, 100 mg/L.
RCRA Land Ban Waste.
RCRA Universal Treatment Standards: Wastewater (mg/L), 0.089; Nonwastewater (mg/kg), 7.4.
RCRA Ground Water Monitoring List. Suggested test method(s) (PQL µg/L): 8040(5); 8270(50)
SAFE DRINKING WATER ACT: MCL, 0.001 mg/L; MCLG, zero; Regulated chemical (47 FR 9352).
EPCRA Section 304 Reportable Quantity (RQ): CERCLA, 10 lbs. (4.54 kgs.).
EPCRA Section 313 Form R de minimus concentration reporting level: 1.0%. Note: Threshold determinations should be made individually and separately from the chlorophenol category.
MARINE POLLUTANT (49CFR, Subchapter 172.101, Appendix B): Severe pollutant.
CALIFORNIA'S PROPOSITION 65: Carcinogen.

EPA NAME: PENTADECYLAMINE
CAS: 2570-26-5
SYNONYMS: 1-PENTADECANAMINE; n-PENTADECYLAMINE; 1-PENTADECYLAMINE

EPCRA Section 302 Extremely Hazardous Substances: TPQ = 100/10,000 lbs. (45.4/4,540 kgs.).
EPCRA Section 304 Reportable Quantity (RQ): EHS, 100 lbs. (45.4 kgs.).

EPA NAME: 1,3-PENTADIENE
CAS: 504-60-9
SYNONYMS: trans-1,3-PENTADIENE; PENTADIENE, 1,3-

CLEAN AIR ACT: Accidental Release Prevention/Flammable substances (Section 112[r], Table 3), TQ = 10,000 lbs. (4540 kgs.).

EPA HAZARDOUS WASTE NUMBER (RCRA No.): U186.
RCRA Section 261 Hazardous Constituents.
EPCRA Section 304 Reportable Quantity (RQ): CERCLA, 100 lbs. (45.4 kgs.).

EPA NAME: PENTANE
CAS: 109-66-0
SYNONYMS: AMYL HYDRIDE; n-PENTANE

CLEAN AIR ACT: Accidental Release Prevention/Flammable substances (Section 112[r], Table 3), TQ = 10,000 lbs. (4540 kgs.).

EPA NAME: 1-PENTENE
CAS: 109-67-1
SYNONYMS: α-n-AMYLENE; PROPYLETHYLENE

CLEAN AIR ACT: Accidental Release Prevention/Flammable substances (Section 112[r], Table 3), TQ = 10,000 lbs. (4540 kgs.).

EPA NAME: 2-PENTENE, (E)-
CAS: 646-04-8
SYNONYMS: β-AMYLENE-trans

CLEAN AIR ACT: Accidental Release Prevention/Flammable substances (Section 112[r], Table 3), TQ = 10,000 lbs. (4540 kgs.).

EPA NAME: 2-PENTENE, (Z)-
CAS: 627-20-3
SYNONYMS: β-AMYLENE-cis

CLEAN AIR ACT: Accidental Release Prevention/Flammable substances (Section 112[r], Table 3), TQ = 10,000 lbs. (4540 kgs.).

EPA NAME: PENTOBARBITOL SODIUM
CAS: 57-33-0
SYNONYMS: BARBITURIC ACID, 5-ETHYL-5-sec-PENTYL-, SODIUM SALT; BARPENTAL; NEMBUTAL SODIUM; SODIUM PENTOBARBITURATE

EPCRA Section 313 Form R de minimus concentration reporting level: 1.0%.
CALIFORNIA'S PROPOSITION 65: Reproductive toxin.

EPA NAME: PERACETIC ACID
CAS: 79-21-0
SYNONYMS: ETHANEPEROXOIC ACID; PEROXOACETIC ACID

CLEAN AIR ACT: Accidental Release Prevention/Flammable substances (Section 112[r], Table 3), TQ = 10,000 lbs. (4540 kgs.).
EPCRA Section 302 Extremely Hazardous Substances: TPQ = 500 lbs. (227 kgs.).
EPCRA Section 304 Reportable Quantity (RQ): EHS, 500 lbs. (227 kgs.).
EPCRA Section 313 Form R de minimus concentration reporting level: 1.0%.

EPA NAME: PERCHLORETHYLENE
[see TETRACHLOROETHYLENE]
CAS: 127-18-4

EPA NAME: PERCHLOROMETHYL MERCAPTAN
CAS: 594-42-3
SYNONYMS: METHANESULFENYL CHLORIDE, TRICHLORO-; PERCHLOROMETHYLMERCAPTAN; TRICHLOROMETHANESULFURYL CHLORIDE; TRICHLOROMETHYLSULFENYL CHLORIDE

CLEAN AIR ACT: Accidental Release Prevention/Flammable substances (Section 112[r], Table 3), TQ = 10,000 lbs. (4540 kgs.).
EPCRA Section 302 Extremely Hazardous Substances: TPQ = 500 lbs. (227 kgs.).
EPCRA Section 304 Reportable Quantity (RQ): EHS/CERCLA, 100 lbs. (45.4 kgs.).
EPCRA Section 313 Form R de minimus concentration reporting level: 1.0%.
MARINE POLLUTANT (49CFR, Subchapter 172.101, Appendix B).

EPA NAME: PERMETHRIN
CAS: 52645-53-1
SYNONYMS: (3-(2,2-DICHLOROETHENYL)-2,2-DIMETHYLCYCLOPROPANE CARBOXYLIC ACID, (3-PHENOXYPHENYL)METHYL ESTER); EPA PESTICIDE CHEMICAL CODE 109701; 3-PHENOXYBENZYL (1RS)-cis-trans-3-(2,2-DICHLOROVINYL)-2,2-DIMETHYLCYCLOPROPANECARBOXYLATE

EPCRA Section 313 Form R de minimus concentration reporting level: 1.0%.

EPA NAME: PHENACETIN
CAS: 62-44-2
SYNONYMS: ACETAMIDE, N-(4-ETHOXYPHENYL)-(9CI); para-ACETOPHENETIDIDE; 4-ETHOXYACETANILIDE

EPA HAZARDOUS WASTE NUMBER (RCRA No.): U187.
RCRA Section 261 Hazardous Constituents.
RCRA Land Ban Waste.
RCRA Universal Treatment Standards: Wastewater (mg/L), 0.081; Nonwastewater (mg/kg), 16.
RCRA Ground Water Monitoring List. Suggested test method(s) (PQL µg/L): 8270(10).
EPCRA Section 304 Reportable Quantity (RQ): CERCLA, 100 lbs. (45.4 kgs.).
CALIFORNIA'S PROPOSITION 65: Carcinogen.

EPA NAME: PHENANTHRENE
CAS: 85-01-8
SYNONYMS: PHENANTHRIN; PHENANTRIN

CLEAN WATER ACT: Section 307 Priority Pollutants.

RCRA Land Ban Waste.
RCRA Universal Treatment Standards: Wastewater (mg/L), 0.059; Nonwastewater (mg/kg), 5.6.
RCRA Ground Water Monitoring List. Suggested test method(s) (PQL µg/L): 8100(200); 8270(10).
EPCRA Section 304 Reportable Quantity (RQ): CERCLA, 5,000 lbs. (2270 kgs.).
EPCRA Section 313 Form R de minimus concentration reporting level: 1.0%.

EPA NAME: PHENOL
CAS: 108-95-2
SYNONYMS: CARBOLIC ACID; PHENYLIC ACID

CLEAN AIR ACT: Hazardous Air Pollutants (Title I, Part A, Section 112).
CLEAN WATER ACT: Section 311 Hazardous Substances/RQ (same as CERCLA); Section 307 Toxic Pollutants; Section 307 Priority Pollutants; Section 313 Priority Chemicals.
EPA HAZARDOUS WASTE NUMBER (RCRA No.): U188.
RCRA Section 261 Hazardous Constituents.
RCRA Land Ban Waste.
RCRA Universal Treatment Standards: Wastewater (mg/L), 0.039; Nonwastewater (mg/kg), 6.2.
RCRA Ground Water Monitoring List. Suggested test method(s) (PQL µg/L): 8040(1); 8270(10).
EPCRA Section 302 Extremely Hazardous Substances: TPQ = 500/10,000 lbs. (227/4,540 kgs.).
EPCRA Section 304 Reportable Quantity (RQ): EHS/CERCLA, 1,000 lbs. (454 kgs.).
EPCRA Section 313 Form R de minimus concentration reporting level: 1.0%.

EPA NAME: PHENOL, 3-(1-METHYLETHYL)-, METHYLCARBAMATE
CAS: 64-00-6
SYNONYMS: m-ISOPROPYLPHENOL-N-METHYLCARBAMATE; m-ISOPROPYLPHENYL METHYLCARBAMATE

RCRA Land Ban Waste.
RCRA Universal Treatment Standards: Wastewater (mg/L), 0.056; Nonwastewater (mg/kg), 1.4 as m-cumenyl methylcarbamate.
EPCRA Section 302 Extremely Hazardous Substances: TPQ = 500/10,000 lbs. (227/4,540 kgs.).
EPCRA Section 304 Reportable Quantity (RQ): EHS, 1 lb. (0.454 kg.).

EPA NAME: PHENOL, 2-(1-METHYLETHOXY)-, METHYLCARBAMATE
[see PROPOXUR]
CAS: 114-26-1

EPA NAME: PHENOL, 2,2′-THIOBIS(4-CHLORO-6-METHYL)-
CAS: 4418-66-0

SYNONYMS: 2,2′-DIHYDROXY-3,3′-DIMETHYL-5,5′-DICHLORODIPHENYL SULFIDE; PHENOL, 2,2′-THIOBIS[4-CHLORO-6-METHYL- (9CI)-

EPCRA Section 302 Extremely Hazardous Substances: TPQ= 100/10,000 lbs. (45.4/4,540 kgs.).
EPCRA Section 304 Reportable Quantity (RQ): EHS, 100 lbs. (45.4 kgs.).

EPA NAME: PHENOTHRIN
CAS: 26002-80-2
SYNONYMS: CYCLOPROPANECAR BOXYLIC ACID, 2,2-DIMETHYL-3-(2-METHYLPROPENYL)-, m-PHENOXYBENZYL ESTER; 2,2-DIMETHYL-3-(2-METHYL-1-PROPENYL)CYCLOPROPANECARBOXYLIC ACID (3-PHENOXYPHENYL)-METHYL ESTER; EPA PESTICIDE CHEMICAL CODE 069005; m-PHENOXYBENZYL 2,2-DIMETHYL-3-(2-METHYLPROPENYL)CYCLOPROPANECARBOXYLATE

EPCRA Section 313 Form R de minimus concentration reporting level: 1.0%.

EPA NAME: PHENOXARSINE, 10,10′-OXYDI-
CAS: 58-36-6
SYNONYMS: BIS(10-PHENOXARSYL)OXIDE; 10-10′-OXYBISPHENOXYARSINE; PHENARSAZINE OXIDE; PHENOXARSINE, 10,10′-OXYDI-

EPCRA Section 302 Extremely Hazardous Substances: TPQ= 500/10,000 lbs. (227/4,540 kgs.).
EPCRA Section 304 Reportable Quantity (RQ): EHS, 500 lbs. (227 kgs.).

EPA NAME: [2-(4-PHENOXY-PHENOXY)ETHYL]CARBAMIC ACID ETHYL ESTER
[see FENOXYCARB]
CAS: 72490-01-8

EPA NAME: PHENYLDICHLOROARSINE
[see DICHLOROPHENYLARSINE]
CAS: 696-28-6

EPA NAME: [1,2-PHENYLENEBIS (IMINOCARBONOTHIOYL)] BISCARBAMIC ACID DIETHYL ESTER
[see THIOPHANATE ETHYL]
CAS: 23564-06-9

EPA NAME: 1,2-PHENYLENEDIAMINE
CAS: 95-54-5
SYNONYMS: o-BENZENEDIAMINE; 1,2-BENZENEDIAMINE; 1,2-DIAMINOBENZENE; ORTHOPHENYLENEDIAMINE; o-PHENYLENEDIAMINE; PHENYLENEDIAMINE, ortho-

RCRA Land Ban Waste.
RCRA Universal Treatment Standards: Wastewater (mg/L), 0.056; Nonwastewater (mg/kg), 5.6.

EPCRA Section 313 Form R de minimus concentration reporting level: 1.0%.

EPA NAME: p-PHENYLENEDIAMINE

CAS: 106-50-3

SYNONYMS: p-BENZENEDIAMINE; 1,4-BENZENEDIAMINE; PARAPHENYLENEDIAMINE; 1,4-PHENYLENEDIAMINE; PHENYLENEDIAMINE, para-

CLEAN AIR ACT: Hazardous Air Pollutants (Title I, Part A, Section 112).

RCRA Ground Water Monitoring List. Suggested test method(s) (PQL μg/L): 8270(10).

EPCRA Section 304 Reportable Quantity (RQ): CERCLA, 5,000 lbs. (2,270 kgs.).

EPCRA Section 313 Form R de minimus concentration reporting level: 1.0%.

EPA NAME: 1,3-PHENYLENEDIAMINE

CAS: 108-45-2

SYNONYMS: m-BENZENEDIAMINE; 1,3-BENZENEDIAMINE; METAPHENYLENEDIAMINE; 3-PHENYLENEDIAMINE; m-PHENYLENEDIAMINE; PHENYLENEDIAMINE, meta

EPCRA Section 313 Form R de minimus concentration reporting level: 1.0%.

EPA NAME: 1,2-PHENYLENEDIAMINE DIHYDROCHLORIDE

CAS: 615-28-1

SYNONYMS: 1,2-BENZENEDIAMINE DIHYDROCHLORIDE; o-PHENYLENEDIAMINE DIHYDROCHLORIDE

EPCRA Section 313 Form R de minimus concentration reporting level: 1.0%.

EPA NAME: 1,4-PHENYLENEDIAMINE DIHYDROCHLORIDE

CAS: 624-18-0

SYNONYMS: 1,4-BENZENEDIAMINE DIHYDROCHLORIDE; p-BENZENEDIAMINE DIHYDROCHLORIDE; p-PHENYLENEDIAMINE DIHYDROCHLORIDE

EPCRA Section 313 Form R de minimus concentration reporting level: 1.0%.

EPA NAME: 1,3-PHENYLENE DIISOCYANATE

[see also DIISOCYANATES]

CAS: 123-61-5

SYNONYMS: BENZENE, 1,3-DIISOCYANATE; BENZENE, 1,3-DIISOCYANATO-

EPCRA Section 313 Form R de minimus concentration reporting level: 1.0%.

EPA NAME: 1,4-PHENYLENE DIISOCYANATE

[see also DIISOCYANATES]

CAS: 104-49-4

SYNONYMS: BENZENE, 1,4-DIISOCYANATO-; p-PHENYLENE DIISOCYANATE

EPCRA Section 313 Form R de minimus concentration reporting level: 1.0%.

EPA NAME: PHENYLHYDRAZINE HYDROCHLORIDE
CAS: 59-88-1
SYNONYMS: PHENYLHYDRAZINE MONOHYDROCHLORIDE; PHENYLHYDRAZINIUM CHLORIDE

EPCRA Section 302 Extremely Hazardous Substances: TPQ = 1,000/10,000 lbs. (454/4,540 kgs.).
EPCRA Section 304 Reportable Quantity (RQ): EHS, 1,000 lbs. (454 kgs.).
CALIFORNIA'S PROPOSITION 65: Carcinogen as phenylhydrazine and its salts.

EPA NAME: PHENYLMERCURIC ACETATE
[see PHENYLMERCURY ACETATE]
CAS: 62-38-4

EPA NAME: PHENYLMERCURY ACETATE
CAS: 62-38-4
SYNONYMS: ACETOXYPHENYLMERCURY; MERCURY, (ACETO-O)PHENYL-; PHENYLMERCURIC ACETATE

EPA HAZARDOUS WASTE NUMBER (RCRA No.): P092.
RCRA Section 261 Hazardous Constituents.
EPCRA Section 302 Extremely Hazardous Substances: TPQ = 500/10,000 lbs. (227/4,540 kgs.).
EPCRA Section 304 Reportable Quantity (RQ): EHS/CERCLA, 100 lbs. (45.4 kgs.).
MARINE POLLUTANT (49CFR, Subchapter 172.101, Appendix B): Severe pollutant.
CALIFORNIA'S PROPOSITION 65: Reproductive toxin.

EPA NAME: 5-(PHENYLMETHYL)-3-FURANYL]METHYL 2,2-DIMETHYL-3-(2-METHYL-1-PROPENYL)CYCLOPROPANECARBOXYLATE
[see RESMETHRIN]
CAS: 10453-86-8

EPA NAME: 2-PHENYLPHENOL
CAS: 90-43-7
SYNONYMS: (1,1'-BIPHENYL)-2-OL; ORTHOPHENYLPHENOL; o-PHENYLPHENOL

EPCRA Section 313 Form R de minimus concentration reporting level: 1.0%.

EPA NAME: PHENYLSILATRANE
CAS: 2097-19-0

SYNONYMS: PHENYL-2,8,9-TRIOXA-5-AZA-1-SILABICYCLO(3,3,3)UNDECANE

EPCRA Section 302 Extremely Hazardous Substances: TPQ = 100/10,000 lbs. (45.4/4,540 kgs.).
EPCRA Section 304 Reportable Quantity (RQ): EHS, 100 lbs. (45.4 kgs.).

EPA NAME: PHENYLTHIOUREA
CAS: 103-85-5
SYNONYMS: α-PHENYLTHIOUREA; 1-PHENYL-2-THIOUREA; 1-PHENYLTHIOUREA

EPA HAZARDOUS WASTE NUMBER (RCRA No.): P093.
RCRA Section 261 Hazardous Constituents.
EPCRA Section 302 Extremely Hazardous Substances: TPQ = 100/10,000 lbs. (45.4/4,540 kgs.).
EPCRA Section 304 Reportable Quantity (RQ): EHS/CERCLA, 100 lbs. (45.4 kgs.).

EPA NAME: PHENYTOIN
CAS: 57-41-0
SYNONYMS: DIPHENYLAN; 5,5-DIPHENYLHYDANTOIN; 2,4-IMIDAZOLIDINEDIONE, 5,5-DIPHENYL-; DIPHENYLHYDANTOIN

CLEAN AIR ACT: Accidental Release Prevention/Flammable substances (Section 112[r], Table 3), TQ = 15,000 lbs. (6810 kgs.).
EPA HAZARDOUS WASTE NUMBER (RCRA No.): U098.
RCRA Section 261 Hazardous Constituents.
EPCRA Section 313 Form R de minimus concentration reporting level: 0.1%.
EPCRA Section 302 Extremely Hazardous Substances: TPQ = 1,000 lbs. (454 kgs.).
EPCRA Section 304 Reportable Quantity (RQ): CERCLA, 10 lbs. (4.54 kgs.).
CALIFORNIA'S PROPOSITION 65: Carcinogen; reproductive toxin.

EPA NAME: PHORATE
CAS: 298-02-2
SYNONYMS: THIMET; THEMET(R); O,O-DIETHYLETHYLTHIOMETHYL PHOSPHORODITHIOATE; PHOSPHORODITHIOIC ACID, O,O-DIETHYL S-[(ETHYLTHIO)METHYL] ESTER

EPA HAZARDOUS WASTE NUMBER (RCRA No.): P094.
RCRA Section 261 Hazardous Constituents.
RCRA Land Ban Waste.
RCRA Universal Treatment Standards: Wastewater (mg/L), 0.021; Nonwastewater (mg/kg), 4.6.
RCRA Ground Water Monitoring List. Suggested test method(s) (PQL µg/L): 8140(2); 8270(10).
EPCRA Section 302 Extremely Hazardous Substances: TPQ = 10 lbs. (4.54 kgs.).

EPCRA Section 304 Reportable Quantity (RQ): EHS/CERCLA, 10 lbs. (4.54 kgs.).

MARINE POLLUTANT (49CFR, Subchapter 172.101, Appendix B): Severe pollutant.

EPA NAME: PHOSACETIM

CAS: 4104-14-7

SYNONYMS: PHOSAZETIM; PHOSPHORAMIDOTHIOIC ACID, (1-IMINOETHYL)-, O,O-BIS(4-CHLOROPHENYL)ESTER

EPCRA Section 302 Extremely Hazardous Substances: TPQ = 100/10,000 lbs. (45.4/4,540 kgs.).

EPCRA Section 304 Reportable Quantity (RQ): EHS, 100 lbs. (45.4 kgs.).

EPA NAME: PHOSFOLAN

CAS: 947-02-4

SYNONYMS: CYCLIC ETHYLENE P,P-DIETHYLPHOSPHONODITHIOIMIDOCARBONATE; PHOSPHOLAN

EPCRA Section 302 Extremely Hazardous Substances: TPQ = 100/10,000 lbs. (45.4/4,540 kgs.).

EPCRA Section 304 Reportable Quantity (RQ): EHS, 100 lbs. (45.4 kgs.).

EPA NAME: PHOSGENE

CAS: 75-44-5

SYNONYMS: CARBON DICHLORIDE OXIDE; CARBONIC DICHLORIDE; CARBON OXYCHLORIDE; CARBONYL CHLORIDE

CLEAN AIR ACT: Hazardous Air Pollutants (Title I, Part A, Section 112); List of high risk pollutants (Section 63.74); Accidental Release Prevention/Flammable substances (Section 112[r], Table 3), TQ = 500 lbs. (227 kgs.).

CLEAN WATER ACT: Section 311 Hazardous Substances/RQ (same as CERCLA); Section 313 Priority Chemicals.

EPA HAZARDOUS WASTE NUMBER (RCRA No.): P095.

RCRA Section 261 Hazardous Constituents.

EPCRA Section 302 Extremely Hazardous Substances: TPQ = 10 lbs. (4.54 kgs.).

EPCRA Section 304 Reportable Quantity (RQ): EHS/CERCLA, 10 lbs. (4.54 kgs.).

EPCRA Section 313 Form R de minimus concentration reporting level: 1.0%.

CALIFORNIA'S PROPOSITION 65: Carcinogen.

EPA NAME: PHOSMET

CAS: 732-11-6

SYNONYMS: (O,O-DIMETHYL-PHTHALIMIDIOMETHYL-DITHIOPHOSPHATE); PHTHALIMIDOMETHYL O,O-DIMETHYLPHOSPHORODITHIOATE

EPCRA Section 302 Extremely Hazardous Substances: TPQ = 10/10,000 lbs. (4.54/4,540 kgs.).

EPCRA Section 304 Reportable Quantity (RQ): EHS, 10 lbs. (4.54 kgs.).
MARINE POLLUTANT (49CFR, Subchapter 172.101, Appendix B).

EPA NAME: PHOSPHAMIDON
CAS: 13171-21-6
SYNONYMS: 2-CHLORO-2-DIETHYLCARBAMOYL-1-METHYL-VINYL DIMETHYLPHOSPHATE; PHOSPHORIC ACID, 2-CHLORO-3-(DIETHYLAMINO)-1-METHYL-3-OXO-1-PROPENYL DIMETHYL ESTER

EPCRA Section 302 Extremely Hazardous Substances: TPQ = 100 lbs. (45.4 kgs.).
EPCRA Section 304 Reportable Quantity (RQ): EHS, 100 lbs. (45.4 kgs.).
MARINE POLLUTANT (49CFR, Subchapter 172.101, Appendix B): Severe pollutant.

EPA NAME: PHOSPHINE
CAS: 7803-51-2
SYNONYMS: HYDROGEN PHOSPHIDE; PHOSPHORATED HYDROGEN

CLEAN AIR ACT: Hazardous Air Pollutants (Title I, Part A, Section 112); Accidental Release Prevention/Flammable substances (Section 112[r], Table 3), TQ = 5,000 lbs. (2270 kgs.).
EPA HAZARDOUS WASTE NUMBER (RCRA No.): P096.
RCRA Section 261 Hazardous Constituents.
EPCRA Section 302 Extremely Hazardous Substances: TPQ = 500 lbs. (227 kgs.).
EPCRA Section 304 Reportable Quantity (RQ): EHS/CERCLA, 100 lbs. (45.4 kgs.).
EPCRA Section 313 Form R de minimus concentration reporting level: 1.0%.

EPA NAME: PHOSPHONIC ACID, (2,2,2-TRICHLORO-1-HYDROXYETHYL)-, DIMETHYL ESTER
[see TRICHLORFON]
CAS: 52-68-6

EPA NAME: PHOSPHONOTHIOIC ACID, METHYL-, O-ETHYL O-(4-(METHYLTHIO)PHENYL) ESTER
CAS: 2703-13-1
SYNONYMS: METHYLPHOSPHONOTHIOIC ACID-O-ETHYL O-(4-(METHYLTHIO)PHENYL) ESTER (9CI)

EPCRA Section 302 Extremely Hazardous Substances: TPQ = 500 lbs. (227 kgs.).
EPCRA Section 304 Reportable Quantity (RQ): EHS, 500 lbs. (227 kgs.).

EPA NAME: PHOSPHONOTHIOIC ACID, METHYL-, S(2-(BIS(1-METHYLETHYL)AMINO)ETHYL) O-ETHYL ESTER
CAS: 50782-69-9
SYNONYMS: O-ETHYL-S-DIISOPROPYLAMINOETHYL METHYLPHOSPHONOTHIOATE

EPCRA Section 302 Extremely Hazardous Substances: TPQ = 100 lbs. (45.4 kgs.).
EPCRA Section 304 Reportable Quantity (RQ): EHS, 100 lbs. (45.4 kgs.).

EPA NAME: PHOSPHONOTHIOIC ACID, METHYL-, O-(4-NITROPHENYL) O-PHENYL ESTER
CAS: 2665-30-7
SYNONYMS: METHYLPHOSPHONOTHIOIC ACID-O-(4-NITROPHENYL)-O-PHENYL ESTER

EPCRA Section 302 Extremely Hazardous Substances: TPQ = 100 lbs. (45.4 kgs.).
EPCRA Section 304 Reportable Quantity (RQ): EHS, 500 lbs. (227 kgs.).

EPA NAME: PHOSPHORIC ACID
CAS: 7664-38-2
SYNONYMS: ORTHOPHOSPHORIC ACID

CLEAN WATER ACT: Section 311 Hazardous Substances/RQ (same as CERCLA); Section 313 Priority Chemicals.
EPCRA Section 304 Reportable Quantity (RQ): CERCLA, 5,000 lbs. (2270 kgs.).
EPCRA Section 313 Form R de minimus concentration reporting level: 1.0%.

EPA NAME: PHOSPHORIC ACID, 2-CHLORO-1-(2,3,5-TRICHLOROPHENYL) ETHENYL DIMETHYL ESTER
[see TETRACHLORVINPHOS]
CAS: 961-11-5

EPA NAME: PHOSPHORIC ACID, 2-DICHLOROETHENYL DIMETHYL ESTER
[see DICHLORVOS]
CAS: 62-73-7

EPA NAME: PHOSPHORIC ACID, DIMETHYL 4-(METHYLTHIO)PHENYL ESTER
CAS: 3254-63-5
SYNONYMS: O,O-DIMETHYL O-(4-METHYLMERCAPTOPHENYL)PHOSPHATE; PHOSPHORIC ACID, DIMETHYL p-(METHYLTHIO)PHENYL ESTER

EPCRA Section 302 Extremely Hazardous Substances: TPQ = 500 lbs. (227 kgs.).
EPCRA Section 304 Reportable Quantity (RQ): EHS, 500 lbs. (227 kgs.).

EPA NAME: PHOSPHORODITHIOIC ACID O-ETHYL S,S-DIPROPYL ESTER
[see ETHOPROP]
CAS: 13194-48-4

EPA NAME: PHOSPHOROTHIOIC ACID, O,O-DIETHYL-O-(4-NITROPHENYL) ESTER
[see PARATHION]

CAS: 56-38-2

EPA NAME: PHOSPHOROTHIOIC ACID, O,O-DIMETHYL-5-(2-(METHYLTHIO)ETHYL)ESTER

CAS: 2587-90-8

SYNONYMS: 2-(METHYLTHIO)-ETHANETHIOL-O,O-DIMETHYL PHOSPHOROTHIOATE

EPCRA Section 302 Extremely Hazardous Substances: TPQ = 500 lbs. (227 kgs.).

EPCRA Section 304 Reportable Quantity (RQ): EHS, 500 lbs. (227 kgs.).

EPA NAME: PHOSPHORUS

CAS: 7723-14-0

SYNONYMS: PHOSPHORUS ELEMENTAL, WHITE; PHOSPHORUS YELLOW; PHOSPHORUS (YELLOW OR WHITE); WHITE PHOSPHORUS; YELLOW PHOSPHORUS

CLEAN AIR ACT: Hazardous Air Pollutants (Title I, Part A, Section 112).

CLEAN WATER ACT: Section 311 Hazardous Substances/RQ (same as CERCLA); Section 313 Priority Chemicals.

EPCRA Section 302 Extremely Hazardous Substances: TPQ = 100 lbs. (45.4 kgs.).

EPCRA Section 304 Reportable Quantity (RQ): EHS/CERCLA, 1 lb. (0.454 kg.).

EPCRA Section 313 (yellow or white) Form R de minimus concentration reporting level: 1.0%.

MARINE POLLUTANT (49CFR, Subchapter 172.101, Appendix B): Severe pollutant, white, yellow dry, molten, or in solution.

EPA NAME: PHOSPHORUS OXYCHLORIDE

CAS: 10025-87-3

SYNONYMS: PHOSPHORIC CHLORIDE; PHOSPHORUS CHLORIDE OXIDE; PHOSPHORYL CHLORIDE; PHOSPHORYL TRICHLORIDE

CLEAN AIR ACT: Accidental Release Prevention/Flammable substances (Section 112[r], Table 3), TQ = 5,000 lbs. (2270 kgs.).

CLEAN WATER ACT: Section 311 Hazardous Substances/RQ (same as CERCLA).

EPCRA Section 302 Extremely Hazardous Substances: TPQ = 500 lbs. (227 kgs.).

EPCRA Section 304 Reportable Quantity (RQ): EHS/CERCLA, 1,000 lbs. (454 kgs.).

EPA NAME: PHOSPHORUS PENTACHLORIDE

CAS: 10026-13-8

SYNONYMS: PHOSPHORANE, PENTACHLORO-; PHOSPHORUS PERCHLORIDE

EPCRA Section 302 Extremely Hazardous Substances: TPQ = 500 lbs. (227 kgs.).

EPCRA Section 304 Reportable Quantity (RQ): EHS, 500 lbs. (227 kgs.).

EPA NAME: PHOSPHORUS PENTOXIDE
CAS: 1314-56-3
SYNONYMS: PHOSPHORIC ANHYDRIDE; PHOSPHORUS(V) OXIDE

Removed from EHS list (FR Vol. 61, No. 89, page 20477).

EPA NAME: PHOSPHOROUS TRICHLORIDE
CAS: 7719-12-2
SYNONYMS: CHLORIDE OF PHOSPHORUS; PHOSPHOROUS CHLORIDE; PHOSPHORUS TRICHLORIDE

CLEAN AIR ACT: Accidental Release Prevention/Flammable substances (Section 112[r], Table 3), TQ = 15,000 lbs. (6810 kgs.).
CLEAN WATER ACT: Section 311 Hazardous Substances/RQ (same as CERCLA).
EPCRA Section 302 Extremely Hazardous Substances: TPQ = 1,000 lbs. (454 kgs.).
EPCRA Section 304 Reportable Quantity (RQ): EHS/CERCLA, 1,000 lbs. (454 kgs.).

EPA NAME: PHOSPHORYL CHLORIDE
CAS: 10025-87-3
SYNONYMS: PHOSPHORIC CHLORIDE; PHOSPHORUS CHLORIDE OXIDE; PHOSPHORUS OXYCHLORIDE; PHOSPHORUS OXYTRICHLORIDE

CLEAN AIR ACT: Accidental Release Prevention/Flammable substances (Section 112[r], Table 3), TQ = 5,000 lbs. (2270 kgs.).
EPCRA Section 302 Extremely Hazardous Substances: TPQ = 500 lbs. (227 kgs.).
EPCRA Section 304 Reportable Quantity (RQ): CERCLA, 1,000 lbs. (454 kgs.).

EPA NAME: PHTHALATE ESTERS
CLEAN WATER ACT: Section 307 Toxic Pollutants.

EPA NAME: PHTHALIC ACID
CAS: 100-21-0
RCRA Land Ban Waste.
RCRA Universal Treatment Standards: Wastewater (mg/L), 0.055; Nonwastewater (mg/kg), 28.

EPA NAME: PHTHALIC ANHYDRIDE
CAS: 85-44-9
SYNONYMS: 1,2-BENZENEDICARBOXYLIC ANHYDRIDE; 1,3-ISOBENZOFURANDIONE

CLEAN AIR ACT: Hazardous Air Pollutants (Title I, Part A, Section 112).
EPA HAZARDOUS WASTE NUMBER (RCRA No.): U190.
RCRA Section 261 Hazardous Constituents.
RCRA Land Ban Waste.
RCRA Universal Treatment Standards: Wastewater (mg/L), 0.055; Nonwastewater (mg/kg), 28.

EPCRA Section 304 Reportable Quantity (RQ): CERCLA, 5,000 lbs. (2270 kgs.).
EPCRA Section 313 Form R de minimus concentration reporting level: 1.0%.

EPA NAME: PHYSOSTIGMINE
CAS: 57-47-6
SYNONYMS: METHYL-CARBAMIC ACID, ESTER with ESEROLINE; PHYSOSTOL

EPA HAZARDOUS WASTE NUMBER (RCRA No.): P204.
RCRA Section 261 Hazardous Constituents.
RCRA Land Ban Waste.
RCRA Universal Treatment Standards: Wastewater (mg/L), 0.056; Nonwastewater (mg/kg), 1.4.
EPCRA Section 302 Extremely Hazardous Substances: TPQ = 100/10,000 lbs. (45.4/4,540 kgs.).
EPCRA Section 304 Reportable Quantity (RQ): EHS, 1 lb. (0.454 kg.).

EPA NAME: PHYSOSTIGMINE, SALICYLATE (1:1)
CAS: 57-64-7
SYNONYMS: PHSOSTOL SALICYLATE SALICYLIC ACID with PHYSOSTIGMINE (1:1)

EPA HAZARDOUS WASTE NUMBER (RCRA No.): P188.
RCRA Section 261 Hazardous Constituents.
RCRA Land Ban Waste.
RCRA Universal Treatment Standards: Wastewater (mg/L), 0.056; Nonwastewater (mg/kg), 1.4.
EPCRA Section 302 Extremely Hazardous Substances: TPQ = 100/10,000 lbs. (45.4/4,540 kgs.).
EPCRA Section 304 Reportable Quantity (RQ): EHS, 1 lb. (0.454 kg.).

EPA NAME: PICLORAM
CAS: 1918-02-1
SYNONYMS: 4-AMINOTRICHLOROPICOLINIC ACID; 4-AMINO-3,5,6-TRICHLORO-2-PYRIDINECARBOXYLIC ACID; 4-AMINO-3,5,6-TRICHLOROPYRIDINE-2-CARBOXYLIC ACID; EPA PESTICIDE CHEMICAL CODE 005101; PICHLORAM; 2-PYRIDINE CARBOXYLIC ACID, 4-AMINO-3,5,6-TRICHLORO-; TORDON

SAFE DRINKING WATER ACT: MCL, 0.5 mg/L; MCLG, 0.5 mg/L; Regulated chemical (47 FR 9352) as pichloram.
EPCRA Section 313 Form R de minimus concentration reporting level: 1.0%.

EPA NAME: 2-PICOLINE
[see 2-METHYLPYRIDINE]
CAS: 109-06-8

EPA NAME: PICRIC ACID
CAS: 88-89-1

SYNONYMS: PHENOL TRINITRATE; PHENOL, 2,4,6-TRINITRO-; TRINITROPHENOL; 2,4,6-TRINITOPHENOL

EPCRA Section 313 Form R de minimus concentration reporting level: 1.0%.

EPA NAME: PICROTOXIN
CAS: 124-87-8
SYNONYMS: COCCULIN; COCCULUS

EPCRA Section 302 Extremely Hazardous Substances: TPQ = 500/10,000 lbs. (227/4,540 kgs.).
EPCRA Section 304 Reportable Quantity (RQ): EHS, 500 lbs. (227 kgs.).

EPA NAME: N,N′-[1,4-PIPERAZINEDIYLBIS(2,2,2-TRICHLOROETHYLIDENE)]BISFORMAMIDE
[see TRIFORINE]
CAS: 26644-46-2

EPA NAME: PIPERIDINE
CAS: 110-89-4
SYNONYMS: AZACYCLOHEXANE; CYCLOPENTIMINE; CYPENTIL; HEXAHYDROPYRIDINE; HEXAZANE; PENTAMETHYLENEIMINE

CLEAN AIR ACT: Accidental Release Prevention/Flammable substances (Section 112[r], Table 3), TQ = 15,000 lbs. (6810 kgs.).
EPCRA Section 302 Extremely Hazardous Substances: TPQ = 1,000 lbs. (454 kgs.).
EPCRA Section 304 Reportable Quantity (RQ): EHS, 1,000 lbs. (454 kgs.).

EPA NAME: PIPERONYL BUTOXIDE
[see also GLYCOL ETHERS]
CAS: 51-03-6
SYNONYMS: 1,3-BENZODIOXOLE, 5-((2-(2-BUTOXYETHOXY)ETHOXY)METHYL)-6-PROPYL-; EPA PESTICIDE CHEMICAL CODE 067501; ETHANOL BUTOXIDE; (3,4-METHYLENEDIOXY-6-PROPYLBENZYL) (BUTYL) DIETHYLENE GLYCOL ETHER

EPCRA Section 313 Form R de minimus concentration reporting level: 1.0%.

EPA NAME: PIRIMFOS-ETHYL
CAS: 23505-41-1
SYNONYMS: O-(2-(DIETHYLAMINO)-6-METHYL-4-PYRIMIDINYL)O,O-DIETHYL PHOSPHOROTHIOATE

EPCRA Section 302 Extremely Hazardous Substances: TPQ = 1,000 lbs. (454 kgs.).
EPCRA Section 304 Reportable Quantity (RQ): EHS, 1,000 lbs. (454 kgs.).

MARINE POLLUTANT (49CFR, Subchapter 172.101, Appendix B): Severe pollutant.

EPA NAME: PIRIMIPHOS METHYL

CAS: 29232-93-7

SYNONYMS: O-(2-(DIETHYLAMINO)-6-METHYL-4-PYRIMIDINYL)-O,O-DIMETHYL PHOSPHOROTHIOATE; O-(2-DIETHYLAMINO-6-METHYLPYRIMIDIN-4-YL) O,O-DIMETHYL PHOSPHOROTHIOATE; EPA PESTICIDE CHEMICAL CODE 108102; PHOSPHOROTHIOIC ACID, O-(2-(DIETHYLAMINO)-6-METHYL-4-PYRIMIDINYL)O,O-DIMETHYL ESTER

EPCRA Section 313 Form R de minimus concentration reporting level: 1.0%.

EPA NAME: PLUMBANE, TETRAMETHYL-

[see TETRAMETHYL LEAD]

CAS: 75-74-1

EPA NAME: POLYBROMINATED BIPHENYLS (PBBs)

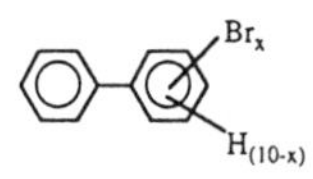

Where x = 1 to 10

CAS: 36355-01-8

RCRA Section 261 Hazardous Constituents, waste number not listed.

RCRA Ground Water Monitoring List. Suggested test method(s) (PQL μg/L): 8080(50); 8250(100).

EPCRA Section 313 Form R de minimus concentration reporting level: 0.1%. Form R Toxic Chemical Category Code: N575.

CALIFORNIA'S PROPOSITION 65: Carcinogen; reproductive toxin.

EPA NAME: POLYCHLORINATED ALKANES (C10 to C13)

$C_xH_{2x+2-y}Cl$ where x = 10 to 13; y=3 to 12; and the average chlorine content ranges from 40-70% with the limiting molecular formulas $C_{10}H_{19}Cl$ and $C_{13}H_{16}Cl_{12}$

EPCRA Section 313 Form R de minimus concentration reporting level: 1.0% (except for those members of the category that have an average chain length of 12 carbons and contain an average chlorine content of 60 percent by weight, which is subject to the 0.1% de minimus). Form R Toxic Chemical Category Code: N583.

EPA NAME: POLYCHLORINATED BIPHENYLS

[see also AROCLORS]

CAS: 1336-36-3

SYNONYMS: AROCLORs; 1,1′-BIPHENYL, CHLORO DERIVATIVES; BIPHENYL, CHLORINATED; PCBs

CLEAN AIR ACT: Hazardous Air Pollutants (Title I, Part A, Section 112).

CLEAN WATER ACT: Section 311 Hazardous Substances/RQ (same as CERCLA); Section 307 Toxic Pollutants; Section 307 Priority Pollutants; Section 313 Priority Chemicals.

RCRA Land Ban Waste.

RCRA Universal Treatment Standards: Wastewater (mg/L), 0.10; Nonwastewater (mg/kg), 10, total PCBs, sum of all PCB isomers, or all AROCLORs.

RCRA Ground Water Monitoring List. Suggested test method(s) (PQL μg/L): 8080(50); 8250(100).

SAFE DRINKING WATER ACT: MCL, 0.0005 mg/L; MCLG, zero; Regulated chemical (47 FR 9352).

EPCRA Section 304 Reportable Quantity (RQ): CERCLA, 1 lb. (0.454 kg.).

EPCRA Section 313 Form R de minimus concentration reporting level: 0.1%.

MARINE POLLUTANT (49CFR, Subchapter 172.101, Appendix B): Severe pollutant.

CALIFORNIA'S PROPOSITION 65: Carcinogen; reproductive toxin.

EPA NAME: POLYCYCLIC AROMATIC COMPOUNDS

EPCRA Section 313: This category includes only 19 chemicals that appear in this book and contains the notation, "also see POLYCYCLIC AROMATIC COMPOUNDS (PACs)". Form R de minimus concentration reporting level: 0.1%. Form R Toxic Chemical Category Code: N590.

EPA NAME: POLYMERIC DIPHENYLMETHANE DIISOCYANATE

[see also DIISOCYANATES]

CAS: 9016-87-9

SYNONYMS: ISOCYANIC ACID, POLYMETHYLENEPOLYPHENYLENE ESTER; POLYMERIC MDI; POLYMETHYLENE POLYPHENYLENE ISOCYANATE; POLYMETHYLENE POLYPHENYL POLYISOCYANATE; POLYMETHYL POLYPHENYL POLYISOCYANATE

EPCRA Section 313 Form R de minimus concentration reporting level: 1.0%.

EPA NAME: POLYNUCLEAR AROMATIC HYDROCARBONS

CLEAN WATER ACT: Section 307 Toxic Pollutants including naphthalene, benzonthracenes, benzopyrenes, benzofluoranthene, chrysenes, dibenzanthracenes, fluorene, and indenopyrenes.

EPA NAME: POTASSIUM ARSENATE

[see also ARSENIC COMPOUNDS]

CAS: 7784-41-0

SYNONYMS: ARSENIC ACID, MONOPOTASSIUM SALT; POTASSIUM DIHYDROGEN ARSENATE; POTASSIUM HYDROGEN ARSENATE

CLEAN WATER ACT: Section 311 Hazardous Substances/RQ (same as CERCLA); Section 313 Priority Chemicals.
EPCRA Section 304 Reportable Quantity (RQ): CERCLA, 1 lb. (0.454 kg.).

EPA NAME: POTASSIUM ARSENITE
[see also ARSENIC COMPOUNDS]
CAS: 10124-50-2
SYNONYMS: ARSENOUS ACID, POTASSIUM SALT; ARSONIC ACID, POTASSIUM SALT

CLEAN WATER ACT: Section 311 Hazardous Substances/RQ (same as CERCLA); Section 313 Priority Chemicals.
EPCRA Section 302 Extremely Hazardous Substances: TPQ = 500/10,000 lbs. (227/4,540 kgs.).
EPCRA Section 304 Reportable Quantity (RQ): EHS/CERCLA, 1 lb. (0.454 kg.).

EPA NAME: POTASSIUM BICHROMATE
[see also CHROMIUM COMPOUNDS]
CAS: 7778-50-9
SYNONYMS: CHROMIC ACID, DIPOTASSIUM SALT; POTASSIUM DICHROMATE

EPCRA Section 304 Reportable Quantity (RQ): CERCLA, 10 lbs. (4.54 kgs).

EPA NAME: POTASSIUM BROMATE
CAS: 7758-01-2
SYNONYMS: BROMIC ACID, POTASSIUM SALT

EPCRA Section 313 Form R de minimus concentration reporting level: 0.1%.
CALIFORNIA'S PROPOSITION 65: Carcinogen.

EPA NAME: POTASSIUM CHROMATE
[see also CHROMIUM COMPOUNDS]
CAS: 7789-00-6
SYNONYMS: CHROMIC ACID, DIPOTASSIUM SALT; DIPOTASSIUM CHROMATE

CLEAN WATER ACT: Section 311 Hazardous Substances/RQ same as CERCLA; Section 313 Priority Chemicals.
EPCRA Section 304 Reportable Quantity (RQ): CERCLA, 10 lbs. (4.54 kgs.).

EPA NAME: POTASSIUM CYANIDE
[see also CHROMIUM COMPOUNDS]
CAS: 151-50-8
SYNONYMS: HYDROCYANIC ACID, POTASSIUM SALT

CLEAN WATER ACT: Section 311 Hazardous Substances/RQ (same as CERCLA); Section 313 Priority Chemicals.
EPA HAZARDOUS WASTE NUMBER (RCRA No.): P098.

RCRA Section 261 Hazardous Constituents.
EPCRA Section 302 Extremely Hazardous Substances: TPQ = 100 lbs. (45.4 kgs.).
EPCRA Section 304 Reportable Quantity (RQ): EHS/CERCLA, 10 lbs. (4.54 kgs.).
MARINE POLLUTANT (49CFR, Subchapter 172.101, Appendix B).

EPA NAME: POTASSIUM DIMETHYLDITHIOCARBAMATE
CAS: 128-03-0
SYNONYMS: CARBAMIC ACID, DIMETHYLDITHIO-, POTASSIUM SALT, HYDRATE; CARBAMODITHIOIC ACID, DIMETHYL-, POTASSIUM SALT; EPA PESTICIDE CHEMICAL CODE 034803 EPA HAZARDOUS WASTE NUMBER (RCRA No.): U383.

RCRA Section 261 Hazardous Constituents.
EPCRA Section 304 Reportable Quantity (RQ): CERCLA, 1 lb. (0.454 kg.).
EPCRA Section 313 Form R de minimus concentration reporting level: 1.0%.

EPA NAME: POTASSIUM HYDROXIDE
CAS: 1310-58-3
SYNONYMS: POTASSIUM HYDRATE

CLEAN WATER ACT: Section 311 Hazardous Substances/RQ (same as CERCLA).
EPCRA Section 304 Reportable Quantity (RQ): CERCLA, 1,000 lbs. (454 kgs.).

EPA NAME: POTASSIUM N-METHYLDITHIOCARBAMATE
CAS: 137-41-7
SYNONYMS: CARBAMIC ACID, N-METHYLDITHIO-, POTASSIUM SALT; EPA PESTICIDE CHEMICAL CODE 039002; POTASSIUM METHYLDITHIOCARBAMATE

EPA HAZARDOUS WASTE NUMBER (RCRA No.): U377.
RCRA Section 261 Hazardous Constituents.
EPCRA Section 304 Reportable Quantity (RQ): CERCLA, 1 lb. (0.454 kg.).
EPCRA Section 313 Form R de minimus concentration reporting level: 1.0%.

EPA NAME: POTASSIUM PERMANGANATE
CAS: 7722-64-7
SYNONYMS: PERMANGANIC ACID, POTASSIUM SALT

CLEAN WATER ACT: Section 311 Hazardous Substances/RQ (same as CERCLA).
EPCRA Section 304 Reportable Quantity (RQ): CERCLA, 100 lbs. (45.4 kgs.).

EPA NAME: POTASSIUM SILVER CYANIDE
[see also CYANIDE COMPOUNDS]
CAS: 506-61-6

SYNONYMS: SILVER POTASSIUM CYANIDE

EPA HAZARDOUS WASTE NUMBER (RCRA No.): P099.
RCRA Section 261 Hazardous Constituents.
EPCRA Section 302 Extremely Hazardous Substances: TPQ = 500 lbs. (227 kgs.).
EPCRA Section 304 Reportable Quantity (RQ): EHS/CERCLA, 1 lb. (0.454 kg.).

EPA NAME: PROFENOFOS
CAS: 41198-08-7
SYNONYMS: O-(4-BROMO-2-CHLOROPHENYL)-O-ETHYL-S-PROPYLPHOSPHOROTHIOATE; EPA PESTICIDE CHEMICAL CODE 111401; PHOSPHOROTHIOIC ACID, O-(4-BROMO-2-CHLOROPHENYL)-O-ETHYL-S-PROPYL ESTER

EPCRA Section 313 Form R de minimus concentration reporting level: 1.0%.

EPA NAME: PROMECARB
CAS: 2631-37-0
SYNONYMS: 3-METHYL-5-ISOPROPYLPHENYL-N-METHYL CARBAMATE

EPA HAZARDOUS WASTE NUMBER (RCRA No.): P201.
RCRA Section 261 Hazardous Constituents.
RCRA Land Ban Waste.
RCRA Universal Treatment Standards: Wastewater (mg/L), 0.056; Non-wastewater (mg/kg), 1.4.
EPCRA Section 302 Extremely Hazardous Substances: TPQ = 500/10,000 lbs. (227/4,540 kgs.).
EPCRA Section 304 Reportable Quantity (RQ): EHS, 1 lb. (0.454 kg.).
MARINE POLLUTANT (49CFR, Subchapter 172.101, Appendix B).

EPA NAME: PROMETHRYN
CAS: 7287-19-6
SYNONYMS: 2,4-BIS(ISOPROPYLAMINO)-6-(METHYLMERCAPTO)-s-TRIAZINE; N,N′-BIS(1-METHYLETHYL)-6-METHYLTHIO-1,3,5-TRIAZINE-2,4-DIAMINE; N,N-DI-ISOPROPYL-6-METHYLTHIO-1,3,5-TRIAZINE-2,4-DIAMINE; EPA PESTICIDE CHEMICAL CODE 080805; 1,3,5-TRIAZINE-2,4-DIAMINE, N,N′-BIS(1-METHYLETHYL)-6-(METHYLTHIO)-

EPCRA Section 313 Form R de minimus concentration reporting level: 1.0%.

EPA NAME: PRONAMIDE
CAS: 23950-58-5
SYNONYMS: BENZAMIDE, 3,5-DICHLORO-N-(1,1-DIMETHYL-2-PROPYNYL); 3,5-DICHLORO-N-(1,1-DIMETHYL-2-PROPYNYL)BENZAMIDE; EPA PESTICIDE CHEMICAL CODE 101701

EPA HAZARDOUS WASTE NUMBER (RCRA No.): U192.
RCRA Section 261 Hazardous Constituents.

RCRA Land Ban Waste.

RCRA Universal Treatment Standards: Wastewater (mg/L), 0.093; Non-wastewater (mg/kg), 1.5.

RCRA Ground Water Monitoring List. Suggested test method(s) (PQL μg/L): 8270(10).

EPCRA Section 304 Reportable Quantity (RQ): CERCLA, 5,000 lbs. (2,270 kgs.).

EPCRA Section 313 Form R de minimus concentration reporting level: 1.0%.

EPA NAME: PROPACHLOR

CAS: 1918-16-7

SYNONYMS: ACETAMIDE, 2-CHLORO-N-(1-METHYLETHYL)-N-PHENYL-; 2-CHLORO-N-(1-METHYLETHYL)-N-PHENYLACETAMIDE; EPA PESTICIDE CHEMICAL CODE 019101

EPCRA Section 313 Form R de minimus concentration reporting level: 1.0%.

EPA NAME: 1,2-PROPADIENE

CAS: 463-49-0

SYNONYMS: ALLENE; PROPADIENE

CLEAN AIR ACT: Accidental Release Prevention/Flammable substances (Section 112[r], Table 3), TQ=10,000 lbs. (4540 kgs.).

EPA NAME: PROPADIENE

[see 1,2-PROPADIENE]

CAS: 463-49-0

EPA NAME: 2-PROPANAMINE

[see ISOPROPYLAMINE]

CAS: 75-31-0

EPA NAME: PROPANE

CAS: 74-98-6

SYNONYMS: DIMETHYL METHANE

CLEAN AIR ACT: Accidental Release Prevention/Flammable substances (Section 112[r], Table 3), TQ=10,000 lbs. (4540 kgs.).

EPA NAME: PROPANE, 1,2-DICHLORO-

[see 1,2-DICHLOROPROPANE]

CAS: 78-87-5

EPA NAME: 1,3-PROPANE SULTONE

CAS: 1120-71-4

SYNONYMS: 3-HYDROXY-1-PROPANESULPHONIC ACID SULTONE; 1,2-OXATHROLANE 2,2-DIOXIDE; PROPANE SULTONE

CLEAN AIR ACT: Hazardous Air Pollutants (Title I, Part A, Section 112).

EPA HAZARDOUS WASTE NUMBER (RCRA No.): U193.

RCRA Section 261 Hazardous Constituents.

EPCRA Section 304 Reportable Quantity (RQ): CERCLA, 10 lbs. (4.54 kgs.).
EPCRA Section 313 Form R de minimus concentration reporting level: 0.1%.

EPA NAME: PROPANE SULTONE
[see 1,3-PROPANE SULTONE]
CAS: 1120-71-4

EPA NAME: PROPANE, 2-CHLORO
[see ISOPROPYL CHLORIDE]
CAS: 75-29-6

EPA NAME: PROPANE, 2,2-DIMETHYL-
[see 2,2-DIMETHYLPROPANE]
CAS: 463-82-1

EPA NAME: PROPANE, 2-METHYL
[see ISOBUTANE]
CAS: 75-28-5

EPA NAME: PROPIONITRILE
[see also CYANIDE COMPOUNDS]
CAS: 107-12-0
SYNONYMS: ETHYL CYANIDE; PROPIONIC NITRILE

CLEAN AIR ACT: Hazardous Air Pollutants (Title I, Part A, Section 112) as cyanide compounds; Accidental Release Prevention/Flammable substances (Section 112[r], Table 3), TQ = 10,000 lbs. (4540 kgs.).
EPA HAZARDOUS WASTE NUMBER (RCRA No.): P101.
RCRA Section 261 Hazardous Constituents.
RCRA Land Ban Waste.
EPCRA Section 302 Extremely Hazardous Substances: TPQ = 500 lbs. (227 kgs.).
EPCRA Section 304 Reportable Quantity (RQ): EHS/CERCLA, 10 lbs. (4.54 kgs.).
MARINE POLLUTANT (49CFR, Subchapter 172.101, Appendix B) as cyanide mixtures, cyanide solutions or cyanides, inorganic, n.o.s.

EPA NAME: PROPANENITRILE, 2-METHYL-
[see ISOBUTYRONITRILE]
CAS: 78-82-0

EPA NAME: 1,3-PROPANE SULTONE
CAS: 1120-71-4
SYNONYMS: 3-HYDROXY-1-PROPANESULFONIC ACID SULTONE; 1,2-OXATHROLANE 2,2-DIOXIDE; PROPANE SULTONE

EPA HAZARDOUS WASTE NUMBER (RCRA No.): U193.
RCRA Section 261 Hazardous Constituents.
EPCRA Section 304 Reportable Quantity (RQ): CERCLA, 10 lbs. (4.54 kgs.).

EPCRA Section 313 Form R de minimus concentration reporting level: 1.0%.

CALIFORNIA'S PROPOSITION 65: Carcinogen.

EPA NAME: PROPANIL

CAS: 709-98-8

SYNONYMS: DCPA; EPA PESTICIDE CHEMICAL CODE 028201; N-(3,4-DICHLOROPHENYL)PROPANAMIDE; PROPIONANILIDE, 3′,4′-DICHLORO-; PROPIONIC ACID 3,4-DICHLOROANILIDE

SAFE DRINKING WATER ACT: Priority List (55 FR 1470) as DCPA (and its acid metabolites).

EPCRA Section 313 Form R de minimus concentration reporting level: 1.0%.

EPA NAME: PROPARGITE

CAS: 2312-35-8

SYNONYMS: EPA PESTICIDE CHEMICAL CODE 097601; SULFUROUS ACID, 2-(4-(1,1-DIMETHYLETHYL)PHENOXY) CYCLOHEXYL 2-PROPYNYL ESTER; 2-(4-(1,1-DIMETHYLETHYL)PHENOXY)CYCLOHEXYL 2-PROPYNYL SULFUROUS ACID; PROPARGIL

CLEAN WATER ACT: Section 311 Hazardous Substances/RQ (same as CERCLA).

EPCRA Section 304 Reportable Quantity (RQ): CERCLA, 10 lbs. (4.54 kgs.).

EPCRA Section 313 Form R de minimus concentration reporting level: 1.0%.

CALIFORNIA'S PROPOSITION 65: Carcinogen.

EPA NAME: PROPARGYL ALCOHOL

CAS: 107-19-7

SYNONYMS: 3-HYDROXY-1-PROPYNE; 2-PROPYN-1-OL; PROP-2-YN-1-OL

EPA HAZARDOUS WASTE NUMBER (RCRA No.): P102.

RCRA Section 261 Hazardous Constituents.

EPCRA Section 304 Reportable Quantity (RQ): CERCLA, 1,000 lbs. (454 kgs.).

EPCRA Section 313 Form R de minimus concentration reporting level: 1.0%.

EPA NAME: PROPARGYL BROMIDE

CAS: 106-96-7

SYNONYMS: 3-BROMO-1-PROPYNE; 1-PROPYENE, 3-BROMO-

EPCRA Section 302 Extremely Hazardous Substances: TPQ = 10 lbs. (4.54 kgs.).

EPCRA Section 304 Reportable Quantity (RQ): EHS, 10 lbs. (4.54 kgs.).

EPA NAME: 2-PROPENAL

[see ACROLEIN]

CAS: 107-02-8

EPA NAME: 2-PROPEN-1-AMINE
[see ALLYLAMINE]
CAS: 107-11-9

EPA NAME: PROPENE
[see PROPYLENE]
CAS: 115-07-1

EPA NAME: 1-PROPENE
[see PROPYLENE]
CAS: 115-07-1

EPA NAME: 1-PROPENE, 1-CHLORO-
[see 1-CHLOROPROPYLENE]
CAS: 590-21-6

EPA NAME: 1-PROPENE, 2-CHLORO-
[see 2-CHLOROPROPYLENE]
CAS: 557-98-2

EPA NAME: 1-PROPENE, 2-METHYL
[see 2-METHYLPROPENE]
CAS: 115-11-7

EPA NAME: 2-PROPENENITRILE, 2-METHYL-
[see METHACRYLONITRILE]
CAS: 126-98-7

EPA NAME: 2-PROPEN-1-OL
[see ALLYL ALCOHOL]
CAS: 107-18-6

EPA NAME: 2-PROPENOYL CHLORIDE
[see ACRYLYL CHLORIDE]
CAS: 814-68-6

EPA NAME: PROPETAMPHOS
CAS: 31218-83-4
SYNONYMS: 2-BUTENOIC ACID, 3-(((ETHYLAMINO)METHOXYPHOSPHINOTHIOYL)OXY)-, 1-METHYLETHYL ESTER, (E)-; EPA PESTICIDE CHEMICAL CODE 113601; 3-[(ETHYLAMINO)METHOXYPHOSPHINOTHIOYL]OXY]-2-BUTENOIC ACID, 1-METHYLETHYL ESTER

EPCRA Section 313 Form R de minimus concentration reporting level: 1.0%.

EPA NAME: PROPICONAZOLE
CAS: 60207-90-1

SYNONYMS: 1-[2-(2,4-DICHLOROPHENYL)-4-PROPYL-1,3-DIOXOLAN-2-YL]-METHYL-1H-1,2,4,-TRIAZOLE; EPA PESTICIDE CHEMICAL CODE 122101; 1H-1,2,4-TRIAZOLE, 1-((2-(2,4-DICHLOROPHENYL)-4-PROPYL-1,3-DIOXOLAN-2-YL)METHYL)-

EPCRA Section 313 Form R de minimus concentration reporting level: 1.0%.

EPA NAME: β-PROPIOLACTONE

CAS: 57-57-8

SYNONYMS: 2-OXETANONE; PROPIOLACTONE, β-; 3-PROPIOLACTONE

CLEAN AIR ACT: Hazardous Air Pollutants (Title I, Part A, Section 112).
EPCRA Section 302 Extremely Hazardous Substances: TPQ = 500 lbs. (227 kgs.).
EPCRA Section 304 Reportable Quantity (RQ): EHS/CERCLA, 10 lbs. (4.54 kgs.).
EPCRA Section 313 Form R de minimus concentration reporting level: 0.1%.
CALIFORNIA'S PROPOSITION 65: Carcinogen.

EPA NAME: PROPIONALDEHYDE

CAS: 123-38-6

SYNONYMS: n-PROPANAL; 1-PROPANAL; PROPANAL

CLEAN AIR ACT: Hazardous Air Pollutants (Title I, Part A, Section 112).
EPCRA Section 304 Reportable Quantity (RQ): CERCLA, 1,000 lbs. (454 kgs.).
EPCRA Section 313 Form R de minimus concentration reporting level: 1.0%.

EPA NAME: PROPIONIC ACID

CAS: 79-09-4

SYNONYMS: METHYLACETIC ACID; PSEUDOACETIC ACID

CLEAN WATER ACT: Section 311 Hazardous Substances/RQ (same as CERCLA).
EPCRA Section 304 Reportable Quantity (RQ): CERCLA, 5,000 lbs. (2270 kgs.).

EPA NAME: PROPIONIC ANHYDRIDE

CAS: 123-62-6

SYNONYMS: METHYLACETIC ANHYDRIDE; PROPIONIC ACID ANHYDRIDE

CLEAN WATER ACT: Section 311 Hazardous Substances/RQ (same as CERCLA).
EPCRA Section 304 Reportable Quantity (RQ): CERCLA, 5,000 lbs. (2270 kgs.).

EPA NAME: 2-PROPENITRILE

[see ACRYLONITRILE]

CAS: 107-13-1

EPA NAME: PROPIONITRILE
CAS: 107-12-0
SYNONYMS: CYANOETHANE; ETHYL CYANIDE; PROPANE-NITRILE; PROPIONIC NITRILE

CLEAN AIR ACT: Accidental Release Prevention/Flammable substances (Section 112[r], Table 3), TQ = 10,000 lbs. (4540 kgs.).
EPA HAZARDOUS WASTE NUMBER (RCRA No.): P101.
RCRA Section 261 Hazardous Constituents.
RCRA Ground Water Monitoring List. Suggested test method(s) (PQL µg/L): 8015(60); 8240(5).
EPCRA Section 302 Extremely Hazardous Substances: TPQ = 500 lbs. (227 kgs.).
EPCRA Section 304 Reportable Quantity (RQ): EHS/CERCLA, 10 lbs. (4.54 kgs.).

EPA NAME: PROPIONITRILE, 3-CHLORO-
[see 3-CHLOROPROPIONITRILE]
CAS: 542-76-7

EPA NAME: PROPIOPHENONE, 4′-AMINO-
CAS: 70-69-9
SYNONYMS: p-AMINOPROPIOPHENONE; 1-(4-AMINOPHENYL)-1-PROPANONE

EPCRA Section 302 Extremely Hazardous Substances: TPQ = 100/10,000 lbs. (45.4/4,540 kgs.).
EPCRA Section 304 Reportable Quantity (RQ): EHS, 100 lbs. (45.4 kgs.).

EPA NAME: PROPOXUR
CAS: 114-26-1
SYNONYMS: BAYGON; DDVP; 2-ISOPROPOXYPHENYL N-METHYLCARBAMATE; 2-ISOPROPOXYPHENYL METHYL-CARBAMATE; PHENOL, 2-(1-METHYLETHOXY)-, METHYL-CARBAMATE

CLEAN AIR ACT: Hazardous Air Pollutants (Title I, Part A, Section 112).
EPA HAZARDOUS WASTE NUMBER (RCRA No.): U411.
RCRA Section 261 Hazardous Constituents.
RCRA Land Ban Waste.
RCRA Universal Treatment Standards: Wastewater (mg/L), 0.056; Non-wastewater (mg/kg), 1.4.
EPCRA Section 304 Reportable Quantity (RQ): CERCLA, 100 lbs. (45.4 kgs.).
EPCRA Section 313 Form R de minimus concentration reporting level: 1.0%.
MARINE POLLUTANT (49CFR, Subchapter 172.101, Appendix B).

EPA NAME: n-PROPYLAMINE
CAS: 107-10-8

SYNONYMS: 1-AMINOPROPANE; 1-IODOPROPANE; PROPANAMINE

EPA HAZARDOUS WASTE NUMBER (RCRA No.): U194.
RCRA Section 261 Hazardous Constituents.
EPCRA Section 304 Reportable Quantity (RQ): CERCLA, 5,000 lbs. (2270 kgs.).

EPA NAME: PROPYL CHLOROFORMATE
CAS: 109-61-5
SYNONYMS: CARBONOCHLORIDIC ACID, PROPYLESTER; CHLOROFORMIC ACID PROPYL ESTER; PROPYL CHLOROCARBONATE; n-PROPYL CHLOROFORMATE

CLEAN AIR ACT: Accidental Release Prevention/Flammable substances (Section 112[r], Table 3), TQ = 15,000 lbs. (6810 kgs.).
EPCRA Section 302 Extremely Hazardous Substances: TPQ = 500 lbs. (227 kgs.).
EPCRA Section 304 Reportable Quantity (RQ): EHS, 500 lbs. (227 kgs.).

EPA NAME: PROPYLENE
CAS: 115-07-1
SYNONYMS: METHYLETHYLENE; PROPENE; 1-PROPENE; 1-PROPYLENE

CLEAN AIR ACT: Accidental Release Prevention/Flammable substances (Section 112[r], Table 3), TQ = 10,000 lbs. (4540 kgs.).
EPCRA Section 313 Form R de minimus concentration reporting level: 1.0%.

EPA NAME: PROPYLENEIMINE
CAS: 75-55-8
SYNONYMS: AZIRIDINE, 2 METHYL; 2-METHYLAZACYCLOPROPANE; 2-METHYLAZIRIDINE; 2-METHYLETHYLENIMINE; 2-METHYLETHYLEN IMINE; 1,2-PROPYLENIMINE

CLEAN AIR ACT: Hazardous Air Pollutants (Title I, Part A, Section 112); Accidental Release Prevention/Flammable substances (Section 112[r], Table 3), TQ = 10,000 lbs. (4540 kgs.).
EPA HAZARDOUS WASTE NUMBER (RCRA No.): P067.
RCRA Section 261 Hazardous Constituents.
EPCRA Section 302 Extremely Hazardous Substances: TPQ = 10,000 lbs. (4,540 kgs.).
EPCRA Section 304 Reportable Quantity (RQ): EHS/CERCLA, 1 lb. (0.454 kg.).
EPCRA Section 313 Form R de minimus concentration reporting level: 0.1%.

EPA NAME: PROPYLENE OXIDE
CAS: 75-56-9
SYNONYMS: 1,2-EPOXYPROPANE; METHYLOXIRANE; OXIRANE, METHYL-; PROPANE, 1,2-EPOXY-; PROPENE OXIDE; PROPYLENE EPOXIDE

CLEAN AIR ACT: Hazardous Air Pollutants (Title I, Part A, Section 112); Accidental Release Prevention/Flammable substances (Section 112[r], Table 3), TQ = 10,000 lbs. (4540 kgs.).
CLEAN WATER ACT: Section 311 Hazardous Substances/RQ (same as CERCLA); Section 313 Priority Chemicals.
EPCRA Section 302 Extremely Hazardous Substances: TPQ = 10,000 lbs. (4,540 kgs.).
EPCRA Section 304 Reportable Quantity (RQ): CERCLA, 100 lbs. (45.4 kgs.).
EPCRA Section 313 Form R de minimus concentration reporting level: 0.1%.
CALIFORNIA'S PROPOSITION 65: Carcinogen.

EPA NAME: 1-PROPYNE
CAS: 74-99-7
SYNONYMS: METHYL ACETYLENE; PROPINE; PROPYNE

CLEAN AIR ACT: Accidental Release Prevention/Flammable substances (Section 112[r], Table 3), TQ = 10,000 lbs. (4540 kgs.).

EPA NAME: PROPYNE
[see 1-PROPYNE]
CAS: 74-99-7

EPA NAME: PROTHOATE
CAS: 2275-18-5
SYNONYMS: O,O-DIETHYLDITHIOPHOSPHORYLACETIC ACID-N-MONOISOPROPYLAMIDE; O,O-DIETHYL-S-(N-ISOPROPYLCARBAMOYLMETHYL)DITHIOPHOSPHATE

EPCRA Section 302 Extremely Hazardous Substances: TPQ = 100/10,000 lbs. (45.4/4,540 kgs.).
EPCRA Section 304 Reportable Quantity (RQ): EHS, 100 lbs. (45.4 kgs.).
MARINE POLLUTANT (49CFR, Subchapter 172.101, Appendix B).

EPA NAME: PYRENE
[see also POLYNUCLEAR AROMATIC HYDROCARBONS]
CAS: 129-00-0
SYNONYMS: BENZO(def)PHENANTHRENE; β-PYRENE

RCRA Land Ban Waste.
RCRA Universal Treatment Standards: Wastewater (mg/L), 0.067; Nonwastewater (mg/kg), 8.2.
RCRA Ground Water Monitoring List. Suggested test method(s) (PQL μg/L): 8100(200); 8270(10).
EPCRA Section 302 Extremely Hazardous Substances: TPQ = 1,000/10,000 lbs. (454/4,540 kgs.).
EPCRA Section 304 Reportable Quantity (RQ): EHS/CERCLA, 5,000 lbs. (2270 kgs.).

EPA NAME: PYRETHRINS
CAS: 121-21-1

SYNONYMS: CYCLOPROPANECARBOXYLIC ACID, 2,2-DIMETHYL-3-(2-METHYL-1-PROPENYL)-,2-METHYL-4-OXO-3-(2,4-PENTADIENYL)-2-CYCLOPENTEN-1-YL ESTER, [1R[1α[S*O],3β]-; PYRETHRIN I

CLEAN WATER ACT: Section 311 Hazardous Substances/RQ (same as CERCLA).

EPA HAZARDOUS WASTE NUMBER (RCRA No.): P008.

RCRA Section 261 Hazardous Constituents.

EPCRA Section 304 Reportable Quantity (RQ): CERCLA, 1 lb. (0.454 kg.).

EPA NAME: PYRETHRINS

CAS: 121-29-9

SYNONYMS: CYCLOPROPANECARBOXYLIC ACID, 3-(3-METHOXY-2-METHYL-3-OXO-1-PROPENYL)-2,2-DIMETHYL-,2-METHYL-4-OXO-3-(2,4-PENTADIENYL)-2-CYCLOPENTEN-1-YL ESTER, [R[1α[3*(Z)],3β(E)]-; PYRETRIN II

CLEAN WATER ACT: Section 311 Hazardous Substances/RQ (same as CERCLA).

EPCRA Section 304 Reportable Quantity (RQ): CERCLA, 1 lb. (0.454 kg.).

EPA NAME: PYRETHRINS

CAS: 8003-34-7

SYNONYMS: PYRETHRUM; PYRETHROIDS

CLEAN WATER ACT: Section 311 Hazardous Substances/RQ (same as CERCLA).

EPCRA Section 304 Reportable Quantity (RQ): CERCLA, 1 lb. (0.454 kg.).

EPA NAME: PYRIDINE

CAS: 110-86-1

SYNONYMS: AZABENZENE; AZINE

EPA HAZARDOUS WASTE NUMBER (RCRA No.): U196, DO38.

RCRA Section 261 Hazardous Constituents.

RCRA Toxicity Characteristic (Section 261.24), Maximum Concentration of Contaminants, regulatory level, 5.0 mg/L.

RCRA Land Ban Waste.

RCRA Universal Treatment Standards: Wastewater (mg/L), 0.014; Nonwastewater (mg/kg), 16.

RCRA Ground Water Monitoring List. Suggested test method(s) (PQL μg/L): 8240(5); 8270(10).

EPCRA Section 304 Reportable Quantity (RQ): CERCLA, 1,000 lbs. (454 kgs.).

EPCRA Section 313 Form R de minimus concentration reporting level: 1.0%.

EPA NAME: PYRIDINE, 4-AMINO-

[see 4-AMINOPYRIDINE]

CAS: 504-24-5

EPA NAME: PYRIDINE, 3-(1-METHYL-2-PYRROLIDINYL)-, (S)
[see NICOTINE]
CAS: 54-11-5

EPA NAME: PYRIDINE, 2-METHYL-5-VINYL
CAS: 140-76-1
SYNONYMS: 5-ETHENYL-2-METHYLPYRIDINE; 2-METHYL-5-VINYLPYRIDINE

EPCRA Section 302 Extremely Hazardous Substances: TPQ = 500 lbs. (227 kgs.).
EPCRA Section 304 Reportable Quantity (RQ): EHS, 500 lbs. (227 kgs.).

EPA NAME: PYRIDINE, 4-NITRO-, 1-OXIDE
CAS: 1124-33-0
SYNONYMS: 4-NITROPYRIDINE-1-OXIDE

EPCRA Section 302 Extremely Hazardous Substances: TPQ = 500/10,000 lbs. (227/4,540 kgs.).
EPCRA Section 304 Reportable Quantity (RQ): EHS, 500 lbs. (227 kgs.).

EPA NAME: 2,4-(1H,3H)-PYRIMIDINEDIONE,5-BROMO-6-METHYL-3(1-METHYLPROPYL), LITHIUM SALT
[see BROMACIL, LITHIUM SALT]
CAS: 53404-19-6

EPA NAME: PYRIMINIL
CAS: 53558-25-1
SYNONYMS: N-(4-NITROPHENYL)-N′-(3-PYRIDINYLMETHYL)UREA; N-3-PYRIDYLMETHYL-N′-p-NITROPHENYLUREA

EPCRA Section 302 Extremely Hazardous Substances: TPQ = 100/10,000 lbs. (45.4/4,540 kgs.).
EPCRA Section 304 Reportable Quantity (RQ): EHS, 100 lbs. (45.4 kgs.).

- Q -

EPA NAME: QUINOLINE
CAS: 91-22-5
SYNONYMS: 1-BENZAZINE; BENZOPYRIDINE; BENZO[B]PYRIDINE

CLEAN AIR ACT: Hazardous Air Pollutants (Title I, Part A, Section 112).
CLEAN WATER ACT: Section 311 Hazardous Substances/RQ (same as CERCLA); Section 313 Priority Chemicals.
EPCRA Section 304 Reportable Quantity (RQ): CERCLA, 5,000 lbs. (2270 kgs.).
EPCRA Section 313 Form R de minimus concentration reporting level: 1.0%.

EPA NAME: QUINONE
CAS: 106-51-4
SYNONYMS: BENZOQUINONE; p-BENZOQUINONE; 1,4-CYCLOHEXADIENEDIONE; 2,5-CYCLOHEXADIENE-1,4-DIONE; 1,4-DIOXYBENZENE

CLEAN AIR ACT: Hazardous Air Pollutants (Title I, Part A, Section 112).
EPA HAZARDOUS WASTE NUMBER (RCRA No.): U197.
RCRA Section 261 Hazardous Constituents.
EPCRA Section 304 Reportable Quantity (RQ): CERCLA, 10 lbs. (4.54 kgs.).
EPCRA Section 313 Form R de minimus concentration reporting level: 1.0%.

EPA NAME: QUINTOZENE
[see also CHLORINATED BENZENES]
CAS: 82-68-8
SYNONYMS: BENZENE, PENTACHLORONITRO-; NITROPENTACHLOROBENZENE; PCNB; PENTACHLORONITROBENZENE; QUINTOCENE

EPA HAZARDOUS WASTE NUMBER (RCRA No.): U185.
RCRA Section 261 Hazardous Constituents.
RCRA Land Ban Waste.
RCRA Universal Treatment Standards: Wastewater (mg/L), 0.055; Nonwastewater (mg/kg), 4.8.
RCRA Ground Water Monitoring List. Suggested test method(s) (PQL µg/L): 8270(10).
EPCRA Section 304 Reportable Quantity (RQ): CERCLA, 100 lbs. (45.4 kgs.).
EPCRA Section 313 Form R de minimus concentration reporting level: 1.0%.

EPA NAME: QUIZALOFOP-ETHYL
CAS: 76578-14-8
SYNONYMS: 2-[4-[(6-CHLORO-2-QUINOXALINYL)OXY]PHENOXY]PROPANOIC ACID ETHYL ESTER; EPA PESTICIDE CHEMICAL CODE 128201; ETHYL 2-(4-(6-CHLORO-2-QUINOXALINYLOXY)PHENOXY)PROPANOATE

EPCRA Section 313 Form R de minimus concentration reporting level: 1.0%.

- R -

EPA NAME: RESPERINE

CAS: 50-55-5

SYNONYMS: AUSTRAPINE; 3,4,5,-TRIMETHOXYBENZOIC ACID; YOHIMBAN-16-CARBOXYLIC ACID, 11,17-DIMETHOXY-18-(3,4,5-TRIMETHOXYBENXOYL)OXY-, METHYL ESTER,(3β, 16β, 17α, 18β, 20α)-

EPA HAZARDOUS WASTE NUMBER (RCRA No.): U200.

RCRA Section 261 Hazardous Constituents.

EPCRA Section 304 Reportable Quantity (RQ): CERCLA, 5,000 lbs. (2270 kgs.).

CALIFORNIA'S PROPOSITION 65: Carcinogen.

EPA NAME: RESMETHRIN

CAS: 10453-86-8

SYNONYMS: (5-BENZYL-3-FURYL)METHYL CHRYSANTHEMATE; 5-BENZYL-3-FURYLMETHYL (+-)-cis-trans-CHRYSANTHEMATE; EPA PESTICIDE CHEMICAL CODE 097801; 5-(PHENYLMETHYL)-3-FURANYL]METHYL 2,2-DIMETHYL-3-(2-METHYL-1-PROPENYL)CYCLOPROPANECARBOXYLATE

EPCRA Section 313 Form R de minimus concentration reporting level: 1.0%.

EPA NAME: RESORCINOL

CAS: 108-46-3

SYNONYMS: m-BENZENEDIOL; 1,3-BENZENEDIOL; m-HYDROXYPHENOL; 3-HYDROXYPHENOL

CLEAN WATER ACT: Section 311 Hazardous Substances/RQ (same as CERCLA).

EPA HAZARDOUS WASTE NUMBER (RCRA No.): U201.

RCRA Section 261 Hazardous Constituents.

EPCRA Section 304 Reportable Quantity (RQ): CERCLA, 5,000 lbs. (2270 kgs.).

- S -

EPA NAME: SACCHARIN (MANUFACTURING)

CAS: 81-07-2

SYNONYMS: 3-BENZISOTHIAZOLINONE 1,1-DIOXIDE; 1,2-BENZISOTHIAZOL-3(2H)-ONE, 1,1-DIOXIDE; 1,2-DIHYDRO-2-KETOBENZISOSULFONAZOLE

EPA HAZARDOUS WASTE NUMBER (RCRA No.): U202.

RCRA Section 261 Hazardous Constituents.

EPCRA Section 304 Reportable Quantity (RQ): CERCLA, 100 lbs. (45.4 kgs.).

EPCRA Section 313 Form R de minimus concentration reporting level: 0.1%.

CALIFORNIA'S PROPOSITION 65: Carcinogen.

EPA NAME: SACCHARIN AND SALTS

CAS: 81-07-2

EPA HAZARDOUS WASTE NUMBER (RCRA No.): U202.

RCRA Section 261 Hazardous Constituents.

EPCRA Section 313 (as saccharin) Form R de minimus concentration reporting level: 0.1%.

CALIFORNIA'S PROPOSITION 65: Carcinogen, as saccaharin.

EPA NAME: SAFROLE

CAS: 94-59-7

SYNONYMS: 5-ALLYL-1,3-BENZODIOXOLE; 4-ALLYL-1,2-(METHYLENEDIOXY)BENZENE; 1,3-BENZODIOXOLE, 5-(2-PROPENYL)-; 3,4-METHYLENEDIOXY-ALLYLBENZENE

EPA HAZARDOUS WASTE NUMBER (RCRA No.): U203.

RCRA Section 261 Hazardous Constituents.

RCRA Land Ban Waste.

RCRA Universal Treatment Standards: Wastewater (mg/L), 0.081; Nonwastewater (mg/kg), 22.

RCRA Ground Water Monitoring List. Suggested test method(s) (PQL µg/L): 8270(10).

EPCRA Section 304 Reportable Quantity (RQ): CERCLA, 100 lbs. (45.4 kgs.).

EPCRA Section 313 Form R de minimus concentration reporting level: 0.1%.

CALIFORNIA'S PROPOSITION 65: Carcinogen.

EPA NAME: SALCOMINE

CAS: 14167-18-1

SYNONYMS: BIS(SALICYALDEHYDE)ETHYLENEDIIMINE COBALT(II)

EPCRA Section 302 Extremely Hazardous Substances: TPQ = 500 lbs. (227 kgs.).

EPCRA Section 304 Reportable Quantity (RQ): EHS, 500 lbs. (227 kgs.).

EPA NAME: SARIN

CAS: 107-44-8

SYNONYMS: ISOPROPYHYL METHYLPHOSPHONOFLUORIDATE; METHYLPHOSPHONOFLUORIDIC ACID ISOPROPYL ESTER

EPCRA Section 302 Extremely Hazardous Substances: TPQ = 10 lbs. (4.54 kgs.).
EPCRA Section 304 Reportable Quantity (RQ): EHS, 10 lbs. (4.54 kgs.).

EPA NAME: SELENIOUS ACID

[see also SELENIUM and SELENIUM COMPOUNDS]

CAS: 7783-00-8

SYNONYMS: SELENIUM DIOXIDE

CLEAN WATER ACT: Section 307 Toxic Pollutants as selenium and compounds.
EPA HAZARDOUS WASTE NUMBER (RCRA No.): U204.
RCRA Section 261 Hazardous Constituents, as selenium compounds, n.o.s., waste number not listed.
EPCRA Section 302 Extremely Hazardous Substances: TPQ = 1,000/10,000 lbs. (454/4,540 kgs.).
EPCRA Section 304 Reportable Quantity (RQ): EHS/CERCLA, 10 lbs. (4.54 kgs.).
EPCRA Section 313 Form R de minimus concentration reporting level: 1.0%. Form R Toxic Chemical Category Code: N725.

EPA NAME: SELENIOUS ACID, DITHALLIUM(1+) SALT

[see also SELENIUM and SELENIUM COMPOUNDS]

CAS: 12039-52-0

SYNONYMS: THALLIUM MONOSELENIDE; THALLIUM SELENIDE

CLEAN WATER ACT: Section 307 Toxic Pollutants as selenium and compounds.
RCRA Section 261 Hazardous Constituents, as selenium compounds, n.o.s., waste number not listed.
EPA HAZARDOUS WASTE NUMBER (RCRA No.): P114.
EPCRA Section 304 Reportable Quantity (RQ): CERCLA, 1,000 lbs. (454 kgs.).
EPCRA Section 313 Form R de minimus concentration reporting level: 1.0%. Form R Toxic Chemical Category Code: N725.

EPA NAME: SELENIUM

[see also SELENIUM COMPOUNDS]

CAS: 7782-49-2

SYNONYMS: ELEMENTAL SELENIUM; SELENIUM ELEMENTAL

CLEAN WATER ACT: Section 307 Priority Pollutants; Section 313 Priority Chemicals; Section 307 Toxic Pollutants.
EPA HAZARDOUS WASTE NUMBER (RCRA No.): D010.
RCRA Section 261 Hazardous Constituents, waste number not listed.
RCRA Land Ban Waste.

RCRA Toxicity Characteristic (Section 261.24), Maximum Concentration of Contaminants, regulatory level, 1.0 mg/L.
RCRA Universal Treatment Standards: Wastewater (mg/L), 0.82; Nonwastewater (mg/L), 0.16 TCLP.
RCRA Ground Water Monitoring List. Suggested test method(s) (PQL μg/L): (total) 6010(750); 7740(20); 7741(20).
SAFE DRINKING WATER ACT: MCL, 0.05 mg/L; MCLG, 0.05 mg/L; Regulated chemical (47 FR 9352).
EPCRA Section 304 Reportable Quantity (RQ): CERCLA, 100 lbs. (45.4 kgs.).
EPCRA Section 313 Form R de minimus concentration reporting level: 1.0%. Form R Toxic Chemical Category Code: N725.

EPA NAME: SELENIUM COMPOUNDS
[see also SELENIUM]
CLEAN WATER ACT: Section 307 Toxic Pollutants as selenium and compounds.
RCRA Section 261 Hazardous Constituents, as selenium compounds, n.o.s., waste number not listed.
EPCRA Section 313: Includes any unique chemical substance that contains selenium as part of that chemical's infrastructure. Form R de minimus concentration reporting level: 1.0%. Form R Toxic Chemical Category Code: N725.

EPA NAME: SELENIUM DIOXIDE
[see also SELENIUM and SELENIUM COMPOUNDS]
CAS: 7446-08-4
SYNONYMS: SELENIUM(IV) DIOXIDE (1:2); SELENIUM OXIDE

CLEAN WATER ACT: Section 311 Hazardous Substances/RQ same as CERCLA; Section 313 Priority Chemicals as selenium oxide; Section 307 Toxic Pollutants as selenium and compounds.
EPA HAZARDOUS WASTE NUMBER (RCRA No.): U204.
RCRA Section 261 Hazardous Constituents.
EPCRA Section 304 Reportable Quantity (RQ): CERCLA, 10 lbs. (4.54 kgs.).
EPCRA Section 313 Form R de minimus concentration reporting level: 1.0%. Form R Toxic Chemical Category Code: N725.

EPA NAME: SELENIUM OXYCHLORIDE
[see also SELENIUM and SELENIUM COMPOUNDS]
CAS: 7791-23-3
SYNONYMS: SELENIUM CHLORIDE OXIDE

CLEAN WATER ACT: Section 307 Toxic Pollutants as selenium and compounds.
RCRA Section 261 Hazardous Constituents, as selenium compounds, n.o.s., waste number not listed.
EPCRA Section 302 Extremely Hazardous Substances: TPQ = 500 lbs. (227 kgs.).
EPCRA Section 304 Reportable Quantity (RQ): EHS, 500 lbs. (227 kgs.).

EPCRA Section 313 Form R de minimus concentration reporting level: 1.0%. Form R Toxic Chemical Category Code: N725.

EPA NAME: SELENIUM SULFIDE

[see also SELENIUM and SELENIUM COMPOUNDS]

CAS: 7488-56-4

SYNONYMS: SELENIUM DISULFIDE; SELENIUM(IV) DISULFIDE; SULFUR SELENIDE

CLEAN WATER ACT: Section 307 Toxic Pollutants as selenium and compounds.

EPA HAZARDOUS WASTE NUMBER (RCRA No.): U205.

RCRA Section 261 Hazardous Constituents.

EPCRA Section 304 Reportable Quantity (RQ): CERCLA, 10 lbs. (4.54 kgs.).

EPCRA Section 313 Form R de minimus concentration reporting level: 1.0%. Form R Toxic Chemical Category Code: N725.

CALIFORNIA'S PROPOSITION 65: Carcinogen.

EPA NAME: SELENOUREA

[see also SELENIUM and SELENIUM COMPOUNDS]

CAS: 630-10-4

SYNONYMS: CARBAMIMIDOSELENOIC ACID; UREA, SELENO-

CLEAN WATER ACT: Section 307 Toxic Pollutants as selenium and compounds.

EPA HAZARDOUS WASTE NUMBER (RCRA No.): P103.

RCRA Section 261 Hazardous Constituents.

EPCRA Section 304 Reportable Quantity (RQ): CERCLA, 1,000 lbs. (454 kgs.).

EPCRA Section 313 Form R de minimus concentration reporting level: 1.0%. Form R Toxic Chemical Category Code: N725.

EPA NAME: SEMICARBAZIDE HYDROCHLORIDE

CAS: 563-41-7

SYNONYMS: AMIDOUREA HYDROCHLORIDE; CARBAMYLHYDRAZINE HYDROCHLORIDE; HYDRAZINECARBOXAMIDE MONOHYDROCHLORIDE

EPCRA Section 302 Extremely Hazardous Substances: TPQ = 1,000/10,000 lbs. (454/4,540 kgs.).

EPCRA Section 304 Reportable Quantity (RQ): EHS, 1,000 lbs. (454 kgs.).

EPA NAME: SETHOXYDIM

CAS: 74051-80-2

SYNONYMS: 2-CYCLOHEXEN-1-ONE,2-(1-(ETHOXYIMINO)BUTYL)-5-(2-(ETHYLTHIO)PROPYL)-3-HYDROXY-; CYETHOXYDIM; 2-[1-(ETHOXYIMINO) BUTYL]-5-[2-(ETHYLTHIO)PROPYL]-3-HYDROXYL-2-CYCLOHEXEN-1-ONE; EPA PESTICIDE CHEMICAL CODE 121001

EPCRA Section 313 Form R de minimus concentration reporting level: 1.0%.

EPA NAME: SILANE
CAS: 7803-62-5
SYNONYMS: SILICANE; SILICON TETRAHYDRIDE

CLEAN AIR ACT: Accidental Release Prevention/Flammable substances (Section 112[r], Table 3), TQ = 10,000 lbs. (4540 kgs.).

EPA NAME: SILANE, (4-AMINOBUTYL)DIETHOXYMETHYL-
CAS: 3037-72-7
SYNONYMS: (4-AMINOBUTYL)DIETHYOXYMETHYLSILANE; δ-AMINOBUTYLMETHYLDIETHOXYSILANE

EPCRA Section 302 Extremely Hazardous Substances: TPQ = 1,000 lbs. (454 kgs.).
EPCRA Section 304 Reportable Quantity (RQ): EHS, 1,000 lbs. (454 kgs.).

EPA NAME: SILANE, CHLOROTRIMETHYL-
[see TRIMETHYLCHLOROSILANE]
CAS: 75-77-4

EPA NAME: SILANE, DICHLORO-
[see DICHLOROSILANE]
CAS: 4109-96-0

EPA NAME: SILANE, DICHLORODIMETHYL-
[see DIMETHYLDICHLOROSILANE]
CAS: 75-78-5

EPA NAME: SILANE, TRICHLORO-
CAS: 10025-78-2
SYNONYMS: TRICHLOROMONOSILANE; TRICHLOROSILANE

CLEAN AIR ACT: Accidental Release Prevention/Flammable substances (Section 112[r], Table 3), TQ = 10,000 lbs. (4540 kgs.).

EPA NAME: SILANE, TRICHLOROMETHYL-
[see METHYLTRICHLOROSILANE]
CAS: 75-79-6

EPA NAME: SILANE, TETRAMETHYL-
[see TETRAMETHYLSILANE]
CAS: 75-76-3

EPA NAME: SILVER
[see also SILVER COMPOUNDS]
CAS: 7440-22-4
SYNONYMS: SILVER ELEMENTAL; SILVER METAL

CLEAN WATER ACT: Section 307 Priority Pollutants; Section 313 Priority Chemicals; Section 307 Toxic Pollutants as silver and compounds.
EPA HAZARDOUS WASTE NUMBER (RCRA No.): D011.

RCRA Section 261 Hazardous Constituents, waste number not listed.
RCRA Toxicity Characteristic (Section 261.24), Maximum Concentration of Contaminants, regulatory level, 5.0 mg/L. Land Ban chemical.
RCRA Universal Treatment Standards: Wastewater (mg/L), 0.43; Non-wastewater (mg/L), 0.30 TCLP.
RCRA Ground Water Monitoring List. Suggested test method(s) (PQL µg/L): 6010(70); 7760(100).
SAFE DRINKING WATER ACT: SMCL, 0.1 mg/L.
EPCRA Section 304 Reportable Quantity (RQ): CERCLA, 1,000 lbs. (454 kgs.).
EPCRA Section 313 Form R de minimus concentration reporting level: 1.0%. Form R Toxic Chemical Category Code: N740.

EPA NAME: SILVER COMPOUNDS

[see also SILVER]
CLEAN WATER ACT: Section 307 Toxic Pollutants as silver and compounds.
RCRA Section 261 Hazardous Constituents, as silver compounds, n.o.s., waste number not listed. Land Ban chemical.
EPCRA Section 313: Includes any unique chemical substance that contains silver as part of that chemical's infrastructure. Form R de minimus concentration reporting level: 1.0%. Form R Toxic Chemical Category Code: N740.

EPA NAME: SILVER CYANIDE

[see also CYANIDE COMPOUNDS, SILVER COMPOUNDS]
CAS: 506-64-9
CLEAN WATER ACT: Section 307 Toxic Pollutants as silver and compounds.
EPA HAZARDOUS WASTE NUMBER (RCRA No.): P104.
RCRA Section 261 Hazardous Constituents. Land Ban chemical.
EPCRA Section 304 Reportable Quantity (RQ): CERCLA, 1 lb. (0.454 kg.).
EPCRA Section 313 Form R de minimus concentration reporting level: 1.0%. Form R Toxic Chemical Category Code: N740.
MARINE POLLUTANT (49CFR, Subchapter 172.101, Appendix B).

EPA NAME: SILVER NITRATE

[see also SILVER COMPOUNDS]
CAS: 7761-88-8
SYNONYMS: NITRIC ACID, SILVER(I) SALT; SILVER(I) NITRATE
CLEAN WATER ACT: Section 311 Hazardous Substances/RQ (same as CERCLA); Section 313 Priority Chemicals; Section 307 Toxic Pollutants as silver and compounds.
RCRA Section 261 Hazardous Constituents, as silver compounds, n.o.s., waste number not listed. Land Ban chemical.
EPCRA Section 304 Reportable Quantity (RQ): CERCLA, 1 lb. (0.454 kg.).

EPCRA Section 313 Form R de minimus concentration reporting level: 1.0%. (silver); 1.0% (nitrate compounds, water dissociable; reportable only when in aqueous solution). Molecular weight 169.87. Form R Toxic Chemical Category Code: N740 (silver); N511 (nitrate compounds).

EPA NAME: SILVEX (2,4,5-TP)

CAS: 93-72-1

SYNONYMS: PROPANOIC ACID, 2-(2,4,5-TRICHLOROPHENOXY)-; SILVEX; 2-(2,4,5-TRICHLOROPHENOXY) PROPANOIC ACID; 2,4,5-TP; 2,4,5-TP ACID

CLEAN WATER ACT: Section 311 Hazardous Substances/RQ (same as CERCLA) as 2,4,5-TP acid.

EPA HAZARDOUS WASTE NUMBER (RCRA No.): U233.

RCRA Section 261 Hazardous Constituents.

RCRA Land Ban Waste.

RCRA Ground Water Monitoring List. Suggested test method(s) (PQL μg/L): 8150(2).

SAFE DRINKING WATER ACT: MCL, 0.05 mg/L; MCLG, 0.05 mg/L; Regulated chemical (47 FR 9352) as 2,4,5-TP.

RCRA Universal Treatment Standards: Wastewater (mg/L), 0.72; Nonwastewater (mg/kg), 7.9.

EPCRA Section 304 Reportable Quantity (RQ): CERCLA, 100 lbs. (45.4 kgs.).

EPA NAME: SIMAZINE

CAS: 122-34-9

SYNONYMS: 2,4-BIS(ETHYLAMINO)-6-CHLORO-S-TRIAZINE; S-TRIAZINE, 2-CHLORO-4,6-BIS(ETHYLAMINO)-; EPA PESTICIDE CHEMICAL CODE 080807

SAFE DRINKING WATER ACT: MCL, 0.004 mg/L; MCLG, 0.004 mg/L; Regulated chemical (47 FR 9352).

EPCRA Section 313 Form R de minimus concentration reporting level: 1.0%.

EPA NAME: SODIUM

CAS: 7440-23-5

SYNONYMS: ELEMENTAL SODIUM; SODIUM ELEMENTAL; SODIUM METAL

CLEAN WATER ACT: Section 311 Hazardous Substances/RQ (same as CERCLA).

EPCRA Section 304 Reportable Quantity (RQ): CERCLA, 10 lbs. (4.54 kgs.).

EPA NAME: SODIUM ARSENATE

CAS: 7631-89-2

SYNONYMS: ARSENIC ACID, SODIUM SALT

CLEAN WATER ACT: Section 311 Hazardous Substances/RQ (same as CERCLA); Section 313 Priority Chemicals.

EPCRA Section 302 Extremely Hazardous Substances: TPQ = 1,000/10,000 lbs. (454/4,540 kgs.).
EPCRA Section 304 Reportable Quantity (RQ): EHS/CERCLA, 1 lb. (0.454 kg.).

EPA NAME: SODIUM ARSENITE
CAS: 7784-46-5
SYNONYMS: ARSENOUS ACID, SODIUM SALT; SODIUM META-ARSENITE

CLEAN WATER ACT: Section 311 Hazardous Substances/RQ (same as CERCLA); Section 313 Priority Chemicals.
EPCRA Section 302 Extremely Hazardous Substances: TPQ = 500/10,000 lbs. (227/4,540 kgs.).
EPCRA Section 304 Reportable Quantity (RQ): EHS/CERCLA, 1 lb. (0.454 kg.).

EPA NAME: SODIUM AZIDE ($Na(N_3)$)
CAS: 26628-22-8
SYNONYMS: HYDRAZOIC ACID, SODIUM SALT; EPA PESTICIDE CHEMICAL CODE 107701

EPA HAZARDOUS WASTE NUMBER (RCRA No.): P105.
RCRA Section 261 Hazardous Constituents.
EPCRA Section 302 Extremely Hazardous Substances: TPQ = 500 lbs. (227 kgs.).
EPCRA Section 304 Reportable Quantity (RQ): EHS/CERCLA, 1,000 lbs. (454 kgs.).
EPCRA Section 313 Form R de minimus concentration reporting level: 1.0%.

EPA NAME: SODIUM BICHROMATE
CAS: 10588-01-9
SYNONYMS: CHROMIC ACID, DISODIUM SALT; SODIUM DICHROMATE(VI); SODIUM DICHROMATE

CLEAN WATER ACT: Section 311 Hazardous Substances/RQ (same as CERCLA); Section 313 Priority Chemicals.
EPCRA Section 304 Reportable Quantity (RQ): CERCLA, 10 lbs. (4.54 kgs.).

EPA NAME: SODIUM BIFLUORIDE
CAS: 1333-83-1
SYNONYMS: HYDROFLUORIC ACID, SODIUM SALT (2:1); SODIUM FLUORIDE ($Na(HF_2)$); SODIUM HYDROGEN DIFLUORIDE; SODIUM HYDROGEN FLUORIDE

CLEAN WATER ACT: Section 311 Hazardous Substances/RQ (same as CERCLA).
EPCRA Section 304 Reportable Quantity (RQ): CERCLA, 100 lbs. (45.4 kgs.).

EPA NAME: SODIUM BISULFITE
CAS: 7631-90-5

SYNONYMS: SODIUM DISULFITE; SODIUM HYDROGEN SULFITE; SULFUROUS ACID, MONOSODIUM SALT

CLEAN WATER ACT: Section 311 Hazardous Substances/RQ (same as CERCLA).

EPCRA Section 304 Reportable Quantity (RQ): CERCLA, 5,000 lbs. (2270 kgs.).

EPA NAME: SODIUM CACODYLATE

CAS: 124-65-2

SYNONYMS: CACODYLIC ACID SODIUM SALT; HYDRODIMETHYLARSINE OXIDE, SODIUM SALT

EPCRA Section 302 Extremely Hazardous Substances: TPQ = 100/10,000 lbs. (45.4/4,540 kgs.).

EPCRA Section 304 Reportable Quantity (RQ): EHS, 100 lbs. (45.4 kgs.).

EPA NAME: SODIUM CHROMATE

[see also CHROMIUM COMPOUNDS]

CAS: 7775-11-3

SYNONYMS: CHROMIC ACID, DISODIUM SALT; SODIUM CHROMATE(VI)

CLEAN WATER ACT: Section 311 Hazardous Substances/RQ (same as CERCLA); Section 313 Priority Chemicals.

EPCRA Section 304 Reportable Quantity (RQ): CERCLA, 10 lbs. (4.54 kgs.).

EPA NAME: SODIUM CYANIDE (Na(CN))

[see also CYANIDE COMPOUNDS]

CAS: 143-33-9

SYNONYMS: HYDROCYANIC ACID, SODIUM SALT

CLEAN WATER ACT: Section 311 Hazardous Substances/RQ (same as CERCLA); Section 313 Priority Chemicals.

EPA HAZARDOUS WASTE NUMBER (RCRA No.): P106.

RCRA Section 261 Hazardous Constituents.

EPCRA Section 302 Extremely Hazardous Substances: TPQ = 100 lbs. (45.4 kgs.).

EPCRA Section 304 Reportable Quantity (RQ): EHS/CERCLA, 10 lbs. (4.54 kgs.).

EPCRA Section 313: See Cyanide Compounds.

MARINE POLLUTANT (49CFR, Subchapter 172.101, Appendix B).

EPA NAME: SODIUM DICAMBA

CAS: 1982-69-0

SYNONYMS: BENZOIC ACID, 3,6-DICHLORO-2-METHOXY-, SODIUM SALT; 3,6-DICHLORO-2-METHOXYBENZOIC ACID, SODIUM SALT; SODIUM 3,6-DICHLORO-2-METHOXYBENZOATE; SODIUM 2-METHOXY-3,6-DICHLOROBENZOATE

EPCRA Section 313 Form R de minimus concentration reporting level: 1.0%.

EPA NAME: SODIUM DIMETHYLDITHIOCARBAMATE
CAS: 128-04-1
SYNONYMS: CARBAMIC ACID, DIMETHYLDITHIO-, SODIUM SALT; EPA PESTICIDE CHEMICAL CODE 034804; SODIUM DIMETHYLAMINOCARBODITHIOATE

EPA HAZARDOUS WASTE NUMBER (RCRA No.): U382.
RCRA Section 261 Hazardous Constituents.
EPCRA Section 304 Reportable Quantity (RQ): CERCLA, 1 lb. (0.454 kgs.).
EPCRA Section 313 Form R de minimus concentration reporting level: 1.0%.

EPA NAME: SODIUM DODECYLBENZENESULFONATE
CAS: 25155-30-0
SYNONYMS: BENZENE SULFONIC ACID, DODECYL-, SODIUM SALT; DODECYLBENZENESULPHONATE, SODIUM SALT

CLEAN WATER ACT: Section 311 Hazardous Substances/RQ (same as CERCLA).
EPCRA Section 304 Reportable Quantity (RQ): CERCLA, 1,000 lbs. (454 kgs.).

EPA NAME: SODIUM FLUORIDE
CAS: 7681-49-4
SYNONYMS: SODIUM HYDROFLUORIDE; TRISODIUM TRIFLUORIDE

CLEAN WATER ACT: Section 311 Hazardous Substances/RQ (same as CERCLA).
EPCRA Section 304 Reportable Quantity (RQ): CERCLA, 1,000 lbs. (454 kgs.).

EPA NAME: SODIUM FLUOROACETATE
CAS: 62-74-8
SYNONYMS: ACETIC ACID, FLUORO-, SODIUM SALT; FLUOROACETIC ACID, SODIUM SALT; SODIUM FLUOACETIC ACID; EPA PESTICIDE CHEMICAL CODE 075003

EPA HAZARDOUS WASTE NUMBER (RCRA No.): P058.
RCRA Section 261 Hazardous Constituents.
EPCRA Section 302 Extremely Hazardous Substances: TPQ = 10/10,000 lbs. (4.54/4,540 kgs.).
EPCRA Section 304 Reportable Quantity (RQ): EHS/CERCLA, 10 lbs. (4.54 kgs.).
EPCRA Section 313 Form R de minimus concentration reporting level: 1.0%.

EPA NAME: SODIUM HYDROSULFIDE
CAS: 16721-80-5
SYNONYMS: SODIUM BISULFIDE; SODIUM SULFIDE; SODIUM SULFHYDRATE

CLEAN WATER ACT: Section 311 Hazardous Substances/RQ (same as CERCLA).

EPCRA Section 304 Reportable Quantity (RQ): CERCLA, 5,000 lbs. (2270 kgs.).

EPA NAME: SODIUM HYDROXIDE

CAS: 1310-73-2

SYNONYMS: SODIUM HYDRATE; SODA LYE

CLEAN WATER ACT: Section 311 Hazardous Substances/RQ (same as CERCLA).

EPCRA Section 304 Reportable Quantity (RQ): CERCLA, 1,000 lbs. (454 kgs.).

EPA NAME: SODIUM HYPOCHLORITE

CAS: 7681-52-9

SYNONYMS: HYPOCHLOROUS ACID, SODIUM SALT

CLEAN WATER ACT: Section 311 Hazardous Substances/RQ (same as CERCLA).

EPCRA Section 304 Reportable Quantity (RQ): CERCLA, 100 lbs. (45.4 kgs.).

EPA NAME: SODIUM METHYLATE

CAS: 124-41-4

SYNONYMS: METHANOL, SODIUM SALT; SODIUM METHOXIDE

CLEAN WATER ACT: Section 311 Hazardous Substances/RQ (same as CERCLA).

EPCRA Section 304 Reportable Quantity (RQ): CERCLA, 1,000 lbs. (454 kgs.).

EPA NAME: SODIUM METHYLDITHIOCARBAMATE

[see METHAM SODIUM]

CAS: 137-42-8

EPA NAME: SODIUM NITRITE

CAS: 7632-00-0

SYNONYMS: EPA PESTICIDE CHEMICAL CODE 076204; NITROUS ACID, SODIUM SALT

CLEAN WATER ACT: Section 311 Hazardous Substances/RQ (same as CERCLA).

EPCRA Section 304 Reportable Quantity (RQ): CERCLA, 100 lbs. (45.4 kgs.).

EPCRA Section 313 Form R de minimus concentration reporting level: 1.0%.

EPA NAME: SODIUM PENTACHLOROPHENATE

CAS: 131-52-2

SYNONYMS: EPA PESTICIDE CHEMICAL CODE 063003; PENTACHLOROPHENATE SODIUM; PENTACHLOROPHENOL, SODIUM SALT; SODIUM PENTACHLOROPHENOL

EPA HAZARDOUS WASTE NUMBER (RCRA No.): listed as "None."
RCRA Section 261 Hazardous Constituents.
EPCRA Section 313 Form R de minimus concentration reporting level: 1.0%.
MARINE POLLUTANT (49CFR, Subchapter 172.101, Appendix B).

EPA NAME: SODIUM O-PHENYLPHENOXIDE
CAS: 132-27-4
SYNONYMS: (1,1′-BIPHENYL)-2-OL, SODIUM SALT; o-PHENYLPHENOL SODIUM; EPA PESTICIDE CHEMICAL CODE 064104; 2-PHENYLPHENOL SODIUM SALT; o-PHENYLPHENOL, SODIUM SALT; SODIUM o-PHENYLPHENATE; SODIUM 2-PHENYLPHENATE

EPCRA Section 313 Form R de minimus concentration reporting level: 0.1%.

EPA NAME: SODIUM PHOSPHATE, DIBASIC
CAS: 7558-79-4
SYNONYMS: DIBASIC SODIUM PHOSPHATE; DISODIUM PHOSPHATE; PHOSPHORIC ACID, DISODIUM SALT

CLEAN WATER ACT: Section 311 Hazardous Substances/RQ (same as CERCLA).
EPCRA Section 304 Reportable Quantity (RQ): CERCLA, 5,000 lbs. (2270 kgs.).

EPA NAME: SODIUM PHOSPHATE, DIBASIC
CAS: 10039-32-4
SYNONYMS: PHOSPHORIC ACID, DISODIUM SALT, DODECAHYDRATE

EPCRA Section 304 Reportable Quantity (RQ): CERCLA, 5,000 lbs. (2270 kgs.).

EPA NAME: SODIUM PHOSPHATE, DIBASIC
CAS: 10140-65-5
SYNONYMS: PHOSPHORIC ACID, DISODIUM SALT, HYDRATE; SODIUM PHOSPHATE, DIBASIC MONOHYDRATE

EPCRA Section 304 Reportable Quantity (RQ): CERCLA, 5,000 lbs. (2270 kgs.).

EPA NAME: SODIUM PHOSPHATE, TRIBASIC
CAS: 7601-54-9
SYNONYMS: PHOSPHORIC ACID, TRISODIUM SALT; TRIBASIC SODIUM PHOSPHATE; TRISODIUM PHOSPHATE

CLEAN WATER ACT: Section 311 Hazardous Substances/RQ (same as CERCLA).
EPCRA Section 304 Reportable Quantity (RQ): CERCLA, 5,000 lbs. (2270 kgs.).

EPA NAME: SODIUM PHOSPHATE, TRIBASIC
CAS: 7758-29-4

SYNONYMS: TRIPHOSPHORIC ACID, PENTASODIUM SALT

CLEAN WATER ACT: Section 311 Hazardous Substances/RQ (same as CERCLA).

EPCRA Section 304 Reportable Quantity (RQ): CERCLA, 5,000 lbs. (2270 kgs.).

EPA NAME: SODIUM PHOSPHATE, TRIBASIC

CAS: 7785-84-4

SYNONYMS: METAPHOSPHORIC ACID, TRISODIUM SALT; SODIUM TRIMETAPHOSPHATE

CLEAN WATER ACT: Section 311 Hazardous Substances/RQ (same as CERCLA).

EPCRA Section 304 Reportable Quantity (RQ): CERCLA, 5,000 lbs. (2270 kgs.).

EPA NAME: SODIUM PHOSPHATE, TRIBASIC

CAS: 10101-89-0

SYNONYMS: PHOSPHORIC ACID, TRISODIUM SALT, DODECAHYDRATE; TRIBASIC SODIUM PHOSPHATE DODECAHYDRATE

CLEAN WATER ACT: Section 311 Hazardous Substances/RQ (same as CERCLA).

EPCRA Section 304 Reportable Quantity (RQ): CERCLA, 5,000 lbs. (2270 kgs.).

EPA NAME: SODIUM PHOSPHATE, TRIBASIC

CAS: 10124-56-8

SYNONYMS: HEXAMETAPHOSPHATE, SODIUM SALT; METAPHOSPHORIC ACID, HEXASODIUM SALT; PHOSPHATE, SODIUM HEXAMETA-; SODIUM HEXAMETAPHOSPHATE

CLEAN WATER ACT: Section 311 Hazardous Substances/RQ (same as CERCLA).

EPCRA Section 304 Reportable Quantity (RQ): CERCLA, 5,000 lbs. (2270 kgs.).

EPA NAME: SODIUM PHOSPHATE, TRIBASIC

CAS: 10361-89-4

SYNONYMS: PHOSPHORIC ACID, TRISODIUM SALT, DECAHYDRATE; SODIUM PHOSPHATE, TRISODIUM SALT, DECAHYDRATE; TRIBASIC SODIUM PHOSPHATE

CLEAN WATER ACT: Section 311 Hazardous Substances/RQ (same as CERCLA).

EPCRA Section 304 Reportable Quantity (RQ): CERCLA, 5,000 lbs. (2270 kgs.).

EPA NAME: SODIUM SELENATE

CAS: 13410-01-0

SYNONYMS: DISODIUM SELENATE

EPCRA Section 302 Extremely Hazardous Substances: TPQ = 100/10,000 lbs. (45.4/4,540 kgs.).
EPCRA Section 304 Reportable Quantity (RQ): EHS, 100 lbs. (45.4 kgs.).

EPA NAME: SODIUM SELENITE
CAS: 10102-18-8; 7782-82-3
SYNONYMS: DISODIUM SELENITE; SELENIOUS ACID, DISODIUM SALT

CLEAN WATER ACT: Section 311 Hazardous Substances/RQ (same as CERCLA); Section 313 Priority Chemicals.
EPCRA Section 302 Extremely Hazardous Substances: TPQ = 100/10,000 lbs. (45.4/4,540 kgs.).
EPCRA Section 304 Reportable Quantity (RQ): EHS/CERCLA, 100 lbs. (45.4 kgs.).

EPA NAME: SODIUM TELLURITE
CAS: 10102-20-2
SYNONYMS: SODIUM TELLURATE(IV); TELLUROUS ACID, DISODIUM SALT

EPCRA Section 302 Extremely Hazardous Substances: TPQ = 500/10,000 lbs. (227/4,540 kgs.).
EPCRA Section 304 Reportable Quantity (RQ): EHS, 500 lbs. (227 kgs.).

EPA NAME: STANNANE, ACETOXYTRIPHENYL-
CAS: 900-95-8
SYNONYMS: ACETOTRIPHENYLSTANNANE; ACETOTRIPHENYLSTANNINE; FENTIN ACETATE; TRIPHENYLTIN ACETATE

EPCRA Section 302 Extremely Hazardous Substances: TPQ = 500/10,000 lbs. (227/4,540 kgs.).
EPCRA Section 304 Reportable Quantity (RQ): EHS, 500 lbs. (227 kgs.).

EPA NAME: STRONTIUM CHROMATE
[see also CHROMIUM COMPOUNDS]
CAS: 7789-06-2
SYNONYMS: CHROMIC ACID, STRONTIUM SALT (1:1); STRONTIUM CHROMATE (VI)

CLEAN WATER ACT: Section 311 Hazardous Substances/RQ (same as CERCLA).
EPCRA Section 304 Reportable Quantity (RQ): CERCLA, 10 lbs. (4.54 kgs.).

EPA NAME: STRYCHNINE
[see also STRYCHNINE and SALTS]
CAS: 57-24-9
SYNONYMS: STRYCHNIDIN-10-ONE

CLEAN WATER ACT: Section 311 Hazardous Substances/RQ (same as CERCLA).
EPA HAZARDOUS WASTE NUMBER (RCRA No.): P108.

RCRA Section 261 Hazardous Constituents.
EPCRA Section 302 Extremely Hazardous Substances: TPQ= 100/10,000 lbs. (45.4/4,540 kgs.).
EPCRA Section 304 Reportable Quantity (RQ): EHS/CERCLA, 10 lbs. (4.54 kgs.).
EPCRA Section 313 Form R de minimus concentration reporting level: 1.0%. Form R Toxic Chemical Category Code: N746.
MARINE POLLUTANT (49CFR, Subchapter 172.101, Appendix B).

EPA NAME: STRYCHNINE AND SALTS
[see also STRYCHNINE]
EPA HAZARDOUS WASTE NUMBER (RCRA No.): P108.
RCRA Section 261 Hazardous Constituents. as strychnine salts.
EPCRA Section 313: Includes any unique chemical substance that contains strychnine or a strychnine salt as part of that chemical's infrastructure. Form R de minimus concentration reporting level: 1.0%. Form R Toxic Chemical Category Code: N746.
MARINE POLLUTANT (49CFR, Subchapter 172.101, Appendix B).

EPA NAME: STRYCHNINE, SULFATE
[see also STRYCHNINE and STRYCHNINE AND SALTS]
CAS: 60-41-3
SYNONYMS: STRYCHININE SULFATE; STRYCHNINE SULFATE (2:1); STRYCHNIDIN-10-ONE SULFATE (2:1)

EPCRA Section 302 Extremely Hazardous Substances: TPQ= 100/10,000 lbs. (45.4/4,540 kgs.).
EPCRA Section 304 Reportable Quantity (RQ): EHS, 10 lbs. (4.54 kgs.).
EPCRA Section 313 Form R de minimus concentration reporting level: 1.0%. Form R Toxic Chemical Category Code: N746.

EPA NAME: STYRENE
CAS: 100-42-5
SYNONYMS: BENZENE, ETHENYL-; PHENYLETHYLENE; STYRENE MONOMER; VINYLBENZENE

CLEAN AIR ACT: Hazardous Air Pollutants (Title I, Part A, Section 112).
CLEAN WATER ACT: Section 311 Hazardous Substances/RQ (same as CERCLA); Section 313 Priority Chemicals.
RCRA Ground Water Monitoring List. Suggested test method(s) (PQL µg/L): 8020(1); 8240(5).
SAFE DRINKING WATER ACT: MCL, 0.1 mg/L; MCLG, 0.1 mg/L.
EPCRA Section 304 Reportable Quantity (RQ): CERCLA, 1,000 lbs. (454 kgs.).
EPCRA Section 313 Form R de minimus concentration reporting level: 0.1%.
MARINE POLLUTANT (49CFR, Subchapter 172.101, Appendix B).

EPA NAME: STYRENE OXIDE
CAS: 96-09-3
SYNONYMS: BENZENE, (EPOXYETHYL)-; α,β-EPOXYSTYRENE; OXIRANE, PHENYL-; 2-PHENYLOXIRANE; PHENYLOXIRANE

CLEAN AIR ACT: Hazardous Air Pollutants (Title I, Part A, Section 112).
EPCRA Section 304 Reportable Quantity (RQ): CERCLA, 100 lbs. (45.4 kgs.).
EPCRA Section 313 Form R de minimus concentration reporting level: 0.1%.
CALIFORNIA'S PROPOSITION 65: Carcinogen.

EPA NAME: SULFATE
SAFE DRINKING WATER ACT: SMCL, 250 mg/L.

EPA NAME: SULFOTEP
CAS: 3689-24-5
SYNONYMS: PYROPHOSPHORODITHIOIC ACID, TETRAETHYL ESTER; TEDP; TETRAETHYL DITHIOPYROPHOSPHATE; TETRAETHYLDITHIONOPYROPHOSPHATE; THIODIPHOSPHORIC ACID, TETRAETHYL ESTER

EPA HAZARDOUS WASTE NUMBER (RCRA No.): P109.
RCRA Section 261 Hazardous Constituents.
RCRA Ground Water Monitoring List. Suggested test method(s) (PQL µg/L): 8270(10).
EPCRA Section 302 Extremely Hazardous Substances: TPQ = 500 lbs. (227 kgs.).
EPCRA Section 304 Reportable Quantity (RQ): EHS/CERCLA, 100 lbs. (45.4 kgs.).
MARINE POLLUTANT (49CFR, Subchapter 172.101, Appendix B).

EPA NAME: SULFOXIDE, 3-CHLOROPROPYL OCTYL
CAS: 3569-57-1
SYNONYMS: 3-CHLOROPROPYL-N-OCTYLSULFOXIDE

EPCRA Section 302 Extremely Hazardous Substances: TPQ = 500 lbs. (227 kgs.).
EPCRA Section 304 Reportable Quantity (RQ): EHS, 500 lbs. (227 kgs.).

EPA NAME: SULFUR DIOXIDE
[see also SULFUR DIOXIDE (ANHYDROUS)]
CAS: 7446-09-5
SYNONYMS: BISULFITE; SULFUROUS OXIDE

EPCRA Section 302 Extremely Hazardous Substances: TPQ = 500 lbs. (227 kgs.).
EPCRA Section 304 Reportable Quantity (RQ): EHS, 500 lbs. (227 kgs.).

EPA NAME: SULFUR DIOXIDE (ANHYDROUS)
[see also SULFUR DIOXIDE]
CAS: 7446-09-5
SYNONYMS: SULFUROUS OXIDE

CLEAN AIR ACT: Accidental Release Prevention/Flammable substances (Section 112[r], Table 3), TQ = 5,000 lbs. (2270 kgs.).
EPCRA Section 302 Extremely Hazardous Substances: TPQ = 500 lbs. (227 kgs.).
EPCRA Section 304 Reportable Quantity (RQ): EHS, 500 lbs. (227 kgs.).

EPA NAME: SULFUR FLUORIDE (SF4), (T-4)-
CAS: 7783-60-0
SYNONYMS: SULFUR TETRAFLUORIDE

CLEAN AIR ACT: Accidental Release Prevention/Flammable substances (Section 112[r], Table 3), TQ=2,500 lbs. (1135 kgs.).
EPCRA Section 302 Extremely Hazardous Substances: TPQ=100 lbs. (45.4 kgs.).
EPCRA Section 304 Reportable Quantity (RQ): EHS, 100 lbs. (45.4 kgs.).

EPA NAME: SULFURIC ACID
CAS: 7664-93-9
SYNONYMS: HYDROGEN SULFATE

CLEAN WATER ACT: Section 311 Hazardous Substances/RQ (same as CERCLA); Section 313 Priority Chemicals.
EPCRA Section 302 Extremely Hazardous Substances: TPQ=1,000 lbs. (454 kgs.).
EPCRA Section 304 Reportable Quantity (RQ): EHS/CERCLA, 1,000 lbs. (454 kgs.).
EPCRA Section 313 (acid aerosols including mists, vapors, gas, fog, and other airborne species of any particle size) Form R de minimus concentration reporting level: 0.1%.

EPA NAME: SULFURIC ACID (FUMING)
[see OLEUM (FUMING SULFURIC ACID)]
CAS: 8014-95-7

EPA NAME: SULFURIC ACID, MIXTURE WITH SULFUR TRIOXIDE
[see OLEUM (FUMING SULFURIC ACID)]
CAS: 8014-95-7

EPA NAME: SULFUR MONOCHLORIDE
CAS: 12771-08-3
SYNONYMS: DISULFUR DICHLORIDE; SULFUR CHLORIDE; SULFUR CHLORIDE (DI)

CLEAN WATER ACT: Section 311 Hazardous Substances/RQ (same as CERCLA).
EPCRA Section 304 Reportable Quantity (RQ): CERCLA, 1,000 lbs. (454 kgs.).

EPA NAME: SULFUR PHOSPHIDE
CAS: 1314-80-3
SYNONYMS: DIPHOSPHOROUS PENTASULPHIDE; PHOSPHORUS PENTASULFIDE; PHOSPHORUS SULFIDE

EPA HAZARDOUS WASTE NUMBER (RCRA No.): U189.
RCRA Section 261 Hazardous Constituents.
EPCRA Section 304 Reportable Quantity (RQ): CERCLA, 100 lbs. (45.4 kgs.).

EPA NAME: SULFUR TETRAFLUORIDE
[see SULFUR FLUORIDE (SF4), (T-4)-]
CAS: 7783-60-0

EPA NAME: SULFUR TRIOXIDE
CAS: 7446-11-9
SYNONYMS: SULFURIC OXIDE

CLEAN AIR ACT: Accidental Release Prevention/Flammable substances (Section 112[r], Table 3), TQ = 10,000 lbs. (4540 kgs.).
EPCRA Section 302 Extremely Hazardous Substances: TPQ = 100 lbs. (45.4 kgs.).
EPCRA Section 304 Reportable Quantity (RQ): EHS, 100 lbs. (45.4 kgs.).

EPA NAME: SULFURYL FLUORIDE
CAS: 2699-79-8
SYNONYMS: EPA PESTICIDE CHEMICAL CODE 078003; SULFONYL FLUORIDE; SULPHURYL DIFLUORIDE; VIKANE

EPCRA Section 313 Form R de minimus concentration reporting level: 1.0%.

EPA NAME: SULPROFOS
CAS: 35400-43-2
SYNONYMS: BOLSTAR; O-ETHYL O-(4-(METHYLMERCAPTO)PHENYL)-S-N-PROPYLPHOSPHOROTHIONOTHIOLATE; O-ETHYL O-[4-(METHYLTHIO)PHENYL]PHOSPHORODITHIOIC ACID S-PROPYL ESTER; O-ETHYL O-(4-(METHYLTHIO)PHENYL)PHOSPHORODITHIOIC ACID S-PROPYL ESTER; PHOSPHORODITHIOIC ACID, O-ETHYL O-(4-(METHYLTHIO)PHENYL) S-PROPYL ESTER

EPCRA Section 313 Form R de minimus concentration reporting level: 1.0%.
MARINE POLLUTANT (49CFR, Subchapter 172.101, Appendix B): Severe pollutant.

- T -

EPA NAME: 2,4,5-T ACID

CAS: 93-76-5

SYNONYMS: ACETIC ACID, (2,4,5-TRICHLOROPHYNOXY)-; 2,4,5-T; 2,4,5-TRICHLOROPHENOXYACETIC ACID

CLEAN WATER ACT: Section 311 Hazardous Substances/RQ (same as CERCLA).

EPA HAZARDOUS WASTE NUMBER (RCRA No.): U232; F027.

RCRA Section 261 Hazardous Constituents.

RCRA Land Ban Waste.

RCRA Universal Treatment Standards: Wastewater (mg/L), 0.72; Non-wastewater (mg/kg), 7.9.

RCRA Ground Water Monitoring List. Suggested test method(s) (PQL µg/L): 8150(2).

SAFE DRINKING WATER ACT: Priority List (55 FR 1470).

EPCRA Section 304 Reportable Quantity (RQ): CERCLA, 1,000 lbs. (454 kgs.).

EPA NAME: 2,4,5-T AMINES

CAS: 1319-72-8

SYNONYMS: ACETIC ACID, (2,4,5-TRICHLOROPHYNOXY)-COMP. WITH 1-AMINO-2-PROPANOL (1:1); 2,4,5-T ACID AMINE

EPCRA Section 304 Reportable Quantity (RQ): CERCLA, 5,000 lbs. (2270 kgs.).

EPA NAME: 2,4,5-T AMINES

CAS: 2008-46-0

SYNONYMS: ACETIC ACID, (2,4,5-TRICHLOROPHYNOXY)-COMP. WITH N,N-DIETHYLETHANAMINE; 1-AMINO-2-PROPANOL (1:1); 2,4,5-T ACID AMINE

EPCRA Section 304 Reportable Quantity (RQ): CERCLA, 5,000 lbs. (2270 kgs.).

EPA NAME: 2,4,5-T AMINES

CAS: 3813-14-7

SYNONYMS: ACETIC ACID, (2,4,5-TRICHLOROPHYNOXY)-COMP. WITH 2,2′,2″-NITRILOTRIS(ETHANOL)(1:1); 2,4,5-T ACID AMINE

CLEAN WATER ACT: Section 311 Hazardous Substances/RQ (same as CERCLA).

EPCRA Section 304 Reportable Quantity (RQ): CERCLA, 5,000 lbs. (2270 kgs.).

EPA NAME: 2,4,5-T AMINES

CAS: 6369-96-6

SYNONYMS: ACETIC ACID, (2,4,5-TRICHLOROPHYNOXY)-COMP. WITH TRIMETHYLAMINE (1:1); 2,4,5-T ACID AMINE; 2,4,5-T SODIUM SALT

CLEAN WATER ACT: Section 311 Hazardous Substances/RQ (same as CERCLA).

EPCRA Section 304 Reportable Quantity (RQ): CERCLA, 5,000 lbs. (2270 kgs.).

EPA NAME: 2,4,5-T AMINES

CAS: 6369-97-7

SYNONYMS: ACETIC ACID, (2,4,5-TRICHLOROPHYNOXY)-COMP. WITH N-METHYLMETHANAMINE; 2,4,5-T ACID AMINE

CLEAN WATER ACT: Section 311 Hazardous Substances/RQ (same as CERCLA).

EPCRA Section 304 Reportable Quantity (RQ): CERCLA, 5,000 lbs. (2270 kgs.).

EPA NAME: 2,4,5-T ESTERS

CAS: 93-79-8

SYNONYMS: ACETIC ACID, (2,4,5-TRICHLOROPHYNOXY)-, BUTYL ESTER; BUTYL-, 2,4,5-T; BUTYLATE-2,4,5-T; 2,4,5-T-N-BUTYL ESTER

CLEAN WATER ACT: Section 311 Hazardous Substances/RQ (same as CERCLA).

EPCRA Section 304 Reportable Quantity (RQ): CERCLA, 1,000 lbs. (454 kgs.).

EPA NAME: 2,4,5-T ESTERS

CAS: 1928-47-8

SYNONYMS: ACETIC ACID, (2,4,5-TRICHLOROPHENOXY)-, 2-ETHYLHEXYL ESTER; ETHYLHEXYL-2,4,5-T; 2,4,5-T ETHYLHEXYL ESTER

CLEAN WATER ACT: Section 311 Hazardous Substances/RQ (same as CERCLA).

EPCRA Section 304 Reportable Quantity (RQ): CERCLA, 1,000 lbs. (454 kgs.).

EPA NAME: 2,4,5-T ESTERS

CAS: 2545-59-7

SYNONYMS: ACETIC ACID, (2,4,5-TRICHLOROPHENOXY)-, 2-BUTOXYETHYL ESTER

CLEAN WATER ACT: Section 311 Hazardous Substances/RQ (same as CERCLA).

EPCRA Section 304 Reportable Quantity (RQ): CERCLA, 1,000 lbs. (454 kgs.).

EPA NAME: 2,4,5-T ESTERS

CAS: 25168-15-4

SYNONYMS: ACETIC ACID, (2,4,5-TRICHLOROPHENOXY)-, ISOOCTYL ESTER; 2,4,5-T-ISOOCTYL ESTER

CLEAN WATER ACT: Section 311 Hazardous Substances/RQ (same as CERCLA).

EPCRA Section 304 Reportable Quantity (RQ): CERCLA, 1,000 lbs. (454 kgs.).

EPA NAME: 2,4,5-T ESTERS

CAS: 61792-07-2

SYNONYMS: ACETIC ACID, (2,4,5-TRICHLOROPHENOXY)-, 1-METHYL PROPYL ESTER

CLEAN WATER ACT: Section 311 Hazardous Substances/RQ (same as CERCLA).

EPCRA Section 304 Reportable Quantity (RQ): CERCLA, 1,000 lbs. (454 kgs.).

EPA NAME: 2,4,5-T SALTS

CAS: 13560-99-1

SYNONYMS: ACETIC ACID, (2,4,5-TRICHLOROPHYNOXY)-, SODIUM SALT; (2,4,5-TRICHLOROPHENOXY)-, ACETIC ACID SODIUM SALT

CLEAN WATER ACT: Section 311 Hazardous Substances/RQ (same as CERCLA).

EPCRA Section 304 Reportable Quantity (RQ): CERCLA, 1,000 lbs. (454 kgs.).

EPA NAME: TABUN

[see also CYANIDE COMPOUNDS]

CAS: 77-81-6

SYNONYMS: ETHYL DIMETHYLPHOSPHORAMIDOCYANIDATE

EPCRA Section 302 Extremely Hazardous Substances: TPQ = 10 lbs. (4.54 kgs.).

EPCRA Section 304 Reportable Quantity (RQ): EHS, 10 lbs. (4.54 kgs.).

EPA NAME: TEBUTHIURON

CAS: 34014-18-1

SYNONYMS: N-[5-(1,1-DIMETHYLETHYL)-1,3,4-THIADIAZOL-2-YL)-N,N′-DIMETHYLUREA; EPA PESTICIDE CHEMICAL CODE 105501; 1-(5-tert-BUTYL-1,3,4-THIADIAZOL-2-YL)-1,3-DIMETHYLUREA; UREA, 2-(5-tert-BUTYL-1,3,4-THIADIAZOL-2-YL)-1,3-DIMETHYL-

EPCRA Section 313 Form R de minimus concentration reporting level: 1.0%.

EPA NAME: TELLURIUM

CAS: 13494-80-9

SYNONYMS: TELLURIUM, ELEMENTAL

Removed from EHS list (FR Vol. 61, No. 89, page 20477).

EPA NAME: TELLURIUM HEXAFLUORIDE
CAS: 7783-80-4
SYNONYMS: TELLURIUM FLUORIDE

EPCRA Section 302 Extremely Hazardous Substances: TPQ = 1,000 lbs. (454 kgs.).
EPCRA Section 304 Reportable Quantity (RQ): EHS, 100 lbs. (45.4 kgs.).

EPA NAME: TEMEPHOS
CAS: 3383-96-8
SYNONYMS: ABATE; EPA PESTICIDE CHEMICAL CODE 059001; PHOSPHOROTHIOIC ACID, O,O′-(THIODI-4,1-PHENYLENE)O,O,O′,O′-TETRAMETHYL ESTER; O,O,O′,O′-TETRAMETHYL O,O′-THIODI-p-PHENYLENE PHOSPHOROTHIOATE; TETRAMETHYL O,O′-THIO-DI-p-PHENYLENE-PHOSPHOROTHIOATE

EPCRA Section 313 Form R de minimus concentration reporting level: 1.0%.
MARINE POLLUTANT (49CFR, Subchapter 172.101, Appendix B).

EPA NAME: TEPP
CAS: 107-49-3
SYNONYMS: BIS-O,O-DIETHYLPHOSPHORIC ANHYDRIDE; DIPHOSPHORIC ACID, TETRAETHYL ESTER; TETRAETHYL PYROPHOSPHATE; O,O,O,O-TETRAETHYL-DIPHOSPHATE

CLEAN WATER ACT: Section 311 Hazardous Substances/RQ (same as CERCLA).
EPA HAZARDOUS WASTE NUMBER (RCRA No.): P111.
EPCRA Section 302 Extremely Hazardous Substances: TPQ = 100 lbs. (45.4 kgs.).
EPCRA Section 304 Reportable Quantity (RQ): EHS/CERCLA, 10 lbs. (4.54 kgs.).
MARINE POLLUTANT (49CFR, Subchapter 172.101, Appendix B).

EPA NAME: TERBACIL
CAS: 5902-51-2
SYNONYMS: 5-CHLORO-3-(1,1-DIMETHYLETHYL)-6-METHYL-2,4(1H,3H)-PYRIMIDINEDIONE; 5-CHLORO-3-tert-BUTYL-6-METHYLURACIL; EPA PESTICIDE CHEMICAL CODE 012701; 2,4(1H,3H)-PYRIMIDINEDIONE,5-CHLORO-3-(1,1-DIMETHYL)-6-METHYL-

EPCRA Section 313 Form R de minimus concentration reporting level: 1.0%.

EPA NAME: TERBUFOS
CAS: 13071-79-9
SYNONYMS: PHOSPHORODITHIOIC ACID S-((tert-BUTYLTHIO)METHYL)-O,O-DIETHYLESTER

EPCRA Section 302 Extremely Hazardous Substances: TPQ = 100 lbs. (45.4 kgs.).

EPCRA Section 304 Reportable Quantity (RQ): EHS, 100 lbs. (45.4 kgs.).
MARINE POLLUTANT (49CFR, Subchapter 172.101, Appendix B): Severe pollutant.

EPA NAME: 1,2,4,5-TETRACHLOROBENZENE
[see also CHLORINATED BENZENES]
CAS: 95-94-3
SYNONYMS: BENZENE, 1,2,4,5-TETRACHLORO-; TETRACHLOROBENZENE

EPA HAZARDOUS WASTE NUMBER (RCRA No.): U207.
RCRA Section 261 Hazardous Constituents.
RCRA Land Ban Waste.
RCRA Universal Treatment Standards: Wastewater (mg/L), 0.055; Nonwastewater (mg/kg), 14.
RCRA Ground Water Monitoring List. Suggested test method(s) (PQL µg/L): 8270(10).
EPCRA Section 304 Reportable Quantity (RQ): CERCLA, 5,000 lbs. (2270 kgs.).

EPA NAME: 2,3,7,8-TETRACHLORODIBENZO-p-DIOXIN (TCDD)
CAS: 1746-01-6
SYNONYMS: DIBENZO[B,E][1,4]DIOXIN, 2,3,7,8-TETRACHLORO-; 2,3,7,8-TCDD; 2,3,6,7-TETRACHLORODIBENZO-para-DIOXIN

CLEAN AIR ACT: Hazardous Air Pollutants (Title I, Part A, Section 112).
CLEAN WATER ACT: Section 307 Priority Pollutants; Section 307 Toxic Pollutants.
RCRA Section 261 Hazardous Constituents, waste number not listed.
RCRA Land Ban Waste.
SAFE DRINKING WATER ACT: MCL, 0.000000003 mg/L; MCLG, zero; Regulated chemical (47 FR 9352).
RCRA Universal Treatment Standards: Wastewater (mg/L), 0.000063; Nonwastewater (mg/kg), 0.001.
RCRA Ground Water Monitoring List. Suggested test method(s) (PQL µg/L): 8250(0.005).
EPCRA Section 304 Reportable Quantity (RQ): CERCLA, 1 lb. (0.454 kg.).
CALIFORNIA'S PROPOSITION 65: Carcinogen; reproductive toxin.

EPA NAME: 1,1,1,2-TETRACHLOROETHANE
[see also CHLORINATED ETHANES]
CAS: 630-20-6
SYNONYMS: ETHANE, 1,1,1,2-TETRACHLORO-

EPA HAZARDOUS WASTE NUMBER (RCRA No.): U208.
RCRA Section 261 Hazardous Constituents.
RCRA Land Ban Waste.
RCRA Universal Treatment Standards: Wastewater (mg/L), 0.057; Nonwastewater (mg/kg), 6.0.

RCRA Ground Water Monitoring List. Suggested test method(s) (PQL µg/L): 8010(5); 8240(5).
SAFE DRINKING WATER ACT: Priority List (55 FR 1470).
EPCRA Section 304 Reportable Quantity (RQ): CERCLA, 100 lbs. (45.4 kgs.).
EPCRA Section 313 Form R de minimus concentration reporting level: 1.0%.
MARINE POLLUTANT (49CFR, Subchapter 172.101, Appendix B).

EPA NAME: 1,1,2,2-TETRACHLOROETHANE

[see also CHLORINATED ETHANES]
CAS: 79-34-5
SYNONYMS: ACETYLENE TETRACHLORIDE; 1,1-DICHLORO-2,2-DICHLOROETHANE; ETHANE, 1,1,2,2-TETRACHLORO-; TETRACHLORETHANE

CLEAN AIR ACT: Hazardous Air Pollutants (Title I, Part A, Section 112).
EPA HAZARDOUS WASTE NUMBER (RCRA No.): U209.
RCRA Land Ban Waste.
RCRA Universal Treatment Standards: Wastewater (mg/L), 0.057; Non-wastewater (mg/kg), 6.0.
RCRA Ground Water Monitoring List. Suggested test methods (PQL µg/L): 8010(0.5); 8240(5).
SAFE DRINKING WATER ACT: Priority List (55 FR 1470).
EPCRA Section 304 Reportable Quantity (RQ): CERCLA, 100 lbs. (45.4 kgs.).
EPCRA Section 313 Form R de minimus concentration reporting level: 1.0%.
MARINE POLLUTANT (49CFR, Subchapter 172.101, Appendix B).
CALIFORNIA'S PROPOSITION 65: Carcinogen.

EPA NAME: TETRACHLOROETHYLENE

CAS: 127-18-4
SYNONYMS: ETHENE, TETRACHLORO-; ETHYLENE TETRA-CHLORIDE; PERCHLORETHYLENE; 1,1,2,2-TETRACHLO-ROETHENE; 1,1,2,2,-TETRACHLOROETHYLENE

CLEAN AIR ACT: Hazardous Air Pollutants (Title I, Part A, Section 112); Section 307 Toxic Pollutants.
CLEAN WATER ACT: Section 307 Priority Pollutants.
EPA HAZARDOUS WASTE NUMBER (RCRA No.): U210.
RCRA Section 261 Hazardous Constituents.
RCRA Land Ban Waste.
RCRA Universal Treatment Standards: Wastewater (mg/L), 0.056; Non-wastewater (mg/kg), 6.0.
RCRA Ground Water Monitoring List. Suggested test method(s) (PQL µg/L): 8010(0.5); 8240(5).
SAFE DRINKING WATER ACT: MCL, 0.005 mg/L; MCLG, zero; regulated chemical (47 FR 9352).
EPCRA Section 304 Reportable Quantity (RQ): CERCLA, 100 lbs. (45.4 kgs.).

EPCRA Section 313 Form R de minimus concentration reporting level: 0.1%.

MARINE POLLUTANT (49CFR, Subchapter 172.101, Appendix B).

CALIFORNIA'S PROPOSITION 65: Carcinogen.

EPA NAME: 1,1,2,2-TETRACHLORO-2-FLUOROETHANE

CAS: 354-11-0

SYNONYMS: ETHANE, 1,1,1,2-TETRACHLORO-1-FLUORO-; HCFC-121a; 1-FLUORO-1,1,1,2-TETRACHLOROETHANE

EPCRA Section 313 Form R de minimus concentration reporting level: 1.0%.

EPA NAME: 1,1,2,2-TETRACHLORO-1-FLUOROETHANE

[see also CHLORINATED ETHANES]

CAS: 354-14-3

SYNONYMS: ETHANE, 1,1,2,2-TETRACHLORO-1-FLUORO-; 1-FLUORO-1,1,2,2-TETRACHLOROETHANE; HCFC-121

CLEAN AIR ACT: Stratospheric ozone protection (40CFR, Part 82), Class II, Ozone Depletion Potential = listed as "reserved."

EPCRA Section 313 Form R de minimus concentration reporting level: 1.0%.

EPA NAME: 2,3,4,6-TETRACHLOROPHENOL

[see also CHLORINATED PHENOLS]

CAS: 58-90-2

SYNONYMS: PHENOL, 2,3,4,6-TETRACHLORO-

EPA HAZARDOUS WASTE NUMBER (RCRA No.): U212.

RCRA Section 261 Hazardous Constituents.

RCRA Land Ban Waste.

RCRA Universal Treatment Standards: Wastewater (mg/L), 0.030; Non-wastewater (mg/kg), 7.4.

RCRA Ground Water Monitoring List. Suggested test method(s) (PQL µg/L): 8270(10).

EPCRA Section 304 Reportable Quantity (RQ): CERCLA, 10 lbs. (4.54 kgs.).

MARINE POLLUTANT (49CFR, Subchapter 172.101, Appendix B).

EPA NAME: TETRACHLORVINPHOS

CAS: 961-11-5

SYNONYMS: BENZYL ALCOHOL, 2,4,5-TRICHLORO-α-(CHLOROMETHYLENE)-, DIMETHYL PHOSPHATE; PHOSPHORIC ACID, 2-CHLORO-1-(2,3,5-TRICHLOROPHENYL) ETHENYL DIMETHYL ESTER

EPCRA Section 313 Form R de minimus concentration reporting level: 1.0%.

EPA NAME: TETRACYCLINE HYDROCHLORIDE

CAS: 64-75-5

SYNONYMS: 2-NAPHTHACENECARBOXAMIDE,4-(DIMETHYL-AMINO)-1,4,4A,5,5A,6,11,12A-OCTAHYDRO-3,6,10,12,12A-PENTAHYDROXY-6-METHYL-1,11-DIOXO-, MONOHYDRO-CHLORIDE, (4S-(4α,4α,5α,6β,12α))-

EPCRA Section 313 Form R de minimus concentration reporting level: 1.0%.

EPA NAME: TETRAETHYLDITHIONOPYROPHOSPHATE
[see SULFOTEP]
CAS: 3689-24-5

EPA NAME: TETRAETHYL LEAD
[see also LEAD and LEAD COMPOUNDS]
CAS: 78-00-2
SYNONYMS: LEAD TETRAETHYL; PLUMBANE, TETRAETHYL-

CLEAN WATER ACT: Section 311 Hazardous Substances/RQ (same as CERCLA).
EPA HAZARDOUS WASTE NUMBER (RCRA No.): P110.
RCRA Section 261 Hazardous Constituents.
EPCRA Section 302 Extremely Hazardous Substances: TPQ = 100 lbs. (45.4 kgs.).
EPCRA Section 304 Reportable Quantity (RQ): EHS/CERCLA, 10 lbs. (4.54 kgs.).
MARINE POLLUTANT (49CFR, Subchapter 172.101, Appendix B) (liquid).

EPA NAME: TETRAETHYL PYROPHOSPHATE
[see TEPP]
CAS: 107-49-3

EPA NAME: TETRAETHYLTIN
CAS: 597-64-8
SYNONYMS: STANNANE, TETRAETHYL-; TIN, TETRAETHYL-

EPCRA Section 302 Extremely Hazardous Substances: TPQ = 100 lbs. (45.4 kgs.).
EPCRA Section 304 Reportable Quantity (RQ): EHS, 100 lbs. (45.4 kgs.).

EPA NAME: TETRAFLUOROETHYLENE
CAS: 116-14-3
SYNONYMS: ETHENE, TETRAFLUORO-; 1,1,2,2-TETRAFLUORO-ETHYLENE

CLEAN AIR ACT: Accidental Release Prevention/Flammable substances (Section 112[r], Table 3), TQ = 10,000 lbs. (4540 kgs.).

EPA NAME: TETRAHYDRO-5,5-DIMETHYL-2(1H)-PYRIMIDI-NONE[3-[4-(TRIFLUOROMETHYL)PHENYL]-1-[2-[4-(TRI-FLUOROMETHYL)PHENYL]ETHENYL]-2-PROPENYLI-DENE]HYDRAZONE
[see HYDRAMETHYLNON]
CAS: 67485-29-4

EPA NAME: TETRAHYDRO-3,5-DIMETHYL-2H-1,3,5-THIA-DIAZINE-2-THIONE
[see DAZOMET]
CAS: 533-74-4

EPA NAME: TETRAHYDRO-3,5-DIMETHYL-2H-1,3,5-THIA-DIAZINE-2-THIONE, ION(1-), SODIUM
[see DAZOMET, SODIUM SALT]
CAS: 53404-60-7

EPA NAME: TETRAMETHRIN
CAS: 7696-12-0
SYNONYMS: 2,2-DIMETHYL-3-(2-METHYL-1-PROPENYL) CYCLOPROPANECARBOXYLIC ACID (1,3,4,5,6,7-HEXAHYDRO-1,3-DIOXO-2H-ISOINDOL-2-YL)METHYL ESTER; EPA PESTICIDE CHEMICAL CODE 069003; NEOPYNAMIN

EPCRA Section 313 Form R de minimus concentration reporting level: 1.0%.

EPA NAME: 2,2,3,3-TETRAMETHYLCYCLOPROPANE CARBOXYLIC ACID CYANO(3-PHENOXYPHENYL)METHYL ESTER
[see FENPROPATHRIN]
CAS: 39515-41-8

EPA NAME: TETRAMETHYL LEAD
[see also LEAD and LEAD COMPOUNDS]
CAS: 75-74-1
SYNONYMS: LEAD TETRAMETHYL; PLUMBANE, TETRAMETHYL-; TETRAMETHYLPLUMBANE

CLEAN AIR ACT: Accidental Release Prevention/Flammable substances (Section 112[r], Table 3), TQ = 10,000 lbs. (4540 kgs.).
EPCRA Section 302 Extremely Hazardous Substances: TPQ = 100 lbs. (45.4 kgs.).
EPCRA Section 304 Reportable Quantity (RQ): EHS, 100 lbs. (45.4 kgs.).
EPCRA Section 313 (as organic lead compound) Form R de minimus concentration reporting level: 1.0%. Form R Toxic Chemical Category Code: N420.
MARINE POLLUTANT (49CFR, Subchapter 172.101, Appendix B).

EPA NAME: TETRAMETHYLSILANE
CAS: 75-76-3
SYNONYMS: SILANE, TETRAMETHYL-

CLEAN AIR ACT: Accidental Release Prevention/Flammable substances (Section 112[r], Table 3), TQ = 10,000 lbs. (4540 kgs.).

EPA NAME: TETRANITROMETHANE
CAS: 509-14-8
SYNONYMS: METHANE, TETRANITRO-; TETAN

CLEAN AIR ACT: Accidental Release Prevention/Flammable substances (Section 112[r], Table 3), TQ=10,000 lbs. (4540 kgs.).
EPA HAZARDOUS WASTE NUMBER (RCRA No.): P112.
RCRA Section 261 Hazardous Constituents.
EPCRA Section 302 Extremely Hazardous Substances: TPQ=500 lbs. (227 kgs.).
EPCRA Section 304 Reportable Quantity (RQ): EHS/CERCLA, 10 lbs. (4.54 kgs.).

EPA NAME: THALLIC OXIDE
[see also THALLIUM and THALLIUM COMPOUNDS]
CAS: 1314-32-5
SYNONYMS: THALLIUM(III) OXIDE

CLEAN WATER ACT: Section 307 Toxic Pollutants.
EPA HAZARDOUS WASTE NUMBER (RCRA No.): P113.
RCRA Section 261 Hazardous Constituents.
EPCRA Section 304 Reportable Quantity (RQ): CERCLA, 100 lbs. (45.4 kgs.).
EPCRA Section 313: Includes any unique chemical substance that contains thallium as part of that chemical's infrastructure. Form R de minimus concentration reporting level: 1.0%. Form R Toxic Chemical Category Code: N750.
MARINE POLLUTANT (49CFR, Subchapter 172.101, Appendix B) as thallium compounds, n.o.s.

EPA NAME: THALLIUM
[see also THALLIUM COMPOUNDS]
CAS: 7440-28-0
SYNONYMS: THALLIUM, ELEMENTAL

CLEAN WATER ACT: Section 307 Toxic Pollutants; Section 307 Priority Pollutants.
RCRA Section 261 Hazardous Constituents, waste number not listed.
RCRA Land Ban Waste.
RCRA Section 261 Hazardous Constituents.
RCRA Universal Treatment Standards: Wastewater (mg/L), 1.4; Non-wastewater (mg/L), 0.78 TCLP.
RCRA Ground Water Monitoring List. Suggested test method(s) (PQL µg/L): (total) 6010(400); 7840(1,000); 7841(10).
SAFE DRINKING WATER ACT: MCL, 0.002 mg/L; MCLG, 0.005 mg/L; Regulated chemical (47 FR 9352).
EPCRA Section 304 Reportable Quantity (RQ): CERCLA, 1,000 lbs. (454 kgs.).
EPCRA Section 313 Form R de minimus concentration reporting level: 1.0%. Form R Toxic Chemical Category Code: N750.
MARINE POLLUTANT (49CFR, Subchapter 172.101, Appendix B) as thallium compounds, n.o.s.

EPA NAME: THALLIUM(I) ACETATE
[see also THALLIUM COMPOUNDS]
CAS: 563-68-8

SYNONYMS: CARBONIC ACID, DITHALLIUM(1+) SALT; THALLOUS ACETATE

CLEAN WATER ACT: Section 307 Toxic Pollutants.
EPA HAZARDOUS WASTE NUMBER (RCRA No.): U214.
RCRA Section 261 Hazardous Constituents.
EPCRA Section 304 Reportable Quantity (RQ): EHS/CERCLA, 100 lbs. (45.4 kgs.).
EPCRA Section 313 Form R de minimus concentration reporting level: 1.0%. Form R Toxic Chemical Category Code: N750.
MARINE POLLUTANT (49CFR, Subchapter 172.101, Appendix B) as thallium compounds, n.o.s.

EPA NAME: THALLIUM(I) CARBONATE
[see also THALLIUM COMPOUNDS]
CAS: 6533-73-9
SYNONYMS: CARBONIC ACID, DITHALLIUM(I) SALT; THALLOUS CARBONATE

CLEAN WATER ACT: Section 307 Toxic Pollutants.
EPA HAZARDOUS WASTE NUMBER (RCRA No.): U215.
RCRA Section 261 Hazardous Constituents.
EPCRA Section 302 Extremely Hazardous Substances: TPQ = 100/10,000 lbs. (45.4/4,540 kgs.).
EPCRA Section 304 Reportable Quantity (RQ): CERCLA, 100 lbs. (45.4 kgs.).
EPCRA Section 313 Form R de minimus concentration reporting level: 1.0%. Form R Toxic Chemical Category Code: N750.
MARINE POLLUTANT (49CFR, Subchapter 172.101, Appendix B) as thallium compounds, n.o.s.

EPA NAME: THALLIUM CHLORIDE (TlCl)
[see also THALLIUM COMPOUNDS]
CAS: 7791-12-0
SYNONYMS: THALLIUM(I) CHLORIDE; THALLOUS CHLORIDE

CLEAN WATER ACT: Section 307 Toxic Pollutants.
EPA HAZARDOUS WASTE NUMBER (RCRA No.): U216.
RCRA Section 261 Hazardous Constituents.
EPCRA Section 302 Extremely Hazardous Substances: TPQ = 100/10,000 lbs. (45.4/4,540 kgs.).
EPCRA Section 304 Reportable Quantity (RQ): EHS/CERCLA, 100 lbs. (45.4 kgs.).
EPCRA Section 313 Form R de minimus concentration reporting level: 1.0%. Form R Toxic Chemical Category Code: N750.
MARINE POLLUTANT (49CFR, Subchapter 172.101, Appendix B) as thallium compounds, n.o.s.

EPA NAME: THALLIUM COMPOUNDS
[see also THALLIUM]
CLEAN WATER ACT: Section 307 Toxic Pollutants as thallium and compounds.
RCRA Section 261 Hazardous Constituents, waste number not listed.

RCRA Section 313 Form R de minimus concentration reporting level: 1.0%. Form R Toxic Chemical Category Code: N750.
MARINE POLLUTANT (49CFR, Subchapter 172.101, Appendix B) as thallium compounds, n.o.s.; thallium compounds (pesticides).

EPA NAME: THALLIUM(I) NITRATE
[see also THALLIUM COMPOUNDS]
CAS: 10102-45-1
SYNONYMS: NITRIC ACID, THALLIUM(1+) SALT; THALLOUS NITRATE

CLEAN WATER ACT: Section 307 Toxic Pollutants.
EPA HAZARDOUS WASTE NUMBER (RCRA No.): U217.
RCRA Section 261 Hazardous Constituents.
EPCRA Section 304 Reportable Quantity (RQ): CERCLA, 100 lbs. (45.4 kgs.).
EPCRA Section 313 Form R de minimus concentration reporting level: 1.0%. Form R Toxic Chemical Category Code: N750.
MARINE POLLUTANT (49CFR, Subchapter 172.101, Appendix B).

EPA NAME: THALLIUM(I) SULFATE
[see also THALLIUM COMPOUNDS]
CAS: 7446-18-6
SYNONYMS: SULFURIC ACID, DITHALLIUM(I) SALT (8CI,9CI); THALLIUM(1+) SULFATE (2:1); THALLOUS SULFATE

CLEAN WATER ACT: Section 311 Hazardous Substances/RQ (same as CERCLA); Section 307 Toxic Pollutants.
EPA HAZARDOUS WASTE NUMBER (RCRA No.): P115.
RCRA Section 261 Hazardous Constituents.
EPCRA Section 313 Form R de minimus concentration reporting level: 1.0%. Form R Toxic Chemical Category Code: N750.
EPCRA Section 302 Extremely Hazardous Substances: TPQ = 100/10,000 lbs. (45.4/4,540 kgs.).
EPCRA Section 304 Reportable Quantity (RQ): EHS/CERCLA, 100 lbs. (45.4 kgs.).
MARINE POLLUTANT (49CFR, Subchapter 172.101, Appendix B).

EPA NAME: THALLIUM SULFATE
[see also THALLIUM COMPOUNDS]
CAS: 10031-59-1
SYNONYMS: SULFURIC ACID, THALLIUM SALT; THALLOUS SULFATE

CLEAN WATER ACT: Section 311 Hazardous Substances/RQ (same as CERCLA); Section 307 Toxic Pollutants.
EPCRA Section 302 Extremely Hazardous Substances: TPQ = 100/10,000 lbs. (45.4/4,540 kgs.).
EPCRA Section 304 Reportable Quantity (RQ): EHS/CERCLA, 100 lbs. (45.4 kgs.).
EPCRA Section 313 Form R de minimus concentration reporting level: 1.0%. Form R Toxic Chemical Category Code: N750.
MARINE POLLUTANT (49CFR, Subchapter 172.101, Appendix B).

EPA NAME: THALLOUS CARBONATE
[see THALLIUM(I) CARBONATE]
CAS: 6533-73-9

EPA NAME: THALLOUS CHLORIDE
[see THALLIUM CHLORIDE TlCl]
CAS: 7791-12-0

EPA NAME: THALLOUS MALONATE
[see also THALLIUM COMPOUNDS]
CAS: 2757-18-8
SYNONYMS: PROPANEDIOIC ACID, DITHALLIUM SALT; THALLIUM MALONITE

CLEAN WATER ACT: Section 307 Toxic Pollutants.
EPCRA Section 302 Extremely Hazardous Substances: TPQ = 100/10,000 lbs. (45.4/4,540 kgs.).
EPCRA Section 304 Reportable Quantity (RQ): EHS, 100 lbs. (45.4 kgs.).
EPCRA Section 313 Form R de minimus concentration reporting level: 1.0%. Form R Toxic Chemical Category Code: N750.
MARINE POLLUTANT (49CFR, Subchapter 172.101, Appendix B) as thallium compounds, n.o.s.

EPA NAME: THALLOUS SULFATE
[see THALLIUM SULFATE]
CAS: 7446-18-6

EPA NAME: THIABENDAZOLE
CAS: 148-79-8
SYNONYMS: BENZIMIDAZOLE, 2-(4-THIAZOLYL)-; 2-THIAZOLE-4-YLBENZIMIDAZOLE; EPA PESTICIDE CHEMICAL CODE 060101; 2-(THIAZOL-4-YL)BENZIMIDAZOLE; 2-(4-THIAZOLYL)-1H-BENZIMIDAZOLE

EPCRA Section 313 Form R de minimus concentration reporting level: 1.0%.

EPA NAME: 2-(4-THIAZOLYL)-1H-BENZIMIDAZOLE
[see THIABENDAZOLE]
CAS: 148-79-8

EPA NAME: THIOACETAMIDE
CAS: 62-55-5
SYNONYMS: ACETOTHIOAMIDE; ETHANETHIOAMIDE

EPA HAZARDOUS WASTE NUMBER (RCRA No.): U218.
RCRA Section 261 Hazardous Constituents.
EPCRA Section 304 Reportable Quantity (RQ): CERCLA, 10 lbs. (4.54 kgs.).
EPCRA Section 313 Form R de minimus concentration reporting level: 0.1%.

EPA NAME: THIOBENCARB
CAS: 28249-77-6

SYNONYMS: BOLERO; CARBAMIC ACID, DIETHYLTHIO-, S-(P-CHLOROBENZYL) ESTER; CARBAMOTHIOIC ACID, DIETHYL-, S-(CHLOROPHENYL)METHYL)ESTER; EPA PESTICIDE CHEMICAL CODE 108401

EPCRA Section 313 Form R de minimus concentration reporting level: 1.0%.

EPA NAME: THIOCARBAZIDE

CAS: 2231-57-4

SYNONYMS: CARBONOTHIOIC DIHYDRAZINE; THIOCARBONOHYDRAZIDE

EPCRA Section 302 Extremely Hazardous Substances: TPQ = 1,000/10,000 lbs. (454/4,540 kgs.).

EPCRA Section 304 Reportable Quantity (RQ): EHS, 1,000 lbs. (454 kgs.).

EPA NAME: THIOCYANIC ACID, METHYL ESTER

CAS: 556-64-9

SYNONYMS: METHYL SULFOCYANATE; METHYL THIOCYANATE

CLEAN AIR ACT: Accidental Release Prevention/Flammable substances (Section 112[r], Table 3), TQ = 20,000 lbs. (9080 kgs.).

EPCRA Section 302 Extremely Hazardous Substances: TPQ = 10,000 lbs. (4,540 kgs.).

EPCRA Section 304 Reportable Quantity (RQ): EHS, 1 lb. (0.454 kg.).

EPA NAME: 4,4′-THIODIANILINE

CAS: 139-65-1

SYNONYMS: BENZENAMINE, 4,4′-THIOBIS-; 4,4′-DIAMINOPHENYL SULFIDE

EPCRA Section 313 Form R de minimus concentration reporting level: 0.1%.

EPA NAME: THIODICARB

CAS: 59669-26-0

SYNONYMS: DIMETHYL N,N′-(THIOBIS((METHYLIMINO) CARBONYLOXY)) BIS(THIOIMIDOACETATE); EPA PESTICIDE CHEMICAL CODE 114501

EPA HAZARDOUS WASTE NUMBER (RCRA No.): U410.

RCRA Section 261 Hazardous Constituents.

RCRA Land Ban Waste.

RCRA Universal Treatment Standards: Wastewater (mg/L), 0.019; Nonwastewater (mg/kg), 1.4.

SAFE DRINKING WATER ACT: Priority List (55 FR 1470).

EPCRA Section 304 Reportable Quantity (RQ): CERCLA, 1 lb. (0.454 kg.).

EPCRA Section 313 Form R de minimus concentration reporting level: 1.0%.

EPA NAME: THIOFANOX

CAS: 39196-18-4

SYNONYMS: 3,3-DIMETHYL-1-(METHYLTHIO)-2-BUTANONE-o-((METHYLAMINO)CARBONYL) OXIME

EPA HAZARDOUS WASTE NUMBER (RCRA No.): P045.

RCRA Section 261 Hazardous Constituents.

EPCRA Section 302 Extremely Hazardous Substances: TPQ = 100/10,000 lbs. (45.4/4,540 kgs.).

EPCRA Section 304 Reportable Quantity (RQ): EHS/CERCLA, 100 lbs. (45.4 kgs.).

EPA NAME: THIOMETHANOL

[see METHYL MERCAPTAN]

CAS: 74-93-1

EPA NAME: THIONAZIN

[see ZINOPHOS]

CAS: 297-97-2

EPA NAME: THIOPHANATE ETHYL

CAS: 23564-06-9

SYNONYMS: ALLOPHANIC ACID, 4,4′-O-PHENYLENEBIS(3-THIO-, DIETHYL ESTER; 1,2-BIS(ETHOXYCARBONYL-THIOUREIDO)BENZENE; 1,2-BIS(3-(ETHOXYCARBONYL)-2-THIOUREIDO)BENZENE; EPA PESTICIDE CHEMICAL CODE 103401; [1,2-PHENYLENEBIS (IMINOCARBONOTHIOYL)] BIS-CARBAMIC ACID DIETHYL ESTER

EPA HAZARDOUS WASTE NUMBER (RCRA No.): U409.

RCRA Section 261 Hazardous Constituents.

EPCRA Section 304 Reportable Quantity (RQ): CERCLA, 1 lb. (0.454 kg.).

EPCRA Section 313 Form R de minimus concentration reporting level: 1.0%.

EPA NAME: THIOPHANATE-METHYL

CAS: 23564-05-8

SYNONYMS: 1,2-BIS(METHOXYCARBONYLTHIO-UREIDO)BENZENE; 1,2-BIS(3-(METHOXYCARBONYL)-2-THIOUREIDO)BENZENE; EPA PESTICIDE CHEMICAL CODE 102001

RCRA Land Ban Waste.

RCRA Universal Treatment Standards: Wastewater (mg/L), 0.056; Non-wastewater (mg/kg), 1.4.

EPCRA Section 313 Form R de minimus concentration reporting level: 1.0%.

EPA NAME: THIOPHENOL

[see BENZENETHIOL]

CAS: 108-98-5

EPA NAME: THIOSEMICARBAZIDE
CAS: 79-19-6
SYNONYMS: 1-AMINO-2-THIOUREA; 1-AMINOTHIOUREA; HYDRAZINE CARBOTHIOAMIDE; 3-THIOSEMICARBAZIDE

EPA HAZARDOUS WASTE NUMBER (RCRA No.): P116.
RCRA Section 261 Hazardous Constituents.
EPCRA Section 302 Extremely Hazardous Substances: TPQ= 100/10,000 lbs. (45.4/4,540 kgs.).
EPCRA Section 304 Reportable Quantity (RQ): EHS/CERCLA, 100 lbs. (45.4 kgs.).
EPCRA Section 313 Form R de minimus concentration reporting level: 1.0%.

EPA NAME: THIOUREA
CAS: 62-56-6
SYNONYMS: ISOTHIOUREA; 2-THIOUREA

EPA HAZARDOUS WASTE NUMBER (RCRA No.): U219.
RCRA Section 261 Hazardous Constituents.
EPCRA Section 304 Reportable Quantity (RQ): CERCLA, 10 lbs. (4.54 kgs.).
EPCRA Section 313 Form R de minimus concentration reporting level: 0.1%.
CALIFORNIA'S PROPOSITION 65: Carcinogen.

EPA NAME: THIOUREA, (2-CHLOROPHENYL)-
CAS: 5344-82-1
SYNONYMS: 2-CHLOROPHENYL THIOUREA

EPA HAZARDOUS WASTE NUMBER (RCRA No.): P026.
RCRA Section 261 Hazardous Constituents.
EPCRA Section 302 Extremely Hazardous Substances: TPQ= 100/10,000 lbs. (45.4/4,540 kgs.).
EPCRA Section 304 Reportable Quantity (RQ): EHS/CERCLA, 100 lbs. (45.4 kgs.).

EPA NAME: THIOUREA, (2-METHYLPHENYL)-
CAS: 614-78-8
SYNONYMS: o-TOLYL THIOUREA

EPCRA Section 302 Extremely Hazardous Substances: TPQ= 500/10,000 lbs. (227/4,540 kgs.).
EPCRA Section 304 Reportable Quantity (RQ): EHS, 500 lbs. (227 kgs.).
EPCRA Section 313 Form R de minimus concentration reporting level: 1.0%.

EPA NAME: THIOUREA, 1-NAPHTHALENYL-
[see ANTU]
CAS: 86-88-4

EPA NAME: THIRAM
CAS: 137-26-8

SYNONYMS: BIS(DIMETHYLTHIOCARBAMOYL) DISULFIDE; α,α′-DITHIOBIS(DIMETHYLTHIO)FORMAMIDE; EPA PESTICIDE CHEMICAL CODE 079801; THIOPEROXYDICARBONIC DIAMIDE (((H_2N)C(S))$_2$$S_2$), TETRAMETHYL-; THIOPEROXYDICARBONIC DIAMIDE, TETRAMETHYL-; THIURAM

EPA HAZARDOUS WASTE NUMBER (RCRA No.): U244.
RCRA Section 261 Hazardous Constituents.
EPCRA Section 304 Reportable Quantity (RQ): CERCLA, 10 lbs. (4.54 kgs.).
EPCRA Section 313 Form R de minimus concentration reporting level: 1.0%.

EPA NAME: THORIUM DIOXIDE
CAS: 1314-20-1
SYNONYMS: THORIUM(IV) OXIDE; THORIUM OXIDE

EPCRA Section 313 Form R de minimus concentration reporting level: 1.0%.

EPA NAME: TIN
CAS: 7440-31-5
RCRA Ground Water Monitoring List. Suggested test method(s) (PQL µg/L): (total) 7870(8,000).

EPA NAME: TITANIUM CHLORIDE ($TiCl_4$) (T-4)-
[see TITANIUM TETRACHLORIDE]
CAS: 7550-45-0

EPA NAME: TITANIUM TETRACHLORIDE
CAS: 7550-45-0
SYNONYMS: TITANIUM CHLORIDE; TITANIUM(IV) CHLORIDE; TITANIUM CHLORIDE ($TiCl_4$) (T-4)-

CLEAN AIR ACT: Hazardous Air Pollutants (Title I, Part A, Section 112); Accidental Release Prevention/Flammable substances (Section 112[r], Table 3), TQ = 2,500 lbs. (1135 kgs.).
EPCRA Section 302 Extremely Hazardous Substances: TPQ = 100 lbs. (45.4 kgs.).
EPCRA Section 304 Reportable Quantity (RQ): CERCLA, 1,000 lbs. (454 kgs.).
EPCRA Section 313 Form R de minimus concentration reporting level: 1.0%.

EPA NAME: o-TOLIDINE
CAS: 119-93-7
SYNONYMS: BENZIDINE, 3,3′-DIMETHYL-; (1,1′-BIPHENYL)-4,4′-DIAMINE,3,3′-DIMETHYL-; 3,3′-DIMETHYLBENZIDINE; 3,3′-DIMETHYL-4,4′-BIPHENYLDIAMINE

EPA HAZARDOUS WASTE NUMBER (RCRA No.): U095.
RCRA Section 261 Hazardous Constituents.
EPCRA Section 304 Reportable Quantity (RQ): CERCLA, 10 lbs. (4.54 kgs.).

EPCRA Section 313 Form R de minimus concentration reporting level: 1.0%.

EPA NAME: o-TOLIDINE DIHYDROCHLORIDE
[see 3,3′-DIMETHYLBENZIDINE DIHYDROCHLORIDE]
CAS: 612-82-8

EPA NAME: o-TOLIDINE DIHYDROFLUORIDE
[see 3,3-DIMETHYLBENZIDINE DIHYDROFLUORIDE]
CAS: 41766-75-0

EPA NAME: TOLUENE
CAS: 108-88-3
SYNONYMS: BENZENE, METHYL-; METHANE, PHENYL-; METHYLBENZENE

CLEAN AIR ACT: Hazardous Air Pollutants (Title I, Part A, Section 112).
CLEAN WATER ACT: Section 311 Hazardous Substances/RQ (same as CERCLA); Section 307 Toxic Pollutants; Section 307 Priority Pollutants.
EPA HAZARDOUS WASTE NUMBER (RCRA No.): U220.
RCRA Section 261 Hazardous Constituents.
RCRA Land Ban Waste.
RCRA Universal Treatment Standards: Wastewater (mg/L), 0.080; Nonwastewater (mg/kg), 10.
RCRA Ground Water Monitoring List. Suggested test method(s) (PQL µg/L): 8020(2); 8240(5).
SAFE DRINKING WATER ACT: MCL, 1.0 mg/L; MCLG, 1.0 mg/L; Regulated chemical (47 FR 9352).
EPCRA Section 304 Reportable Quantity (RQ): CERCLA, 1,000 lbs. (454 kgs.).
EPCRA Section 313 Form R de minimus concentration reporting level: 1.0%.
CALIFORNIA'S PROPOSITION 65: Reproductive toxin.

EPA NAME: TOLUENE 2,4-DIISOCYANATE
CAS: 584-84-9
SYNONYMS: BENZENE, 2,4-DIISOCYANATO-1-METHYL-; BENZENE 2,4-DIISOCYANATOMETHYL-; BENZENE,2,4-DIISOCYANATO-1-METHYL-; CRESORCINOL DIISOCYANATE; DESMODUR T80; DI-ISO-CYANATOLUENE; 2,4-DIISOCYANATO-1-METHYLBENZENE (9CI); 2,4-DIISOCYANATOTOLUENE; ISOCYANIC ACID, METHYLPHENYLENE ESTER; ISOCYANIC ACID, 4-METHYL-M-PHENYLENE ESTER; 4-METHYL-PHENYLENE DIISOCYANATE; 4-METHYL-PHENYLENE ISOCYANATE; TOLUENE DIISOCYANATE; 2,4-TOLUENE DIISOCYANATE

CLEAN AIR ACT: Hazardous Air Pollutants (Title I, Part A, Section 112); Accidental Release Prevention/Flammable substances (Section 112[r], Table 3), TQ=10,000 lbs. (4540 kgs.).
EPA HAZARDOUS WASTE NUMBER (RCRA No.): U223.

EPCRA Section 302 Extremely Hazardous Substances: TPQ = 500 lbs (127.5 kgs).

EPCRA Section 304 Reportable Quantity (RQ): EHS/CERCLA, 100 lbs. (45.4 kgs.).

EPCRA Section 313 Form R de minimus concentration reporting level: 0.1%.

EPA NAME: TOLUENE-2,6-DIISOCYANATE

CAS: 91-08-7

SYNONYMS: BENZENE, 1,3-DIISOCYANATO-2-METHYL-; 2,6-DIISOCYANATOTOLUENE; ISOCYANIC ACID, 2-METHYL-m-PHENYLENE ESTER; 2-METHYL-m-PHENYLENE ISOCYANATE; 2,6-TDI; TOLYLENE 2,6-DIISOCYANATE; m-TOLYENE DIISOCYANATE

CLEAN AIR ACT: Accidental Release Prevention/Flammable substances (Section 112[r], Table 3), TQ = 10,000 lbs (4,540 kgs).

EPCRA Section 302 Extremely Hazardous Substances: TPQ =, 100 lbs (45.4 kgs).

EPCRA Section 304 Reportable Quantity (RQ): EHS/CERCLA, 100 lbs. (45.4 kgs.).

EPCRA Section 313 Form R de minimus concentration reporting level: 0.1%.

EPA NAME: TOLUENEDIISOCYANATE (MIXED ISOMERS)

CAS: 26471-62-5

SYNONYMS: BENZENE, 1,3-DIISOCYANATOMETHYL-; DIISOCYANATOTOLUENE; METHYL-m-PHENYLENE ISOCYANATE; METHYLPHENYLENE ISOCYANATE; TOLUENEDIISOCYANATE (UNSPECIFIED ISOMERS); TOLYENE DIISOCYANATE; TOLYLENE ISOCYANATE

CLEAN AIR ACT: Accidental Release Prevention/Flammable substances (Section 112[r], Table 3), TQ = 10,000 lbs (4,540 kgs).

EPA HAZARDOUS WASTE NUMBER (RCRA No.): U223.

RCRA Section 261 Hazardous Constituents.

EPCRA Section 304 Reportable Quantity (RQ): CERCLA, 100 lbs. (45.4 kgs.).

EPCRA Section 313 Form R de minimus concentration reporting level: 0.1%.

CALIFORNIA'S PROPOSITION 65: Carcinogen.

EPA NAME: TOLUENEDIISOCYANATE (UNSPECIFIED ISOMERS)

[see TOLUENEDIISOCYANATE (MIXED ISOMERS)]

CAS: 26471-62-5

EPA NAME: o-TOLUIDINE

CAS: 95-53-4

SYNONYMS: 1-AMINO-2-METHYLBENZENE; BENZENAMINE,2-METHYL- (9CI); o-METHYLANILINE; 2-METHYLANILINE; ortho-TOLUIDINE; TOLUIDINE, ortho-

CLEAN AIR ACT: Hazardous Air Pollutants (Title I, Part A, Section 112).
EPA HAZARDOUS WASTE NUMBER (RCRA No.): U328.
RCRA Section 261 Hazardous Constituents.
RCRA Ground Water Monitoring List. Suggested test method(s) (PQL µg/L): 8270(10).
EPCRA Section 304 Reportable Quantity (RQ): CERCLA, 100 lbs. (45.4 kgs.).
EPCRA Section 313 Form R de minimus concentration reporting level: 0.1%.
CALIFORNIA'S PROPOSITION 65: Carcinogen.

EPA NAME: p-TOLUIDINE
CAS: 106-49-0
SYNONYMS: 4-AMINO-1-METHYLBENZENE; p-METHYLANILINE; 4-METHYLANILINE; para-TOLUIDINE; 4-TOLUIDINE; TOLUIDINE, para-

EPA HAZARDOUS WASTE NUMBER (RCRA No.): U353.
RCRA Section 261 Hazardous Constituents.
EPCRA Section 304 Reportable Quantity (RQ): CERCLA, 100 lbs. (45.4 kgs.).
CALIFORNIA'S PROPOSITION 65: Carcinogen.

EPA NAME: o-TOLUIDINE HYDROCHLORIDE
CAS: 636-21-5
SYNONYMS: 1-AMINO-2-METHYLBENZENE HYDROCHLORIDE; BENZENAMINE, 2-METHYL-, HYDROCHLORIDE; ortho-TOLUIDINE HYDROCHLORIDE; TOLUIDINE HYDROCHLORIDE, ortho-

EPA HAZARDOUS WASTE NUMBER (RCRA No.): U222.
RCRA Section 261 Hazardous Constituents.
EPCRA Section 304 Reportable Quantity (RQ): CERCLA, 100 lbs. (45.4 kgs.).
EPCRA Section 313 Form R de minimus concentration reporting level: 0.1%.
CALIFORNIA'S PROPOSITION 65: Carcinogen.

EPA NAME: TOXAPHENE
CAS: 8001-35-2
SYNONYMS: CAMPHENE, OCTACHLORO-; CAMPHECHLOR; POLYCHLORINATED CAMPHENE; TECHNICAL CHLORINATED CAMPHENE, 67-69% CHLORINE

CLEAN AIR ACT: Hazardous Air Pollutants (Title I, Part A, Section 112).
CLEAN WATER ACT: Section 311 Hazardous Substances/RQ (same as CERCLA); Toxic Pollutant (Section 401.15); Section 307 Priority Pollutants; Section 313 Priority Chemicals.
EPA HAZARDOUS WASTE NUMBER (RCRA No.): P123; D015.
RCRA Section 261 Hazardous Constituents.
RCRA Toxicity Characteristic (Section 261.24), Maximum Concentration of Contaminants, regulatory level, 0.5 mg/L.
RCRA Land Ban Waste.

RCRA Universal Treatment Standards: Wastewater (mg/L), 0.0095; Nonwastewater (mg/kg), 2.6.

RCRA Ground Water Monitoring List. Suggested test method(s) (PQL μg/L): 8080(2); 8250(10).

SAFE DRINKING WATER ACT: MCL, 0.003 mg/L; MCLG, zero; Regulated chemical (47 FR 9352).

EPCRA Section 302 Extremely Hazardous Substances: TPQ = 500/10,000 lbs. (227/4,540 kgs.).

EPCRA Section 304 Reportable Quantity (RQ): EHS/CERCLA, 1 lb. (0.454 kg.).

EPCRA Section 313 Form R de minimus concentration reporting level: 0.1%.

MARINE POLLUTANT (49CFR, Subchapter 172.101, Appendix B).

CALIFORNIA'S PROPOSITION 65: Carcinogen.

EPA NAME: 2,4,5-TP ESTERS

CAS: 32534-95-5

SYNONYMS: PROPANOIC ACID, 2-(2,4,5-TRICHLOROPHENOXY)-, ISOOCTYL ESTER; 2,4,5-TP ACID ESTERS; α-(2,4,5-TRICHLOROPHENOXY)PROPIONIC ACID

EPA HAZARDOUS WASTE NUMBER (RCRA No.): U233.

RCRA Section 261 Hazardous Constituents.

EPCRA Section 304 Reportable Quantity (RQ): CERCLA, 100 lbs. (45.4 kgs.).

EPA NAME: TRIADIMEFON

CAS: 43121-43-3

SYNONYMS: 2-BUTANONE, 1-(4-CHLOROPHENOXY)-3,3-DIMETHYL-1-(1-H-1,2,4-TRIAZOL-1-YL)-; (1-(4-CHLOROPHENOXY)-3,3-DIMETHYL-1-(1H-1,2,4-TRIAZOL-1-YL)-2-BUTANONE; EPA PESTICIDE CHEMICAL CODE 109901; 1-(1,2,4-TRIAZOYL-1)-1-(4-CHLORO-PHENOXY)3,3-DIMETHYLBUTANONE

EPCRA Section 313 Form R de minimus concentration reporting level: 1.0%.

EPA NAME: TRIALLATE

CAS: 2303-17-5

SYNONYMS: CARBAMIC ACID, DIISOPROPYLTHIO-, S-(2,3,3-TRICHLOROALLYL) ESTER; 2,3,3-TRICHLOROALLYL N,N-DIISOPROPYLTHIOCARBAMATE

EPA HAZARDOUS WASTE NUMBER (RCRA No.): U389.

RCRA Section 261 Hazardous Constituents.

RCRA Land Ban Waste.

RCRA Universal Treatment Standards: Wastewater (mg/L), 0.003; Nonwastewater (mg/kg), 1.4.

EPCRA Section 304 Reportable Quantity (RQ): CERCLA, 1 lb. (0.454 kg.).

EPCRA Section 313 Form R de minimus concentration reporting level: 1.0%.

EPA NAME: TRIAMIPHOS

CAS: 1031-47-6

SYNONYMS: 5-AMINO-1-BIS(DIMETHYLAMIDE)PHOSPHORYL-3-PHENYL-1,2,4-TRIAZOLE; 3-PHENYL-5-AMINO-1,2,4-TRIAZOYL-(1)-(N,N'-TETRAMETHYL)DIAMIDOPHOSPHONATE

EPCRA Section 302 Extremely Hazardous Substances: TPQ = 500/10,000 lbs. (227/4,540 kgs.).

EPCRA Section 304 Reportable Quantity (RQ): EHS, 500 lbs. (227 kgs.).

EPA NAME: TRIAZIQUONE

CAS: 68-76-8

SYNONYMS: 2,5-CYCLOHEXADIENE-1,4-DIONE, 2,3,5-TRIS(1-AZIRIDINYL)-; TRIS(AZIRIDINO)-1,4-BENZOQUINONE; 2,3,5-TRIS(AZIRIDINYL)-p-BENZOQUINONE

EPCRA Section 313 Form R de minimus concentration reporting level: 1.0%.

EPA NAME: TRIAZOFOS

CAS: 24017-47-8

SYNONYMS: O,O-DIETHYLO-(1-PHENYL-1H-1,2,4-TRIAZOL-3-YL)PHOSPHOROTHIOATE 1-PHENYL-3-(O,O-DIETHYL-THIONOPHOPHORYL)-1,2,4-TRIAZOLE

EPCRA Section 302 Extremely Hazardous Substances: TPQ = 500 lbs. (227 kgs.).

EPCRA Section 304 Reportable Quantity (RQ): EHS, 500 lbs. (227 kgs.).

MARINE POLLUTANT (49CFR, Subchapter 172.101, Appendix B).

EPA NAME: TRIBENURON METHYL

CAS: 101200-48-0

SYNONYMS: BENZOIC ACID, 2-(((((4-METHOXY-6-METHYL-1,3,5-TRIAZIN-2-YL)METHYLAMINO)CARBONYL) AMINO)SULFONYL)-, METHYL ESTER; 2-(4-METHOXY-6-METHYL-1,3,5-TRIAZIN-2-YL)-METHYLAMINO)-CARBONYL)AMINO)SULFONYL) -, METHYL ESTER

EPCRA Section 313 Form R de minimus concentration reporting level: 1.0%.

EPA NAME: TRIBROMOMETHANE

[see BROMOFORM]

CAS: 75-25-2

EPA NAME: TRIBUTYLTIN FLUORIDE

[see also TIN]

CAS: 1983-10-4

SYNONYMS: EPA PESTICIDE CHEMICAL CODE 083112; STANNANE, FLUOROTRIBUTYL-; STANNANE, TRIBUTYLFLUORO-

EPCRA Section 313 Form R de minimus concentration reporting level: 1.0%.

MARINE POLLUTANT (49CFR, Subchapter 172.101, Appendix B) as tributyltin compounds.

EPA NAME: TRIBUTYLTIN METHACRYLATE
[see also TIN]
CAS: 2155-70-6
SYNONYMS: EPA PESTICIDE CHEMICAL CODE 083120; ((METHACRYLOYL)OXY)TRIBUTYLSTANNANE; STANNANE, TRIBUTYL(METHACRYLOYLOXY)-; TRI-N-BUTYLSTANNYL-METHACRYLATE

EPCRA Section 313 Form R de minimus concentration reporting level: 1.0%.
MARINE POLLUTANT (49CFR, Subchapter 172.101, Appendix B) as tributyltin compounds.

EPA NAME: S,S,S-TRIBUTYLTRITHIOPHOSPHATE
CAS: 78-48-8
SYNONYMS: BUTYL PHOSPHOROTRITHIOATE; S,S,S-TRIBUTYL PHOSPHOROTRITHIOATE; S,S,S-TRIBUTYL TRITHIOPHOSPHATE; DEF; EPA PESTICIDE CHEMICAL CODE 074801

EPCRA Section 313 Form R de minimus concentration reporting level: 1.0%.

EPA NAME: TRICHLORFON
CAS: 52-68-6
SYNONYMS: CHLOROFOS; DIPTEREX; DYLOX; PHOSPHONIC ACID, (2,2,2-TRICHLORO-1-HYDROXYETHYL)-, DIMETHYL ESTER; TRICHLOROFON

CLEAN WATER ACT: Section 311 Hazardous Substances/RQ (same as CERCLA).
EPCRA Section 304 Reportable Quantity (RQ): CERCLA, 100 lbs. (45.4 kgs.).
EPCRA Section 313 Form R de minimus concentration reporting level: 1.0%.
MARINE POLLUTANT (49CFR, Subchapter 172.101, Appendix B).

EPA NAME: TRICHLOROACETYL CHLORIDE
CAS: 76-02-8
SYNONYMS: ACETYL CHLORIDE, TRICHLORO-

EPCRA Section 302 Extremely Hazardous Substances: TPQ = 500.
EPCRA Section 304 Reportable Quantity (RQ): EHS, 500 lbs. (227 kgs.).
EPCRA Section 313 Form R de minimus concentration reporting level: 1.0%.

EPA NAME: 1,2,4-TRICHLOROBENZENE
[see also CHLORINATED BENZENES]
CAS: 120-82-1
SYNONYMS: BENZENE, 1,2,4-TRICHLORO-

RCRA Section 261 Hazardous Constituents, waste number not listed.

RCRA Land Ban Waste.
RCRA Universal Treatment Standards: Wastewater (mg/L), 0.055; Non-wastewater (mg/kg), 19.
RCRA Ground Water Monitoring List. Suggested test method(s) (PQL μg/L): 8270(10).
SAFE DRINKING WATER ACT: Regulated chemical (47 FR 9352) as tri-chlorobenzene.
SAFE DRINKING WATER ACT: MCL, 0.07 mg/L; MCLG, 0.7 mg/L.
EPCRA Section 304 Reportable Quantity (RQ): CERCLA, 100 lbs. (45.4 kgs.).
EPCRA Section 313 Form R de minimus concentration reporting level: 1.0%.
MARINE POLLUTANT (49CFR, Subchapter 172.101, Appendix B) as tri-chlorobenzenes, liquid.

EPA NAME: TRICHLORO(CHLOROMETHYL)SILANE
CAS: 1558-25-4
SYNONYMS: (CHLOROMETHYL)TRICHLOROSILANE

EPCRA Section 302 Extremely Hazardous Substances: TPQ = 100 lbs. (45.4 kgs.).
EPCRA Section 304 Reportable Quantity (RQ): EHS, 100 lbs. (45.4 kgs.).

EPA NAME: TRICHLORO(DICHLOROPHENYL)SILANE
CAS: 27137-85-5
SYNONYMS: (DICHLOROPHENYL)TRICHLOROSILANE; DI-CHLOROPHENYL(TRICHLORO)SILANE

EPCRA Section 302 Extremely Hazardous Substances: TPQ = 500 lbs. (227 kgs.).
EPCRA Section 304 Reportable Quantity (RQ): EHS, 500 lbs. (45.4 kgs.).

EPA NAME: 1,1,1-TRICHLOROETHANE
[see also CHLORINATED ETHANES]
CAS: 71-55-6
SYNONYMS: ETHANE, 1,1,1-TRICHLORO-; METHYL CHLOROFORM; α-TRICHLOROETHANE; TRICHLOROETHANE, 1,1,1-; TRICHLOROMETHYLMETHANE

CLEAN AIR ACT: Hazardous Air Pollutants (Title I, Part A, Section 112); Stratospheric ozone protection (Title VI, Subpart A, Appendix A), Class I, Ozone Depletion Potential = 0.1, all isomers except 1,1,2-tri-chlorethane.
CLEAN WATER ACT: Toxic Pollutant (Section 401.15) as chlorinated ethanes.
EPA HAZARDOUS WASTE NUMBER (RCRA No.): U226.
RCRA Section 261 Hazardous Constituents.
RCRA Land Ban Waste.
RCRA Universal Treatment Standards: Wastewater (mg/L), 0.054; Non-wastewater (mg/kg), 6.0.
RCRA Ground Water Monitoring List. Suggested test method(s) (PQL μg/L): 8240(5).

SAFE DRINKING WATER ACT: MCL, 0.2 mg/L; MCLG, 0.20 mg/L; Regulated chemical (47 FR 9352).
EPCRA Section 304 Reportable Quantity (RQ): CERCLA, 1,000 lbs. (454 kgs.).
EPCRA Section 313 Form R de minimus concentration reporting level: 1.0%.

EPA NAME: 1,1,2-TRICHLOROETHANE

[see also CHLORINATED ETHANES]
CAS: 79-00-5
SYNONYMS: ETHANE, 1,1,2-TRICHLORO-; TRICHLOROETHANE, 1,1,2-; VINYL TRICHLORIDE

CLEAN AIR ACT: Hazardous Air Pollutants (Title I, Part A, Section 112).
EPA HAZARDOUS WASTE NUMBER (RCRA No.): U227.
RCRA Section 261 Hazardous Constituents.
RCRA Land Ban Waste.
RCRA Universal Treatment Standards: Wastewater (mg/L), 0.054; Nonwastewater (mg/kg), 6.0.
RCRA Ground Water Monitoring List. Suggested test method(s) (PQL µg/L): 8010(0.2); 8240(5).
SAFE DRINKING WATER ACT: MCL, 0.005 mg/L; MCLG, 0.003 mg/L; Regulated chemical (47 FR 9352).
EPCRA Section 304 Reportable Quantity (RQ): CERCLA, 100 lbs. (45.4 kgs.).
EPCRA Section 313 Form R de minimus concentration reporting level: 1.0%.
CALIFORNIA'S PROPOSITION 65: Carcinogen.

EPA NAME: TRICHLOROETHYLENE

CAS: 79-01-6
SYNONYMS: ACETYLENE TRICHLORIDE; ETHENE, TRICHLORO-; ETHYLENE, TRICHLORO-

CLEAN AIR ACT: Hazardous Air Pollutants (Title I, Part A, Section 112).
CLEAN WATER ACT: Section 311 Hazardous Substances/RQ (same as CERCLA); Section 307 Toxic Pollutants; Section 307 Priority Pollutants.
EPA HAZARDOUS WASTE NUMBER (RCRA No.): U228; D040.
RCRA Toxicity Characteristic (Section 261.24), Maximum Concentration of Contaminants, regulatory level, 0.5 mg/L.
RCRA Section 261 Hazardous Constituents.
RCRA Land Ban Waste.
RCRA Universal Treatment Standards: Wastewater (mg/L), 0.054; Nonwastewater (mg/kg), 6.0.
RCRA Ground Water Monitoring List. Suggested test method(s) (PQL µg/L): 8010(1); 8240(5).
SAFE DRINKING WATER ACT: MCL, 0.005 mg/L; MCLG, zero; Regulated chemical (47 FR 9352).
EPCRA Section 304 Reportable Quantity (RQ): CERCLA, 100 lbs. (45.4 kgs.).

EPCRA Section 313 Form R de minimus concentration reporting level: 1.0%.
CALIFORNIA'S PROPOSITION 65: Carcinogen.

EPA NAME: TRICHLOROETHYLSILANE
CAS: 115-21-9
SYNONYMS: ETHYL TRICHLOROETHYLSILANE; ETHYL SILICON TRICHLORIDE; SILANE, TRICHLOROETHYL-

EPCRA Section 302 Extremely Hazardous Substances: TPQ = 500 lbs. (227 kgs.).
EPCRA Section 304 Reportable Quantity (RQ): EHS, 500 lbs. (227 kgs.).

EPA NAME: TRICHLOROFLUOROMETHANE
CAS: 75-69-4
SYNONYMS: CFC-11; FLUOROTRICHLOROMETHANE; FREON 11; METHANE, TRICHLOROFLUORO-; MONOFLUROTRICHLOROMETHANE; TRICHLOROMONOFLUOROMETHANE

CLEAN AIR ACT: Stratospheric ozone protection (Title VI, Subpart A, Appendix A), Class I, Ozone Depletion Potential = 1.0.
EPA HAZARDOUS WASTE NUMBER (RCRA No.): U121.
RCRA Section 261 Hazardous Constituents.
RCRA Land Ban Waste.
RCRA Universal Treatment Standards: Wastewater (mg/L), 0.020; Nonwastewater (mg/kg), 30.
RCRA Ground Water Monitoring List. Suggested test method(s) (PQL µg/L): 8010(10); 8240(5).
EPCRA Section 304 Reportable Quantity (RQ): CERCLA, 5,000 lbs. (2270 kgs.).
EPCRA Section 313 Form R de minimus concentration reporting level: 1.0%.

EPA NAME: TRICHLOROMETHYLSULFENYL CHLORIDE
[see PERCHLOROMETHYL MERCAPTAN]
CAS: 594-42-3

EPA NAME: TRICHLOROMONOFLUOROMETHANE
[see TRICHLOROFLUOROMETHANE]
CAS: 75-69-4

EPA NAME: TRICHLORONATE
CAS: 327-98-0
SYNONYMS: O-ETHYL-O-2,4,5-TRICHLOROPHENYL ETHYLPHOSPHONOTHIOATE; ETHYL TRICHLOROPHENYLETHYLPHOSPHONOTHIOATE; TRICHLORONAT

EPCRA Section 302 Extremely Hazardous Substances: TPQ = 500 lbs. (227 kgs.).
EPCRA Section 304 Reportable Quantity (RQ): EHS, 500 lbs. (227 kgs.).
MARINE POLLUTANT (49CFR, Subchapter 172.101, Appendix B) as trichloronat

EPA NAME: TRICHLOROPHENOL
[see also CHLOROPHENOLS]
CAS: 25167-82-2
SYNONYMS: OMAL; PHENACHLOR; PHENOL, TRICHLORO-

CLEAN WATER ACT: Section 311 Hazardous Substances/RQ (same as CERCLA).
EPCRA Section 304 Reportable Quantity (RQ): CERCLA, 10 lbs. (4.54 kgs.).

EPA NAME: 2,3,4-TRICHLOROPHENOL
[see also CHLOROPHENOLS]
CAS: 15950-66-0
SYNONYMS: PHENOL, 2,3,4-TRICHLORO-

EPCRA Section 304 Reportable Quantity (RQ): CERCLA, 10 lbs. (4.54 kgs.).

EPA NAME: 2,3,5-TRICHLOROPHENOL
[see also CHLOROPHENOLS]
CAS: 933-78-8
SYNONYMS: PHENOL, 2,3,5-TRICHLORO-

EPCRA Section 304 Reportable Quantity (RQ): CERCLA, 10 lbs. (4.54 kgs.).

EPA NAME: 2,3,6-TRICHLOROPHENOL
[see also CHLOROPHENOLS]
CAS: 933-75-5
SYNONYMS: PHENOL, 2,3,6-TRICHLORO-

EPCRA Section 304 Reportable Quantity (RQ): CERCLA, 10 lbs. (4.54 kgs.).

EPA NAME: 2,4,5-TRICHLOROPHENOL
CAS: 95-95-4
SYNONYMS: PHENOL, 2,4,5-TRICHLORO-

CLEAN AIR ACT: Hazardous Air Pollutants (Title I, Part A, Section 112).
EPA HAZARDOUS WASTE NUMBER (RCRA No.): U230; D041; F027.
RCRA Toxicity Characteristic (Section 261.24), Maximum Concentration of Contaminants, regulatory level, 400.0 mg/L.
RCRA Section 261 Hazardous Constituents.
RCRA Land Ban Waste.
RCRA Universal Treatment Standards: Wastewater (mg/L), 0.18; Nonwastewater (mg/kg), 7.4.
RCRA Ground Water Monitoring List. Suggested test method(s) (PQL µg/L): 8270(10).
EPCRA Section 304 Reportable Quantity (RQ): CERCLA, 10 lbs. (4.54 kgs.).
EPCRA Section 313 Form R de minimus concentration reporting level: 1.0%. Note: Threshold determinations should be made individually and separately from chlorophenols category.

EPA NAME: 2,4,6-TRICHLOROPHENOL

[see also CHLORINATED PHENOLS]

CAS: 88-06-2

SYNONYMS: 1,3,5-TRICHLORO-2-HYDROXYBENZENE; PHENOL, 2,4,6-TRICHLORO-

CLEAN AIR ACT: Hazardous Air Pollutants (Title I, Part A, Section 112).

EPA HAZARDOUS WASTE NUMBER (RCRA No.): U231; D042; F027.

RCRA Toxicity Characteristic (Section 261.24), Maximum Concentration of Contaminants, regulatory level, 2.0 mg/L.

RCRA Section 261 Hazardous Constituents.

RCRA Land Ban Waste.

RCRA Universal Treatment Standards: Wastewater (mg/L), 0.035; Nonwastewater (mg/kg), 7.4.

RCRA Ground Water Monitoring List. Suggested test method(s) (PQL µg/L): 8040(5); 8270(10).

EPCRA Section 304 Reportable Quantity (RQ): CERCLA, 10 lbs. (4.54 kgs.).

EPCRA Section 313 Form R de minimus concentration reporting level: 0.1%. Note: Threshold determinations should be made individually and separately from chlorophenols category.

CALIFORNIA'S PROPOSITION 65: Carcinogen.

EPA NAME: 3,4,5-TRICHLOROPHENOL

[see also CHLOROPHENOLS]

CAS: 609-19-8

SYNONYMS: PHENOL, 3,4,5-TRICHLORO-

EPCRA Section 304 Reportable Quantity (RQ): CERCLA, 10 lbs. (4.54 kgs.).

EPA NAME: TRICHLOROPHENYLSILANE

CAS: 98-13-5

SYNONYMS: PHENYL TRICHLOROSILANE; SILANE, TRICHLOROPHENYL-

EPCRA Section 302 Extremely Hazardous Substances: TPQ = 500 lbs. (227 kgs.).

EPCRA Section 304 Reportable Quantity (RQ): EHS, 500 lbs. (227 kgs.).

EPA NAME: 1,2,3-TRICHLOROPROPANE

CAS: 96-18-4

SYNONYMS: ALLYL TRICHLORIDE; PROPANE, 1,2,3-TRICHLORO-

RCRA Section 261 Hazardous Constituents, waste number not listed.

RCRA Land Ban Waste.

RCRA Universal Treatment Standards: Wastewater (mg/L), 0.85; Nonwastewater (mg/kg), 30.

RCRA Ground Water Monitoring List. Suggested test method(s) (PQL µg/L): 8010(10); 8240(5).

SAFE DRINKING WATER ACT: Priority List (55 FR 1470).

EPCRA Section 313 Form R de minimus concentration reporting level: 1.0%.
CALIFORNIA'S PROPOSITION 65: Carcinogen.

EPA NAME: TRICHLOROSILANE
CAS: 10025-78-2
SYNONYMS: SILANE, TRICHLORO-; SILICOCHLOROFORM

CLEAN AIR ACT: Accidental Release Prevention/Flammable substances (Section 112[r], Table 3), TQ = 10,000 lbs. (4540 kgs.).

EPA NAME: TRICLOPYR TRIETHYLAMMONIUM SALT
CAS: 57213-69-1
SYNONYMS: ACETIC ACID, ((3,5,6-TRICHLORO-2-PYRIDINYL)OXY)-, compd. with N,N-DIETHYLETHANAMINE (1:1); EPA PESTICIDE CHEMICAL CODE 116002; ((3,5,6-TRICHLORO-2-PYRIDYL)OXY)ACETIC ACID, COMPOUND WITH TRIETHYLAMINE (1:1)

EPCRA Section 313 Form R de minimus concentration reporting level: 1.0%.

EPA NAME: TRIETHANOLAMINE DODECYLBENZENE SULFONATE
CAS: 27323-41-7
SYNONYMS: BENZENESULFONIC ACID, DODECYL-, compd. with 2,2′,2″-NITRILOTRIS[ETHANOL](1:1); DODECYLBENZENESULFONIC ACID, TRIETHANOLAMINE SALT; 2-2′,2″-NITRILOTRIS-DODECYLBENZENESULFONATE (SALT); TRIETHANOLAMINE DODECYLBENZENE SULFONATE

CLEAN WATER ACT: Section 311 Hazardous Substances/RQ (same as CERCLA).
EPCRA Section 304 Reportable Quantity (RQ): CERCLA, 1,000 lbs. (454 kgs.).

EPA NAME: TRIETHOXYSILANE
CAS: 998-30-1
SYNONYMS: SILANE, TRIETHOXY-

EPCRA Section 302 Extremely Hazardous Substances: TPQ = 500 lbs. (227 kgs.).
EPCRA Section 304 Reportable Quantity (RQ): EHS, 500 lbs. (227 kgs.).

EPA NAME: TRIETHYLAMINE
CAS: 121-44-8
SYNONYMS: (DIETHYLAMINO)ETHANE; N,N-DIETHYLETHANEAMINE; ETHANAMINE,N,N-DIETHYL-

CLEAN AIR ACT: Hazardous Air Pollutants (Title I, Part A, Section 112).
CLEAN WATER ACT: Section 311 Hazardous Substances/RQ (same as CERCLA).
EPA HAZARDOUS WASTE NUMBER (RCRA No.): U404.
RCRA Section 261 Hazardous Constituents.

EPCRA Section 304 Reportable Quantity (RQ): CERCLA, 5,000 lbs. (2270 kgs.).
EPCRA Section 313 Form R de minimus concentration reporting level: 1.0%.

EPA NAME: TRIFLUOROCHLOROETHYLENE
CAS: 79-38-9
SYNONYMS: CHLOROTRIFLUOROETHYLENE; ETHENE, CHLOROTRIFLUORO-; 1,1,2-TRIFLUORO-2-CHLOROETHYLENE; TRIFLUOROMONOCHLOROETHYLENE

CLEAN AIR ACT: Accidental Release Prevention/Flammable substances (Section 112[r], Table 3), TQ = 10,000 lbs. (4540 kgs.).

EPA NAME: 2-[4-[[5-(TRIFLUOROMETHYL)-2-PYRIDINYL]OXY]-PHENOXY]PROPANOIC ACID, BUTYL ESTER
[see FLUAZIFOP-BUTYL]
CAS: 69806-50-4

EPA NAME: TRIFLURALIN
CAS: 1582-09-8
SYNONYMS: BENZENEAMINE, 2,6-DINITRO-N,N-DIPROPYL-4-(TRIFLUOROMETHYLANILINE); 2,6-DINITRO-N,N-DIPROPYL-4-(TRIFLUOROMETHYL)BENZENAMINE; 2,6-DINITRO-N,N-DI-N-PROPYL-α,α,α-TRIFLURO-p-TOLUIDINE; 4-(DI-N-PROPYLAMINO)-3,5-DINITRO-1-TRIFLUOROMETHYLBENZENE; N,N-DI-N-PROPYL-2,6-DINITRO-4-TRIFLUOROMETHYLANILINE; N,N-DIPROPYL-4-TRIFLUOROMETHYL-2,6-DINITROANILINE; 2,6-DINITRO-N,N-DIPROPYL-4-(TRIFLUOROMETHYL)ANILINE; p-TOLUIDINE, α,α,α-TRIFLUORO-2,6-DINITRO-N,N-DIPROPYL-; α,α,α-TRIFLUORO-2,6-DINITRO-N,N-DIPROPYL-p-TOLUIDINE

CLEAN AIR ACT: Hazardous Air Pollutants (Title I, Part A, Section 112).
SAFE DRINKING WATER ACT: Priority List (55 FR 1470).
EPCRA Section 304 Reportable Quantity (RQ): CERCLA, 10 lbs. (4.54 kgs.).
EPCRA Section 313 Form R de minimus concentration reporting level: 1.0%.

EPA NAME: TRIFORINE
CAS: 26644-46-2
SYNONYMS: 1,1′-PIPERAZINE-1,4-DIYLDI-(N-(2,2,2-TRICHLOROETHYL)FORMAMIDE); N,N′-[1,4-PIPERAZINEDIYLBIS(2,2,2-TRICHLOROETHYLIDENE)] BISFORMAMIDE

EPCRA Section 313 Form R de minimus concentration reporting level: 1.0%.

EPA NAME: TRIMETHYLAMINE
CAS: 75-50-3
SYNONYMS: N,N-DIMETHYLMETHANAMINE; METHAMINE, N,N-DIMETHYL-

CLEAN AIR ACT: Accidental Release Prevention/Flammable substances (Section 112[r], Table 3), TQ=10,000 lbs. (4540 kgs.).
CLEAN WATER ACT: Section 311 Hazardous Substances/RQ (same as CERCLA).
EPCRA Section 304 Reportable Quantity (RQ): CERCLA, 100 lbs. (45.4 kgs.).

EPA NAME: 1,2,4-TRIMETHYLBENZENE
CAS: 95-63-6
SYNONYMS: BENZENE, 1,2,4-TRIMETHYL-; PSEUDOCUMENE; 1,3,4-TRIMETHYLBENZENE

EPCRA Section 313 Form R de minimus concentration reporting level: 1.0%.

EPA NAME: TRIMETHYLCHLOROSILANE
CAS: 75-77-4
SYNONYMS: CHLOROTRIMETHYLSILANE; SILANE, CHLOROTRIMETHYL-; TRIMETHYLSILYL CHLORIDE

CLEAN AIR ACT: Accidental Release Prevention/Flammable substances (Section 112[r], Table 3), TQ=10,000 lbs. (4540 kgs.).
EPCRA Section 302 Extremely Hazardous Substances: TPQ=1,000 lbs. (454 kgs.).
EPCRA Section 304 Reportable Quantity (RQ): EHS, 1,000 lbs. (454 kgs.).
EPCRA Section 313 Form R de minimus concentration reporting level: 1.0%.

EPA NAME: 2,2,4-TRIMETHYLHEXAMETHYLENE DIISOCYANATE
CAS: 16938-22-0
[see also DIISOCYANATES]
SYNONYMS: 1,6-DIISOCYANATO-2,2,4-TRIMETHYLHEXANE; HEXANE, 1,6-DIISOCYANATO-2,2,4-TRIMETHYL-

EPCRA Section 313 Form R de minimus concentration reporting level: 1.0%.

EPA NAME: 2,4,4-TRIMETHYLHEXAMETHYLENE DIISOCYANATE
[see also DIISOCYANATES]
CAS: 15646-96-5
SYNONYMS: 1,6-DIISOCYANATO-2,4,4-TRIMETHYLHEXANE; HEXANE, 1,6-DIISOCYANATO-2,4,4-TRIMETHYL-

EPCRA Section 313 Form R de minimus concentration reporting level: 1.0%.

EPA NAME: TRIMETHYLOLPROPANE PHOSPHITE
CAS: 824-11-3
SYNONYMS: 4-ETHYL-PHOSPHA-2,6,7-TRIOXABICYCLO(2,2,2)OCTANE

EPCRA Section 302 Extremely Hazardous Substances: TPQ= 100/10,000 lbs. (45.4/4,540 kgs.).
EPCRA Section 304 Reportable Quantity (RQ): EHS, 100 lbs. (45.4 kgs.).

EPA NAME: 2,2,4-TRIMETHYLPENTANE
CAS: 540-84-1
SYNONYMS: ISOOCTANE; PENTANE, 2,2,4-TRIMETHYL-

CLEAN AIR ACT: Hazardous Air Pollutants (Title I, Part A, Section 112).
EPCRA Section 304 Reportable Quantity (RQ): CERCLA, 1,000 lbs. (454 kgs.).

EPA NAME: 2,3,5-TRIMETHYLPHENYL METHYLCARBAMATE
CAS: 2655-15-4
EPCRA Section 313 Form R de minimus concentration reporting level: 1.0%.

EPA NAME: TRIMETHYLTIN CHLORIDE
CAS: 1066-45-1
SYNONYMS: CHLOROTRIMETHYLTIN; TRIMETHYLSTANNYL CHLORIDE

EPCRA Section 302 Extremely Hazardous Substances: TPQ= 500/10,000 lbs. (227/4,540 kgs.).
EPCRA Section 304 Reportable Quantity (RQ): EHS, 500 lbs. (227 kgs.).

EPA NAME: 1,3,5-TRINITROBENZENE
CAS: 99-35-4
SYNONYMS: BENZENE, 1,3,5-TRINITRO-; TRINITROBENZENE; sym-TRINITROBENZENE

EPA HAZARDOUS WASTE NUMBER (RCRA No.): U234.
RCRA Section 261 Hazardous Constituents.
RCRA Ground Water Monitoring List. Suggested test method(s) (PQL µg/L): 8270(10).
EPCRA Section 304 Reportable Quantity (RQ): CERCLA, 10 lbs. (4.54 kgs.).

EPA NAME: TRIPATE
CAS: 26419-73-8
SYNONYMS: CARBAMIC ACID, METHYL-,O-(((2,4-DIMETHYL-1,3-DITHIOLAN-2-YL)METHYLENE)AMINO)-; 2,4-DIMETHYL-1,3-DITHIOLANE-2-CARBOXALDEHYDE O-(METHYLCARBAMOYL)OXIME; 1,3-DITHOLANE-2-CARBOXYALDEHYDE,2,4-DIMETHYL,O-(METHYLCARBAMOYL)OXIME; 2,4-DIMETHYL-1,3-DITHIOLANE-2-CARBOXALDEHYDE O-((METHYLAMINO)CARBONYL)-OXIME

EPA HAZARDOUS WASTE NUMBER (RCRA No.): P185 as tripate.
RCRA Section 261 Hazardous Constituents.
RCRA Land Ban Waste.
RCRA Universal Treatment Standards: Wastewater (mg/L), 0.056; Nonwastewater (mg/kg), 0.28.

EPCRA Section 302 Extremely Hazardous Substances: TPQ = 100/10,000 lbs. (45.4/4,540 kgs.).
EPCRA Section 304 Reportable Quantity (RQ): EHS, 1 lb. (0.454 kg.).

EPA NAME: TRIPHENYLTIN CHLORIDE
CAS: 639-58-7
SYNONYMS: EPA PESTICIDE CHEMICAL CODE 496500; FENTIN CHLORIDE; STANNANE, CHLOROTRIPHENYL-

EPCRA Section 302 Extremely Hazardous Substances: TPQ = 500/10,000 lbs. (227/4,540 kgs.).
EPCRA Section 304 Reportable Quantity (RQ): EHS, 500 lbs. (227 kgs.).
EPCRA Section 313 Form R de minimus concentration reporting level: 1.0%.
MARINE POLLUTANT (49CFR, Subchapter 172.101, Appendix B) as triphenyltin compounds.

EPA NAME: TRIPHENYLTIN HYDROXIDE
CAS: 76-87-9
SYNONYMS: EPA PESTICIDE CHEMICAL CODE 083601; FENTIN; FENTIN HYDROXIDE

EPCRA Section 313 Form R de minimus concentration reporting level: 1.0%.
MARINE POLLUTANT (49CFR, Subchapter 172.101, Appendix B) as triphenyltin compounds.
CALIFORNIA'S PROPOSITION 65: Carcinogen.

EPA NAME: TRIS(2-CHLOROETHYL)AMINE
CAS: 555-77-1
SYNONYMS: TRICHLORMETHINE; TRI-(2-CHLORO-ETHYL)AMINE; TRIMUSTINE HYDROCHLORIDE

EPCRA Section 302 Extremely Hazardous Substances: TPQ = 100 lbs. (45.4 kgs.).
EPCRA Section 304 Reportable Quantity (RQ): EHS, 100 lbs. (45.4 kgs.).

EPA NAME: TRIS(2,3-DIBROMOPROPYL) PHOSPHATE
CAS: 126-72-7
SYNONYMS: PHOSPHORIC ACID TRIS(2,3-DIBROMOPROPYL)ESTER; 1-PROPANOL, 2,3-DIBROMO-, PHOSPHATE (3:1)

EPA HAZARDOUS WASTE NUMBER (RCRA No.): U235.
RCRA Section 261 Hazardous Constituents.
RCRA Land Ban Waste.
RCRA Universal Treatment Standards: Wastewater (mg/L), 0.11; Nonwastewater (mg/kg), 0.10.
EPCRA Section 304 Reportable Quantity (RQ): CERCLA, 10 lbs. (4.54 kgs.).
EPCRA Section 313 Form R de minimus concentration reporting level: 0.1%.

EPA NAME: TRIS(DIMETHYLCARBAMODITHIOATO-S,S′)IRON

[see FERBAM]

CAS: 14484-64-1

EPA NAME: TRYPAN BLUE

CAS: 72-57-1

SYNONYMS: 3,3′-((3,3′-DIMETHYL(1,1′-BIPHENYL)-4,4′-DIYL)BIS(AZO))BIS(5-AMINO-4-HYDROXYNAPHTHALENE-2,7-DISULPHONATE); 2,7-NAPHTHALENEDISULFONIC ACID, 3,3′-((3,3′-DIMETHYL(1,1′-BIPHENYL)-4,4′-DIYL)BIS(AZO))BIS(5-AMINO-4-HYDROXY-, TETRASODIUM SALT; 2,7-NAPHTHALENEDISULFONIC ACID, 3,3′-((3,3′-DIMETHYL-4,4′-BIPHENYLYLENE)BIS(AZO))BIS(5-AMINO-4-HYDROXY-, TETRASODIUM SALT; C.I. DIRECT BLUE 14, TETRASODIUM SALT

EPA HAZARDOUS WASTE NUMBER (RCRA No.): U236.

RCRA Section 261 Hazardous Constituents.

EPCRA Section 304 Reportable Quantity (RQ): CERCLA, 10 lbs. (4.54 kgs.).

EPCRA Section 313 Form R de minimus concentration reporting level: 0.1%.

CALIFORNIA'S PROPOSITION 65: Carcinogen.

- U -

EPA NAME: URACIL MUSTARD

CAS: 66-75-1

SYNONYMS: 5-(BIS(2-CHLOROETHYL)AMINO)-2,4 (1H,3H)-PYRIMIDINEDIONE; 5-N,N-BIS(2-CHLOROETHYL)AMINO-URACIL; 2,4(1H,3H)-PYRIMIDINEDIONE, 5-[BIS(2-CHLORO-ETHYL)AMINO]-

EPA HAZARDOUS WASTE NUMBER (RCRA No.): U237.

RCRA Section 261 Hazardous Constituents.

EPCRA Section 304 Reportable Quantity (RQ): CERCLA, 10 lbs. (4.54 kgs.).

CALIFORNIA'S PROPOSITION 65: Carcinogen; reproductive toxin (male, female).

EPA NAME: URANYL ACETATE

CAS: 541-09-3

SYNONYMS: URANIUM ACETATE; URANIUM BIS(ACETO-O)DIOXO-; URANIUM OXYACETATE

CLEAN WATER ACT: Section 311 Hazardous Substances/RQ (same as CERCLA).

EPCRA Section 304 Reportable Quantity (RQ): CERCLA, 100 lbs. (45.4 kgs.).

EPA NAME: URANYL NITRATE

CAS: 10102-06-4

SYNONYMS: BIS(NITRATO-O,O')DIOXO URANIUM (solid); URANIUM BIS(NITRATO-O)DIOXO-, (T-4)

CLEAN WATER ACT: Section 311 Hazardous Substances/RQ (same as CERCLA).

EPCRA Section 304 Reportable Quantity (RQ): CERCLA, 100 lbs. (45.4 kgs.).

EPA NAME: URANYL NITRATE

CAS: 36478-76-9

SYNONYMS: URANIUM, BIS(NITRATO-O,O')DIOXO-,(OC-6-11)-

CLEAN WATER ACT: Section 311 Hazardous Substances/RQ (same as CERCLA).

EPCRA Section 304 Reportable Quantity (RQ): CERCLA, 100 lbs. (45.4 kgs.).

EPA NAME: UREA, N,N-DIMETHYL-N'-[3-(TRIFLUOROMETHYL)PHENYL]-

[see FLUOMETURON]

CAS: 2164-17-2

EPA NAME: URETHANE

CAS: 51-79-6

SYNONYMS: CARBAMIC ACID, ETHYL ESTER; ETHYLCARBAMATE

CLEAN AIR ACT: Hazardous Air Pollutants (Title I, Part A, Section 112).
EPA HAZARDOUS WASTE NUMBER (RCRA No.): U238.
RCRA Section 261 Hazardous Constituents.
EPCRA Section 304 Reportable Quantity (RQ): CERCLA, 100 lbs. (45.4 kgs.).
EPCRA Section 313 Form R de minimus concentration reporting level: 0.1%.
CALIFORNIA'S PROPOSITION 65: Carcinogen; reproductive toxin.

- V -

EPA NAME: VALINOMYCIN
CAS: 2001-95-8
SYNONYMS: ANTIBIOTIC N-329 B; VALINOMICIN

EPCRA Section 302 Extremely Hazardous Substances: TPQ = 1,000/10,000 lbs. (454/4,540 kgs.).
EPCRA Section 304 Reportable Quantity (RQ): EHS, 1,000 lbs. (454 kgs.).

EPA NAME: VANADIUM (FUME OR DUST)
CAS: 7440-62-2
SYNONYMS: VANADIUM, ELEMENTAL

CLEAN WATER ACT: Section 313 Priority Chemicals.
RCRA Land Ban Waste.
RCRA Universal Treatment Standards: Wastewater (mg/L), 4.3; Non-wastewater (mg/L), 0.23.
RCRA Ground Water Monitoring List. Suggested test method(s) (PQL µg/L): (total) 6010(80); 7910(2,000); 7911(40).
SAFE DRINKING WATER ACT: Priority List (55 FR 1470) as vanadium.
EPCRA Section 313 Form R de minimus concentration reporting level: 1.0%..

EPA NAME: VANADIUM PENTOXIDE
[see also VANADIUM (FUME OR DUST)]
CAS: 1314-62-1
SYNONYMS: DIVANADIUM PENTOXIDE; VANADIUM OXIDE; VANADIUM(V) OXIDE FUME

CLEAN WATER ACT: Section 311 Hazardous Substances/RQ (same as CERCLA).
EPA HAZARDOUS WASTE NUMBER (RCRA No.): P120.
RCRA Section 261 Hazardous Constituents.
RCRA Ground Water Monitoring List. Suggested test method(s) (PQL µg/L): (total) 6010(80); 7910(2,000); 7911(40).
EPCRA Section 302 Extremely Hazardous Substances: TPQ = 100/10,000 lbs. (45.4/4,540 kgs.).
EPCRA Section 304 Reportable Quantity (RQ): EHS/CERCLA, 1,000 lbs. (454 kgs.).

EPA NAME: VANADYL SULFATE
CAS: 27774-13-6
SYNONYMS: OXYSULFATOVANADIUM; VANADIUM, OXO[SULFATO(2-)-O]-

CLEAN WATER ACT: Section 311 Hazardous Substances/RQ (same as CERCLA).
EPCRA Section 304 Reportable Quantity (RQ): CERCLA, 1,000 lbs. (454 kgs.).

EPA NAME: VIKANE
[see SULFURYL FLUORIDE]
CAS: 2699-79-8

EPA NAME: VINCLOZOLIN
CAS: 50471-44-8
SYNONYMS: 3-(3,5-DICHLOROPHENYL)-5-ETHENYL-5-METHYL-2,4-OXAZOLIDINEDIONE; EPA PESTICIDE CHEMICAL CODE 113201; 2,4-OXAZOLIDINEDIONE, 3-(3,5-DICHLOROPHENYL)-5-METHYL-5-VINYL-

EPCRA Section 313 Form R de minimus concentration reporting level: 1.0%.

EPA NAME: VINYL ACETATE
CAS: 108-05-4
SYNONYMS: ACETIC ACID, ETHENYL ESTER; ACETIC ACID, VINYL ESTER; ETHENYL ETHANOATE; VINYL ACETATE MONOMER

CLEAN AIR ACT: Hazardous Air Pollutants (Title I, Part A, Section 112); Accidental Release Prevention/Flammable substances (Section 112[r], Table 3), TQ = 15,000 lbs (6825 kgs).
CLEAN WATER ACT: Section 311 Hazardous Substances/RQ (same as CERCLA); Priority Pollutants (40CFR PART 423); Section 313 Priority Chemicals.
RCRA Ground Water Monitoring List. Suggested test method(s) (PQL µg/L): 8240(5). CERCLA Section 302, Extremely Hazardous Substances, TPQ = 1,000 lbs. (455 kgs.).
EPCRA Section 304 Reportable Quantity (RQ): EHS/CERCLA, 5,000 lbs. (2270 kgs.). CERCLA Section 313 Form R de minimus concentration reporting level: 1.0%.

EPA NAME: VINYL ACETYLENE
CAS: 689-97-4
SYNONYMS: 1-BUTEN-3-YNE

CLEAN AIR ACT: Accidental Release Prevention/Flammable substances (Section 112[r], Table 3), TQ = 10,000 lbs. (4540 kgs.).

EPA NAME: VINYL BROMIDE
CAS: 593-60-2
SYNONYMS: BROMOETHYLENE; ETHENE, BROMO-; ETHYLENE, BROMO-

CLEAN AIR ACT: Hazardous Air Pollutants (Title I, Part A, Section 112).
EPCRA Section 304 Reportable Quantity (RQ): CERCLA, 100 lbs. (45.4 kgs.).
EPCRA Section 313 Form R de minimus concentration reporting level: 0.1%.
CALIFORNIA'S PROPOSITION 65: Carcinogen.

EPA NAME: VINYL CHLORIDE
CAS: 75-01-4

SYNONYMS: CHLOROETHYLENE; CHLOROETHENE; ETHENE, CHLORO-

CLEAN AIR ACT: Hazardous Air Pollutants (Title I, Part A, Section 112); Accidental Release Prevention/Flammable substances (Section 112[r], Table 3), TQ = 10,000 lbs. (4540 kgs.).

CLEAN WATER ACT: Section 313 Priority Chemicals; Section 307 Toxic Pollutants.

EPA HAZARDOUS WASTE NUMBER (RCRA No.): U043; D043.

RCRA Toxicity Characteristic (Section 261.24), Maximum Concentration of Contaminants, regulatory level, 0.2 mg/L.

RCRA Section 261 Hazardous Constituents.

RCRA Land Ban Waste.

RCRA Universal Treatment Standards: Wastewater (mg/L), 0.27; Nonwastewater (mg/kg), 6.0.

RCRA Ground Water Monitoring List. Suggested test method(s) (PQL µg/L): 8010(2); 8240(10).

SAFE DRINKING WATER ACT: MCL, 0.002 mg/L; MCLG, zero; Regulated chemical (47 FR 9352).

EPCRA Section 304 Reportable Quantity (RQ): CERCLA, 1 lb. (0.454 kg.).

EPCRA Section 313 Form R de minimus concentration reporting level: 0.1%.

CALIFORNIA'S PROPOSITION 65: Carcinogen.

EPA NAME: VINYL ETHYL ETHER

CAS: 109-92-2

SYNONYMS: ETHENE, ETHOXY-; ETHER, VINYL ETHYL

CLEAN AIR ACT: Accidental Release Prevention/Flammable substances (Section 112[r], Table 3), TQ = 10,000 lbs. (4540 kgs.).

EPA NAME: VINYL FLUORIDE

CAS: 75-02-5

SYNONYMS: ETHENE, FLUORO-; FLUOROETHENE; MONOFLUOROETHYLENE

CLEAN AIR ACT: Accidental Release Prevention/Flammable substances (Section 112[r], Table 3), TQ = 10,000 lbs. (4540 kgs.).

EPA NAME: VINYLIDENE CHLORIDE

CAS: 75-35-4

SYNONYMS: DICHLOROETHYLENE; 1,1-DICHLOROETHYLENE; 1,1-DICHLOROETHENE; ETHENE, 1,1 DICHLORO-; ETHYLENE, 1,1-DICHLORO-

CLEAN AIR ACT: Hazardous Air Pollutants (Title I, Part A, Section 112); Accidental Release Prevention/Flammable substances (Section 112[r], Table 3), TQ = 10,000 lbs. (4540 kgs.).

CLEAN WATER ACT: Section 311 Hazardous Substances/RQ (same as CERCLA); Section 307 Priority Pollutants; Section 313 Priority Chemicals; Toxic Pollutant (Section 401.15).

EPA HAZARDOUS WASTE NUMBER (RCRA No.): U078; D029.

RCRA Section 261 Hazardous Constituents.
RCRA Toxicity Characteristic (Section 261.24), Maximum Concentration of Contaminants, regulatory level, 0.7 mg/L.
RCRA Land Ban Waste.
RCRA Universal Treatment Standards: Wastewater (mg/L), 0.025; Nonwastewater (mg/kg), 6.0.
RCRA Ground Water Monitoring List. Suggested test method(s) (PQL µg/L): 8010(1); 8240(5).
SAFE DRINKING WATER ACT: MCL, 0.007 mg/L; MCLG, 0.007 mg/L; Regulated chemical (47 FR 9352).
EPCRA Section 304 Reportable Quantity (RQ): CERCLA, 100 lbs. (45.4 kgs.).
EPCRA Section 313 Form R de minimus concentration reporting level: 1.0%.
MARINE POLLUTANT (49CFR, Subchapter 172.101, Appendix B).

EPA NAME: VINYLIDENE FLUORIDE

CAS: 75-38-7
SYNONYMS: 1,1-DIFLUOROETHYLENE; ETHENE, 1,1-DIFLUORO-

CLEAN AIR ACT: Accidental Release Prevention/Flammable substances (Section 112[r], Table 3), TQ = 10,000 lbs. (4540 kgs.).

EPA NAME: VINYL METHYL ETHER

CAS: 107-25-5
SYNONYMS: ETHENE, METHOXY-; METHYL VINYL ETHER

CLEAN AIR ACT: Accidental Release Prevention/Flammable substances (Section 112[r], Table 3), TQ = 10,000 lbs. (4540 kgs.).

- W -

EPA NAME: WARFARIN

[see also WARFARIN AND SALTS]

CAS: 81-81-2

SYNONYMS: 3-(α-ACETONYLBENZYL)-4-HYDROXYCOUMARIN; 2H-1-BENZOPYRAN-2-ONE,4-HYDROXY-3-(3-OXO-1-PHENYLBUTYL)-; COUMARIN, 3-(α-ACETONYLBENZYL)-4-HYDROXY-

EPA HAZARDOUS WASTE NUMBER (RCRA No.): P001; U248.

RCRA Section 261 Hazardous Constituents.

EPCRA Section 302 Extremely Hazardous Substances: TPQ = 500/10,000 lbs. (227/4,540 kgs.).

EPCRA Section 304 Reportable Quantity (RQ): CERCLA, 100 lbs. (45.4 kgs.).

EPCRA Section 313 Form R de minimus concentration reporting level: 1.0%. Form R Toxic Chemical Category Code: N874.

MARINE POLLUTANT (49CFR, Subchapter 172.101, Appendix B).

CALIFORNIA'S PROPOSITION 65: Reproductive toxin.

EPA NAME: WARFARIN AND SALTS

[see also WARFARIN]

SYNONYMS: WARFARIN SALTS; WARFARIN SALTS, conc. <0.3%; WARFARIN SALTS, conc. >0.3%

EPA HAZARDOUS WASTE NUMBER (RCRA No.): U248, warfarin salts when present at concentrations less than 0.3%; P001, warfarin salts when present at concentrations greater than 0.3%

RCRA Section 261 Hazardous Constituents.

EPCRA Section 304 Reportable Quantity (RQ): CERCLA, 100 lbs. (45.4 kgs.), when present at concentrations greater than 0.3%.

EPCRA Section 313: Includes any unique chemical substance that contains warfarin or a warfarin salt as part of that chemical's infrastructure. Form R de minimus concentration reporting level: 1.0%. Form R Toxic Chemical Category Code: N874.

MARINE POLLUTANT (49CFR, Subchapter 172.101, Appendix B).

CALIFORNIA'S PROPOSITION 65: Reproductive toxin, as warfarin.

EPA NAME: WARFARIN SODIUM

[see also WARFARIN and WARFARIN AND SALTS]

CAS: 129-06-0

SYNONYMS: 3-(α-ACETONYLBENZYL)-4-HYDROXY-COUMARIN SODIUM SALT; 4-HYDROXY-3-(3-OXO-1-PHENYLBUTYL)-2H-1-BENZOPYRAN-2-ONE SODIUM SALT (9CI)

EPCRA Section 302 Extremely Hazardous Substances: TPQ = 100/10,000 lbs. (45.4/4,540 kgs.).

EPCRA Section 304 Reportable Quantity (RQ): EHS, 100 lbs. (45.4 kgs.).

EPCRA Section 313 Form R de minimus concentration reporting level: 1.0%. Form R Toxic Chemical Category Code: N874.

MARINE POLLUTANT (49CFR, Subchapter 172.101, Appendix B).

- X -

EPA AME: m-XYLENE
[see also XYLENE (MIXED ISOMERS)]
CAS: 108-38-3
SYNONYMS: BENZENE, m-DIMETHYL-; BENZENE-1,3-DIMETHYL-; m-DIMETHYLBENZENE; 1,3-DIMETHYLBENZENE; m-METHYLTOLUENE; 1,3-XYLENE; m-XYLOL

CLEAN AIR ACT: Hazardous Air Pollutants (Title I, Part A, Section 112).
CLEAN WATER ACT: Section 313 Priority Chemicals.
EPA HAZARDOUS WASTE NUMBER (RCRA No.): U239.
RCRA Section 261 Hazardous Constituents.
EPCRA Section 304 Reportable Quantity (RQ): CERCLA, 1,000 lbs. (454.0 kgs.).
EPCRA Section 313 Form R de minimus concentration reporting level: 1.0%.

EPA NAME: o-XYLENE
[see also XYLENE (MIXED ISOMERS)]
CAS: 95-47-6
SYNONYMS: BENZENE-o-DIMETHYL; BENZENE-1,2-DIMETHYL-; o-DIMETHYLBENZENE; 1,2-DIMETHYLBENZENE; o-METHYLTOLUENE; 1,2-METHYLTOLUENE; 1,2-XYLENE

CLEAN AIR ACT: Hazardous Air Pollutants (Title I, Part A, Section 112).
CLEAN WATER ACT: Section 313 Priority Chemicals.
EPA HAZARDOUS WASTE NUMBER (RCRA No.): U239.
RCRA Section 261 Hazardous Constituents.
EPCRA Section 304 Reportable Quantity (RQ): CERCLA, 1,000 lbs. (454.0 kgs.).
EPCRA Section 313 Form R de minimus concentration reporting level: 1.0%.

EPA NAME: p-XYLENE
[see also XYLENE (MIXED ISOMERS)]
CAS: 106-42-3
SYNONYMS: BENZENE-p-DIMETHYL; BENZENE-1,4-DIMETHYL; 1,4-DIMETHYLBENZENE; 4-METHYLTOLUENE; p-METHYLTOLUENE; 1,4-XYLENE

CLEAN AIR ACT: Hazardous Air Pollutants (Title I, Part A, Section 112).
CLEAN WATER ACT: Section 313 Priority Chemicals.
EPA HAZARDOUS WASTE NUMBER (RCRA No.): U239.
RCRA Section 261 Hazardous Constituents.
EPCRA Section 304 Reportable Quantity (RQ): CERCLA, 100 lbs. (45.4 kgs.).
EPCRA Section 313 Form R de minimus concentration reporting level: 1.0%.

EPA NAME: XYLENE (MIXED ISOMERS)
CAS: 1330-20-7

SYNONYMS: BENZENE, DIMETHYL; DIMETHYLBENZENE; METHYL TOLUENE; XYLENE; XYLENES, N.O.S.; XYLENES (TOTAL)

CLEAN AIR ACT: Hazardous Air Pollutants (Title I, Part A, Section 112) as xylenes (isomers and mixtures.)

CLEAN WATER ACT: Section 311 Hazardous Substances/RQ 100 lbs. (45.4 kgs.); Section 313 Priority Chemicals.

EPA HAZARDOUS WASTE NUMBER (RCRA No.): U239.

RCRA Section 261 Hazardous Constituents.

RCRA Land Ban Waste.

RCRA Universal Treatment Standards: Wastewater (mg/L), 0.32; Non-wastewater (mg/kg), 30.

RCRA Ground Water Monitoring List. Suggested methods (PQL μg/L): (total) 8020(5); 8240(5).

SAFE DRINKING WATER ACT: Xylenes (total), MCL, 10 mg/L; MCLG, 10 mg/L; Regulated chemical (47 FR 9352) as xylene.

EPCRA Section 304 Reportable Quantity (RQ): CERCLA, 100 lbs. (45.4 kgs.).

EPCRA Section 313 Form R de minimus concentration reporting level: 1.0%.

EPA NAME: XYLENOL

CAS: 1300-71-6

SYNONYMS: DIMETHYLPHENOL; PHENOL, DIMETHYL-; XYLENOLS, MIXED

CLEAN WATER ACT: Section 311 Hazardous Substances/RQ (same as CERCLA).

EPCRA Section 304 Reportable Quantity (RQ): CERCLA, 1,000 lbs. (454 kgs.).

MARINE POLLUTANT (49CFR, Subchapter 172.101, Appendix B) as xylenols.

EPA NAME: 2,6-XYLIDINE

CAS: 87-62-7

SYNONYMS: 2-AMINO-1,3-XYLENE; BENZENAMINE, 2,6-DIMETHYL-; 2,6-DIMETHYLANILINE; 2,6-DIMETHYLBENZENAMINE

EPCRA Section 313 Form R de minimus concentration reporting level: 1.0%.

CALIFORNIA'S PROPOSITION 65: Carcinogen.

EPA NAME: XYLYLENE DICHLORIDE

CAS: 28347-13-9

SYNONYMS: BENZENE, BIS(CHLOROMETHYL)-; BIS (CHLOROMETHYL)BENZENE

EPCRA Section 302 Extremely Hazardous Substances: TPQ = 100/10,000 lbs. (45.4/4,540 kgs.).

EPCRA Section 304 Reportable Quantity (RQ): EHS, 100 lbs. (45.4 kgs.).

- Z -

EPA NAME: ZINC
[see also ZINC COMPOUNDS]
CAS: 7440-66-6
SYNONYMS: ZINC, METALLIC

CLEAN WATER ACT: Section 307 Toxic Pollutants; Section 307 Priority Pollutants.
RCRA Universal Treatment Standards: Wastewater (mg/L), 2.61; Non-wastewater (mg/L), 5.3 TCLP Note: these constituents are not "underlying hazardous constituents" in characteristic wastes, according to the definition at Section 268.2(i).
RCRA Ground Water Monitoring List. Suggested test method(s) (PQL µg/L): (total) 6010(20); 7950(50).
SAFE DRINKING WATER ACT: SMCL, 5 mg/L; Priority List (55 FR 1470).
EPCRA Section 304 Reportable Quantity (RQ): CERCLA, 1,000 lbs. (454 kgs.).
EPCRA Section 313 Form R de minimus concentration reporting level: 1.0%. Form R Toxic Chemical Category Code: N982.

EPA NAME: ZINC (FUME OR DUST)
[see also ZINC and ZINC COMPOUNDS]
CAS: 7440-66-6
CLEAN WATER ACT: Section 307 Toxic Pollutants; Section 313 Priority Chemicals.
EPCRA Section 304 Reportable Quantity (RQ): CERCLA, 1,000 lbs. (454 kgs.).
EPCRA Section 313 Form R de minimus concentration reporting level: 1.0%. Form R Toxic Chemical Category Code: N982.

EPA NAME: ZINC ACETATE
[see also ZINC and ZINC COMPOUNDS]
CAS: 557-34-6
SYNONYMS: ACETIC ACID, ZINC SALT; ZINC DIACETATE

CLEAN WATER ACT: Section 311 Hazardous Substances/RQ (same as CERCLA); Section 307 Toxic Pollutants; Section 313 Priority Chemicals.
EPCRA Section 304 Reportable Quantity (RQ): CERCLA, 1,000 lbs. (454 kgs.).
EPCRA Section 313 Form R de minimus concentration reporting level: 1.0%. Form R Toxic Chemical Category Code: N982.

EPA NAME: ZINC AMMONIUM CHLORIDE
[see also ZINC and ZINC COMPOUNDS]
CAS: 14639-97-5
SYNONYMS: ZINCATE(2-), TETRACHLORO-, DIAMMONIUM, (T-4)-

CLEAN WATER ACT: Section 311 Hazardous Substances/RQ (same as CERCLA); Section 307 Toxic Pollutants; Section 313 Priority Chemicals.

EPCRA Section 304 Reportable Quantity (RQ): CERCLA, 1,000 lbs. (454 kgs.).

EPCRA Section 313 Form R de minimus concentration reporting level: 1.0%. Form R Toxic Chemical Category Code: N982.

EPA NAME: ZINC AMMONIUM CHLORIDE
[see also ZINC and ZINC COMPOUNDS]
CAS: 14639-98-6
SYNONYMS: ZINCATE(3-), TETRACHLORO-, TRIAMMONIUM, (T-4)-

CLEAN WATER ACT: Section 311 Hazardous Substances/RQ (same as CERCLA); Section 307 Toxic Pollutants; Section 313 Priority Chemicals.

EPCRA Section 304 Reportable Quantity (RQ): CERCLA, 1,000 lbs. (454 kgs.).

EPCRA Section 313 Form R de minimus concentration reporting level: 1.0%. Form R Toxic Chemical Category Code: N982.

EPA NAME: ZINC AMMONIUM CHLORIDE
[see also ZINC and ZINC COMPOUNDS]
CAS: 52628-25-8
SYNONYMS: AMMONIUM ZINC CHLORIDE; ZINCATE (2-), TETRACHLORO-, DIAMMONIUM SALT

CLEAN WATER ACT: Section 311 Hazardous Substances/RQ (same as CERCLA); Section 307 Toxic Pollutants; Section 313 Priority Chemicals.

EPCRA Section 304 Reportable Quantity (RQ): CERCLA, 1,000 lbs. (454 kgs.).

EPCRA Section 313 Form R de minimus concentration reporting level: 1.0%. Form R Toxic Chemical Category Code: N982.

EPA NAME: ZINC BORATE
[see also ZINC and ZINC COMPOUNDS]
CAS: 1332-07-6
SYNONYMS: BORIC ACID, ZINC SALT

CLEAN WATER ACT: Section 311 Hazardous Substances/RQ (same as CERCLA); Section 307 Toxic Pollutants; Section 313 Priority Chemicals.

EPCRA Section 304 Reportable Quantity (RQ): CERCLA, 1,000 lbs. (454 kgs.).

EPCRA Section 313 Form R de minimus concentration reporting level: 1.0%. Form R Toxic Chemical Category Code: N982.

EPA NAME: ZINC BROMIDE
[see also ZINC and ZINC COMPOUNDS]
CAS: 7699-45-8
SYNONYMS: ZINC DIBROMIDE

CLEAN WATER ACT: Section 311 Hazardous Substances/RQ (same as CERCLA); Section 307 Toxic Pollutants; Section 313 Priority Chemicals.

EPCRA Section 304 Reportable Quantity (RQ): CERCLA, 1,000 lbs. (454 kgs.).

EPCRA Section 313 Form R de minimus concentration reporting level: 1.0%. Form R Toxic Chemical Category Code: N982.

MARINE POLLUTANT (49CFR, Subchapter 172.101, Appendix B).

EPA NAME: ZINC CARBONATE

[see also ZINC and ZINC COMPOUNDS]

CAS: 3486-35-9

SYNONYMS: CARBONIC ACID, ZINC SALT (1:1)

CLEAN WATER ACT: Section 311 Hazardous Substances/RQ (same as CERCLA); Section 307 Toxic Pollutants; Section 313 Priority Chemicals.

EPCRA Section 304 Reportable Quantity (RQ): CERCLA, 1,000 lbs. (454 kgs.).

EPCRA Section 313 Form R de minimus concentration reporting level: 1.0%. Form R Toxic Chemical Category Code: N982.

EPA NAME: ZINC CHLORIDE

[see also ZINC and ZINC COMPOUNDS]

CAS: 7646-85-7

SYNONYMS: ZINC CHLORIDE FUME

CLEAN WATER ACT: Section 311 Hazardous Substances/RQ (same as CERCLA); CLEAN WATER ACT: Section 307 Toxic Pollutants; Section 313 Priority Chemicals.

EPCRA Section 304 Reportable Quantity (RQ): CERCLA, 1,000 lbs. (454 kgs.).

EPCRA Section 313 Form R de minimus concentration reporting level: 1.0%. Form R Toxic Chemical Category Code: N982.

EPA NAME: ZINC COMPOUNDS

[see also ZINC]

CLEAN WATER ACT: Section 307 Toxic Pollutants as zinc and compounds.

EPCRA Section 313: Includes any unique chemical substance that contains zinc as part of that chemical's infrastructure. Form R de minimus concentration reporting level: 1.0%. Form R Toxic Chemical Category Code: N982.

EPA NAME: ZINC CYANIDE

[see also CYANIDE, CYANIDE COMPOUNDS, ZINC, ZINC COMPOUNDS]

CAS: 557-21-1

SYNONYMS: ZINC DICYANIDE

CLEAN WATER ACT: Section 311 Hazardous Substances/RQ (same as CERCLA); Section 307 Toxic Pollutants; Section 313 Priority Chemicals.

EPA HAZARDOUS WASTE NUMBER (RCRA No.): P121.
RCRA Section 261 Hazardous Constituents.
EPCRA Section 304 Reportable Quantity (RQ): CERCLA, 10 lbs. (4.54 kgs.).
EPCRA Section 313 (as zinc or cyanide compound) Form R de minimus concentration reporting level: 1.0%. Form R Toxic Chemical Category Code: N982 (zinc). Form R Toxic Chemical Category Code: N016 (cyanide).
MARINE POLLUTANT (49CFR, Subchapter 172.101, Appendix B).

EPA NAME: ZINC, DICHLORO(4,4-DIMETHYL-5((((METHYLAMINO)CARBONYL)OXY)IMINO)PENTANENITRILE)-,(T-4)-

[see also ZINC and ZINC COMPOUNDS]
CAS: 58270-08-9
SYNONYMS: ETHIENOCARB

CLEAN WATER ACT: Section 307 Toxic Pollutants.
EPCRA Section 302 Extremely Hazardous Substances: TPQ = 100/10,000 lbs. (45.4/4,540 kgs.).
EPCRA Section 304 Reportable Quantity (RQ): EHS, 100 lbs. (45.4 kgs.).
EPCRA Section 313 Form R de minimus concentration reporting level: 1.0%. Form R Toxic Chemical Category Code: N982.

EPA NAME: ZINC FLUORIDE

[see also ZINC and ZINC COMPOUNDS]
CAS: 7783-49-5
SYNONYMS: ZINC FLUORIDE (ZnF_2)

CLEAN WATER ACT: Section 311 Hazardous Substances/RQ (same as CERCLA); Section 307 Toxic Pollutants; Section 313 Priority Chemicals.
EPCRA Section 304 Reportable Quantity (RQ): CERCLA, 1,000 lbs. (454 kgs.).
EPCRA Section 313 Form R de minimus concentration reporting level: 1.0%. Form R Toxic Chemical Category Code: N982.

EPA NAME: ZINC FORMATE

[see also ZINC and ZINC COMPOUNDS]
CAS: 557-41-5
SYNONYMS: FORMIC ACID, ZINC SALT; ZINC DIFORMATE

CLEAN WATER ACT: Section 311 Hazardous Substances/RQ (same as CERCLA); Section 307 Toxic Pollutants; Section 313 Priority Chemicals.
EPCRA Section 304 Reportable Quantity (RQ): CERCLA, 1,000 lbs. (454 kgs.).
EPCRA Section 313 Form R de minimus concentration reporting level: 1.0%. Form R Toxic Chemical Category Code: N982.

EPA NAME: ZINC HYDROSULFITE

[see also ZINC and ZINC COMPOUNDS]
CAS: 7779-86-4

SYNONYMS: DITHIONOUS ACID, ZINC SALT (1:1); ZINC DITHIONITE

CLEAN WATER ACT: Section 311 Hazardous Substances/RQ (same as CERCLA); Section 307 Toxic Pollutants; Section 313 Priority Chemicals.

EPCRA Section 304 Reportable Quantity (RQ): CERCLA, 1,000 lbs. (454 kgs.).

EPCRA Section 313 Form R de minimus concentration reporting level: 1.0%. Form R Toxic Chemical Category Code: N982.

EPA NAME: ZINC NITRATE

[see also NITRATE COMPOUNDS, ZINC, and ZINC COMPOUNDS]

CAS: 7779-88-6

SYNONYMS: NITRIC ACID, ZINC SALT

CLEAN WATER ACT: Section 311 Hazardous Substances/RQ (same as CERCLA); Section 307 Toxic Pollutants; Section 313 Priority Chemicals.

EPCRA Section 304 Reportable Quantity (RQ): CERCLA, 1,000 lbs. (454 kgs.).

EPCRA Section 313 Form R de minimus concentration reporting level: 1.0%. Form R Toxic Chemical Category Code: N982 (zinc); N511 (nitrate compounds, water dissociable; reportable only in aqueous solution).

EPA NAME: ZINC PHENOLSULFONATE

[see also ZINC and ZINC COMPOUNDS]

CAS: 127-82-2

SYNONYMS: BENZENESULFONIC ACID, 4-HYDROXY-, ZINC SALT (2:1); BENZENESULFONIC ACID, p-HYDROXY-, ZINC SALT (2:1)

CLEAN WATER ACT: Section 311 Hazardous Substances/RQ (same as CERCLA); Section 307 Toxic Pollutants; Section 313 Priority Chemicals.

EPCRA Section 304 Reportable Quantity (RQ): CERCLA, 5,000 lbs. (2270 kgs.).

EPCRA Section 313 Form R de minimus concentration reporting level: 1.0%. Form R Toxic Chemical Category Code: N982.

EPA NAME: ZINC PHOSPHIDE

[see also ZINC and ZINC COMPOUNDS]

CAS: 1314-84-7

SYNONYMS: ZINC PHOSPHIDE (conc. >10%); ZINC PHOSPHIDE (Zn_3P_2), when present at concentrations greater than 10%

CLEAN WATER ACT: Section 311 Hazardous Substances/RQ (same as CERCLA); Section 307 Toxic Pollutants; Section 313 Priority Chemicals.

EPA HAZARDOUS WASTE NUMBER (RCRA No.): P122.

RCRA Section 261 Hazardous Constituents.

EPCRA Section 302 Extremely Hazardous Substances: TPQ = 500 lbs. (227 kgs.).

EPCRA Section 304 Reportable Quantity (RQ): EHS/CERCLA, 100 lbs. (45.4 kgs.).

EPCRA Section 313 Form R de minimus concentration reporting level: 1.0%. Form R Toxic Chemical Category Code: N982.

EPA NAME: ZINC PHOSPHIDE (conc. ≤ 10%)

[see also ZINC and ZINC COMPOUNDS]

CAS: 1314-84-7

SYNONYMS: ZINC PHOSPHIDE (Zn_3P_2) when present at concentrations of 10% or less

CLEAN WATER ACT: Section 311 Hazardous Substances/RQ (same as CERCLA); Section 307 Toxic Pollutants; Section 313 Priority Chemicals.

EPCRA Section 302 Extremely Hazardous Substances: TPQ = 500 lbs. (227 kgs.).

EPCRA Section 304 Reportable Quantity (RQ): CERCLA, 100 lbs. (45.4 kgs.).

EPA HAZARDOUS WASTE NUMBER (RCRA No.): P249.

RCRA Section 261 Hazardous Constituents.

EPCRA Section 313 Form R de minimus concentration reporting level: 1.0%. Form R Toxic Chemical Category Code: N982.

EPA NAME: ZINC SILICOFLUORIDE

[see also ZINC and ZINC COMPOUNDS]

CAS: 16871-71-9

SYNONYMS: SILICATE(2-), HEXAFLUORO-, ZINC (1:1); ZINC FLUOSILICATE; ZINC HEXAFLUOROSILICATE

CLEAN WATER ACT: Section 311 Hazardous Substances/RQ (same as CERCLA); Section 307 Toxic Pollutants; Section 313 Priority Chemicals.

EPCRA Section 304 Reportable Quantity (RQ): CERCLA, 5,000 lbs. (2270 kgs.).

EPCRA Section 313 Form R de minimus concentration reporting level: 1.0%. Form R Toxic Chemical Category Code: N982.

EPA NAME: ZINC SULFATE

[see also ZINC and ZINC COMPOUNDS]

CAS: 7733-02-0

SYNONYMS: SULFURIC ACID, ZINC SALT (1:1)

CLEAN WATER ACT: Section 311 Hazardous Substances/RQ (same as CERCLA); Section 307 Toxic Pollutants; Section 313 Priority Chemicals.

EPCRA Section 304 Reportable Quantity (RQ): CERCLA, 1,000 lbs. (454 kgs.).

EPCRA Section 313 Form R de minimus concentration reporting level: 1.0%. Form R Toxic Chemical Category Code: N982.

EPA NAME: ZINEB
[see also ZINC COMPOUNDS and ETHYLENEBISDITHIOCARBAMIC ACID, SALTS and ESTERS (EBDCs)]
CAS: 12122-67-7
SYNONYMS: CARBAMODITHIOIC ACID, 1,2-ETHANEDIYLBIS-, ZINC COMPLEX; 1,2-ETHANEDIYLBIS(CARBAMODITHIOATO)ZINC; 1,2-ETHANEDIYLBISCARBAMOTHIOIC ACID, ZINC SALT

EPCRA Section 313 Form R de minimus concentration reporting level: 1.0%. Form R Toxic Chemical Category Code: N982.
CALIFORNIA'S PROPOSITION 65: Carcinogen.

EPA NAME: ZINOPHOS
CAS: 297-97-2
SYNONYMS: O,O-DIETHYL-O-PARAZINYL PHOSPHOROTHIOATE; O,O-DIETHYL-O,2-PYRAZINYL PHOSPHOROTHIOATE; PHOSPHOROTHIOIC ACID,O,O-DIETHYL O-PYRAZINYL ESTER; THIONAZIN

EPA HAZARDOUS WASTE NUMBER (RCRA No.): P040.
RCRA Section 261 Hazardous Constituents.
RCRA Ground Water Monitoring List. Suggested test method(s) (PQL μg/L): 8270(10).
EPCRA Section 302 Extremely Hazardous Substances: TPQ = 500 lbs. (227 kgs.).
EPCRA Section 304 Reportable Quantity (RQ): EHS/CERCLA, 100 lbs. (45.4 kgs.).

EPA NAME: ZIRCONIUM NITRATE
CAS: 13746-89-9
SYNONYMS: NITRIC ACID, ZIRCONIUM(IV) SALT

CLEAN WATER ACT: Section 311 Hazardous Substances/RQ (same as CERCLA).
EPCRA Section 304 Reportable Quantity (RQ): CERCLA, 5,000 lbs. (2270 kgs.).

EPA NAME: ZIRCONIUM POTASSIUM FLUORIDE
CAS: 16923-95-8
SYNONYMS: POTASSIUM FLUOZIRCONATE; ZIRCONATE(2-), HEXAFLUORO-, DIPOTASSIUM, (OC-6-11)-

CLEAN WATER ACT: Section 311 Hazardous Substances/RQ (same as CERCLA).
EPCRA Section 304 Reportable Quantity (RQ): CERCLA, 1,000 lbs. (454 kgs.).

EPA NAME: ZIRCONIUM SULFATE
CAS: 14644-61-2
SYNONYMS: DISULFATOZIRCONIC ACID; SULFURIC ACID, ZIRCONIUM(IV) SALT (2:1); ZIRCONIUM(IV) SULFATE (1:2)

CLEAN WATER ACT: Section 311 Hazardous Substances/RQ (same as CERCLA).

EPCRA Section 304 Reportable Quantity (RQ): CERCLA, 5,000 lbs. (2270 kgs.).

EPA NAME: ZIRCONIUM TETRACHLORIDE

CAS: 10026-11-6

SYNONYMS: ZIRCONIUM CHLORIDE; ZIRCONIUM CHLORIDE (ZRCl4), (T-4)-; ZIRCONIUM(IV) CHLORIDE (1:4)

CLEAN WATER ACT: Hazardous Substances (Section 311)/RQ (same as CERCLA).

EPCRA Section 304 Reportable Quantity (RQ): CERCLA, 5,000 lbs. (2,270 kgs.).

Chemical Substance Cross Reference Index

2-AAF see 2-ACETYLAMINOFLUORENE

ABATE see TEMEPHOS

ACENAPHTHYLENE, 1,2-DIHYDRO- see ACENAPHTHENE

ACETALDEHYDE, CHLORO- see CHLOROACETALDEHYDE

ACETAMIDE, N-(AMINOTHIOXOMETHYL)- see 1-ACETYL-2-THIOREA

ACETAMIDE, 2-CHLORO-N-(1-METHYLETHYL)-N-PHENYL- see PROPACHLOR 2-CHLORO-N-(1-METHYLETHYL)-N-PHENYLACETAMIDE

ACETAMIDE, 2-CHLORO-N-(2,6-DIETHYLPHENYL)-N-(METHOXYMETHYL)- see ALACHLOR

ACETAMIDE, N-(4-ETHOXYPHENYL)-(9CI) see PHENACETIN

ACETAMIDE, N-FLUOREN-2-YL- see 2-ACETYLAMINOFLUORENE

ACETAMIDE, N-9H-FLUOREN-2-YL see 2-ACETYLAMINOFLUORENE

ACETAMIDE, 2-FLUORO see FLUOROACETAMIDE

ACETAMIDE, N-[4-[(2-HYDROXY-5-METHYLPHENYL)AZO]PHENYL- see C.I. DISPERSE YELLOW 3

ACETATO(2-METHOXYETHYL)MERCURY see METHOXYETHYLMERCURIC ACETATE

ACETEHYDE see ACETALDEHYDE

ACETENE see ETHYLENE

ACETIC ACID, AMIDE see ACETAMIDE

ACETIC ACID, AMMONIUM SALT see AMMONIUM ACETATE

ACETIC ACID, ANHYDRIDE see ACETIC ANHYDRIDE

ACETIC ACID, 2-BUTOXY ESTER see sec-BUTYL ACETATE

ACETIC ACID, BUTYL ESTER see BUTYL ACETATE

ACETIC ACID, n-BUTYL ESTER see BUTYL ACETATE

ACETIC ACID, tert-BUTYL ESTER- see tert-BUTYL ACETATE

ACETIC ACID, CADMIUM SALT see CADMIUM ACETATE

ACETIC ACID, CHLORIDE see ACETYL CHLORIDE

ACETIC ACID, (4-CHLORO-2-METHYLPHENOXY)-, SODIUM SALT see METHOXONE, SODIUM SALT

ACETIC ACID, 4-CHLORO-α-(4-CHLOROPHENYL)-α-HYDROXY-, ETHYL ESTER see CHLOROBENZILATE

ACETIC ACID (4-CHLORO-2-METHYLPHENOXY)- see METHOXONE

ACETIC ACID, ((4-CHLORO-o-TOLYL)OXY)- SODIUM SALT see METHOXONE, SODIUM SALT

ACETIC ACID, CHLORO- see CHLOROACETIC ACID

ACETIC ACID, CHROMIUM(3+) SALT see CHROMIC ACETATE

ACETIC ACID, COPPER(2+) SALT see CUPRIC ACETATE

ACETIC ACID (2,4-DICHLOROPHENOXY)- see 2,4-D

ACETIC ACID, 2,4-DICHLOROPHENOXY-, BUTOXYMETHYLETHYL ESTER see 2,4-D PROPYLENE GLYCOL BUTYL ETHER ESTER

ACETIC ACID, (2,4-DICHLOROPHENOXY)-,2-BUTOXYETHYL ESTER see 2,4-D 2-ETHYLHEXYL ESTER

ACETIC ACID, 2,4-DICHLOROPHENOXY-, BUTOXYPROPYL ESTER see 2,4-D PROPYLENE GLYCOL BUTYL ETHER ESTER

ACETIC ACID, (2,4-DICHLOROPHENOXY)-, BUTYL ESTER see 2,4-D BUTYL ESTER

ACETIC ACID, (2,4-DICHLOROPHENOXY)-, 4-CHLORO-2-BUTENYL ESTER see 2,4-D CHLOROCROTYL ESTER

ACETIC ACID, (2,4-DICHLOROPHENOXY)-, 2-ETHYL-4-METHYLPENTYL ESTER see 2,4-D 2-ETHYL-4-METHYLPENTYL ESTER

ACETIC ACID, (2,4-DICHLOROPHENOXY)-, ISOPROPYL ESTER see 2,4-D ISOPROPYL ESTER

ACETIC ACID (2,4-DICHLOROPHENOXY)-, ISOOCTYL ESTER see 2,4-D ISOOCTYL ESTER

ACETIC ACID, (2,4-DICHLOROPHENOXY)-, METHYL ESTER see 2,4-D METHYL ESTER

ACETIC ACID, (2,4-DICHLOROPHENOXY)-, 1-METHYLETHYL ESTER see 2,4-D ISOPROPYL ESTER

ACETIC ACID, (2,4-DICHLOROPHENOXY), 1-METHYL PROPYL ESTER see 2,4-D sec-BUTYL ESTER

ACETIC ACID, (2,4-DICHLOROPHENOXY)-,PROPYL ESTER see 2,4-D PROPYL ESTER

ACETIC ACID, O,O-DIMETHYLDITHIOPHOSPHORYL-, N-MONOMETHYLAMIDE SALT see DIMETHOATE

ACETIC ACID, 1,1-DIMETHYLETHYL ESTER (9CI) see tert-BUTYL ACETATE

ACETIC ACID, (ETHYLENEDINITRILO)TETRA- see ETHYLENEDIAMINE-TETRAACETIC ACID (EDTA)

ACETIC ACID, ETHENYL ESTER see VINYL ACETATE

ACETIC ACID, ETHYL ESTER see ETHYL ACETATE

ACETIC ACID, FLUORO-, SODIUM SALT see SODIUM FLUOROACETATE

ACETIC ACID, ISOBUTYL ESTER see ISO-BUTYL ACETATE

ACETIC ACID, LEAD(II) SALT see LEAD ACETATE

ACETIC ACID-2-METHYL-2-PROPENE-1,1-DIOL DIESTER see METHACROLEIN DIACETATE

ACETIC ACID, 2-METHYLPROPYL ESTER see ISO-BUTYL ACETATE

ACETIC ACID, PENTYL ESTER see AMYL ACETATE

ACETIC ACID, (2,4,5-TRICHLOROPHYNOXY)- see 2,4,5-T ACID (93-76-5)

ACETIC ACID, (2,4,5-TRICHLOROPHENOXY)-, 2-BUTOXYETHYL ESTER see 2,4,5-T ESTERS (2545-59-7)

ACETIC ACID, (2,4,5-TRICHLOROPHYNOXY)-, BUTYL ESTER see 2,4,5-T ESTERS (93-79-8)

ACETIC ACID, (2,4,5-TRICHLOROPHYNOXY)- comp. with 1-AMINO-2-PROPANOL (1:1) see 2,4,5-T AMINES (1319-72-8)

ACETIC ACID, (2,4,5-TRICHLOROPHYNOXY)- comp. with N,N-DIETHYLETHANAMINE see 2,4,5-T AMINES (2008-46-0)

ACETIC ACID, (2,4,5-TRICHLOROPHYNOXY)- comp. with N-METHYLMETHANAMINE see 2,4,5-T AMINES (6369-97-7)

ACETIC ACID, (2,4,5-TRICHLOROPHYNOXY)- comp. with TRIMETHYLAMINE (1:1) see 2,4,5-T AMINES (6369-96-6)

ACETIC ACID, ((3,5,6-TRICHLORO-2-PYRIDINYL)OXY)-, comp. with N,N-DIETHYLETHANAMINE (1:1) see TRICLOPYR TRIETHYLAMMONIUM SALT

ACETIC ACID, (2,4,5-TRICHLOROPHYNOXY)-, SODIUM SALT see 2,4,5-T SALTS

ACETIC ACID, (2,4,5-TRICHLOROPHENOXY)-, 2-ETHYLHEXYL ESTER see 2,4,5-T ESTERS (1928-47-8)

ACETIC ACID, (2,4,5-TRICHLOROPHENOXY)-, ISOOCTYL ESTER see 2,4,5-T ESTERS (25168-15-4)

ACETIC ACID, (2,4,5-TRICHLOROPHENOXY)-, 1-METHYL PROPYL ESTER see 2,4,5-T ESTERS (61792-07-2)

ACETIC ACID, (2,4,5-TRICHLOROPHYNOXY)- comp. WITH 2,2′,2″-NITRILOTRIS(ETHANOL)(1:1) see 2,4,5-T AMINES (3813-14-7)

ACETIC ACID, VINYL ESTER see VINYL ACETATE

ACETIC ACID, ZINC SALT see ZINC ACETATE

ACETIC ALDEHYDE see ACETALDEHYDE

ACETIC BROMIDE see ACETYL BROMIDE

ACETIC CHLORIDE see ACETYL CHLORIDE

ACETIC EHYDE see ACETALDEHYDE

ACETIC ETHER see ETHYL ACETATE

ACETIMIDIC ACID see ACETAMIDE

ACETIMIDIC ACID, THIO-N-(METHYLCARBAMOYL)OXY-,METHYL ESTER see METHOMYL

ACETONE CYANOHYDRIN see 2-METHYLLACTONITRILE

ACETONITRILE see BENZYL CYANIDE

3-(α-ACETONYLBENZYL)-4-HYDROXYCOUMARIN see WARFARIN

3-(α-ACETONYLBENZYL)-4-HYDROXY-COUMARIN SODIUM SALT see WARFARIN SODIUM

para-ACETOPHENETIDIDE see PHENACETIN

ACETOTHIOAMIDE see THIOACETAMIDE

ACETOTRIPHENYLSTANNANE see STANNANE, ACETOXYTRIPHENYL-

ACETOTRIPHENYLSTANNINE see STANNANE, ACETOXYTRIPHENYL-

ACETOXYPHENYLMERCURY see PHENYLMERCURY ACETATE

2-ACETYLAMIDOFLUORENE see 2-ACETYLAMINOFLUORENE

2-ACETYLAMINEFLUARONE see 2-ACETYLAMINOFLUORENE

ACETYL BENZENE see ACETOPHENONE

ACETYL CHLORIDE, FLUORO- see FLUOROACETYL CHLORIDE

ACETYL CHLORIDE, TRICHLORO- see TRICHLOROACETYL CHLORIDE

ACETYLENE DICHLORIDE see 1,2-DICHLOROETHYLENE

trans-ACETYLENE DICHLORIDE see 1,2-DICHLOROETHYLENE

ACETYLENE TETRACHLORIDE see 1,1,2,2-TETRACHLOROETHANE

ACETYLENE TRICHLORIDE see TRICHLOROETHYLENE

ACETYL OXIDE see ACETIC ANHYDRIDE

ACETYLPHOSPHORAMIDOTHIOIC ACID O,S-DIMETHYL ESTER see ACEPHATE

ACETYL THIOUREA see 1-ACETYL-2-THIOREA

ACIDAL PONCEAU G see C.I. FOOD RED 5

ACID,3-(2-CHLORO-3,3,3-TRIFLUORO-1-PROPENYL)-2,2-DIMETHYL-,(2-METHYL(1,1′-BIPHENYL)3-YL)METHYL ESTER,(Z)-see BIFENTHRIN

ACID-1-((DIMETHYLAMINO)CARBONYL)-5-METHYL-1H-PYRAZOL-3-YL ESTER see DIMETILAN

ACIFLUORFEN see ACIFLUORFEN, SODIUM SALT

ACIFLUORFEN SODIUM see ACIFLUORFEN, SODIUM SALT

ACROLEIC ACID see ACRYLIC ACID

ACROLEINE see ACROLEIN

ACRYLALDEHYDE see ACROLEIN

ACRYLIC ACID, BUTYL ESTER see BUTYL ACRYLATE

ACRYLIC ACID n-BUTYL ESTER see BUTYL ACRYLATE

ACRYLIC ACID CHLORIDE see ACRYLYL CHLORIDE

ACRYLIC ACID, ETHYL ESTER see ETHYL ACRYLATE

ACRYLIC ACID, GLACIAL see ACRYLIC ACID

ACRYLIC ACID, METHYL ESTER see METHYL ACRYLATE

ACRYLIC ALDEHYDE see ACROLEIN

ACRYLOYL CHLORIDE see ACRYLYL CHLORIDE

AF RED NO. 5 see C.I. SOLVENT ORANGE 7

AGARIN see MUSCIMOL

ALLENE see 1,2-PROPADIENE

ALLOPHANIC ACID, 4,4′-O-PHENYLENEBIS(3-THIO-, DIETHYL ESTER see THIOPHANATE ETHYL

5-ALLYL-1,3-BENZODIOXOLE see SAFROLE

ALLYL-1-(2,4-DICHLOROPHENYL)-2-IMIDAZOL-1-YLETHYL ETHER see IMAZALIL

4-ALLYL-1,2-(METHYLENEDIOXY)BENZENE see SAFROLE

ALLYL TRICHLORIDE see 1,2,3-TRICHLOROPROPANE

ALPHANAPHTHYL THIOUREA see ANTU

ALUM see ALUMINUM SULFATE

A-ALUMINA see ALUMINUM OXIDE (fibrous forms)

α-ALUMINA see ALUMINUM OXIDE (fibrous forms)

ALUNDUM (Al_2O_3) see ALUMINUM OXIDE (fibrous forms)

ALUMINUM ALUM, ALUMINUM SALT (3:2) see ALUMINUM SULFATE

ALUMINUM, ELEMENTAL see ALUMINUM (fume or dust)

ALUMINUM, METAL AND OXIDE see ALUMINUM (fume or dust)

ALUMINUM MONOPHOSPHIDE see ALUMINUM PHOSPHIDE

ALUNDUM 600 see ALUMINUM OXIDE (fibrous forms)

ALUNDUM see ALUMINUM OXIDE (fibrous forms)

ALVIT see DIELDRIN

AMETYCIN see MITOMYCIN C

AMID KYSELINY OCTOVE see ACETAMIDE

AMIDOUREA HYDROCHLORIDE see SEMICARBAZIDE HYDROCHLORIDE

ortho-AMINOANISOLE see o-ANISIDINE

para-AMINOANISOLE see p-ANISIDINE

β-AMINOANTHRAQUINONE see 2-AMINOANTHRAQUINONE

AMINOAZOBENZENE see 4-AMINOAZOBENZENE

p-AMINOAZOBENZENE see 4-AMINOAZOBENZENE

para-AMINOAZOBENZENE see 4-AMINOAZOBENZENE

o-AMINOAZOTOLUENE see C.I. SOLVENT YELLOW 3

ortho-AMINOAZOTOLUENE see C.I. SOLVENT YELLOW 3

AMINOBENZENE see ANILINE

p-AMINOBIPHENYL see 4-AMINOBIPHENYL

5-AMINO-1-BIS(DIMETHYLAMIDE)PHOSPHORYL-3-PHENYL-1,2,4-TRIAZOLE see TRIAMIPHOS

1-AMINOBUTANE see BUTYLAMINE

AMINOBUTYLMETHYLDIETHOXYSILANE see SILANE, (4-AMINOBUTYL)DIETHOXYMETHYL-

(4-AMINOBUTYL)DIETHYOXYMETHYLSILANE see SILANE, (4-AMINOBUTYL)DIETHOXYMETHYL-

4-AMINO-6-tert-BUTYL-3-(METHYLTHIO)-1,2,4-TRIAZIN-5-ONE see METRIBUZIN

2-((AMINOCARBONYL)OXY)-N,N,N-TRIMETHYLETHANAMINIUM CHLORIDE CARBAMOYLCHOLINE CHLORIDE see CARBACHOL CHLORIDE

1-AMINO-4-CHLOROBENZENE see p-CHLOROANILINE

2-AMINO-5-CHLOROTOLUENE HYDROCHLORIDE see 4-CHLORO-o-TOLUIDINE, HYDROCHLORIDE

4-AMINO-4-DEOXYPTEROYLGLUTAMATE see AMINOPTERIN

3-AMINO-2,5-DICHLOROBENZOIC ACID see CHLORAMBEN

4-AMINODIPHENYL see 4-AMINOBIPHENYL

p-AMINODIPHENYLE see 4-AMINOBIPHENYL

AMINOETHYLENE see ETHYLENEIMINE

AMINOMETHANE see METHANAMINE

2-AMINO-1-METHOXY-4-NITROBENZENE see 5-NITRO-O-ANISIDINE

7-AMINO-9-α-METHOXYMITOSANE see MITOMYCIN C

1-AMINO-2-METHYL-9,10-ANTHRACENEDIONE see 1-AMINO-2-METHYLANTHRAQUINONE

1-AMINO-2-METHYLBENZENE see o-TOLUIDINE

4-AMINO-1-METHYLBENZENE see p-TOLUIDINE

1-AMINO-2-METHYLBENZENE HYDROCHLORIDE see o-TOLUIDINE HYDROCHLORIDE

5-AMINOMETHYL-3-HYDROXYISOXAZOLE see MUSCIMOL

5-(AMINOMETHYL)-3-ISOXAZOLOL see MUSCIMOL

5-(AMINOMETHYL)-3-(2H)ISOXAZOLONE see MUSCIMOL

1-AMINO-2-METHYLPROPANE see ISO-BUTYLAMINE

1-(4-AMINOPHENYL)-1-PROPANONE see PROPIOPHENONE, 4′-AMINO

1-AMINOPROPANE see n-PROPYLAMINE

1-AMINO-2-PROPANOL (1:1) see 2,4,5-T AMINES (2008-46-0)

3-AMINOPROPENE see ALLYLAMINE

3-AMINO-1-PROPENE see ALLYLAMINE

p-AMINOPROPIOPHENONE see PROPIOPHENONE, 4′-AMINO

AMINOPTERIDINE see AMINOPTERIN

4-AMINOPTEROYLGLUTAMIC ACID see AMINOPTERIN

AMINO-4-PYRIDINE see 4-AMINOPYRIDINE

γ-AMINOPYRIDINE see 4-AMINOPYRIDINE

p-AMINOPYRIDINE see 4-AMINOPYRIDINE

1-AMINOTHIOUREA see THIOSEMICARBAZIDE

1-AMINO-2-THIOUREA see THIOSEMICARBAZIDE

AMINOTRIAZOLE see AMITROLE

2-AMINOTRIAZOLE see AMITROLE

2-AMINO-1,3,4-TRIAZOLE see AMITROLE

3-AMINO-1,2,4-TRIAZOLE see AMITROLE

3-AMINOTRIAZOLE see AMITROLE

3-AMINO-S-TRIAZOLE see AMITROLE

3-AMINO-1H-1,2,4-TRIAZOLE see AMITROLE

4-AMINO-3,5,6-TRICHLORO-2-PYRIDINECARBOXYLIC ACID see PICLORAM

4-AMINOTRICHLOROPICOLINIC ACID see PICLORAM

4-AMINO-3,5,6-TRICHLOROPYRIDINE-2-CARBOXYLIC ACID see PICLORAM

AMINOTRIETHANOIC ACID see NITRILOTRIACETIC ACID

2-AMINO-1,3-XYLENE see 2,6-XYLIDINE

AMMATE see AMMONIUM SULFAMATE

AMMONIUM AMIDE see AMMONIA (conc. 20% or greater)

AMMONIUM AMIDOSULPHATE see AMMONIUM SULFAMATE

AMMONIUM AMINOSULFONATE see AMMONIUM SULFAMATE

AMMONIUM BISULFIDE see AMMONIUM SULFIDE

AMMONIUM BOROFLUORIDE see AMMONIUM FLUOBORATE

AMMONIUM DICHROMATE(VI) see AMMONIUM BICHROMATE

AMMONIUM DICHROMATE see AMMONIUM BICHROMATE

AMMONIUM FERRIC CITRATE see FERRIC AMMONIUM CITRATE

AMMONIUM FERRIC OXALATE TRIHYDRATE see FERRIC AMMONIUM OXALATE

AMMONIUM FLUORIDE $(NH_4)(HF_2)$ see AMMONIUM BIFLUORIDE

AMMONIUM HYDROGEN CARBONATE see AMMONIUM BICARBONATE

AMMONIUM HYDROGEN DIFLUORIDE see AMMONIUM BIFLUORIDE

AMMONIUM HYDROGEN SULFITE see AMMONIUM BISULFITE

AMMONIUM HYDROXIDE see AMMONIA (conc. 20% or greater)

AMMONIUM IRON SULFATE see FERROUS AMMONIUM SULFATE

AMMONIUM METAVANADATE see AMMONIUM VANADATE

AMMONIUM MURIATE see AMMONIUM CHLORIDE

AMMONIUM OXALATE HYDRATE see AMMONIUM OXALATE

AMMONIUM OXALATE, MONOHYDRATE see AMMONIUM OXALATE

AMMONIUM RHODANATE see AMMONIUM THIOCYANATE

AMMONIUM RHODANIDE see AMMONIUM THIOCYANATE

AMMONIUM SULFOCYANATE see AMMONIUM THIOCYANATE

AMMONIUM SULFOCYANIDE see AMMONIUM THIOCYANATE

AMMONIUM SULPHAMIDATE see AMMONIUM SULFAMATE

AMMONIUM TARTRATE, DIAMMONIUM SALT see AMMONIUM TARTRATE

AMMONIUM TETRAFLUOBORATE see AMMONIUM FLUOBORATE

AMMONIUM ZINC CHLORIDE see ZINC AMMONIUM CHLORIDE

AMYL HYDRIDE see PENTANE

β-AMYLENE-cis see 2-PENTENE, (Z)-

β-AMYLENE-trans see 2-PENTENE, (E)-

α-n-AMYLENE see 1-PENTENE

ANHYDROUS AMMONIA see AMMONIA

ANHYDROUS CHLORAL see ACETALDEHYDE, TRICHLORO

ANILINE, N,N-DIMETHYL- see N,N-DIMETHYLANILINE

ANILINE, 3,4-DIMETHYL-2,6-DINITRO-N-(1-ETHYLPROPYL)- see PENDIMETHALIN N-(1-ETHYLPROPYL)-3,4-DIMETHYL-2,6-DINITROBENZENAMINE

ANILINE, 4-METHOXY- see p-ANISIDINE

ortho-ANSIDINE see o-ANISIDINE

ortho-ANISIDINE HYDROCHLORIDE see o-ANISIDINE HYDROCHLORIDE

para-ANISIDINE see p-ANISIDINE

9,10-ANTHRACENEDIONE, 1-AMINO-2-METHYL- see 1-AMINO-2-METHYLANTHRAQUINONE

9,10-ANTHRACENEDIONE, 2-AMINO- see 2-AMINOANTHRAQUINONE

ANTIBIOTIC N-329 B see VALINOMYCIN

ANTIMONATE(2-), BIS μ-2,3-DIHYDROXYBUTANEDIOATA (4-)-01,02:03,04DI-, DIPOTASSIUM, TRIHYDRATE, STEREOISOMER see ANTIMONY POTASSIUM TARTRATE

ANTIMONIC CHLORIDE see ANTIMONY PENTACHLORIDE

ANTIMONOUS BROMIDE see ANTIMONY TRIBROMIDE

ANTIMONOUS OXIDE see ANTIMONY TRIOXIDE

ANTIMONY(III) CHLORIDE see ANTIMONY TRICHLORIDE

ANTIMONY(III) FLUORIDE (1:3) see ANTIMONY TRIFLUORIDE

ANTIMONY(III) TRICHLORIDE see ANTIMONY TRICHLORIDE

ANTIMONY(5+) PENTAFLUORIDE see ANTIMONY PENTAFLUORIDE

ANTIMONY(V) PENTAFLUORIDE see ANTIMONY PENTAFLUORIDE

ANTIMONY CHLORIDE ($SbCl_5$) see ANTIMONY PENTACHLORIDE

ANTIMONY, ELEMENTAL see ANTIMONY

ANTIMONY FLUORIDE (SbF_5) see ANTIMONY PENTAFLUORIDE

ANTIMONY OXIDE (Sb_2O_3) see ANTIMONY TRIOXIDE

ANTIMONYL POTASSIUM TARTRATE see ANTIMONY POTASSIUM TARTRATE

ANTIMYCIN A see ANTIMYCIN

ANTIPIRICULLIN see ANTIMYCIN

AROCHLOR 1016 see AROCLOR 1016

AROCHLOR 1221 see AROCLOR 1221

AROCHLOR 1232 see AROCLOR 1232

AROCHLOR 1242 see AROCLOR 1242

AROCHLOR 1248 see AROCLOR 1248

AROCHLOR 1254 see AROCLOR 1254

AROCHLOR 1260 see AROCLOR 1260

AROCLORs see POLYCHLORINATED BIPHENYLS

ARSENIC(III) CHLORIDE see ARSENOUS TRICHLORIDE

ARSENIC(III) OXIDE see ARSENIC TRIOXIDE

ARSENIC(III) TRICHLORIDE see ARSENOUS TRICHLORIDE

ARSENIC(V) OXIDE see ARSENIC PENTOXIDE

ARSENIC ACID (H_3AsO_4) see ARSENIC ACID

ARSENIC ACID, CALCIUM SALT (2:3) see CALCIUM ARSENATE

ARSENIC ACID, LEAD(IV) SALT (3:2) see LEAD ARSENATE (10102-48-4)

ARSENIC ACID, LEAD(II) SALT (1:1) see LEAD ARSENATE (7784-40-9)

ARSENIC ACID, LEAD SALT see LEAD ARSENATE (7645-25-2)

ARSENIC ACID, MONOPOTASSIUM SALT see POTASSIUM ARSENATE

ortho-ARSENIC ACID see ARSENIC ACID

ARSENIC ACID, SODIUM SALT see SODIUM ARSENATE

ARSENIC CHLORIDE see ARSENOUS TRICHLORIDE

ARSENIC, ELEMENTAL see ARSENIC

ARSENIC OXIDE (As_2O_5) see ARSENIC PENTOXIDE

ARSENIC SULFIDE (As_2S_3) see ARSENIC TRISULFIDE

ARSENIC YELLOW see ARSENIC TRISULFIDE

ARSENOUS ACID, CALCIUM SALT see CALCIUM ARSENITE

ARSENOUS ACID, POTASSIUM SALT see POTASSIUM ARSENITE

ARSENOUS ACID, SODIUM SALT see SODIUM ARSENITE

ARSENOUS CHLORIDE see ARSENOUS TRICHLORIDE

ARSENOUS HYDRIDE see ARSINE

ARSENOUS OXIDE ANHYDRIDE see ARSENIC TRIOXIDE

ARSENOUS OXIDE see ARSENIC TRIOXIDE

ARSINE, DICHLOROPHENYL- see DICHLOROPHENYLARSINE

ARSINE, DIETHYL- see DIETHYLARSINE

ARSONIC ACID, CALCIUM SALT(1:1) see CALCIUM ARSENITE

ARSONIC ACID, POTASSIUM SALT see POTASSIUM ARSENITE

ARSONOUS DICHLORIDE, PHENYL- see DICHLOROPHENYLARSINE

AURAMINE BASE see C.I. SOLVENT YELLOW 34

AURAMINE see C.I. SOLVENT YELLOW 34

AUSTRAPINE see RESPERINE

AVERMECTIN B1 see ABAMECTIN

AVERMECTIN B(SUB1) see ABAMECTIN

AVITROL see 4-AMINOPYRIDINE

12-AZABENZ(a)ANTHRACENE see BEN[c]ACRIDINE

AZABENZENE see PYRIDINE

AZACYCLOHEXANE see PIPERIDINE

AZACYCLOPROPANE see ETHYLENEIMINE

1-AZASERINE see AZASERINE

2H-AZEPIN-2-ONE,HEXAHYDRO see CAPROLACTUM

1H-AZEPINE-1-CARBOTHIOIC ACID, HEXAHYDRO-S-ETHYL ESTER see MOLINATE

AZIMETHYLENE see DIAZOMETHANE

AZINE see PYRIDINE

AZINPHOS-METHYL (ISO) see AZINPHOS-METHYL

AZIRANE see ETHYLENEIMINE

AZIRIDINE, 2-METHYL- see PROPYLENEIMINE

AZIRIDINE see ETHYLENEIMINE

AZIRINE see ETHYLENEIMINE

1H-AZIRINE,DIHYDRO- see ETHYLENEIMINE

AZIRINOL(2′,3′:3,4)PYRROLO(1,2-A)INDOLE-4,7-DIONE,6-AMINO-8[[(AMINOCARBONYL)OXY]METHYL]-1,1A,2,8,8A,8B-HEXAHYDRO-8A-METHOXY-5-METHYL-, [1AS-(Aα,8β,8Aα,8Bα)]- see MITOMYCIN C

AZODRIN see MONOCROPTOPHOS

BANANA OIL see tert-AMYL ACETATE

BARBITURIC ACID, 5-ETHYL-5-sec-PENTYL-, SODIUM SALT see PENTOBARBITOL SODIUM

BARIUM DICYANIDE see BARIUM CYANIDE

BARIUM, ELEMENTAL see BARIUM

BARIUM METAL see BARIUM

BARPENTAL see PENTOBARBITOL SODIUM

BAY 23323 see OXYDISULFOTON

BAYGON see PROPOXUR

BAYTEX see FENTHION

BENOMYL(ISO) see BENOMYL

BENSONITRILE, 3,5-DIBROMO-4-HYDROXY- see BROMOXYNIL

BENZ(A)ANTHRACENE, 7,12-DIMETHYL- see 7,12-DIMETHYL-BENZ(A)ANTHRACENE

BENZ(A)ANTHRACENE, 9,10-DIMETHYL- see 7,12-DIMETHYL-BENZ(A)ANTHRACENE

1,2-BENZ(a)ANTRHRACENE see BENZ[a]ANTHRACENE

BENZ(A)PHENANTHRENE see CHRYSENE

3,4-BENZ(a)PYRENE see BENZO(a)PYRENE

BENZ(e)ACEPHENANTHRYLENE see BENZO[b]FLUORANTHENE

BENZ(j)FLUORANTHENE see BENZO[j]FLUORANTHENE

BENZ(k)FLUORANTHENE see BENZO[k]FLUORANTHENE

BENZALDEHYDE, α-CHLORO- see BENZOYL CHLORIDE

BENZAMIDE, 5-(2-CHLORO-4-(TRIFLUOROMETHYL)PHENOXY)-N-(METHYLSULFONYL)-2-NITRO- see FOMESAFEN

BENZAMIDE, 3,5-DICHLORO-N-(1,1-DIMETHYL-2-PROPYNYL) see PRONAMIDE

1,2-BENZANTHRACENE see BENZ[a]ANTHRACENE

1-BENZAZINE see QUINOLINE

BENZELENE see BENZENEPHENYL ARSENIC ACID see BENZENEARSONIC ACID

BENZENAMINE, 4-CHLORO-2-METHYL see p-CHLORO-o-TOLUIDINE

BENZENAMINE, 2,6-DICHLORO-4-NITRO- see DICHLORAN

BENZENAMINE, N,N-DIETHYL- see N,N-DIETHYLANILINE

BENZENAMINE, 2,6-DIMETHYL- see 2,6-XYLIDINE

BENZENAMINE, 3,4-DIMETHYL-2,6-DINITRO-N-(1-ETHYLPROPYL)- see PENDIMETHALIN N-(1-ETHYLPROPYL)-3,4-DIMETHYL-2,6-DINITROBENZENAMINE

BENZENAMINE, N,N-DIMETHYL- see N,N-DIMETHYLANILINE

BENZENAMINE, N,N-DIMETHYL-4-(PHENYLAZO)- see 4-DIMETHYLAMINOAZOBENZENE

BENZENAMINE, 4-ETHOXY-N-(5-NITRO-2FURANYL)METHYLENE- see NITROFEN

BENZENAMINE, 2-METHOXY- see o-ANISIDINE

BENZENAMINE, 4-METHOXY- see p-ANISIDINE

BENZENAMINE, 2-METHOXY-5-NITRO- see 5-NITRO-O-ANISIDINE

BENZENAMINE, 2-METHOXY HYDROCHLORIDE see o-ANISIDINE HYDROCHLORIDE

BENZENAMINE, 2-METHYL- (9CI) see o-TOLUIDINE

BENZENAMINE, 2-METHYL-, HYDROCHLORIDE see o-TOLUIDINE HYDROCHLORIDE

BENZENAMINE, 2-METHYL-4-[(2-METHYLPHENYL)AZO]- see C.I. SOLVENT YELLOW 3

BENZENAMINE, 2-METHYL-5-NITRO- see 5-NITRO-o-TOLUENE

BENZENAMINE, 4,4′-METHYLENEBIS- see 4,4′-METHYLENEDIANILINE

BENZENAMINE, 4,4′-METHYLENEBIS(N,N-DIMETHYL)- see 4,4′-METHYLENEBIS (N,N-DIMETHYL)BENZENAMINE

BENZENAMINE, 2-NITRO- see o-NITROANILINE

BENZENAMINE, 4-NITRO- see p-NITROANILINE

BENZENAMINE, 4-NITROSO-N-PHENYL- see p-NITROSODIPHENYLAMINE

BENZENAMINE, N-NITROSO-N-PHENYL- see N-NITROSODIPHENYLAMINE

BENZENAMINE, 4,4′-OXYBIS- see 4,4′-DIAMINOPHENYL ETHER

BENZENAMINE, N-PHENYL- see DIPHENYLAMINE

BENZENAMINE, 4,4′-THIOBIS- see 4,4′-THIODIANILINE

BENZENEACETIC ACID, 4-CHLORO-α-(1-METHYLETHYL)-,CYANO(3-PHENOXYPHENYL)METHYL ESTER see FENVALERATE

BENZENEAMINE see ANILINE

BENZENEAMINE, 4,4′-CABONIMIDOYLBIS[N-DIMETHYL- see C.I. SOLVENT YELLOW 34

BENZENEAMINE, 4-CHLORO- see p-CHLOROANILINE

BENZENEAMINE, 4-CHLORO-2-METHYL-, HYDROCHLORIDE see 4-CHLORO-o-TOLUIDINE, HYDROCHLORIDE

BENZENEAMINE, 2,6-DINITRO-N,N-DIPROPYL-4-(TRIFLUOROMETHYLANILINE) see TRIFLURALIN

BENZENEAMINE, N-HYDROXY-N-NITROSO, AMMONIUM SALT see CUPFERRON

BENZENEAMINE, 4-4′-METHYLENEBIS(2-CHLORO)- see 4,4′-METHYLENEBIS(2-CHLOROANILINE)

BENZENEAMINE, 2-METHOXY-5-METHYL- see p-CRESIDINE

BENZENEAZO-β-NAPHTHOL see C.I. SOLVENT YELLOW 14

BENZENE, BIS(CHLOROMETHYL)- see XYLYLENE DICHLORIDE

BENZENE, 1-BROMO-4-PHENOXY- see 4-BROMOPHENYL PHENYL ETHER

BENZENE, 2-BROMO-4-PHENOXY- see 4-BROMOPHENYL PHENYL ETHER

BENZENEBUTANOIC ACID, 4-[BIS(2-CHLOROETHYL)AMINO]- see CHLORAMBUCIL

BENZENECARBOXYLIC ACID see BENZOIC ACID

BENZENE, CHLORO- see CHLOROBENZENE

BENZENE, 2-CHLORO-1-(3-ETHOXY-4-NITROPHENOXY)-4-(TRIFLUOROMETHYL)-see OXYFLUORFEN

BENZENE, 1,3-DIISOCYANATE see 1,3-PHENYLENE DIISOCYANATE

BENZENE, 1,3-DIISOCYANATO-2-METHYL- see TOLUENE-2,6-DIISOCYANATE

BENZENE, 2,4-DIISOCYANATO-1-METHYL- see TOLUENE 2,4-DIISOCYANATE

BENZENE, 1,3-DIISOCYANATO- see 1,3-PHENYLENE DIISOCYANATE

BENZENE, 1,4-DIISOCYANATO- see 1,4-PHENYLENE DIISOCYANATE

BENZENE, 2,4-DIISOCYANATO-1-METHYL- see TOLUENE 2,4-DIISOCYANATE

BENZENE, 1,3-DIISOCYANATOMETHYL- see TOLUENEDIISOCYANATE (MIXED ISOMERS)

BENZENE 2,4-DIISOCYANATOMETHYL- see TOLUENE 2,4-DIISOCYANATE

BENZENE-1,2-DIMETHYL- see o-XYLENE

BENZENE-1,3-DIMETHYL- see m-XYLENE

BENZENE-1,4-DIMETHYL see p-XYLENE

BENZENE, DIMETHYL see XYLENE (MIXED ISOMERS)

BENZENE, m-DIMETHYL- see m-XYLENE

BENZENE, o-DIMETHYL see o-XYLENE

BENZENE, p-DIMETHYL see p-XYLENE

BENZENE, DINITRO- see DINITROBENZENE (MIXED ISOMERS)

BENZENE, 1,3-DINITRO- see m-DINITROBENZENE

BENZENE, 1,4-DINITRO- see p-DINITROBENZENE

BENZENE, o-DINITRO- see o-DINITROBENZENE

BENZENE, p-DINITRO- see p-DINITROBENZENE

1,2-BENZENEDIOL see CATECHOL

1,2-BENZENEDIOL-4,3,4-(1-DIHYDROXY-2-[METHYLAMINO]ETHYL)- see EPINEPHRINE

1,3-BENZENEDIOL see RESORCINOL

1,4-BENZENEDIOL see HYDROQUINONE

m-BENZENEDIOL see RESORCINOL

BENZENE, EPOXYETHYL- see STYRENE OXIDE

BENZENE, ETHENYL- see STYRENE

BENZENE, ETHYL- see ETHYLBENZENE

β-BENZENEHEXACHLORIDE see β-HEXACHLOROCYCLOHEXANE

δ-BENZENEHEXACHLORIDE see δ-HEXACHLOROCYCLOHEXANE

BENZENE HEXACHLORIDE see HEXACHLOROCYCLOHEXANE (ALL ISOMERS)

BENZENE, HEXACHLORO- see HEXACHLOROBENZENE

BENZENE, HEXAHYDRO see CYCLOHEXANE

BENZENE, 1-ISOCYANATO-2-((4-ISOCYANATOPHENYL)THIO)- see 4-METHYLDIPHENYLMETHANE-3,4-DIISOCYANATE

BENZENE, 1-ISOCYANATO-2-((4-ISOCYANATOPHENYL)THIO)- see 2,4′-DIISOCYANATODIPHENYL SULFIDE

BENZENE, 2-ISOCYANATO-4((4-ISOCYANATOPHENYL)METHYL)-1-METHYL-3,4′-DIISOCYANATO-4-METHYLDIPHENYLMETHANE see 4-METHYLDIPHENYLMETHANE-3,4-DIISOCYANATE

BENZENEMETHANOL, 4-CHLORO-α-(4-CHLOROPHENYL)-α-(TRICHLOROMETHYL)- see DICOFOL

BENZENE, METHYL- see TOLUENE

BENZENE, 2-METHYL see o-CRESOL

BENZENE, 3-METHYL see m-CRESOL

BENZENE, 1-METHYL-2-NITRO- see o-NITROTOLUENE

BENZENE, 1-METHYL-3-NITRO- see m-NITROTOLUENE

BENZENE, 1-METHYL-4-NITRO- see p-NITROTOLUENE

BENZENE, 2-METHYL-1,3-DINITRO- see 2,6-DINITROTOLUENE

BENZENE, 1-METHYL-2,4-DINITRO- see 2,4-DINITROTOLUENE

BENZENE, 4-METHYL-1,2-DINITRO- see 3,4-DINITROTOLENE

BENZENE, METHYLDINITRO- see DINITROTOLUENE (MIXED ISOMERS)

BENZENE, 1,1′-METHYLENEBIS(4-ISOCYANATO- see METHYLBIS(PHENYLISOCYANATE)

BENZENE, 1,1′-METHYLENEBIS(4-ISOCYANATO-3-METHYL- see 3,3′-DIMETHYLDIPHENYLMETHANE-4,4′-DIISOCYANATE

BENZENE, 1,2-(METHYLENEDIOXY)-4-PROPENYL- see ISOSAFROLE

BENZENE, 1,2-(METHYLENEDIOXY)-4-PROPYL- see DIHYDROSAFROLE

BENZENE, (1-METHYLETHYL-)- see CUMENE

BENZENE, METHYLNITRO- see NITROTOLUENE

BENZENENITRILE see BENZONITRILE

BENZENE, NITRO- see NITROBENZENE

BENZENE, 1,1′-OXYBIS(4-ISOCYANATO)- see 4,4′-DIISOCYANATODIPHENYL ETHER

BENZENE, 1,1′-OXYBIS[2,3,4,5,6-PENTABROMO- see DECABROMODIPHENYL OXIDE

BENZENE, PENTACHLORO- see PENTACHLOROBENZENE

BENZENE, PENTACHLORONITRO- see QUINTOZENE

BENZENESULFONAMIDE, 2-CHLORO-N-(((4-METHOXY-6-METHYL-1,3,5-TRIAZIN-2-YL)AMINO)CARBONYL) see CHLORSULFURON

BENZENESULFONAMIDE, 4-(DIPROPYLAMINO)-3,5-DINITRO- see ORYZALIN

BENZENESULFONIC ACID, 4-HYDROXY-, ZINC SALT (2:1) see ZINC PHENOLSULFONATE

BENZENESULFONIC ACID CHLORIDE see BENZENESULFONYL CHLORIDE

BENZENESULFONIC ACID, DODECYL- see DODECYLBENZENESULFONIC ACID

BENZENESULFONIC ACID, DODECYL-, comp. with 2,2′,2″-NITRILOTRIS[ETHANOL](1:1) see TRIETHANOLAMINE DODECYLBENZENE SULFONATE

BENZENESULFONIC ACID, DODECYL-, CALCIUM SALT see CALCIUM DODECYLBENZENESULFONATE

BENZENE SULFONIC ACID, DODECYL ESTER see DODECYLBENZENE-SULFONIC ACID

BENZENE SULFONIC ACID, DODECYL-, comp. with 1-AMINO-2-PROPANOL (1:1) see ISOPROPANOLAMINE DODECYLBENZENE SULFONATE

BENZENE SULFONIC ACID, DODECLY-, SODIUM SALT see SODIUM DODECYLBENZENESULFONATE

BENZENESULFONIC ACID, p-HYDROXY-, ZINC SALT (2:1) see ZINC PHENOLSULFONATE

BENZENE, 1,2,4,5-TETRACHLORO- see 1,2,4,5-TETRACHLOROBENZENE

BENZENE, 1,2,4-TRICHLORO- see 1,2,4-TRICHLOROBENZENE

BENZENE,1,1′-(2,2,2-TRICHLOROETHYLIDENE)BIS[4-METHOXY-] see METHOXYCHLOR

BENZENE,1,1′-(2,2,2-TRICHLOROETHYLIDENE)BIS(4-CHLORO) see DDT

BENZENE, TRICHLOROMETHYL- see BENZOIC TRICHLORIDE

BENZENE, 1,2,4-TRIMETHYL- see 1,2,4-TRIMETHYLBENZENE

BENZENE, 1,3,5-TRINITRO- see 1,3,5-TRINITROBENZENE

BENZIDINE, 3,3′-DICHLORO- see 3,3′-DICHLOROBENZIDINE

BENZIDINE, 3,3′-DICHLORO-, DIHYDROCHLORIDE see 3,3′-DICHLOROBENZIDINE DIHYDROCHLORIDE

BENZIDINE, 3,3′-DIMETHOXY- see 3,3′-DIMETHOXYBENZIDINE

BENZIDINE, 3,3′-DIMETHOXY-, DIHYDROCHLORIDE see 3,3′-DIMETHOXYBENZIDINE DIHYDROCHLORIDE

BENZIDINE, 3,3′-DIMETHYL-, DIHYDROCHLORIDE see 3,3′-DIMETHYLBENZIDINE DIHYDROCHLORIDE

BENZIDINE, 3,3′-DIMETHYL- see o-TOLIDINE

BENZILIC ACID, 4,4′-DICHLORO-, ETHYL ESTER see CHLOROBENZILATE

BENZIMIDAZOLE, 2-(4-THIAZOLYL)- see THIABENDAZOLE

1,2-BENZISOTHIAZOL-3(2H)-ONE, 1,1-DIOXIDE see SACCHARIN

3-BENZISOTHIAZOLINONE 1,1-DIOXIDE see SACCHARIN

1,2-BENZO(a)ANTHRACENE see BENZ[a]ANTHRACENE

BENZO(d,e,f)CHRYSENE see BENZO(a)PYRENE

1,3-BENZODIOXOLE, 5-((2-(2-BUTOXYETHOXY)ETHOXY)METHYL)-6-PROPYL- see PIPERONYL BUTOXIDE

1,3-BENZODIOXOLE, 5-(2-PROPENYL)- see SAFROLE

2,3-BENZOFLUORANTHENE see BENZO[b]FLUORANTHENE

BENZO(e)FLUORANTHENE see BENZO[b]FLUORANTHENE

BENZO(jk)FLUORENE see FLUORANTHENE

7-BENZOFURANOL, 2,3-DIHYDRO-2,2-DIMETHYL-, METHYLCARBAMATE see CARBOFURAN

BENZOIC ACID, 3-AMINO-2,5-DICHLORO- see CHLORAMBEN

BENZOIC ACID AMIDE see BENZAMIDE

BENZOIC ACID, AMMONIUM SALT see AMMONIUM BENZOATE

BENZOIC ACID, CHLORIDE see BENZOYL CHLORIDE

BENZOIC ACID, 5-(2-CHLORO-4-(TRIFLUOROMETHYL)PHENOXY)-2-NITRO-2-ETHOXY-1-METHYL-2-OXOETHYL ESTER see LACTOFEN

BENZOIC ACID, 3,6-DICHLORO-2-METHOXY- see DICAMBA

BENZOIC ACID, 3,6-DICHLORO-2-METHOXY-, SODIUM SALT see SODIUM DICAMBA

BENZOIC ACID, 2-((ETHOXY((1-METHYLETHYL)AMINO)PHOSPHINOTHIOYL)OXY), 1-METHYLETHYL ESTER see ISOFENPHOS

BENZOIC ACID, 2-(((((4-METHOXY-6-METHYL-1,3,5-TRIAZIN-2-YL) METHYLAMINO)CARBONYL)AMINO)SULFONYL)-, METHYL ESTER see TRIBENURON METHYL

BENZOL see BENZENE

BENZONITRILE, 2,6-DICHLORO- see DICHLOBENIL

1,12-BENZOPERYLENE see BENZO[ghi]PERYLENE

BENZO[g]PERYLENE see BENZO[ghi]PERYLENE

1,2-BENZOPHENANTHRENE see CHRYSENE

BENZO(A)PHENANTHRENE see CHRYSENE

BENZO(def)PHENANTHRENE see PYRENE

BENZOPHENONE, 4,4′-BIS(DIMETHYLAMINO)- see MICHLER'S KETONE

2h-1-BENZOPYRAN-2-ONE,4-HYDROXY-3-(3-OXO-1-PHENYLBUTYL)- see WARFARIN

BENZOPYRIDINE see QUINOLINE

BENZO[B]PYRIDINE see QUINOLINE

p-BENZOQUINONE see QUINONE

BENZOQUINONE see QUINONE

2-BENZOTHIAZOLETHIOL see 2-MERCAPTOBENZOTHIAZOLE

BENZOTRICHLORIDE see BENZOIC TRICHLORIDE

1,12-BENZPERYLENE see BENZO[ghi]PERYLENE

1,2-BENZPHENANTHRENE see CHRYSENE

(5-BENZYL-3-FURYL)METHYL CHRYSANTHEMATE see RESMETHRIN

5-BENZYL-3-FURYLMETHYL (+-)-cis-trans-CHRYSANTHEMATE see RESMETHRIN

BENZYL ALCOHOL, 2,4,5-TRICHLORO-α-(CHLOROMETHYLENE)-, DIMETHYL PHOSPHATE see TETRACHLORVINPHOS

BENZYL ALCOHOL, 2,4-DICHLORO-α-(CHLOROMETHYLENE)-, DIETHYL PHOSPHATE see CHLORFENVINFOS

n-BENZYL BUTYL PHTHALATE see BUTYL BENZYL PHTHALATE

BENZYL DICHLORIDE see BENZAL CHLORIDE

BENZYLIDENE CHLORIDE see BENZAL CHLORIDE

BENZYL ALCOHOL, 2,4-DICHLORO-α-(CHLOROMETHYLENE)-, DIETHYL PHOSPHATE see CHLORFENVINFOS

BENZYL ALCOHOL, 2,4,5-TRICHLORO-α-(CHLOROMETHYLENE)-, DIMETHYL PHOSPHATE see TETRACHLORVINPHOS n-BENZYL BUTYL PHTHALATE see BUTYL BENZYL PHTHALATE

BENZYL DICHLORIDE see BENZAL CHLORIDE

(5-BENZYL-3-FURYL)METHYL CHRYSANTHEMATE see RESMETHRIN

5-BENZYL-3-FURYLMETHYL (+-)-cis-trans-CHRYSANTHEMATE see RESMETHRIN

BENZYLIDENE CHLORIDE see BENZAL CHLORIDE

BERYLLIUM-9 see BERYLLIUM

BERYLLIUM DICHLORIDE see BERYLLIUM CHLORIDE

BERYLLIUM DIFLUORIDE see BERYLLIUM FLUORIDE

BERYLLIUM DINITRATE see BERYLLIUM NITRATE

BERYLLIUM, ELEMENTAL see BERYLLIUM

BERYLLIUM NITRATE (HYDRATED) see BERYLLIUM NITRATE

BERYLLIUM POWDER see BERYLLIUM

BHC see HEXACHLOROCYCLOHEXANE (ALL ISOMERS)

BHC-HEXACHLOROCYCLOHEXANE see HEPTACHLOR EPOXIDE

α-BHC see α-HEXACHLOROCYCLOHEXANE

α-BHC-α see α-HEXACHLOROCYCLOHEXANE

β-BHC see β-HEXACHLOROCYCLOHEXANE

β-BHC-β see β-HEXACHLOROCYCLOHEXANE

δ-BHC see δ-HEXACHLOROCYCLOHEXANE

δ-BHC see δ-HEXACHLOROCYCLOHEXANE

δ-BHC-δ see δ-HEXACHLOROCYCLOHEXANE

δ-BHC (PCB-POLYCHLORINATED BIPHENYLS) see δ-HEXACHLOROCYCLOHEXANE

BICARBURRETTED HYDROGEN see ETHYLENE

BICYCLOPENTADIENE see DICYCLOPENTADIENE

BIDRIN see DICROTOPHOS

(1,1′-BIFENYL)-4,4′-DIAMINE see BENZIDINE

BIOXIRANE see DIEPOXYBUTANE

2,2′-BIOXIRANE see DIEPOXYBUTANE

4-BIPHENYLACETIC ACID, 2-FLUOROETHYL ESTER see FLUENETIL

[1,1′-BIPHENYL]-4-AMINE see 4-AMINOBIPHENYL

1,1′-BIPHENYL see BIPHENYL

BIPHENYL, CHLORINATED see POLYCHLORINATED BIPHENYLS

1,1′-BIPHENYL, CHLORO DERIVATIVES see POLYCHLORINATED BIPHENYLS

(1,1′-BIPHENYL)-4,4′-DIAMINE, 3,3′-DICHLORO-, DIHYDROCHLORIDE see 3,3′-DICHLOROBENZIDINE DIHYDROCHLORIDE

(1,1′-BIPHENYL)-4,4′-DIAMINE, 3,3′-DICHLORO-, SULFATE (1:2) see 3,3′-DICHLOROBENZIDINE SULFATE

(1,1′-BIPHENYL)-4,4′-DIAMINE, 3,3′-DIMETHOXY-, MONOHYDROCHLORIDE see 3,3′-DIMETHOXYBENZIDINE HYDROCHLORIDE

1,1′BIPHENYL]-4,4′-DIAMINE, 3,3′-DIMETHOXY- see 3,3′-DIMETHOXYBENZIDINE

(1,1′-BIPHENYL)-4,4′-DIAMINE, 3,3′-DIMETHYL- see 3,3′-DIMETHYLBENZIDINE

(1,1′-BIPHENYL)-4,4′-DIAMINE,3,3′-DIMETHYL- see o-TOLIDINE

1,1′-BIPHENYL, 4,4′-DIISOCYANATO-3,3′-DIMETHOXY- see 3,3′-DIMETHOXYBENZIDINE-4,4′-DIISOCYANATE

1,1′-BIPHENYL,4,4′-DIISOCYANATO-3,3′-DIMETHYL- see 3,3′-DIMETHYL-4,4′-DIPHENYLENE DIISOCYANATE

BIPHENYL, 4-NITRO- see 4-NITROBIPHENYL

1,1′-BIPHENYL, 4-NITRO- see 4-NITROBIPHENYL

(1,1′-BIPHENYL)-2-OL see 2-PHENYLPHENOL

(1,1′-BIPHENYL)-2-OL, SODIUM SALT see SODIUM O-PHENYLPHENOXIDE

2,2′-BIPHENYLYLEME OXIDE see DIBENZOFURAN

4,4′-BIPYRIDINIUM, 1,1′-DIMETHYL-, DICHLORIDE see PARAQUAT DICHLORIDE

BIPYRIDINIUM, 1,1′-DIMETHYL-4,4′-, DICHLORIDE see PARAQUAT DICHLORIDE

4,4′-BIPYRIDYNIUM, 1,1-DIMETHYLSULFATE, BIS (METHYL SULFATE) see PARAQUAT METHOSULFATE

BIS(ACETATO)TETRAHYDROXYTRILEAD see LEAD SUBACETATE

BIS(4-AMINOPHENYL) ETHER see 4,4′-DIAMINOPHENYL ETHER

S-(1,2-BIS(CARBETHOXY)ETHYL) O,O-DIMETHYLDITHIOPHOSPHATE see MALATHION

BIS(β-CHLORETHYL)FORMAL see BIS(2-CHLOROETHOXY)METHANE

BIS(2-CHLORETHYL)FORMAL see BIS(2-CHLOROETHOXY)METHANE

BIS(2-CHLORO-ISOPROPYL) ETHER see BIS(2-CHLORO-1-METHYLETHYL) ETHER

BIS(2-CHLOROETHOXY)METHANE see BIS(2-CHLOROETHYL)ETHER

BIS(2-CHLOROETHYL)-β-NAPHTHYLAMINE see CHLORNAPHAZINE

N,N-BIS(2-CHLOROETHYL)-2-NAPHTHYLAMINE see CHLORNAPHAZINE

BIS(2-CHLOROETHYL) ETHER see BIS(2-CHLOROETHYL)ETHER

2-(BIS(2-CHLOROETHYL)AMINO)-1-OXA-3-AZA-2-PHOSPHOCYCLOHEXANE 2-OXIDE MONOHYDRATE see CYCLOPHOSPHAMIDE

4-(BIS(2-CHLOROETHYL)AMINO)BENZENEBUTANOIC ACID see CHLORAMBUCIL

4-(BIS(2-CHLOROETHYL)AMINO)-L-PHENYLALANINE see MELPHALAN

5-(BIS(2-3CHLOROETHYL)AMINO)-2,4(1H,3H)PYRIMIDINEDIONE see URACIL MUSTARD

2-BIS(2-CHLOROETHYL)AMINONAPHTHALENE see CHLORNAPHAZINE

5-N,N-BIS(2-CHLOROETHYL)AMINOURACIL see URACIL MUSTARD

N,N-BIS(2-CHLOROETHYL)METHYLAMINE see NITROGEN MUSTARD

BIS(CHLOROMETHYL) ETHER see BIS(2-CHLORO-1-METHYLETHYL) ETHER

BIS(CHLOROMETHYL)BENZENE see XYLYLENE DICHLORIDE

BIS(2-CHLOROMETHYL)ETHER see BIS(CHLOROMETHYL)ETHER

3,3-BIS(CHLOROMETHYL)OXETANE see OXETANE, 3,3-BIS (CHLOROMETHYL)-

1,1-BIS(p-CHLOROPHENYL)- see DDE

2,2-BIS(p-CHLOROPHENYL)-1,1-DICHLOROETHYLENE see DDE

2,2-BIS(p-CHLOROPHENYL)ETHANE see DDE

1,1-BIS(P-CHLOROPHENYL)-2,2,2-TRICHLOROETHANOL see DICOFOL

BIS-O,O-DIETHYLPHOSPHORIC ANHYDRIDE see TEPP

BIS(DIMETHYLAMIDO)FLUOROPHOSPHATE see DIMEFOX

BIS(DIMETHYLTHIOCARBAMOYL) DISULFIDE see THIRAM

BIS(2,3-EPOXYPROPYL)ETHER see DIGLYCIDYL ETHER

1,2-BIS(3-(ETHOXYCARBONYL)-2-THIOUREIDO)BENZENE see THIOPHANATE ETHYL

1,2-BIS(ETHOXYCARBONYLTHIOUREIDO)BENZENE see THIOPHANATE ETHYL

2,4-BIS(ETHYLAMINO)-6-CHLORO-S-TRIAZINE see SIMAZINE

BIS(2-ETHYLHEXYL)-1,2-BENZENEDICARB OXYLATE see DI(2-ETHYLHEXYL)PHTHALATE

BIS(2-ETHYLHEXYL)PHTHALATE see DI(2-ETHYLHEXYL)PHTHALATE

BIS(3-FLUOROSALICYLALDEHYDE)ETHYLENEDIIMINE-COBALT see COBALT,((2,2′-(1,2-ETHANEDIYLBIS(NITRILOMETHYLIDYNE))BIS(6-FLUOROPHENOLATO))(2)-

N,N-BIS(2-HYDROXYETHYL)AMINE see DIETHANOLAMINE

1,3-BIS(ISOCYANATOMETHYL)CYCLOHEXANE see 1,3-BIS(METHYLISOCYANATE)CYCLOHEXANE

2,4-BIS(ISOPROPYLAMINO)-6-(METHYLMERCAPTO)-s-TRIAZINE see PROMETHRYN

N,N′-BIS(1-METHYLETHYL)-6-METHYLTHIO-1,3,5-TRIAZINE-2,4-DIAMINE see PROMETHRYN

BIS(1-METHYLETHYL) CARBAMOTHIOIC ACID, S-(2,3-DICHLORO-2-PROPENYL)ESTER see DIALLATE

1,2-BIS(3-(METHOXYCARBONYL)-2-THIOUREIDO)BENZENE see THIOPHANATE-METHYL

1,2-BIS(METHOXYCARBONYLTHIOUREIDO)BENZENE see THIOPHANATE-METHYL

2,2-BIS(p-METHOXYPHENYL)-1,1,1-TRICHLOROETHANE see METHOXYCHLOR

BIS(NITRATO-O,O′)DIOXO URANIUM (solid) see URANYL NITRATE

P,P′-BISPHENOL A see 4,4′-ISOPROPYLIDENEDIPHENOL

BISPHENOL A see 4,4′-ISOPROPYLIDENEDIPHENOL

BIS(10-PHENOXARSYL)OXIDE see PHENOXARSINE, 10,10′-OXYDI-

BIS(SALICYALDEHYDE)ETHYLENEDIIMINE COBALT(II) see SALCOMINE

BIS(THYOCYANATO)-MERCURY see MERCURIC THIOCYANATE

BIS(3,5,6-TRICHLORO-2-HYDROXYPHENYL)METHANE see HEXACHLOROPHENE

BIS(TRIS(2-METHYL-2-PHENYLPROPYL)TIN)OXIDE see FENBUTATIN OXIDE

BISULFITE see SULFUR DIOXIDE

N,N-BIS(2,4-XYLYLIMINOMETHYL)METHYLAMINE see AMITRAZ

BIURET, 2,4-DITHIO- see DITHIOBIURET

1,1′-BI[ETHYLENE OXIDE] see DIEPOXYBUTANE

BLADEX see CYANAZINE

BOLERO see THIOBENCARB

BOLSTAR see SULPROFOS

BORANE, TRICHLORO- see BORON TRICHLORIDE

BORANE, TRIFLUORO- see BORON TRIFLUORIDE

BORATE (1-), TETRAFLUORO-, LEAD (2+) (2:1) see LEAD FLUOBORATE

BORIC ACID, ZINC SALT see ZINC BORATE

BORON TRIFLUORIDE DIMETHYL ETHERATE see BORON TRIFLUORIDE COMPOUND with METHYL ETHER (1:1)

BORON, TRIFLUORO[OXYBIS[METHANE]]-,(T-4)- see BORON TRIFLUORIDE COMPOUND with METHYL ETHER (1:1)

BROMIC ACID, POTASSIUM SALT see POTASSIUM BROMATE

BROMINE, ELEMENTAL see BROMINE

3-(3-(4′-BROMO(1,1′-BIPHENYL)-4-YL)3-HYDROXY-1-PHENYLPROPYL)-4-HYDROXY-2H-1-BENZOPYRAN-2-ONE see BROMADIOLONE

O-(4-BROMO-2,5-BROMO-2,5-DICHLOROPHENYL)O-METHYL ESTER see LEPTOPHOS

2-BROMO-2-(BROMOETHYL)PENTANEDINITRILE see 1-BROMO-1-(BROMOMETHYL)-1,3-PROPANEDICARBONITRILE

5-BROMO-3-sec-BUTYL-6-METHYLURACIL see BROMACIL

5-BROMO-3-sec-BUTYL-6-METHYLPYRIMIDINE-2,4(1H,3H)-DIONE, LITHIUM SALT see BROMACIL, LITHIUM SALT

o-(4-BROMO-2-CHLOROPHENYL)-O-ETHYL-S-PROPYLPHOSPHOROTHIOATE see PROFENOFOS

BROMODICHLOROMETHANE see DICHLOROBROMOMETHANE

o-(4-BROMO-2,5-DICHLOROPHENYL)O-METHYL PHENYLPHOSPHONOTHIOATE see LEPTOPHOS

BROMOETHYLENE see VINYL BROMIDE

5-BROMO-6-METHYL-3-(1-METHYLPROPYL)-2,4(1H,3H)-PYRIMIDINEDIONE see BROMACIL

2-BROMO-2-NITRO-1,3-PROPANEDIOL see 2-BROMO-2-NITROPROPANE-1,3-DIOL

BROMO-2-PROPANONE see BROMOACETONE

3-BROMO-1-PROPYNE see PROPARGYL BROMIDE

BROMOTRIFLUORETHENE see BROMOTRIFLUORETHYLENE

BRONOPOL see 2-BROMO-2-NITROPROPANE-1,3-DIOL

BUTADIENE see 1,3-BUTADIENE

BUTA-1,3-DIENE see 1,3-BUTADIENE

1,3-BUTADIENE, 2-CHLORO- see CHLOROPRENE

BUTADIENE DIEPOXIDE see DIEPOXYBUTANE

1,3-BUTADIENE DIEPOXIDE see DIEPOXYBUTANE

BUTADIENE DIOXIDE see DIEPOXYBUTANE

BUTADIENE, HEXACHLORO- see HEXACHLORO-1,3-BUTADIENE

1,3-BUTADIENE, 1,1,2,3,4,4-HEXACHLORO- see HEXACHLORO-1,3-BUTADIENE

1,3-BUTADIENE, 2-METHYL see ISOPRENE

BUTAN-2-OL see sec-BUTYL ALCOHOL

BUTAN-1-OL see n-BUTYL ALCOHOL

BUTAN-2-ONE see METHYL ETHYL KETONE

1-BUTANAMINE see BUTYLAMINE

2-BUTANAMINE see sec-BUTYLAMINE

1-BUTANAMINE,N-BUTYL-N-NITROSO- see N-NITROSODI-n-BUTYLAMINE

BUTANE, 1,2:3,4-DIEPOXY- see DIEPOXYBUTANE

BUTANE, 2-METHYL- see ISOPENTANE

BUTANE DIEPOXIDE see DIEPOXYBUTANE

n-BUTANE see BUTANE

1,4-BUTANEDICARBOXYLIC ACID see ADIPIC ACID

cis-BUTENEDIOIC ACID, (Z) see MALEIC ACID

BUTANOIC ACID, 4-(2,4-DICHLOROPHENOXY)- see 2,4-DB

n-BUTANOIC ACID see BUTYRIC ACID

BUTANOIC ACID see BUTYRIC ACID

1-BUTANOL see n-BUTYL ALCOHOL

2-BUTANOL see sec-BUTYL ALCOHOL

sec-BUTANOL see sec-BUTYL ALCOHOL

tert-BUTANOL see tert-BUTYL ALCOHOL

2-BUTANOL, 2-METHYL-, ACETATE see tert-AMYL ACETATE

1-BUTANOL, 3-METHYL-, ACETATE see ISO-AMYL ACETATE

2-BUTANONE, 1-(4-CHLOROPHENOXY)-3,3-DIMETHYL-1-(1-H-1,2,4-TRIAZOL-1-YL)- see TRIADIMEFON

2-BUTANONE see METHYL ETHYL KETONE

2-BUTANONE, PEROXIDE see METHYL ETHYL KETONE PEROXIDE

2-BUTENAL, (e) see CROTONALDEHYDE, (E)

2-BUTENAL PROPYLENE ALDEHYDE see CROTONALDEHYDE

(E)-2-BUTENAL see CROTONALDEHYDE, (E)

2-BUTENAL see CROTONALDEHYDE

2-BUTENE, (z)- see 2-BUTENE-cis

(E)-, 2-BUTENE see 2-BUTENE, (E)-

2-BUTENE, trans see 2-BUTENE, (E)-

trans-2-BUTENE see 2-BUTENE, (E)-

BUTENE-1 see 1-BUTENE

2, BUTENE, 1,4-DICHLORO- see 1,4-DICHLORO-2-BUTENE

1-BUTENE, 2-METHYL see 2-METHYL-1-BUTENE

1-BUTENE, 3-METHYL see 3-METHYL-1-BUTENE

2-BUTENEDIOIC ACID (E)- see FURMARIC ACID

BUTENEDIOIC ACID, (Z)- see MALEIC ACID

(E)-BUTENEDIOIC ACID see FURMARIC ACID

trans-BUTENEDIOIC ACID see FURMARIC ACID

2-BUTENOIC ACID, 3-((DIMETHOXYPHOSPHINYL)OXY)-, METHYL ESTER see MEVINPHOS

2-BUTENOIC ACID, 3-(((ETHYLAMINO)METHOXYPHOSPHINOTHIOYL)OXY)-, 1-METHYLETHYL ESTER, (E)- see PROPETAMPHOS

2-BUTENOIC ACID, 2-METHYL-, 7-((2,3-DIHYDROXY-2-(1-METHOXYETHYL)-3-METHYL-1-OXOBUTOXY)METHYL)-2,3,5,7A-TETRAHYDRO-1H-PYRROLIZIN-1-YL ESTER,(IS(1α (Z),7(2S*,3R*),7Aα)-see LASIOCARPINE

2-BUTENOIC ACID 2-(1-METHYLHEPTYL)-4,6-DINITROPHENYL ESTER see DINOCAP

1-BUTEN-3-YNE see VINYL ACETYLENE

BUTOXYETHYL 2,4-D see 2,4-D BUTOXYETHYL ESTER

BUTOXYETHYL 2,4-DICHLOROPHENOXYACETATE see 2,4-D BUTOXYETHYL ESTER

tert-BUTOXYMETHANE see METHYL tert-BUTYL ETHER

n-BUTYLACRYLATE see BUTYL ACRYLATE

BUTYL ALCOHOL see n-BUTYL ALCOHOL

2-BUTYL ALCOHOL see sec-BUTYL ALCOHOL

BUTYL ALDEHYDE see BUTYRALDEHYDE

n-BUTYLAMINE see BUTYLAMINE

sec-BUTYLAMINE (s-) see sec-BUTYLAMINE

(RS)-sec-BUTYLAMINE see sec-BUTYLAMINE

BUTYLATE-2,4,5-T see 2,4,5-T ESTERS (93-79-8)

α-BUTYL-α-(4-CHLOROPHENYL)-1H-1,2,4-TRIAZOLE-1-PROPANENITRILE see MYCLOBUTANIL

sec-BUTYL, 2,4-D ESTER see 2,4-D sec-BUTYL ESTER

n-BUTYL 2,4-D ESTER see 2,4-D BUTYL ESTER

sec-BUTYL, 2,4-D see 2,4-D sec-BUTYL ESTER

BUTYL 2,4-D see 2,4-D BUTYL ESTER

2-sec-BUTYL-4,6-DINITROPHENOL see DINITROBUTYL PHENOL

BUTYLENE see BUTENE

α-BUTYLENE see BUTENE

BUTYLENE see 1-BUTENE

α-BUTYLENE see 1-BUTENE

β-BUTYLENE see 2-BUTENE

γ-BUYLENE see 2-METHYLPROPENE

2-BUTYLENE DICHLORIDE see trans-1,4-DICHLORO-2-BUTENE

N-BUTYL-N-ETHYL-2,6-DINITRO-4-(TRIFLUROMETHYL)BENZENEAMINE see BENFLURALIN

BUTYLETHYLCARBAMOTHIOIC ACID S-PROPYL ESTER see PEBULATE

n-BUTYL-N-ETHYL-α,α,α-TRIFLUORO-2,6-DINITRO-p-TOLUIDINE see BENFLURALIN

t-BUTYL METHYL ETHER see METHYL tert-BUTYL ETHER

BUTYL PHOSPHOROTRITHIOATE see S,S,S-TRIBUTYLTRITHIOPHOSPHATE

n-BUTYL PHTHALATE see DIBUTYL PHTHALATE

BUTYL-, 2,4,5-T see 2,4,5-T ESTERS (93-79-8)

1-(5-tert-BUTYL-1,3,4-THIADIAZOL-2-YL)-1,3-DIMETHYLUREA see TEBUTHIURON

BUTYL(RS)-2-(4-((5-(TRIFLUOROMETHYL)-2-PYRIDINYL)OXY) PHENOXY)PROPANOATE see FLUAZIFOP-BUTYL

n-BUTYRALDEHYDE see BUTYRALDEHYDE

BUTYRIC ALCOHOL see n-BUTYL ALCOHOL

CACODYLIC ACID SODIUM SALT see SODIUM CACODYLATE

CADMIUM(II) ACETATE see CADMIUM ACETATE

CADMIUM DIBROMIDE see CADMIUM BROMIDE

CADMIUM DICHLORIDE see CADMIUM CHLORIDE

CADMIUM, ELEMENTAL see CADMIUM

CALCIUM CHLOROHYDROCHLORITE see CALCIUM HYPOCHLORITE

CALCYANIDE see CALCIUM CYANIDE

CAMPHECHLOR see TOXAPHENE

CAMPHENE, OCTACHLORO- see TOXAPHENE

CANTHARIDINE see CANTHARIDIN

ε-CAPROLACTAM see CAPROLACTUM

σ-CAPROLACTAM see CAPROLACTUM

CAPTOBENZOTHIAZOL see 2-MERCAPTOBENZOTHIAZOLE

CARBAMATE, 4-DIMETHYLAMINO-3,5-XYLYLN-METHYL- see MEXACARBATE

CARBAMIC ACID, AMMONIUM SALT see AMMONIUM CARBAMATE

CARBAMIC ACID, 1-(BUTYLAMINO)CARBONYL- 1H-BENZIMIDAZOL-2YL, METHYL ESTER see BENOMYL

CARBAMIC ACID, BUTYL-, 3-IODO-2-PROPYNYL ESTER see 3-IODO-2-PROPYNYL BUTYLCARBAMATE

CARBAMIC ACID, DIETHYLTHIO-, S-(P-CHLOROBENZYL) ESTER see THIOBENCARB

CARBAMIC ACID, DIISOPROPYLTHIO-, S-(2,3,3-TRICHLOROALLYL) ESTER see TRIALLATE

CARBAMIC ACID, DIMETHYLDITHIO-, POTASSIUM SALT, HYDRATE see POTASSIUM DIMETHYLDITHIOCARBAMATE

CARBAMIC ACID, DIMETHYLDITHIO-, SODIUM SALT see SODIUM DIMETHYLDITHIOCARBAMATE

CARBAMIC ACID, DIPROPYLTHIO-, S-ETHYL ESTER see ETHYL DIPROPYLTHIOCARBAMATE

CARBAMIC ACID, ETHYL ESTER see URETHANE

CARBAMIC ACID, METHYL-,O-(((2,4-DIMETHYL-1,3-DITHIOLAN-2-YL)METHYLENE)AMINO)- see TRIPATE

CARBAMIC ACID, METHYL-, 3-METHYLPHENYL ESTER see METOLCARB

CARBAMIC ACID, METHYLDITHIO-, MONOSODIUM SALT see METHAM SODIUM

CARBAMIC ACID, N-METHYLDITHIO-, POTASSIUM SALT see POTASSIUM N-METHYLDITHIOCARBAMATE

CARBAMIC ACID, METHYLNITROSO-, ETHYL ESTER see N-NITROSO-N-METHYLURETHANE

CARBAMIC ACID, MONOAMMONIUM SALT see AMMONIUM CARBAMATE

CARBAMIC ACID, (3-(((PHENYLAMINO)CARBONYL)OXY)PHENYL)-, ETHYL ESTER see DESMEDIPHAM

CARBAMIC CHLORIDE, DIMETHYL- see DIMETHYLCARBAMOYL CHLORIDE

CARBAMIMIDOSELENOIC ACID see SELENOUREA

CARBAMODITHIOIC ACID, BIS(1-METHYLETHYL)-, S-(2,3-DICHLORO-2-PROPENYL)ESTER see DIALLATE

CARBAMOTHIOIC ACID, DIETHYL-, S-(CHLOROPHENYL)METHYL) ESTER see THIOBENCARB

CARBAMODITHIOIC ACID, DIMETHYL-, POTASSIUM SALT see POTASSIUM DIMETHYLDITHIOCARBAMATE

CARBAMODITHIOIC ACID, 1,2-ETHANEDIYLBIS-, MANGANESE SALT see MANEB

CARBAMODITHIOIC ACID, 1,2-ETHANEDIYLBIS, SALTS AND ESTERS see ETHYLENEBISDITHIOCARBAMIC ACID

CARBAMODITHIOIC ACID, 1,2-ETHANEDIYLBIS-, ZINC COMPLEX see ZINEB

CARBAMOTHIOIC ACID, BUTYLETHYL-, S-PROPYL ESTER see PEBULATE

CARBAMYLHYDRAZINE HYDROCHLORIDE see SEMICARBAZIDE HYDROCHLORIDE

CARBARYL (ISO) see CARBARYL

CARBARYL (SEVIN) see CARBARYL

CARBIDE, ACETYLENOGEN see CALCIUM CARBIDE

CARBOLIC ACID see PHENOL

α-2-CARBOMETHOXY-1-METHYLVINYL DIMETHYL PHOSPHATE see MEVINPHOS

2-CARBOMETHOXY-1-METHYLVINYL DIMETHYL PHOSPHATE see MEVINPHOS

CARBON BISULFIDE see CARBON DISULFIDE

CARBON DICHLORIDE OXIDE see PHOSGENE
CARBON HYDRIDE NITRIDE (CHN) see HYDROGEN CYANIDE
CARBON NITRIDE see CYANOGEN
CARBON OXIDE SULFIDE (COS) see CARBONYL SULFIDE
CARBON OXYCHLORIDE see PHOSGENE
CARBON OXYFLUORIDE see CARBONIC DIFLUORIDE
CARBON OXYSULFIDE see CARBONYL SULFIDE
CARBONIC ACID, AMMONIUM SALT see AMMONIUM CARBONATE
CARBONIC ACID, DIAMMONIUM SALT see AMMONIUM CARBONATE
CARBONIC ACID, DILITHIUM SALT see LITHIUM CARBONATE
CARBONIC ACID, DITHALLIUM(1+) SALT see THALLIUM(I) ACETATE
CARBONIC ACID, DITHALLIUM(I) SALT see THALLIUM(I) CARBONATE
CARBONIC ACID, DITHIO-, CYCLIC see CHINOMETHIONAT
CARBONIC ACID, ZINC SALT (1:1) see ZINC CARBONATE
CARBONIC DICHLORIDE see PHOSGENE
CARBONOCHLORIDIC ACID-2-CHLOROETHYL ESTER see CHLOROETHYL CHLOROFORMATE
CARBONOCHLORIDIC ACID, ETHYL ESTER see ETHYL CHLOROFORMATE
CARBONOCHLORIDIC ACID, 1-METHYLETHYL ESTER see ISOPROPYL CHLOROFORMATE
CARBONOCHLORIDIC ACID, METHYL ESTER see METHYL CHLOROCARBONATE
CARBONOCHLORIDIC ACID, PROPYLESTER see PROPYL CHLOROFORMATE
CARBONOTHIOIC DIHYDRAZINE see THIOCARBAZIDE
CARBONYL CHLORIDE see PHOSGENE
CARBONYL DIFLUORIDE see CARBONIC DIFLUORIDE
CARBONYL FLUORIDE see CARBONIC DIFLUORIDE
CARBOXYLIC ACID, (3-PHENOXYPHENYL)METHYL ESTER) see PERMETHRIN
CASTRIX see CRIMIDINE
CFC-11 see TRICHLOROFLUOROMETHANE
CFC-13 see CHLOROTRIFLUOROMETHANE
CFC-12 see DICHLORODIFLUOROMETHANE
CFC-114 see DICHLOROTETRAFLUOROETHANE
CFC 113 see FREON 113
CFC-115 see MONOCHLOROPENTAFLUOROETHANE
2-CHLORAETHYLTRIMETHYLAMMONIUM CHLORIDE see CHLORMEQUAT CHLORIDE
CHLORAL see ACETALDEHYDE, TRICHLORO
CHLORDAN see CHLORDANE
CHLORIDE OF PHOSPHORUS see PHOSPHOROUS TRICHLORIDE
CHLORINATED FLUOROCARBON 113 see FREON 113

CHLORINE CYANIDE see CYANOGEN CHLORIDE

CHLORINE MOLECULAR (Cl_2) see CHLORINE

CHLORINE MONOXIDE see CHLORINE OXIDE

CHLORINE OXIDE (ClO_2) see CHLORINE DIOXIDE

CHLORINE(IV) OXIDE see CHLORINE DIOXIDE

2-CHLOROACETOPHENONE see 2-CHLOROACETOPHENONE CHLOROALKYL ESTERS

α-CHLOROACETOPHENONE see 2-CHLOROACETOPHENONE CHLOROALKYL ESTERS

(CHLOROACETYL)-N-(2,6-DIETHYLPHENYL)GLYCINE ETHYL ESTER see DIETHATYL ETHYL

2-CHLOROACRYLIC ACID, METHYL ESTER see METHYL 2-CHLOROACRYLATE

5-CHLORO-2-AMINOTOLUENE see p-CHLORO-o-TOLUIDINE

para-CHLOROANILINE see p-CHLOROANILINE

CHLOROBENZAL see BENZAL CHLORIDE

α-CHLOROBENZALDEHYDE see BENZOYL CHLORIDE

CHLOROBENZENE (MONO) see CHLOROBENZENE

2-(4-((6-CHLORO-2-BENZOXAZOLYLEN)OXY)PHENOXY)PROPANOIC ACID, ETHYL ESTER see FENOXAPROP ETHYL

2-CHLOROBUTA-1,3-DIENE see CHLOROPRENE

2-CHLORO-1,3-BUTADIENE see CHLOROPRENE

5-CHLORO-3-tert-BUTYL-6-METHYLURACIL see TERBACIL

CHLOROCARBONIC ACID ETHYL ESTER see ETHYL CHLOROFORMATE

3-CHLOROCHLORDENE see HEPTACHLOR

2-CHLORO-N-(2-CHLOROETHYL)-N-METHYLETHANAMINE see NITROGEN MUSTARD

para-CHLORO-meta-CRESOL see p-CHLORO-m-CRESOL

CHLOROCRESOL see p-CHLORO-m-CRESOL

4-CHLORO-m-CRESOL see p-CHLORO-m-CRESOL

2-CHLORO-4-((1-CYANO-1-METHYLETHYL)AMINO)-6-(ETHYLAMINO)-s-TRIAZINE see CYANAZINE

2-exo-3-CHLORO-6-endo-CYANO-2-NORBORNANONE-o-(METHYLCARBOMOYL)OXIME, 2-CARBONITRILE see BICYCLO[2.2.1]HEPTANE-2-CARBONITRILE, 5-CHLORO-6-((((METHYAMINO)CARBONYL) OXY)IMINO)-,(1ST-(1-α,2-β,4-α,5-α,6e))-CYCLOPROPANECARBOXYLIC see BIFENTHRIN

3-CHLORO-1,2-DIBROMOPROPANE see 1,2-DIBROMO-3-CHLOROPROPANE

2-CHLORO-2-DIETHYLCARBAMOYL-1-METHYLVINYL DIMETHYLPHOSPHATE see PHOSPHAMIDON

2-CHLORO-N-(2,6-DIETHYLPHENYL)-N-(METHOXYMETHYL) ACETAMIDE see ALACHLOR

CHLORODIFLUOROBROMOMETHANE see BROMOCHLORODIFLUOROMETHANE

CHLORODIFLUOROETHANE see 1-CHLORO-1,1-DIFLUOROETHANE

S-(2-CHLORO-1-(1,3-DIHYDRO-1,3-DIOXO-2H-ISOINDOL-2-YL)ETHYL)-O,O-DIETHYL PHOSPHORODITHIOATE see DIALIFOR

5-CHLORO-3-(1,1-DIMETHYLETHYL)-6-METHYL-2,4(1H,3H)-PYRIMIDINEDIONE see TERBACIL

CHLORODIPHENYL (21% Cl) see AROCLOR 1221

CHLORODIPHENYL (32% Cl) see AROCLOR 1232

CHLORODIPHENYL (42% Cl) see AROCLOR 1242

CHLORODIPHENYL (48% Cl) see AROCLOR 1248

CHLORODIPHENYL (54% Cl) see AROCLOR 1254

CHLORODIPHENYL (60% Cl) see AROCLOR 1260

1-CHLORO-2,3-EPOXYPROPANE see EPICHLOROHYDRIN

3-CHLORO-1,2-EPOXYPROPANE see EPICHLOROHYDRIN

2-CHLORO-1-ETHANAL see CHLOROACETALDEHYDE

2-CHLOROETHANESULFONYL CHLORIDE see ETHANESULFONYL CHLORIDE, 2-CHLORO-

2-CHLOROETHANOL see CHLOROETHANOL

CHLOROETHENE see VINYL CHLORIDE

(2-CHLOROETHENYL)ARSONOUS DICHLORIDE see LEWISITE

(2-CHLOROETHOXY)ETHENE see 2-CHLOROETHYL VINYL ETHER

N,N-DI-2-CHLOROETHYL-γ-p-AMINOPHENYLBUTYRIC ACID see CHLORAMBUCIL

6-CHLORO-N-ETHYL-N′-(1-METHYLETHYL)-1,3,5-TRIAZINE-2,4-DIAMINE see ATRAZINE

2-CHLOROETHYL CHLOROCARBONATE see CHLOROETHYL CHLOROFORMATE

CHLOROETHYLENE see VINYL CHLORIDE

CHLOROFLURAZOLE see BENZIMIDAZOLE, 4,5-DICHLORO-2-(TRIFLUOROMETHYL)-

CHLOROFORMIC ACID ISOPROPYL ESTER see ISOPROPYL CHLOROFORMATE

CHLOROFORMIC ACID PROPYL ESTER see PROPYL CHLOROFORMATE

CHLOROFORMIC DIGITALIN see DIGOXIN

CHLOROFOS see TRICHLORFON

3-CHLORO-7-HYDROXY-4-METHYL-COUMARIN O,O-DIETHYL PHOSPHOROTHIOATE see COUMAPHOS

1-CHLORO-4-ISOCYANATO- see p-CHLOROPHENYL ISOCYANATE

2-CHLORO-N-(((4-METHOXY-6-METHYL-1,3,5-TRIAZIN-2-YL)AMINO)CARBONYL)BENZENESULFONAMIDE see CHLORSULFURON

4-CHLORO-5-METHYLAMINO-2-(3-TRIFLUOROMETHYLPHENYL)PYRIDAZIN-3-ONE see NORFLURAZON

4-CHLORO-5-(METHYLAMINO)-2-[3-(TRIFLUOROMETHYL)PHENYL]-3(2H)-PYRIDAZINONE see NORFLURAZON

4-CHLORO-5-(METHYLAMINO)-2-(α, α, α-TRIFLUORO-m-TOLYL)-3(2H)-PYRIDAZINONE see NORFLURAZON

4-CHLORO-2-METHYLBENZENAMINE see p-CHLORO-o-TOLUIDINE

2-CHLORO-4-METHYL-6-DIMETHYLAMINOPYRIMIDINE see CRIMIDINE

S-(CHLOROMETHYL)-O,O-DIETHYL PHOSPHORODITHIOATE see CHLORMEPHOS

(CHLOROMETHYL)TRICHLOROSILANE see TRICHLORO(CHLOROMETHYL)SILANE

CHLOROMETHYL ETHER see BIS(CHLOROMETHYL)ETHER

4-CHLORO-α-(1-METHYLETHYL)BENZENEACETIC ACID CYANO(3-PHENOXYPHENYL)METHYL ESTER see FENVALERATE

CHLOROMETHYL PHENYL KETONE see 2-CHLOROACETOPHENONE CHLOROALKYL ESTERS

(4-CHLORO-2-METHYLPHENOXY) ACETATE SODIUM SALT see METHOXONE, SODIUM SALT

(4-CHLORO-2-METHYLPHENOXY) ACETIC ACID see METHOXONE

3-CHLORO-2-METHYLPROP-1-ENE see 3-CHLORO-2-METHYL-1-PROPENE

3-CHLORO-2-METHYLPROPENE see 3-CHLORO-2-METHYL-1-PROPENE

β-CHLORONAPHTHALENE see 2-CHLORONAPHTHALENE

α-CHLORO-p-NITROTOLUENE see BENZENE, 1-(CHLOROMETHYL)-4-NITRO-

CHLOROPENTAFLUOROETHANE see MONOCHLOROPENTAFLUOROETHANE

1-CHLORO-1,1,2,2,2-PENTAFLUOROMETHANE see MONOCHLOROPENTAFLUOROETHANE

o-CHLOROPHENOL see 2-CHLOROPHENOL

(1-(4-CHLOROPHENOXY)-3,3-DIMETHYL-1-(1H-1,2,4-TRIAZOL-1-YL)-2-BUTANONE see TRIADIMEFON

N'-(4-(4-CHLOROPHENOXY)PHENYL-N,N-DIMETHYLUREA see CHLOROXURON

2-(α-p-CHLOROPHENYLACETYL)INDANE-1,3-DIONE see CHLOROPHACIONONE

N-(((4-CHLOROPHENYL)AMINO)CARBONYL)-2,6-DIFLUOROBENZAMIDE see DIFLUBENZURON

(2-CHLOROPHENYL)-α-(4-CHLOROPHENYL)-5-PYRIMIDINEMETHANOL see FENARIMOL

α-(2-CHLOROPHENYL)-α-(4-CHLOROPHENYL)-5-PYRIMIDINEMETHANOL see FENARIMOL

3-para-CHLOROPHENYL-1,1-DIMETHYLUREA see MONURON

N-(P-CHLOROPHENYL)-N',N'-DIMETHYLUREA see MONURON

2-CHLORO-1-PHENYLETHANONE see 2-CHLOROACETOPHENONE CHLOROALKYL ESTERS

4-CHLOROPHENYL ISOCYANATE see p-CHLOROPHENYL ISOCYANATE

s-(4-CHLOROPHENYLTHIOMETHYL)DIETHYL PHOSPHOROTHIOLO-THIONATE see CARBOPHENOTHION

2-CHLOROPHENYL THIOUREA see THIOUREA, (2-CHLOROPHENYL)-

2-p-CHLOROPHENYL-2-(1H-1,2,4-TRIAZOLE-1-YLMETHYL)HEX-ANENITRILE see MYCLOBUTANIL

2(4-CHLOROPHENYL)-2-(1H-1,2,4-TRIAZOLE-1-YLMETHYL)HEX-ANENITRILE see MYCLOBUTANIL

CHLOROPHOSPHORIC ACID DIETHYL ESTER see DIETHYL CHLORO-PHOSPHATE

CHLOROPICRINE see CHLOROPICRIN

CHLOROPRENE see 1-CHLOROPROPYLENE

3-CHLOROPRENE see ALLYL CHLORIDE

β-CHLOROPRENE see CHLOROPRENE

2-CHLOROPROPANE see ISOPROPYL CHLORIDE

1-CHLORO-1-PROPENE see 1-CHLOROPROPYLENE

2-CHLORO-1-PROPENE see 2-CHLOROPROPYLENE

2-CHLORO-2-PROPENOIC ACID METHYL ESTER (9CI) see METHYL 2-CHLOROACRYLATE

1-3-CHLORO-2-PROPENYL)-3,5,7-TRIAZA-1-AZONIATRICYCLO(3.3.1) DECANE CHLORIDE see 1-(3-CHLORALLYL)-3,5,7-TRIAZA-1-AZO-NIAADAMANTANE CHLORIDE

β-CHLOROPROPIONITRILE see 3-CHLOROPROPIONITRILE

3-CHLOROPROPYL-N-OCTYLSULFOXIDE see SULFOXIDE, 3-CHLORO-PROPYL OCTYL

3-CHLOROPROPYLENE see ALLYL CHLORIDE

2-[4-[(6-CHLORO-2-QUINOXALINYL)OXY]PHENOXY]PROPANOIC ACID ETHYL ESTER see QUIZALOFOP-ETHYL

CHLOROSULFURIC ACID see CHLOROSULFONIC ACID

CHLOROSULPHONIC ACID see CHLOROSULFONIC ACID

1-CHLORO-1,2,2,2-TETRAFLUOROETHANE see 2-CHLORO-1,1,1,2-TET-RAFLUOROETHANE

1-CHLORO-1,1,2,2-TETRAFLUOROETHANE see CHLOROTETRAFLUO-ROETHANE

(4-CHLORO-o-TOLOXY)ACETIC ACID see METHOXONE

α-CHLOROTOLUENE see BENZYL CHLORIDE

CHLOROTOLUENE see BENZYL CHLORIDE

para-CHLORO-ortho-TOLUIDINE HYDROCHLORIDE see CHLOROSUL-FONIC ACID

4-CHLORO-o-TOLUIDINE, HYDROCHLORIDE see CHLOROSULFONIC ACID

para-CHLORO-ortho-TOLUIDINE see p-CHLORO-o-TOLUIDINE

4-CHLORO-6-(TRICHLOROMETHYL)PYRIDINE see NITRAPYRIN

3-(CHLORO-3,3,3-TRIFLUORO-1-PROPENYL)-2,2-DIMETHYL-CYCLOPROPANECARBOXYLIC ACID CYANO(3-PHENOXYPHE-NYL)METHYL ESTER see CYHALOTHRIN

CHLOROTRIFLUOROETHANE see 2-CHLORO-1,1,1-TRIFLUORO-ETHANE

CHLORO-1,1,1-TRIFLUOROETHANE see 2-CHLORO-1,1,1-TRIFLUO-ROETHANE

CHLOROTRIFLUOROETHYLENE see TRIFLUOROCHLOROETHYLENE

N-[2-CHLORO-4-(TRIFLUOROMETHYL)PHENYL]-DL-VALINE(+)-CYA-NO(3-PHENOXYLPHENYL)METHYL ESTER see FLUVALINATE

5-(2-2-CHLORO-4-(TRIFLUOROMETHYL)PHENOXY)-2-NITROBEN-ZOIC ACID, SODIUM SALT see ACIFLUORFEN, SODIUM SALT

5-(2-CHLORO-4-(TRIFLUOROMETHYL)PHENOXY)-N-METHYLSULFO-NYL)-2-NITROBENZAMIDE see FOMESAFEN

1-CHLORO-3,3,3-TRIFLUOROPROPANE see 3-CHLORO-1,1,1-TRIFLUO-ROPROPANE

2-CHLORO-N,N,N-TRIMETHYLAMMONIUM CHLORIDE see CHLORME-QUAT CHLORIDE

CHLOROTRIMETHYLSILANE see TRIMETHYLCHLOROSILANE

CHLOROTRIMETHYLTIN see TRIMETHYLTIN CHLORIDE

β-CHLOROVINYLBICHLOROARSINE see LEWISITE

CHLORPYRIFOS-ETHYL see CHLORPYRIFOS

CHROMIC(VI) ACID see CHROMIC ACID

CHROMIC ACID, CALCIUM SALT (1:1) see CALCIUM CHROMATE

CHROMIC(II) CHLORIDE see CHROMOUS CHLORIDE

CHROMIC ACID, DIAMMONIUM SALT see AMMONIUM CHROMATE

CHROMIC ACID, DILITHIUM SALT see LITHIUM CHROMATE

CHROMIC ACID, DIPOTASSIUM SALT see POTASSIUM CHROMATE

CHROMIC ACID, DISODIUM SALT see SODIUM BICHROMATE

CHROMIC ACID, ESTER see CHROMIC ACID

CHROMIC ACID, STRONTIUM SALT (1:1) see STRONTIUM CHROMATE

CHROMIUM(III) ACETATE see CHROMIC ACETATE

CHROMIUM CHLORIDE see CHROMOUS CHLORIDE

CHROMIUM(III) CHLORIDE (1:3) see CHROMIC CHLORIDE

CHROMIUM DICHLORIDE see CHROMOUS CHLORIDE

CHROMIUM LITHIUM OXIDE see LITHIUM CHROMATE

CHROMIUM METAL see CHROMIUM

CHROMIUM(III) SULFATE see CHROMIC SULFATE

CHROMIUM TRIACETATE see CHROMIC ACETATE

CHROMIUM TRICHLORIDE see CHROMIC CHLORIDE

d-trans-CHRYSANTHEMIC ACID of ALLETHRONE see d-trans-ALLE-THRIN

d-trans-CHRYSANTHEMUMONOCARBOXYLIC ESTER see d-trans-AL-LETHRIN

CHRYSENE, 5-METHYL- see 5-METHYLCHRYSENE

C.I. ACID RED 114, DISODIUM SALT see C.I. ACID RED 114

C.I. DIRECT BLUE 14, TETRASODIUM SALT see TRYPAN BLUE

C.I. PIGMENT GREEN 21 see CUPRIC ACETOARSENITE

C.I. SOLVENT YELLOW 1 see 4-AMINOAZOBENZENE

CITRIC ACID, DIAMMONIUM SALT see AMMONIUM CITRATE, DIBASIC

COAL TAR CREOSOTE see CREOSOTE

COBALT BROMIDE see COBALTOUS BROMIDE

COBALT(II) BROMIDE see COBALTOUS BROMIDE

COBALT(II), N,N'-ETHYLENEBIS(3-FLUOROSALICYLIDENEIMINATO)- see COBALT,((2,2'-(1,2-ETHANEDIYLBIS(NITRILOMETHYLIDYNE)) BIS(6-FLUOROPHENOLATO))(2)-

COBALT FORMATE see COBALTOUS FORMATE

COBALT(II) FORMATE see COBALTOUS FORMATE

COBALT SULFAMATE see COBALTOUS SULFAMATE

COBALT(II) SULFAMATE see COBALTOUS SULFAMATE

COBALT TETRACARBONYL see COBALT CARBONYL

COCCULIN see PICROTOXIN

COCCULUS see PICROTOXIN

COPPER ACETATE see CUPRIC ACETATE

COPPER ACETOARSENITE see CUPRIC ACETOARSENITE

COPPER AMMONIUM SULFATE see CUPRIC SULFATE, AMMONIATED

COPPER(II) CHLORIDE (1:2) see CUPRIC CHLORIDE

COPPER CHLORIDE see CUPRIC CHLORIDE

COPPER(II) CYANIDE see COPPER CYANIDE

COPPER, [DIHYDROGEN 5-[[4'-[[2,6-DIHYDROXY-3-[(2-HYDROXY-5-SULFOPHENYL)AZO]PHENYL]AZO]-4-BIPHENYLYL]AZO]-2-HYDROXYBENZOATO(4-)]-,DISODIUM see C.I. DIRECT BROWN 95

COPPER, ELEMENTAL see COPPER

COPPER NITRATE see CUPRIC NITRATE

COPPER(II) NITRATE see CUPRIC NITRATE

COPPER OXALATE see CUPRIC OXALATE

COPPER SULFATE see CUPRIC SULFATE

COPPER(II) SULFATE see CUPRIC SULFATE

COPPER(2+), TETRAAMINE-,SULFATE (1:1), MONOHYDRATE see CUPRIC SULFATE, AMMONIATED

COPPER, (μ-((TETRAHYDROGEN, 3,3'-((3,3'-DIHYDROXY-4,4'-BIPHENYLENE)BIS(AZO)BIS(5-AMINO-4-HYDROXY-2,7-NAPHTHALENEDISULFONATO))(8-)))DI-TETRASODIUM SALT see C.I. DIRECT BLUE 218

CO-RAL see COUMAPHOS

COUMARIN, 3-(α-ACETONYLBENZYL)-4-HYDROXY- see WARFARIN

CREOSOTE OIL see CREOSOTE

para-CRESIDINE see p-CRESIDINE

2-CRESOL see o-CRESOL

3-CRESOL see m-CRESOL

meta-CRESOL see m-CRESOL

ortho-CRESOL see o-CRESOL

CYANOGEN MONOBROMIDE see CYANOGEN BROMIDE

CYCLIC ETHYLENE P,P-DIETHYLPHOSPHONODITHIOIMIDOCARBONATE see PHOSFOLAN

2,5-CYCLOHEXADIENE-1,4-DIONE see QUINONE

2,5-CYCLOHEXADIENE-1,4-DIONE, 2,3,5-TRIS(1-AZIRIDINYL)- see TRIAZIQUONE

1,4-CYCLOHEXADIENEDIONE see QUINONE

CYCLOHEXANE, 1,3-BIS(ISOCYANATOMETHYL)- see 1,3-BIS(METHYLISOCYANATE)CYCLOHEXANE

1,3-CYCLOHEXANE BIS(METHYLISOCYANATE) see 1,3-BIS(METHYLISOCYANATE)CYCLOHEXANE

1,4-CYCLOHEXANE BIS(METHYLISOCYANATE) see 1,4-BIS(METHYLISOCYANATE)CYCLOHEXANE

CYCLOHEXANE, 1,4-DIISOCYANATE see 1,4-CYCLOHEXANE DIISOCYANATE

1,4-CYCLOHEXANE DIISOCYANATO- see 1,4-CYCLOHEXANE DIISOCYANATE

CYCLOHEXANE, 1,2,3,4,5,6-HEXACHLORO-,(1α,2α,3β,4α,5α,6β) see LINDANE

CYCLOHEXANE 1,2,3,4,5,6-HEXACHLORO-,(1α,2α,3α,4β,5β,6β)- see δ-HEXACHLOROCYCLOHEXANE

CYCLOHEXANE, 1,2,3,4,5,6-HEXACHLORO- see HEXACHLOROCYCLOHEXANE (ALL ISOMERS)

CYCLOHEXANE 1,2,3,4,5,6-HEXACHLORO-(1α,2β,3α,4β,5α,6β)- see β-HEXACHLOROCYCLOHEXANE

2,5-CYCLOHEXANE,1,2,3,4,5,6-HEXACHLORO-,(1α,2α,3β,4α,5α,6β)- see LINDANE

CYCLOHEXANE see α-HEXACHLOROCYCLOHEXANE 1,2,3,4,5,6-HEXACHLORO-(1α,2α,3β,4α,5β,6β)- see α-HEXACHLOROCYCLOHEXANE

CYCLOHEXANE, 5-ISOCYANATO-1-(ISOCYANATOMETHYL)-1,3,3-TRIMETHYL- see ISOPHORONE DIISOCYANATE

CYCLOHEXANE, 1,1′-METHYLENEBIS(4-ISOCYANATO- see 1,1-METHYLENE BIS(4-ISOCYANATOCYCLOHEXANE)

CYCLOHEXANE, NITRO- see NITROCYCLOHEXANE

2-CYCLOHEXEN-1-ONE,2-(1-(ETHOXYIMINO)BUTYL)-5-(2-(ETHYLTHIO)PROPYL)-3-HYDROXY- see SETHOXYDIM

2-CYCLOHEXEN-1-ONE,3,5,5-TRIMETHYL- see ISOPHRONE

4-CYCLOHEXENE-1,2-DICARBOXIMIDE,N-[(TRICHLOROMETHYL)MERCAPTO see CAPTAN

3-CYCLOHEXYL-6-(DIMETHYLAMINO)-1-METHYL-1,3,5-TRIAZINE-2,4(1H,3H)-DIONE see HEXAZINONE

6-CYCLOHEXYL-2,4-DINITROPHENOL see 2-CYCLOHEXYL-4,6-DINITROPHENOL

CYCLOHEXYL KETONE see CYCLOHEXANONE

CYCLOPENTA(de)NAPHTHALENE see ACENAPHTHYLENE

1,3-CYCLOPENTADIENE, 1,2,3,4,5,5-HEXACHLORO- see HEXACHLOROCYCLOPENTADIENE

4,4-DDE see DDE

P,P′-DDE see DDE

4,4′-DDT see DDT

P,P′-DDT see DDT

DDVP see PROPOXUR

P,P-DDX see DDE

DECABROMOBIPHENYL ETHER see DECABROMODIPHENYL OXIDE

DECARBORON TETRADECAHYDRIDE see DECABORANE(14)

DEF see S,S,S-TRIBUTYLTRITHIOPHOSPHATE

DEGRADATION PRODUCT OF PARATHION see p-NITROPHENOL

DEHP see DI(2-ETHYLHEXYL)PHTHALATE

DEMETON-S METHYL SULFOXIDE see OXYDEMETON METHYL

DEMETON O + DEMETON S see DEMETON

DES see DIETHYLSTILBESTROL

DESMODUR T80 see TOLUENE 2,4-DIISOCYANATE

DFP see DIISOPROPYLFLUOROPHOSPHATE

DGE

DGRE see DIGLYCIDYL RESORCINOL ETHER

N,N-DI-N-BUTYLNITROSAMINE see N-NITROSODI-n-BUTYLAMINE

DI(2-CHLOROETHYL) ETHER see BIS(2-CHLOROETHYL)ETHER

DI(2-ETHYLHEXYL)PHTHALATE see DI(2-ETHYLHEXYL)PHTHALATE

DI(2,3-EPOXY)PROPYL ETHER see DIGLYCIDYL ETHER

DI(5-CHLORO-2-HYDROXYPHENYL)METHANE see DICHLOROPHENE

DI(TRI-(2,2-DIMETHYL-2-PHENYLETHYL)TIN)OXIDE see FENBUTA-TIN OXIDE

2,4-DIACETIC ACID, (CHLOROPHENOXY)-, 1-METHYLETHYL ESTER see 2,4-D ISOPROPYL ESTER

DIALIFOS see DIALIFOR

DIAMINE see HYDRAZINE

1,3-DIAMINO-4-METHOXYBENZENE see 2,4-DIAMINOSOLE

2,4-DIAMINO-1-METHYLBENZENE see 2,4-DIAMINOTOLUENE

4,4′-DIAMINO-1,1′-BIPHENYL see BENZIDINE

4,4′-DIAMINO-3,3′-DIMETHYLBIPHENYL see 3,3′-DIMETHYLBENZI-DINE

2,4-DIAMINOANISOLE SULPHATE see 2,4-DIAMINOSOLE, SULFATE

1,2-DIAMINOBENZENE see 1,2-PHENYLENEDIAMINE

4,4′-DIAMINODIPHENYLMETHANE see 4,4′-METHYLENEDIANILINE

1,2-DIAMINOETHANE, ANHYDROUS see ETHYLENEDIAMINE

2,4-DIAMINOPHENYL METHYL ETHER see 2,4-DIAMINOSOLE

4,4′-DIAMINOPHENYL SULFIDE see 4,4′-THIODIANILINE

DIAMINOTOLUENE, 2,4- see 2,4-DIAMINOTOLUENE

DIAMINOTOLUENE see DIAMINOTOLUENE (MIXED ISOMERS)

2,4-DIAMINOTOLUENE see 2,4-DIAMINOTOLUENE

DIBENZPYRENE see DIBENZ[a,i]PYRENE

1,2:7,8-DIBENZPYRENE see DIBENZ[a,i]PYRENEDIBORANE (6)

3,4:9,10-DIBENZPYRENE see DIBENZ[a,i]PYRENE

3,4,9,10-DIBENZPYRENE see DIBENZ[a,i]PYRENE

DIBORANE HEXANHYDRIDE see DIBORANE

DIBROM see NALED

1,2-DIBROMO-2,4-DICYANOBUTANE see 1-BROMO-1-(BROMOMETHYL)-1,3-PROPANEDICARBONITRILE

2,2-DIBROMO-2-CARBAMOYLACETONITRILE see 2,2-DIBROMO-3-NITRILOPROPIONAMIDE

3,5-DIBROMO-4-HYDROXYBENZONITRILE see BROMOXYNIL

3,5-DIBROMO-4-OCTANOYLOXY-BENZONITRILE see BROMOXYNIL OCTANOATE

DIBROMOCHLOROMETHANE see CHLORODIBROMOMETHANE

DIBROMOCHLOROPROPANE see 1,2-DIBROMO-3-CHLOROPROPANE

DIBROMOMETHANE see METHYLENE BROMIDE

DIBUTYL O-PHTHALATE see DIBUTYL PHTHALATE

N,N′-DIBUTYLHEXAMETHYLENEDIAMINE see HEXAMETHYLENEDIAMINE, N,N′-DIBUTYL-

1,6-N,N′-DIBUTYLHEXANEDIAMINE see HEXAMETHYLENEDIAMINE, N,N′-DIBUTYL-

DI-N-BUTYL PHTHALATE see DIBUTYL PHTHALATEDICAMBRA see DICAMBA

DI-CHLORICIDE see DICHLOROBENZENE (MIXED ISOMERS)

DICHLORINE OXIDE (Cl_2O) see CHLORINE OXIDE

o-(2,5-DICHLORO-4-BROMOPHENYL) O-METHYL PHENYLTHIOPHOSPHONATE see LEPTOPHOS

1,1-DICHLORO-2,2-BIS(P-CHLORO-PHENYL) ETHANE see DDD

1,1-DICHLORO-2,2,3,3,3-PENTAFLUOROPROPANE see 3,3-DICHLORO-1,1,1,2,2-PENTAFLUOROPROPANE

1,1-DICHLORO-2,2-DICHLOROETHANE see 1,1,2,2-TETRACHLOROETHANE

1,2-DICHLORO-1,1,2,2-TETRAFLUOROETHANE see DICHLOROTETRAFLUOROETHANE

1,2-DICHLORO-2,2-DIFLUOROETHANE see 1,2-DICHLORO-1,1-DIFLUOROETHANE

1,2-DICHLORO-4-PHENYL ISOCYANATE see ISOCYANIC ACID, 3,4-DICHLOROPHENYL ESTER

1,3-DICHLORO-2-PROPANONE see BIS(CHLOROMETHYL)KETONE

1,3-DICHLORO-2-PROPENE see 1,3-DICHLOROPROPYLENE

1,3-DICHLORO-1-PROPENE see 1,3-DICHLOROPROPYLENE

1,4-DICHLORO-2-BUTENE see 1,4-DICHLORO-2-BUTENE

2,2′-DICHLORO-DIETHYLETHER see BIS(2-CHLOROETHYL)ETHER

2,2′-DICHLORO-4,4′-METHYLENEDIANILINE see 4,4′-METHYLENEBIS(2-CHLOROANILINE)

2,3-DICHLORO-1-PROPENE see 2,3-DICHLOROPRENE

2,4-DICHLORO-6-(2-CHLOROANILINO)-1,3,5-TRIAZINE see ANILAZINE

2,4-DICHLORO-6-(o-CHLOROANILINO)-s-TRIAZINE see ANILAZINE

2,4-DICHLORO-4′-NITRODIPHENYL ETHER see NITROFEN

2,4-DICHLORO-1-(4-NITROPHENOXY)BENZENE see NITROFEN

2,6-DICHLORO-4-NITROANILINE see DICHLORAN

3,3′-DICHLORO-4,4′-DIAMINODIPHENYLMETHANE see 4,4′-METHYLENEBIS(2-CHLOROANILINE)

3,5-DICHLORO-N-(1,1-DIMETHYL-2-PROPYNYL)BENZAMIDE see PRONAMIDE

3,6-DICHLORO-2-METHOXYBENZOIC ACID see DICAMBA

3,6-DICHLORO-O-ANISIC ACID, comp. with DIMETHYLAMINE (1:1) see DIMETHYLAMINE DICAMBA

4,5-DICHLORO-2-TRIFLUOROMETHYLBENZIMIDAZOLE see BENZIMIDAZOLE, 4,5-DICHLORO-2-(TRIFLUOROMETHYL)-

4,6-DICHLORO-N-(2-CHLOROPHENYL)-1,3,5-TRIAZIN-2-AMINE see ANILAZINE

3,6-DICHLORO-2-METHOXYBENZOIC ACID, SODIUM SALT see SODIUM DICAMBA

3-[2,4-DICHLORO-5-(1-METHYLETHOXY)PHENYL]-5-(1,1-DIMETHYLETHYL)-1,3,4-OXADIAZOL-2(3H)-ONE see OXYDIAZON

3-(2,4-DICHLORO-5-ISOPROPYLOXY-PHENYL)-Δ(SUP4)-5-(TERT-BUTYL)-1,3,4-OXADIAZOLINE-2-ONE see OXYDIAZON

2,2′-DICHLORO ISOPROPYL ETHER see BIS(2-CHLORO-1-METHYLETHYL) ETHER

DICHLOROALLYLDIISOPROPYLTHIOCARBAMATE see DIALLATE

DICHLOROBENZENE, (ORTHO) see o-DICHLOROBENZENE

m-DICHLOROBENZENE see 1,3-DICHLOROBENZENE

p-DICHLOROBENZENE see 1,4-DICHLOROBENZENE

DICHLOROBENZENE see DICHLOROBENZENE (MIXED ISOMERS)

1,2-DICHLOROBENZENE see o-DICHLOROBENZENE

DICHLOROBENZIDINE see 3,3′-DICHLOROBENZIDINE

3,3′-DICHLOROBENZIDENE see 3,3′-DICHLOROBENZIDINE

o-DICHLOROBENZOL see o-DICHLOROBENZENE

2,6-DICHLOROBENZONITRILE see DICHLOBENIL

1,4-DICHLOROBUTENE-2 see 1,4-DICHLORO-2-BUTENE

trans-1,4-DICHLOROBUTENE see trans-1,4-DICHLORO-2-BUTENE

2,2′-DICHLORODIETHYL SULFIDE see MUSTARD GAS

DICHLORODIMETHYLSILANE see DIMETHYLDICHLOROSILANE

P,P′-DICHLORODIPHENYL ETHANE see DDE

DICHLORODIPHENYLDICHLOROETHANE see DDD

P,P′-DICHLORODIPHENYLDICHLOROETHYLENE see DDE

DICHLORODIPHENYLDICHLOROETHYLENE see DDE

DICHLORODIPHENYLTRICHLOROETHANE see DDT

P,P′-DICHLORODIPHENYLTRICHLOROETHANE see DDT

4,4′-DICHLORODIPHENYLTRICHLOROETHANE see DDT

1,1-DICHLOROETHANE see ETHYLIDENE DICHLORIDE

cis & trans-1,2-DICHLOROETHENE see 1,2-DICHLOROETHYLENE

1,1-DICHLOROETHENE see VINYLIDENE CHLORIDE

3-(2,2-DICHLOROETHENYL)-2,2-DIMETHYLCYCLOPROPANE-CARBOXYLIC ACID, CYANO(4-FLUORO-3-PHENOXYPHENYL)METHYL ESTER see CYFLUTHRIN

(3-(2,2-DICHLOROETHENYL)-2,2-DIMETHYLCYCLOPROPANE see PERMETHRIN

(1,1′-DICHLOROETHENYLIDENE)BIS(4-CHLOROBENZENE) see DDE

1,2-DICHLOROETHYL ACETATE see ETHANOL, 1,2-DICHLORO-, ACETATE

DICHLOROETHYL ETHER see BIS(2-CHLOROETHYL)ETHER

1,2-DICHLOROETHYLENE-trans- see 1,2-DICHLOROETHYLENE

DICHLOROETHYLENE see 1,2-DICHLOROETHANE

DICHLOROETHYLENE see VINYLIDENE CHLORIDE

1,1-DICHLOROETHYLENE see VINYLIDENE CHLORIDE

DICHLOROISOPROPYL ETHER see BIS(2-CHLORO-1-METHYLETHYL) ETHER

DICHLOROMETHOXY ETHANE see BIS(2-CHLOROETHOXY)METHANE

(DICHLOROMETHYL)BENZENE see BENZAL CHLORIDE

DICHLOROMETHYL ETHER see BIS(CHLOROMETHYL)ETHER

DICHLOROMETHYLETHANE see ETHYLIDENE DICHLORIDE

3,3-DICHLOROMETHYLOXYCYCLOBUTANE see OXETANE, 3,3-BIS(CHLOROMETHYL)-

DICHLOROMONOFLUOROMETHANE see DICHLOROFLUOROMETHANE

DICHLORONAPHTHOQUINONE see DICHLONE

DICHLOROPHEN see DICHLOROPHENE

4,6-DICHLOROPHENOL see 2,4-DICHLOROPHENOL

2,4-DICHLOROPHENOXY-α-PROPIONIC ACID see 2,4-DP

(2,4-DICHLOROPHENOXY)ACETIC ACID ESTER see 2,4-D ISOPROPYL ESTER

4-(2,4-DICHLOROPHENOXY)BUTYRIC ACID see 2,4-DB

4-(2,4-DICHLOROPHENOXY)NITROBENZENE see NITROFEN

2-(4-(2,4-DICHLOROPHENOXY)PHENOXY)PROPANOIC ACID, METHYL ESTER see DICLOFOP METHYL

2,4-DICHLOROPHENOXY, 2-BUTOXYMETHYLETHYL ESTER see 2,4-D PROPYLENE GLYCOL BUTYL ETHER ESTER

DICHLOROPHENOXYACETIC ACID see 2,4-D

2,4-DICHLOROPHENOXYACETIC ACID ISOOCTYL(2-ETHYL-4-METHYLPENTYL) ESTER see 2,4-D 2-ETHYL-4-METHYLPENTYL ESTER

2,4-DICHLOROPHENOXYACETIC ACID BUTOXYETHYL ESTER see 2,4-D BUTOXYETHYL ESTER

2,4-DICHLOROPHENOXYACETIC ACID ESTER see 2,4-D ESTERS

2,4-DICHLOROPHENOXYACETIC ACID, 4-CHLOROCROTONYL ESTER see 2,4-D CHLOROCROTYL ESTERBENZENE, 1,1′-(2,2-DICHLOROETHYLIDENE)BIS[4-CHLORO]- see DDD

2,4-DICHLOROPHENOXYACETIC ACID, BUTYL ESTER see 2,4-D BUTYL ESTERACETIC ACID, (2,4-DICHLOROPHENOXY), sec-BUTYL ESTER see 2,4-D, sec-BUTYL ESTER

2,4-DICHLOROPHENOXYACETIC ACID, SALTS AND ESTERS see 2,4-D

2,4-DICHLOROPHENOXYACETIC ACID, ISOOCTYL ESTER see 2,4-D ISOOCTYL ESTER

2,4-DICHLOROPHENOXYACETIC ACID, ISOPROPYL ESTER see 2,4-D ISOPROPYL ESTER

2,4-DICHLOROPHENOXYACETIC ACID, SODIUM SALT see 2,4-D SODIUM SALT

2,4-DICHLOROPHENOXYPROPIONIC ACID see 2,4-DP

DICHLOROPHENYL(TRICHLORO)SILANE see TRICHLORO(DICHLOROPHENYL)SILANE

N′-(3,4-DICHLOROPHENYL)-N-METHOXY-N-METHYLUREA see LINURON

1-[2-(2,4-DICHLOROPHENYL)-2-(2-PROPENYLOXY)ETHYL]-1H-IMIDAZOLE see IMAZALIL

2-(3,4-DICHLOROPHENYL)-4-METHYL-1,2,4-OXADIAZOLIDINE-3,5-DIONE see METHAZOLE

1-[2-(2,4-DICHLOROPHENYL)-4-PROPYL-1,3-DIOXOLAN-2-YL]-METHYL-1H-1,2,4,-TRIAZOLE see PROPICONAZOLE

3-(3,5-DICHLOROPHENYL)-5-ETHENYL-5-METHYL-2,4-OXAZOLIDINEDIONE see VINCLOZOLIN

N-(3,4-DICHLOROPHENYL)PROPANAMIDE see PROPANIL

(DICHLOROPHENYL)TRICHLOROSILANE see TRICHLORO(DICHLOROPHENYL)SILANE

2,4-DICHLOROPHENYL 4-NITROPHENYL ETHER see NITROFEN

DICHLOROPHENYL ISOCYANATE see ISOCYANIC ACID, 3,4-DICHLOROPHENYL ESTER

2,4-DICHLOROPHENYL p-NITROPHENYL ETHER see NITROFEN

(E)-1,3-DICHLOROPROPENE see trans-1,3-DICHLOROPROPENE

1,3-DICHLOROPROPENE see 1,3-DICHLOROPROPYLENE

trans-1,3-DICHLOROPROPYLENE see trans-1,3-DICHLOROPROPENE

DICHLOROPROPYLENE see DICHLOROPROPENE

2,3-DICHLOROPROPYLENE see 2,3-DICHLOROPRENE

2,2′-DICHLOROTRIETHYLAMINE see ETHYLBIS(2-CHLOROETHYL)AMINE

2,4-DICHLORPHENOXYACETIC ACID see 2,4-D

DICOBALT CARBONYL see COBALT CARBONYL

1,4-DICYANOBUTANE see ADIPONITRILE

1,3-DICYANOTETRACHLOROBENZENE see CHLOROTHALONIL

DICYCLOHEXYLMETHANE-4,4′-DIISOCYANATE see 1,1-METHYLENE BIS(4-ISOCYANATOCYCLOHEXANE)

DIDRIN see DICROTOPHOS

(DIETHOXYPHOSPHINYLIMINO)-1,3-DITHIETANE see FOSTHIETAN

DI-(2-ETHYLHEXYL)ADIPATE see BIS(2-ETHYLHEXYL)ADIPATE

DI-2-ETHYLHEXYL ADIPATE see BIS(2-ETHYLHEXYL)ADIPATE

o-(2-(DIETHYLAMINO)-6-METHYL-4-PYRIMIDINYL)O,O-DIETHYL PHOSPHOROTHIOATE see PIRIMFOS-ETHYL

O-(2-(DIETHYLAMINO)-6-METHYL-4-PYRIMIDINYL)-O,O-DIMETHYL PHOSPHOROTHIOATE see PIRIMIPHOS METHYL

(DIETHYLAMINO)ETHANE see TRIETHYLAMINE

2-(2-DIETHYLAMINO)ETHYL)-O,O-DIETHYL ESTER, OXALATE (1:1) see AMITON OXALATE

O-(2-DIETHYLAMINO-6-METHYLPYRIMIDIN-4-YL) O,O-DIMETHYL PHOSPHOROTHIOATE see PIRIMIPHOS METHYL

DIETHYL 1,2-BENZENEDICARBOXYLATE see DIETHYL PHTHALATE

DIETHYLCARBAMAZINE ÁCID CITRATE see DIETHYLCARBAMAZINE CITRATE

O,O-DIETHY-S-P-CHLOROPHENYLTHIOMETHYL DITHIOPHOSPHATE see CARBOPHENOTHION

α,α′-DIETHYL-(E)-4,4′-D-STILBENEDIOL see DIETHYLSTILBESTROL

O,O-(DIETHYL-O-2,4,5-DICHLORO(METHYLTHIO)PHENYL THIONO-PHOSPHATE see CHLORTHIOPHOS

DIETHYL-S-2-DIETHYLAMINOETHYL PHOSPHOROTHIOATE see AMI-TON

O,O-DIETHYLDITHIOPHOSPHORYLACETIC ACID-N-MONOISOPRO-PYLAMIDE see PROTHOATE

1,4-DIETHYLENE DIOXIDE see 1,4-DIOXANE

N,N-DIETHYLETHANEAMINE see TRIETHYLAMINE

O,O-DIETHYL S-(2-ETHTHIOETHYL)PHOSPHORODITHIOATE see DI-SULFOTON

O,O-DIETHYL S-((2-ETHTYLTHIO)ETHYL)ESTER see DISULFOTON

O,O-DIETHYL-2-ETHYLMERCAPTOETHYL THIOPHOSPHATE, DIETH-OXYTHIOPHOSPHORIC ACID see DEMETON

O,O-DIEYHYL-S-((ETHYLSULFINYL)ETHYL)PHOSPHORODITHIOATE see OXYDISULFOTON

O,O-DIETHYLETHYLTHIOMETHYL PHOSPHORODITHIOATE see PHO-RATE

1,2-DIETHYLHYDRAZINE see HYDRAZINE, 1,2-DIETHYL-

O,O-DIETHYL O-2-ISOPROPYL-6-METHYLPYRIMIDIN-4-YLPHOSPHO-ROTHIONATE see DIAZINON

O,O-DIETHYL-S-(N-ISOPROPYLCARBAMOYLMETHYL)DITHIO-PHOSPHATE see PROTHOATE

DIETHYL MERCAPTOSUCCINATE, S-ESTER WITH O,O-DIMETHYL PHOSPHORODITHIOATE see MALATHION

N,N-DIETHYL-4-METHYL-1-PIPERAZINE CARBOXAMIDE CITRATE see DIETHYLCARBAMAZINE CITRATE

DIETHYL(4-METHYL-1,3-DITHIOLAN-2-YLIDENE)PHOSPHOROAMIDATE see MEPHOSFOLAN

O,O-DIETHYL O-(6-METHYL-2-(1-METHYLETHYL)-4-PYRIMIDINYL) PHOSPHORTHIOATE see DIAZINON

O,O-DIETHYL O-(P-NITROPHENYL) PHOSPHOROTHIOATE see PARATHION

DIETHYLNITROSAMIDE see N-NITROSODIETHYLAMINE

N,N-DIETHYLNITROSOAMINE see N-NITROSODIETHYLAMINE

O,O-DIETHYLO-(1-PHENYL-1H-1,2,4-TRIAZOL-3-YL)PHOSPHOROTHIOATE 1-PHENYL-3-(O,O-DIETHYL-THIONOPHOPHORYL)-1,2,4-TRIAZOLE see TRIAZOFOS

O,O-DIETHYLO(P-(METHYLSULFINYL)PHENYL)PHOSPHOROTHIOATE see FENSULFOTHION

O,O-DIETHYL-o-PYRAZINYL PHOSPHOROTHIOATE see ZINOPHOS

DIETHYLPHENYLAMINE see N,N-DIETHYLANILINE

O,O-DIETHYL-O,2-PYRAZINYL PHOSPHOROTHIOATE see ZINOPHOS

DIETHYL SULPHATE see DIETHYL SULFATE

DIFLUORO-1-CHLOROETHANE see 1-CHLORO-1,1-DIFLUOROETHANE

DIFLUOROCHLOROMETHANE see CHLORODIFLUOROMETHANE

DIFLUORODICHLOROMETHANE see DICHLORODIFLUOROMETHANE

1,1-DIFLUOROETHYLENE see VINYLIDENE FLUORIDE

DIFLUOROMONOCHLOROETHANE see 1-CHLORO-1,1-DIFLUOROETHANE

DIFLUOROMONOCHLOROMETHANE see CHLORODIFLUOROMETHANE

1,1-DIFLUROETHANE see DIFLUOROETHANE

DIFONATE see FONFOS

1,3-DIGLYCIDYLOXYBENZENE see DIGLYCIDYL RESORCINOL ETHER

DIHYDRO-1-AZIRINE see ETHYLENEIMINE

DIHYDROAZIRINE see ETHYLENEIMINE

9,10-DIHYDRO-8A,10A-DIAZONIAPHENANTHRENE(1,1′-ETHYLENE-2,2′-BIPYRIDYLIUM)DIBROMIDE see DIQUAT

2,3-DIHYDRO-2,2-DIMETHYL-7-BENZOFURANOLMETHYLCARBAMATE see CARBOFURAN

2,3,-DIHYDRO-5,6-DIMETHYL-1,4-DITHIIN-1,1,4,4-TETRAOXIDE see DIMETHIPIN

DIHYDRO-2,5-DIOXOFURAN see MALEIC ANHYDRIDE

5,6-DIHYDRO-DIPYRIDO(1,2A,2,1C)PYRAZINIUM DIBROMIDE see DIQUAT

1,2-DIHYDRO-2-KETOBENZISOSULFONAZOLE see SACCHARIN

1,2-DIHYDRO-3-METHYL-BENZ(j)ACEANTHRYLENE see 3-METHYLCHLOANTHRENE

2,3-DIHYDRO-6-METHYL-1,4-OXATHIIN-5-CARBOXANILIDE see CARBOXIN

5,6-DIHYDRO-2-METHYL-N-PHENYL-1,4-OXATHIIN-3-CARBOXAMIDE see CARBOXIN

3,4-DIHYDRO-4-OXO-3-BENZOTRIAZINYLMETHYL O,O-DIETHYL PHOSPHORODITHIOATE see AZINPHOS-ETHYL

DIHYDROPENTABORANE(9) see PENTABORANE

1,2-DIHYDROPYRIDAZINE-3,6-DIONE see MALEIC HYDRAZIDE

1,2-DIHYDRO-3,6-PYRIDAZINEDIONE see MALEIC HYDRAZIDE

6,7-DIHYDROPYRIDO (1,2-a:2′,1′-c)PYRAZINEDIIUMION see DIQUAT

1,2-DIHYDROXYBENZENE see CATECHOL

p-DIHYDROXYBENZENE see HYDROQUINONE

1,4-DIHYDROXYBENZENE see HYDROQUINONE

2,2′-DIHYDROXY-3,3′-DIMETHYL-5,5′-DICHLORODIPHENYL SULFIDE see PHENOL, 2,2′-THIOBIS(4-CHLORO-6-METHYL)-

1,2-DIHYDROXYETHANE see ETHYLENE GLYCOL

2,2′-DIHYDROXY-N-NITROSODIETHYLAMINE see N-NITROSODIETHANOLAMINE

DI-ISO-CYANATOLUENE see TOLUENE 2,4-DIISOCYANATE

4,4′-DIISOCYANATO-3,3′-DIMETHYL-1,1′-BIPHENYL see 3,3′-DIMETHYL-4,4′-DIPHENYLENE DIISOCYANATE

4,4′-DIISOCYANATO-3,3′-DIMETHYLDIPHENYLMETHANE see 3,3′-DIMETHYLDIPHENYLMETHANE-4,4′-DIISOCYANATE

2,4-DIISOCYANATO-1-METHYLBENZENE (9CI) see TOLUENE 2,4-DIISOCYANATE

1,6-DIISOCYANATO-2,2,4-TRIMETHYLHEXANE see 2,2,4-TRIMETHYL HEXAMETHYLENE DIISOCYANATE

1,6-DIISOCYANATO-2,4,4-TRIMETHYLHEXANE see 2,4,4-TRIMETHYLHEXAMETHYLENE DIISOCYANATE

DIISOCYANATOTOLUENE see TOLUENEDIISOCYANATE (MIXED ISOMERS)

2,4-DIISOCYANATOTOLUENE see TOLUENE 2,4-DIISOCYANATE

2,6-DIISOCYANATOTOLUENE see TOLUENE-2,6-DIISOCYANATE

O,O-DIISOPROPYLFLUOROPHOSPHATE see DIISOPROPYLFLUOROPHOSPHATE

N,N-DI-ISOPROPYL-6-METHYLTHIO-1,3,5-TRIAZINE-2,4-DIAMINE see PROMETHRYN

1,4-DIISOTHIOCYANATOBENZENE see BITOSCANATE

DILITHIUM CHROMATE see LITHIUM CHROMATE

1,4:5,8-DIMETHANONAPHTHALENE,1,2,3,4,10,10-HEXACHLORO-1,4,4a,5,8,8a-HEXAHYDRO-(1α,4α,4β,5α,8α,8β)- see ALDRIN

2,7:3,6-DIMETHANONAPHTH(2,3-B)OXIRENE,3,4,5,6,9,9-HEXACHLORO-1A,2,2A,3,6,6A,7,7A-OCTAHYDRO-,(Aα,2β,2Aβ,2Aβ,3α,6α,6Aβ,7β,7Aα)- see ENDRIN

2,7:3,6-DIMETHANONAPHTH[2,3B]OXIRENE,3,4,5,6,9,9-HEXACHLORO-1A,2,2A,3,6,6A,7,7A-OCTAHYDRO-(1 Aα,2β,2Aα,3β,6β,6Aα,7β,7A α) see DIELDRIN

3,3′-DIMETHOXYBENZIDINE DIHYDROCHLORIDE see 3,3′-DIMETHOXYBENZIDINE HYDROCHLORIDE

3,3′-DIMETHOXYBIPHENYL-4,4′-YLENEDIAMMONIUM DICHLORIDE see 3,3′-DIMETHOXYBENZIDINE DIHYDROCHLORIDE

3,3′-DIMETHOXY-4,4′-DIAMINODIPHENYL see 3,3′-DIMETHOXYBENZIDINE

p,p′-DIMETHOXYDIPHENYLTRICHLOROETHANE see METHOXYCHLOR

3-(DIMETHOXYPHOSPHINYLOXY)N-METHYL-CIS-CROTONAMIDE see MONOCROPTOPHOS

N,N′-DIMETHYHYDRAZINE see HYDRAZINE, 1,2-DIMETHYLDIMETHOXY STRYCHNINE see BRUCINE

O,S-DIMETHYL ACETYLPHOSPHORAMIDOTHIOATE see ACEPHATE

2-(DIMETHYLAMINO)-N(((METHYLAMINO)CARBONYL)OXY)2-OXO-ETHANIMIDOTHIOIC ACID METHYL ESTER see OXAMYL

p-(DIMETHYLAMINO)AZOBENZENE see 4-DIMETHYLAMINOAZOBENZENE

(DIMETHYLAMINO)BENZENE see N,N-DIMETHYLANILINE

m-(((DI-METHYLAMINO)METHYLENE)AMINO)PHENYLCARBAMATE, HYDROCHLORIDE see FORMETANATE HYDROCHLORIDE

o-(4-((DIMETHYLAMINO)SULFONYL)PHENYL) O,O-DIMETHYL PHOSPHOROTHIOATE see FAMPHUR

DIMETHYLAMINOAZOBENZENE see 4-DIMETHYLAMINOAZOBENZENE

2′,3-DIMETHYL-4-AMINOAZOBENZENE see C.I. SOLVENT YELLOW 3

DIMETHYLAMINOCARBONYL CHLORIDE see DIMETHYLCARBAMOYL CHLORIDE

2,6-DIMETHYLANILINE see 2,6-XYLIDINE

DIMETHYLANILINE see N,N-DIMETHYLANILINE

DIMETHYLARSINIC ACID see CACODYLIC ACID

2,6-DIMETHYLBENZENAMINE see 2,6-XYLIDINE

DIMETHYLBENZENE see XYLENE (MIXED ISOMERS)

1,2-DIMETHYLBENZENE see o-XYLENE

1,3-DIMETHYLBENZENE see m-XYLENE

1,4-DIMETHYLBENZENE see p-XYLENE

m-DIMETHYLBENZENE see m-XYLENE

o-DIMETHYLBENZENE see o-XYLENE

DIMETHYL 1,2-BENZENEDICARBOXYLATE see DIMETHYL PHTHALATE

3,3′-DIMETHYLBENZIDINE see o-TOLIDINE

2,2-DIMETHYL-1,3-BENZODIOXOL-4-YL-N-METHYLCARBAMATE see BENDIOCARB

2,2-DIMETHYL-1,3-BENZODIOXOL-4-OL METHYLCARBAMATE see BENDIOCARB

α,α-DIMETHYLBENZYL HYDROPEROXIDE see CUMENE HYDROPEROXIDE

3,3′-((3,3′-DIMETHYL(1,1′-BIPHENYL)-4,4′-DIYL)BIS(AZO))BIS(5-AMINO-4-HYDROXYNAPHTHALENE-2,7-DISULPHONATE) see TRYPAN BLUE

3,3′-DIMETHYL-4,4′-BIPHENYLDIAMINE see o-TOLIDINE

DIMETHYLCARBAMIC see DIMETILAN

DIMETHYLCARBAMOYL CHLORIDE see DIMETHYLCARBAMOYL CHLORIDE

N,N-DIMETHYLCARBAMOYL CHLORIDE see DIMETHYLCARBAMOYL CHLORIDE

DIMETHYLCARBAMYL CHLORIDE see DIMETHYLCARBAMOYL CHLORIDE

DIMETHYLCARBINOL see ISOPROPYL ALCOHOL

DIMETHYLCHLOROETHER see CHLOROMETHYL METHYL ETHER

O,O-DIMETHYL CHLOROTHIONOPHOSPHATE see DIMETHYL CHLOROTHIOPHOSPHATE

O,O-DIMETHYL-O-4-CYANOPHENYL-PHOSPHOROTHIOATE see CYANOPHOS

DIMETHYL 1,2-DIBROMO-2,2-DICHLOROETHYL PHOSPHATE see NALED

O,O-DIMETHYL-O-(1,2-DIBROMO-2,2-DICHLOROETHYL)PHOSPHATE see NALED

DIMETHYL 2,2-DICHLOROETHENYL PHOSPHATE see DICHLORVOS

O,O-DIMETHYL S-(2,5-DICHLOROPHENYLTHIO)METHYL PHOSPHORODITHIOATE see METHYL PHENKAPTON

DIMETHYL O,O-DICHLOROVINYL-2,2-PHOSPHATE see DICHLORVOS

O,O-DIMETHYL 2,2-DICHLOROVINYL PHOSPHATE see DICHLORVOS

N,N-DIMETHYL-2,2-DIPHENYLACETAMIDE see DIPHENAMID

N,N-DIMETHYLDIPHENYLACETAMIDE see DIPHENAMID

2,4-DIMETHYL-1,3-DITHIOLANE-2-CARBOXALDEHYDE O-((METHYLAMINO)CARBONYL)OXIME see TRIPATE

2,4-DIMETHYL-1,3-DITHIOLANE-2-CARBOXALDEHYDE O-(METHYLCARBAMOYL)OXIME see TRIPATE

DIMETHYLENEDIAMINE see ETHYLENEDIAMINE

DIMETHYLENEIMINE see ETHYLENEIMINE

DIMETHYLENIMINE see ETHYLENEIMINE

1,1-DIMETHYLETHANE see ISOBUTANE

DIMETHYL ETHER see METHYL ETHER

O,O-DIMETHYL-S-(2-ETHTHIOETHYL)PHOSPHOROTHIOATE see DEMETON-s-METHYL

N-[5-(1,1-DIMETHYLETHYL)-1,3,4-THIADIAZOL-2-YL])-N,N′-DIMETHYLUREA see TEBUTHIURON

2-(4-(1,1-DIMETHYLETHYL)PHENOXY)CYCLOHEXYL 2-PROPYNYL SULFUROUS ACID see PROPARGITE

1,1-DIMETHYLETHYL METHYL ETHER see METHYL tert-BUTYL ETHER

1,1-DIMETHYLETHYLAMINE see tert-BUTYLAMINE

DIMETHYLFORMALDEHYDE see ACETONE

N,N-DIMETHYLFORMAMIDE see DIMETHYLFORMAMIDE

O,O-DIMETHYL-S-(N-FORMYL-N-METHYLCARBAMOYLMETHYL) PHOSPHORODITHIOATE see FORMOTHION

DIMETHYLHYDRAZINE see 1,1-DIMETHYL HYDRAZINE

asym-DIMETHYLHYDRAZINE see 1,1-DIMETHYL HYDRAZINE

N,N-DIMETHYLHYDRAZINE see 1,1-DIMETHYL HYDRAZINE

1,2-DIMETHYLHYDRAZINE see HYDRAZINE, 1,2-DIMETHYL-

N,N-DIMETHYLMETHANAMIDE see DIMETHYLFORMAMIDE

N,N-DIMETHYLMETHANAMINE see TRIMETHYLAMINE

DIMETHYL METHANE see PROPANE

O,O-DIMETHYL-S-(5-METHOXY-4-OXO-4H-PYRAN-2-YL)PHOSPHOROTHIOATE see ENDOTHION

O,O-DIMETHYL)-S-(2-METHOXY-1,3,4-THIADIAZOLE-5(4H)-ONYL-(4)-METHYL)-PHOSPHORODITHIOATE see METHIDATHION

O,O-DIMETHYL METHYLCARBAMOYLMETHYL PHOSPHORODITHIOATE see DIMETHOATE

N,N-DIMETHYL-N'(2-METHYL-4(((METHYLAMINO)CARBONYL) OXY)PHENYL)METHANIMIDAMIDE see FORMPARANATE

O,O-DIMETHYL-O-(3-METHYL-4-NITROPHENYL)-PHOSPHOROTHIOATE see FENITROTHION

2,2-DIMETHYL-3-(2-METHYL-1-PROPENYL)CYCLOPROPANECARBOXYLIC ACID (1,3,4,5,6,7-HEXAHYDRO-1,3-DIOXO-2H-ISOINDOL-2-YL)METHYL ESTER see TETRAMETHRIN

2,2-DIMETHYL-3-(2-METHYL-1-PROPENYL)CYCLOPROPANECARBOXYLIC ACID (3-PHENOXYPHENYL)METHYL ESTER see PHENOTHRIN

O,O-DIMETHYL O-[3-METHYL-4-(METHYLTHIO)PHENYL] ESTER, PHOSPHOROTHIOIC ACID see FENTHION

3,3-DIMETHYL-1-(METHYLTHIO)-2-BUTANONE-o-((METHYLAMINO)CARBONYL) OXIME see THIOFANOX

3,5-DIMETHYL-4-(METHYLTHIO)PHENOL METHYLCARBAMATE see METHIOCARB

N,N'-((METHYLIMINO)DIMETHYLIDYNE)BIS(2,4-XYLIDINE) see AMITRAZ

O,O-DIMETHYL O-(P-NITROPHENYL) PHOSPHOROTHIOATE see METHYL PARATHION

DIMETHYLNITROSAMINE see N-NITROSODIMETHYLAMINE

2,3-DIMETHYL-7-OXABICYCLO(2,2,1)HEPTANE-2,3-DICARBOXYLIC ANHYDRIDE see CANTHARIDIN

O,O-DIMETHYL-S-(4-OXO-1,2,3-BEZOTRIAZIN-3(4H)-YL METHYL)PHOSPHORODITHIOATE see AZINPHOS-METHYL

O,O-DIMETHYLO-(4-METHYLMERCAPTOPHENYL)PHOSPHATE see PHOSPHORIC ACID, DIMETHYL 4-(METHYLTHIO)PHENYL ESTER

3(2-(3,5-DIMETHYL-2-OXOCYCLOHEXYL)-2-HYDROXYETH-YL)GLUTARIMIDE see CYCLOHEXIMIDE

α,α-DIMETHYLPHENETHYLAMINE see BENZENEETHANAMINE, α,α-DI-METHYL-DIMETHYLPHENOL see XYLENOL

N,N-DIMETHYLPHENYLAMINE see N,N-DIMETHYLANILINE

N,N-DIMETHYL-p-PHENYLENEDIAMINE see DIMETHYL-p-PHENYL-ENEDIAMINE

1,5-DI-(2,4-DIMETHYLPHENYL)-3-METHYL-1,3,5-TRIAZAPENTA-1,4-DIENE see AMITRAZ

1,5-DI-(2,4-DIMETHYLPHENYL)-3-METHYL-1,3,5-TRIAZAPENTA-1,4-DIENE BIURET, 2,4,-DITHIO see DITHIOBIURET

O,S-DIMETHYLPHOSPHORAMIDOTHIOATE see METHAMIDOPHOS

DIMETHYL PHOSPHOROCHLOROTHIOATE see DIMETHYL CHLORO-THIOPHOSPHATE

O,O-DIMETHYL PHOSPHORODITHIOATE, S-ESTER with 4-(MERCAPTO-METHYL)-2-METHOXY-2-OXO-1,3,4-THIADIAZOLIN-5-ONE see METHIDATHION

N,N-DIMETHYLPHTHALATE see DIMETHYL PHTHALATE

DIMETHYL O-PHTHALATE see DIMETHYL PHTHALATE

(O,O-DIMETHYL-PHTHALIMIDIOMETHYL-DITHIOPHOSPHATE) see PHOSMET

N,N-DIMETHYL-N'-2-PYRIDINYL-N'-(2-THIENYLMETHYL)-1,2-ETH-ANEDIAMIDE see METHAPYRILENE

3,5-DIMETHYL-1,2,3,5-TETRAHYDRO-1,3,5-THIADIAZINETHIONE-2 see DAZOMET

DIMETHYL N,N'-(THIOBIS((METHYLIMINO)CARBONYLOXY))BIS (THIOIMIDOACETATE) see THIODICARB

DIMETHYL THIOPHOSPHORYL CHLORIDE see DIMETHYL CHLORO-THIOPHOSPHATE

O,O-DIMETHYL THIOPHOSPHORYL CHLORIDE see DIMETHYL CHLO-ROTHIOPHOSPHATE

O,O-DIMETHYL-O-(3,5,6-TRICHLORO-2-PYRIDYL)PHOSPHORO-THIOATE see CHLORPYRIFOS METHYL

O,O-DIMETHYL O-(3,5,6-TRICHLORO-2-PYRIDINYL) PHOSPHORO-THIOATE see CHLORPYRIFOS

N,N-DIMETHYL-N'-[3-(TRIFLUOROMETHYL)PHENYL]UREA see FLUOMETURON

DINEX see 2-CYCLOHEXYL-4,6-DINITROPHENOL

1,2-DINITROBENZENE see o-DINITROBENZENE

1,3-DINITROBENZENE see m-DINITROBENZENE

DINITROBUTYL PHENOL see DINITROBUTYL PHENOL

2,4-DINITRO-6-tert-BUTYLPHENOL see DINOTERB

DINITRO-ortho-CRESOL see 4,6-DINITRO-o-CRESOL

DINITRO-o-CYCLOHEXYLPHENOL see 2-CYCLOHEXYL-4,6-DINITRO-PHENOL

2,6-DINITRO-N,N-DI-N-PROPYL-α,α,α-TRIFLURO-P-TOLUIDINE see TRIFLURALIN

BENZENE, 1,2-DINITRO- see o-DINITROBENZENE

2,6-DINITRO-N,N-DIPROPYL-4-(TRIFLUOROMETHYL)ANILINE see TRIFLURALIN

2,6-DINITRO-N,N-DIPROPYL-4-(TRIFLUOROMETHYL)BENZENAMINE see TRIFLURALIN

DINITROGEN DIOXIDE see NITROGEN DIOXIDE

DINITROGEN TETROXIDE see NITROGEN DIOXIDE

4,6-DINITRO-2-METHYLPHENOL see 4,6-DINITRO-o-CRESOL

4,6-DINITRO-2-(1-METHYL-N-PROPYL)PHENOL see DINITROBUTYL PHENOL

β-DINITROPHENOL see 2,6-DINITROPHENOL

γ-DINITROPHENOL see 2,5-DINITROPHENOL

DINOSEB see DINITROBUTYL PHENOL

DIOCTYL ADIPATE see BIS(2-ETHYLHEXYL)ADIPATECYCLOHEXANE, 1,4-BIS(ISOCYANATOMETHYL)- see 1,4-BIS(METHYLISOCYANATE) CYCLOHEXANE

DIOCTYL PHTHALATE see DI(2-ETHYLHEXYL)PHTHALATE

p-DIOXANE see 1,4-DIOXANE

2,3-DIOXANEDITHIOL S,S-BIS(O,O-DIETHYLPHOSPHORODITHIOATE) see DIOXATHION

1,4-DIOXIN, TETRAHYDRO- see 1,4-DIOXANE

2,4-DIOXO-5-FLUOROPYRIMIDINE see FLUOROURACIL

1,4-DIOXYBENZENE see QUINONE

DIOXYBUTADIENE see DIEPOXYBUTANE

DIPHENAMIDE see DIPHENAMID

DIPHENYL see BIPHENYL

1,1′-DIPHENYL see BIPHENYL

2-(DIPHENYLACETYL)-1H-INDENE-1,3(2H)-DIONE see DIPHACIONE

N,N-DIPHENYLAMINE see DIPHENYLAMINE

DIPHENYLAMINE, N-NITROSO- see N-NITROSODIPHENYLAMINE

DIPHENYLAN see PHENYTOIN

DIPHENYLHYDANTOIN see PHENYTOIN

5,5-DIPHENYLHYDANTOIN see PHENYTOIN

N,N′-DIPHENYLHYDRAZINE see 1,2-DIPHENYLHYDRAZINE

4,4′-DIPHENYLMETHANE DIISOCYANATE see METHYL-BIS(PHENYLISOCYANATE)

DIPHENYLNITROSAMINE see N-NITROSODIPHENYLAMINE

DIPHOSPHORIC ACID, TETRAETHYL ESTER see TEPP

DIPHOSPHOROUS PENTASULPHIDE see SULFUR PHOSPHIDE

DIPOTASSIUM CHROMATE see POTASSIUM CHROMATE

DIPROPYLAMINE, N-NITROSO- see N-NITROSODI-N-PROPYLAMINE

4-(DI-N-PROPYLAMINO)-3,5-DINITRO-1-TRIFLUOROMETHYLBENZENE see TRIFLURALIN

4-(DIPROPYLAMINO)-3,5-DINITROBENZENESULFONAMIDE) see ORYZALIN

N,N-DI-N-PROPYL-2,6-DINITRO-4-TRIFLUOROMETHYLANILINE see TRIFLURALIN

DIPROPYLNITROSAMINE see N-NITROSODI-N-PROPYLAMINE

DIPROPYL 2,5-PYRIDINEDICARBOXYLATE see DIPROPYL ISOCINCHOMERONATE

N,N-DIPROPYL-4-TRIFLUOROMETHYL-2,6-DINITROANILINE see TRIFLURALIN

DIPTEREX see TRICHLORFON

DIPYRIDO(1,2-A:2′,1′-C)PYRAZINEDIUM, 6,7-DIHYDRO-, DIBROMIDE see DIQUAT

DIPYRIDO(1,2-a:2′,1′-c)PYRAZINEDIIUM, 6,7-DIHYDRO- see DIQUAT

DIQUAT, DIBROMIDE see DIQUAT

DIRECT BLACK 38 see C.I. DIRECT BLACK 38

DIRECT BLUE 6 see C.I. DIRECT BLUE 6

DIRECT BROWN 95 see C.I. DIRECT BROWN 95

DI-SEC-OCTYL PHTHALATE see DI(2-ETHYLHEXYL)PHTHALATE

DISODIUM 8-((3,3′-DIMETHYL-4′-(4-(4-METHYLPHENYLSULPHONYLOXY)PHENYLAZO)(1,1′-BIPHENYL)-4-YL)AZO)-7-HYDROXYNAPHTHALENE-1,3-DISULPHONATE see C.I. ACID RED 114

DISODIUM PHOSPHATE see SODIUM PHOSPHATE, DIBASIC

DISODIUM SELENATE see SODIUM SELINATE

DISODIUM SELENITE see SODIUM SELINITE

DISPERSE YELLOW 3 see C.I. DISPERSE YELLOW 3

DISULFATOZIRCONIC ACID see ZIRCONIUM SULFATE

DISULFUR DICHLORIDE see SULFUR MONOCHLORIDE

DISULPHURIC ACID see OLEUM (FUMING SULFURIC ACID)

DI-SYSTON see DISULFOTON

p-DITHIANE, 2,3-DEHYDRO-2,3-DIMETHYL-, TETROXIDE see DIMETHIPIN

α,α′-DITHIOBIS(DIMETHYLTHIO)FORMAMIDE see THIRAM

2,4-DITHIOBIURET see DITHIOBIURET

DITHIONOUS ACID, ZINC SALT (1:1) see ZINC HYDROSULFITE

1,3-DITHOLANE-2-CARBOXYALDEHYDE,2,4-DIMETHYL,O-(METHYLCARBAMOYL)OXIME see TRIPATE

DIVANADIUM PENTOXIDE see VANADIUM PENTOXIDE

DMBA see 7,12-DIMETHYLBENZ(A)ANTHRACENE

7,12-DMBA see 7,12-DIMETHYLBENZ(A)ANTHRACENE

DMSP see FENSULFOTHION

1,2-DNB see o-DINITROBENZENE

DNOC see 4,6-DINITRO-o-CRESOL

DODECYLBENZENESULFONIC ACID, CALCIUM SALT see CALCIUM DODECYLBENZENESULFONATE

DODECYLBENZENESULFONIC ACID, TRIETHANOLAMINE SALT see TRIETHANOLAMINE DODECYLBENZENE SULFONATE

DODECYLBENZENESULPHONATE, SODIUM SALT see SODIUM DODECYLBENZENESULFONATE

DODECYLGUANIDINE MONOACETATE see DODINE

1-DODECYLGUANIDINIUM ACETATE see DODINE

DTMC see DICOFOL

DURSBAN METHYL see CHLORPYRIFOS METHYL

DURSBAN see CHLORPYRIFOS

DYFONATE see FONFOS

DYLOX see TRICHLORFON

EBDCs see ETHYLENEBISDITHIOCARBAMIC ACID

EGEE see 2-ETHOXYETHANOL

EGME see 2-METHOXYETHANOL

EGMME see 2-METHOXYETHANOL

ELAYL see ETHYLENE

ELEMENTAL SELENIUM see SELENIUM

ELEMENTAL SODIUM see SODIUM

1-EMETINE, DIHYDROCHLORIDE see EMETINE, DIHYDROCHLORIDE

EMETINE, HYDROCHLORIDE see EMETINE, DIHYDROCHLORIDE

3,6-ENDOOXOHEXAHYDROPHTHALIC ACID see ENDOTHALL

ENDOSULFAN I see α-ENDOSULFAN

ENDOSULFAN II see β-ENDOSULFAN

ENDOTHALL DIPOTASSIUM SALT see DIPOTASSIUM ENDOTHALL

ENT PHOSPHOROTHIOATE see FENSULFOTHION

EPA PESTICIDE CHEMICAL Code 004003 see d-trans-ALLETHRIN

EPA PESTICIDE CHEMICAL Code 004401 see AMITROLE

EPA PESTICIDE CHEMICAL Code 005101 see PICLORAM

EPA PESTICIDE CHEMICAL Code 008701 see BROMINE

EPA PESTICIDE CHEMICAL Code 012302 see BROMACIL, LITHIUM SALT

EPA PESTICIDE CHEMICAL Code 012701 see TERBACIL

EPA PESTICIDE CHEMICAL Code 014601 see METIRAM

EPA PESTICIDE CHEMICAL Code 015801 see MEVINPHOS

EPA PESTICIDE CHEMICAL Code 017901 see 1-(3-CHLORALLYL)-3,5,7-TRIAZA-1-AZONIAADAMANTANE CHLORIDE

EPA PESTICIDE CHEMICAL Code 019101 see PROPACHLOR 2-CHLORO-N-(1-METHYLETHYL)-N-PHENYLACETAMIDE

EPA PESTICIDE CHEMICAL Code 028201 see PROPANIL

EPA PESTICIDE CHEMICAL Code 029801 see DICAMBA

EPA PESTICIDE CHEMICAL Code 029802 see DIMETHYLAMINE DICAMBA

EPA PESTICIDE CHEMICAL Code 030004 see 2,4-D SODIUM SALT

EPA PESTICIDE CHEMICAL Code 030053 see 2,4-D BUTOXYETHYL ESTER

EPA PESTICIDE CHEMICAL Code 030063 see 2,4-D 2-ETHYLHEXYL ESTER

EPA PESTICIDE CHEMICAL Code 030064 see 2,4-D 2-ETHYL-4-METHYLPENTYL ESTER

EPA PESTICIDE CHEMICAL Code 030501 see METHOXONE

EPA PESTICIDE CHEMICAL Code 030502 see METHOXONE, SODIUM SALT

EPA PESTICIDE CHEMICAL Code 030801 see 2,4-DB

EPA PESTICIDE CHEMICAL Code 031301 see DICHLORAN

EPA PESTICIDE CHEMICAL Code 031401 see 2,4-DP

EPA PESTICIDE CHEMICAL Code 034801 see FERBAM

EPA PESTICIDE CHEMICAL Code 034803 see POTASSIUM DIMETHYLDITHIOCARBAMATE

EPA PESTICIDE CHEMICAL Code 034804 see SODIUM DIMETHYLDITHIOCARBAMATE

EPA PESTICIDE CHEMICAL Code 035001 see DIMETHOATE

EPA PESTICIDE CHEMICAL Code 035301 see BROMOXYNIL

EPA PESTICIDE CHEMICAL Code 035302 see BROMOXYNIL OCTANOATE

EPA PESTICIDE CHEMICAL Code 035501 see MONURON

EPA PESTICIDE CHEMICAL Code 035505 see DIURON

EPA PESTICIDE CHEMICAL Code 035506 see LINURON

EPA PESTICIDE CHEMICAL Code 035602 see DAZOMET

EPA PESTICIDE CHEMICAL Code 036001 see DINOCAP

EPA PESTICIDE CHEMICAL Code 036601 see DIPHENAMID

EPA PESTICIDE CHEMICAL Code 037505 see DINITROBUTYL PHENOL

EPA PESTICIDE CHEMICAL Code 038501 see DIPHENYLAMINE

EPA PESTICIDE CHEMICAL Code 038904 see DIPOTASSIUM ENDOTHALL

EPA PESTICIDE CHEMICAL Code 039002 see POTASSIUM N-METHYLDITHIOCARBAMATE

EPA PESTICIDE CHEMICAL Code 041101 see ETHOPROP

EPA PESTICIDE CHEMICAL Code 041301 see CYCLOATE

EPA PESTICIDE CHEMICAL Code 041401 see ETHYL DIPROPYLTHIOCARBAMATE

EPA PESTICIDE CHEMICAL Code 041402 see MOLINATE

EPA PESTICIDE CHEMICAL Code 044301 see DODINE

EPA PESTICIDE CHEMICAL Code 047201 see DIPROPYL ISOCINCHOMERONATE

EPA PESTICIDE CHEMICAL Code 053301 see FENTHION

EPA PESTICIDE CHEMICAL Code 053501 see METHYL PARATHION

EPA PESTICIDE CHEMICAL Code 054101 see CHINOMETHIONAT

EPA PESTICIDE CHEMICAL Code 055001 see DICHLOROPHENE

EPA PESTICIDE CHEMICAL Code 057701 see MALATHION

EPA PESTICIDE CHEMICAL Code 057801 see DIAZINON

EPA PESTICIDE CHEMICAL Code 059001 see TEMEPHOS

EPA PESTICIDE CHEMICAL Code 059901 see FAMPHUR

EPA PESTICIDE CHEMICAL Code 060101 see THIABENDAZOLE

EPA PESTICIDE CHEMICAL Code 061601 see PARAQUAT DICHLORIDE

EPA PESTICIDE CHEMICAL Code 063003 see SODIUM PENTACHLOROPHENATE

EPA PESTICIDE CHEMICAL Code 063301 see DISODIUM CYANODITHIOIMIDOCARBONATE

EPA PESTICIDE CHEMICAL Code 064104 see SODIUM O-PHENYLPHENOXIDE

EPA PESTICIDE CHEMICAL Code 066501 see ALUMINUM PHOSPHIDE

EPA PESTICIDE CHEMICAL Code 067501 see PIPERONYL BUTOXIDE

EPA PESTICIDE CHEMICAL Code 068103 see METHYL ISOTHIOCYANATE

EPA PESTICIDE CHEMICAL Code 069003 see TETRAMETHRIN

EPA PESTICIDE CHEMICAL Code 069005 see PHENOTHRIN

EPA PESTICIDE CHEMICAL Code 074801 see S,S,S-TRIBUTYLTRITHIOPHOSPHATE

EPA PESTICIDE CHEMICAL Code 074901 see MERPHOS

EPA PESTICIDE CHEMICAL Code 076204 see SODIUM NITRITE

EPA PESTICIDE CHEMICAL Code 078003 see SULFURYL FLUORIDE

EPA PESTICIDE CHEMICAL Code 079801 see THIRAM

EPA PESTICIDE CHEMICAL Code 080805 see PROMETHRYN

EPA PESTICIDE CHEMICAL Code 080807 see SIMAZINE

EPA PESTICIDE CHEMICAL Code 080811 see ANILAZINE

EPA PESTICIDE CHEMICAL Code 081601 see FOLPET

EPA PESTICIDE CHEMICAL Code 083001 see BIS(TRIBUTYLTIN) OXIDE

EPA PESTICIDE CHEMICAL Code 083112 see TRIBUTYLTIN FLUORIDE

EPA PESTICIDE CHEMICAL Code 083120 see TRIBUTYLTIN METHACRYLATE

EPA PESTICIDE CHEMICAL Code 083601 see TRIPHENYLTIN HYDROXIDE

EPA PESTICIDE CHEMICAL Code 084301 see BENFLURALIN

EPA PESTICIDE CHEMICAL Code 090201 see CARBOXIN

EPA PESTICIDE CHEMICAL Code 090501 see ALACHLOR

EPA PESTICIDE CHEMICAL Code 097601 see PROPARGITE

EPA PESTICIDE CHEMICAL Code 097801 see RESMETHRIN

EPA PESTICIDE CHEMICAL Code 098301 see ALDICARB

EPA PESTICIDE CHEMICAL Code 100101 see CYANAZINE

EPA PESTICIDE CHEMICAL Code 100501 see METHIOCARB

EPA PESTICIDE CHEMICAL Code 101701 see PRONAMIDE

EPA PESTICIDE CHEMICAL Code 101801 see 2,2-DIBROMO-3-NITRILO-PROPIONAMIDE

EPA PESTICIDE CHEMICAL Code 102001 see THIOPHANATE-METHYL

EPA PESTICIDE CHEMICAL Code 103301 see ACEPHATE

EPA PESTICIDE CHEMICAL Code 103401 see THIOPHANATE ETHYL

EPA PESTICIDE CHEMICAL Code 104201 see ORYZALIN

EPA PESTICIDE CHEMICAL Code 104601 see FENBUTATIN OXIDE

EPA PESTICIDE CHEMICAL Code 104801 see DESMEDIPHAM

EPA PESTICIDE CHEMICAL Code 105501 see TEBUTHIURON

EPA PESTICIDE CHEMICAL Code 105801 see NORFLURAZON

EPA PESTICIDE CHEMICAL Code 106001 see METHAZOLE

EPA PESTICIDE CHEMICAL Code 106201 see AMITRAZ

EPA PESTICIDE CHEMICAL Code 107201 see HEXAZINONE

EPA PESTICIDE CHEMICAL Code 107701 see SODIUM AZIDE ($Na(N_3)$)

EPA PESTICIDE CHEMICAL Code 107801 see 3-IODO-2-PROPYNYL BUTYLCARBAMATE

EPA PESTICIDE CHEMICAL Code 108102 see PIRIMIPHOS METHYL

EPA PESTICIDE CHEMICAL Code 108401 see THIOBENCARB

EPA PESTICIDE CHEMICAL Code 108501 see PENDIMETHALIN N-(1-ETHYLPROPYL)-3,4-DIMETHYL-2,6-DINITROBENZENAMINE

EPA PESTICIDE CHEMICAL Code 109001 see OXYDIAZON

EPA PESTICIDE CHEMICAL Code 109301 see FLUVALINATE

EPA PESTICIDE CHEMICAL Code 109301 see FENVALERATE

EPA PESTICIDE CHEMICAL Code 109401 see ISOFENPHOS

EPA PESTICIDE CHEMICAL Code 109701 see PERMETHRIN

EPA PESTICIDE CHEMICAL Code 109901 see TRIADIMEFON

EPA PESTICIDE CHEMICAL Code 110902 see DICLOFOP METHYL

EPA PESTICIDE CHEMICAL Code 111001 see 1-BROMO-1-(BROMOMETHYL)-1,3-PROPANEDICARBONITRILE

EPA PESTICIDE CHEMICAL Code 111401 see PROFENOFOS

EPA PESTICIDE CHEMICAL Code 111601 see OXYFLUORFEN

EPA PESTICIDE CHEMICAL Code 111901 see IMAZALIL

EPA PESTICIDE CHEMICAL Code 113201 see VINCLOZOLIN

EPA PESTICIDE CHEMICAL Code 113601 see PROPETAMPHOS

EPA PESTICIDE CHEMICAL Code 114402 see ACIFLUORFEN, SODIUM SALT

EPA PESTICIDE CHEMICAL Code 114501 see THIODICARB

EPA PESTICIDE CHEMICAL Code 116002 see TRICLOPYR TRIETHYLAMMONIUM SALT

EPA PESTICIDE CHEMICAL Code 118401 see HYDRAMETHYLNON

EPA PESTICIDE CHEMICAL Code 118601 see CHLORSULFURON

EPA PESTICIDE CHEMICAL Code 118901 see DIMETHIPIN

EPA PESTICIDE CHEMICAL Code 121001 see SETHOXYDIM

EPA PESTICIDE CHEMICAL Code 122101 see PROPICONAZOLE

EPA PESTICIDE CHEMICAL Code 122805 see FLUAZIFOP-BUTYL

EPA PESTICIDE CHEMICAL Code 125301 see FENOXYCARB

EPA PESTICIDE CHEMICAL Code 127901 see FENPROPATHRIN

EPA PESTICIDE CHEMICAL Code 128201 see QUIZALOFOP-ETHYL

EPA PESTICIDE CHEMICAL Code 128701 see FENOXAPROP ETHYL

EPA PESTICIDE CHEMICAL Code 128825 see BIFENTHRIN

EPA PESTICIDE CHEMICAL Code 128857 see MYCLOBUTANIL

EPA PESTICIDE CHEMICAL Code 206600 see FENARIMOL

EPA PESTICIDE CHEMICAL Code 216400 see 2-BROMO-2-NITROPROPANE-1,3-DIOL

EPA PESTICIDE CHEMICAL Code 366200 see DIMETHYLFORMAMIDE

EPA PESTICIDE CHEMICAL Code 496500 see TRIPHENYLTIN CHLORIDE

1,2-EPOXYBUTANE see 1,2-BUTYLENE OXIDE

1,2-EPOXYPROPANE see PROPYLENE OXIDE

2,3-EPOXY-1-PROPANAL see GLYCIDYLALDEHYDE

α,β-EPOXYSTYRENE see STYRENE OXIDE

EPTC see ETHYL DIPROPYLTHIOCARBAMATE

ERGOTAMINE BITARTRATE see ERGOTAMINE TARTRATE

ETHANAL see ACETALDEHYDE

ETHANAMIDE see ACETAMIDE

ETHANAMINE, 2-CHLORO-N-(2-CHLOROETHYL)-N-METHYL- see NITROGEN MUSTARD

ETHANAMINE,N,N-DIETHYL- see TRIETHYLAMINE

ETHANE, CHLORO- see CHLOROETHANE

ETHANE, 1-CHLORO-1,1-DIFLUORO- see 1-CHLORO-1,1-DIFLUOROETHANE

ETHANE, 2-CHLORO-1,1,2,2-TETRAFLUORO- see 1-CHLORO-1,1,2,2-TETRAFLUOROETHANE

ETHANE, 2-CHLORO-1,1,1,2-TETRAFLUORO- see 2-CHLORO-1,1,1,2-TETRAFLUOROETHANE

ETHANE, 2-CHLORO-1,1,1-TRIFLUORO- see 2-CHLORO-1,1,1-TRIFLUOROETHANE

1,2-ETHANEDIAMINE, N,N-DIETHYL N'-PYRIDINYL-N'(2-THIENYLMETHYL)- see METHAPYRILENE

1,2-ETHANEDIAMINE see ETHYLENEDIAMINE

ETHANE, 1,2-DIBROMO- see 1,2-DIBROMOETHANE

ETHANE, 1,2-DIBROMOTETRAFLUORO see DIBROMOTETRAFLUOROETHANE

ETHANE, DICHLORO-1,1,2-TRIFLUORO see DICHLORO-1,1,2-TRIFLUOROETHANE

ETHANE, 1,1-DICHLORO-1,2,2-TRIFLUORO see 1,1-DICHLORO-1,2,2-TRIFLUOROETHANE

ETHANE, 1,1-DICHLORO-1-FLUORO- see 1,1-DICHLORO-1-FLUOROETHANE

ETHANE, 1,1-DICHLORO- see ETHYLIDENE DICHLORIDE

ETHANE, 1,2-DICHLORO-1,1-DIFLUORO- see 1,2-DICHLORO-1,1-DIFLUOROETHANE

ETHANE, 1,2-DICHLORO-1,1,2-TRIFLUORO- see 1,2-DICHLORO-1,1,2-TRIFLUOROETHANE

ETHANE, 1,2-DICHLORO- see 1,2-DICHLOROETHANE

ETHANE, 2,2-DICHLORO-1,1,1-TRIFLUORO- see 2,2-DICHLORO-1,1,1-TRIFLUOROETHANE

ETHANE, DICHLOROTRIFLUORO- see DICHLOROTRIFLUROETHANE

ETHANE, 1,1-DIFLUORO- see DIFLUOROETHANE

ETHANEDINITRILE see CYANOGEN

ETHANEDIOIC ACID, AMMONIUM SALT see FERRIC AMMONIUM OXALATE

ETHANEDIOIC ACID, AMMONIUM IRON(3+) SALT see FERRIC AMMONIUM OXALATE

ETHANEDIOIC ACID, AMMONIUM IRON(III) SALT see FERRIC AMMONIUM OXALATE

ETHANEDIOIC ACID, DIAMMONIUM SALT, MONOHYDRATE see AMMONIUM OXALATEOXALIC ACID, AMMONIUM SALT see AMMONIUM OXALATE

ETHANEDIOIC ACID, MONOAMMONIUM SALT, MONOHYDRATE see AMMONIUM OXALATE

1,2-ETHANEDIOL see ETHYLENE GLYCOL

1,2-ETHANEDIYLBIS(CARBAMODITHIOATO)(2-)-MANGANESE see MANEB

1,2-ETHANEDIYLBIS(CARBAMODITHIOATO)ZINC see ZINEB

1,2-ETHANEDIYLBISCARBAMODITHIOIC ACID DISODIUM SALT see NABAM

1,2-ETHANEDIYLBISCARBAMODITHIOIC ACID, MANGANESE COMPLEX see MANEB

1,2-ETHANEDIYLBISCARBAMODITHIOIC ACID see ETHYLENEBISDITHIOCARBAMIC ACID

1,2-ETHANEDIYLBISCARBAMOTHIOIC ACID, ZINC SALT see ZINEB

ETHANE, HEXACHLORO- see HEXACHLOROETHANE

ETHANE,1,1'-(METHYLENEBIS(OXY))BIS(2-CHLORO- see BIS(2-CHLOROETHOXY)METHANE

ETHANE, 1,1'-OXYBIS- see ETHYL ETHER

ETHANE, 1,1'-OXYBIS(2-CHLORO- see BIS(2-CHLOROETHYL)ETHER

ETHANE PENTACHLORIDE see PENTACHLOROETHANE

ETHANE, PENTACHLORO- see PENTACHLOROETHANE

ETHANEPEROXOIC ACID see PERACETIC ACID

ETHANE, 1,1,1,2-TETRACHLORO-1-FLUORO- see 1,1,2,2-TETRACHLORO-2-FLUOROETHANE

ETHANE, 1,1,1,2-TETRACHLORO- see 1,1,1,2-TETRACHLOROETHANE

ETHANE, 1,1,2,2-TETRACHLORO-1-FLUORO- see HCFC-121

ETHANE, 1,1,2,2-TETRACHLORO- see 1,1,2,2-TETRACHLOROETHANE

ETHANETHIOAMIDE see THIOACETAMIDE

ETHANE, 1,1′-THIOBIS[2-CHLORO- see MUSTARD GAS

1-ETHANETHIOL see ETHYL MERCAPTAN

ETHANETHIOL see ETHYL MERCAPTAN

ETHANE, 1,1,1-TRICHLORO- see 1,1,1-TRICHLOROETHANE

ETHANE, 1,1,2-TRICHLORO-1,2,2,-TRIFLUORO- FREON 113

ETHANE, 1,1,2-TRICHLORO- see 1,1,2-TRICHLOROETHANE

ETHANIMIDOTHIOIC ACID, 2-(DIMETHYLAMINO)-N-HYDROXY-2-OXO-, METHYL ESTER see A2213

ETHANIMIDOTHIC ACID, N-[[METHYLAMINO]CARBONYL]OXY]-, METHYL ESTER see METHOMYL

ETHANOIC ACID see ACETIC ACID

ETHANOIC ANHYDRIDE see ACETIC ANHYDRIDE

ETHANOL BUTOXIDE see PIPERONYL BUTOXIDE

ETHANOL, 2-CHLORO- see CHLOROETHANOL

ETHANOL, 2-ETHOXY- see 2-ETHOXYETHANOL

ETHANOL, 2,2′-IMINOBIS- see DIETHANOLAMINE

ETHANOL, 2-METHOXY- see 2-METHOXYETHANOL

ETHANOL, 2,2′-(NITROSOIMINO)BIS- see N-NITROSODIETHANOLAMINE

ETHANONE, 1-PHENYL- see ACETOPHENONE

ETHANOYL CHLORIDE see ACETYL CHLORIDE

ETHENAMINE, N-METHYL-N-NITROSO- see N-NITROSOMETHYLVINYLAMINE

ETHENE see ACETYLENE

ETHENE see ETHYLENE

ETHENE, BROMO- see VINYL BROMIDE

ETHENE, BROMOTRIFLUORO- see BROMOTRIFLUORETHYLENE

ETHENE, CHLORO- see VINYL CHLORIDE

ETHENE, (2-CHLOROETHOXY)- see 2-CHLOROETHYL VINYL ETHER

ETHENE, CHLOROTRIFLUORO- see TRIFLUOROCHLOROETHYLENE

ETHENE, 1,1 DICHLORO- see VINYLIDENE CHLORIDE

ETHENE, 1,2-DICHLORO-, (E) see 1,2-DICHLOROETHYLENE

ETHENE, trans-1,2-DICHLORO- see 1,2-DICHLOROETHYLENE

ETHENE, 1,1-DIFLUORO- see VINYLIDENE FLUORIDE

ETHENE, ETHOXY- see VINYL ETHYL ETHER

ETHENE, FLUORO- see VINYL FLUORIDE

ETHENE, METHOXY- see VINYL METHYL ETHER

ETHENE, TETRACHLORO- see TETRACHLOROETHYLENE

ETHENE, TETRAFLUORO- see TETRAFLUOROETHYLENE

ETHENE, TRICHLORO- see TRICHLOROETHYLENE

ETHENYL ETHANOATE see VINYL ACETATE

5-ETHENYL-2-METHYLPYRIDINE see PYRIDINE, 2-METHYL-5-VINYL

ETHER,2,4-DICHLOROPHENYL p-NITROPHENYL see NITROFEN

ETHER, VINYL ETHYL see VINYL ETHYL ETHER

ETHIENOCARB see ZINC, DICHLORO(4,4-DIMETHYL-5((((METHYL-AMINO)CARBONYL)OXY)IMINO)PENTANENITRILE)-,(T-4)-

ETHINE see ACETYLENE

ETHOPROPHOS see ETHOPROP

4-ETHOXYACETANILIDE see PHENACETIN

2-ETHOXYACETYLENE see 2-ETHOXYETHANOL

ETHOXY CARBONYL ETHYLENE see ETHYL ACRYLATE

2-[1-(ETHOXYIMINO)BUTYL]-5-[2-(ETHYLTHIO)PROPYL]-3-HYDROX-YL-2-CYCLOHEXEN-1-ONE see SETHOXYDIM

2-[[ETHOXYL[(1-METHYLETHYL)AMINO]PHOSPHINOTHIOYL]OXY] BENZOIC ACID 1-METHYLETHYL ESTER see ISOFENPHOS

ETHYL ACETYLENE see 1-BUTYNE

ETHYL ALDEHYDE see ACETALDEHYDE

ETHYLAMINE see ETHANAMINE

3-[(ETHYLAMINO)METHOXYPHOSPHINOTHIOYL]OXY]-2-BUTENOIC ACID, 1-METHYLETHYL ESTER see PROPETAMPHOS

S-ETHYL AZEPANE-1-CARBOTHIOATE see MOLINATE

ETHYLCARBAMATE see URETHANE

ETHYL CHLORIDE see CHLOROETHANE

ETHYL-2-[[[(4-CHLORO-6-METHOXYPYRIMIDIN-2-YL)-CARBONYL]-AMINO]SULFONYL]BENZOATE see CHLORIMURON ETHYL

ETHYL-2-((4-(6-CHLORO-2-BENZOXAZOLYLOXY))-PHENOXY) PROPIONATE see FENOXAPROP ETHYL

ETHYL 2-(4-(6-CHLORO-2-QUINOXALINYLOXY)PHENOXY) PROPANOATE see QUIZALOFOP-ETHYL

ETHYL CYANIDE see PROPIONITRILE

S-ETHYL CYCLOHEXYLETHYLCARBAMOTHIOATE see CYCLOATE

ETHYL 4,4′-DICHLOROBENZILATE see CHLOROBENZILATE

o-ETHYL-S-DIISOPROPYLAMINOETHYL METHYLPHOSPHONO-THIOATE see PHOSPHONOTHIOIC ACID, METHYL-, S[2-[BIS(1-METHYLETHYL)AMINOETHYL) O-ETHYL ESTER

ETHYL DIMETHYL METHANE see ISOPENTANE

ETHYL DIMETHYLPHOSPHORAMIDOCYANIDATE see TABUN

o-ETHYL S,S-DIPROPYL DITHIOPHOSPHATE see ETHOPROP

o-ETHYL S,S-DIPROPYL PHOSPHORODITHIOATE see ETHOPROP

ETHYL N,N-DIPROPYLTHIOLCARBAMATE see ETHYL DIPROPYLTHI-OCARBAMATE

ETHYL EHYDE see ACETALDEHYDE

ETHYLENE BIS(DITHIOCARBAMATE), DISODIUM SALT see NABAM

ETHYLENE, BROMO- see VINYL BROMIDE

ETHYLENE CHLORHYDRIN see CHLOROETHANOL

ETHYLENE CHLORIDE see 1,2-DICHLOROETHANE

ETHYLENE CHLOROHYDRINE see CHLOROETHANOL

ETHYLENE DIBROMIDE see 1,2-DIBROMOETHANE

1,2-ETHYLENE DIBROMIDE see 1,2-DIBROMOETHANE

ETHYLENE DICHLORIDE see 1,2-DICHLOROETHANE

1,2-ETHYLENE DICHLORIDE see 1,2-DICHLOROETHANE

ETHYLENE, 1,1-DICHLORO- see VINYLIDENE CHLORIDE

ETHYLENE, 1,2-DICHLORO- see 1,2-DICHLOROETHYLENE

ETHYLENE DIPROPIONATE (8CI) see CROTONALDEHYDE, (E)

ETHYLENE GLYCOL ETHYL ETHER see 2-ETHOXYETHANOL

ETHYLENE GLYCOL MONOETHYL ETHER see 2-ETHOXYETHANOL

ETHYLENE GLYCOL MONOMETHYL ETHER see 2-METHOXYETHANOL

ETHYLENE, N-METHYL-N-NITROSO- see N-NITROSOMETHYLVINYLAMINE

ETHYLENE TETRACHLORIDE see TETRACHLOROETHYLENE

ETHYLENE, TRICHLORO- see TRICHLOROETHYLENE

ETHYLENEBIS(DITHIOCARBAMIC ACID) see ETHYLENEBISDITHIOCARBAMIC ACID

ETHYLENECARBOXAMIDE see ACRYLAMIDE

cis-1,2-ETHYLENEDICARBOXYLIC ACID, TOXILIC ACID see MALEIC ACID

ETHYLENEDIAMINETETRAACETIC ACID see ETHYLENEDIAMINETETRAACETIC ACID (EDTA)

1,3-ETHYLENETHIOUREA see ETHYLENE THIOUREA

n-ETHYLETHANAMINE see DIETHYLAMINE

3-ETHYL-2-(5-(3-ETHYL-2-BENZOTHIAZOLINYLIDENE)-1,3-PENTADIENYL)BENZOTHIAZOLIUM IODIDE see DITHIAZANINE IODIDE

ETHYLETHYLENE see BUTENE

ETHYLETHYNE see 1-BUTYNE BUTANAL see BUTYRALDEHYDE

ETHYLHEXYL-2,4,5-T see 2,4,5-T ESTERS (1928-47-8)

2-ETHYLHEXYL(2,4-DICHLOROPHENOXY)ACETATE see 2,4-D 2-ETHYLHEXYL ESTER

ETHYL HYDRIDE see ETHANE

ETHYLIC ACID see ACETIC ACID

ETHYLIDENE DICHLORIDE see ETHYLIDENE DICHLORIDE

1,2-ETHYLIDENE DICHLORIDE see ERGOCALCIFEROL

ETHYLIMINE see ETHYLENEIMINE

ETHYL 3-METHYL-4-(METHYLTHIO)PHENYL(1-METHYLETHYL) PHOSPHORAMIDATE see FENAMIPHOS

ETHYL METHYL KETONE see METHYL ETHYL KETONE

N-ETHYL-N′-(1-METHYLETHYL)-6-(METHYLTHIOL)-1,3,5,-TRIAZINE-2,4-DIAMINE see AMETRYN

o-ETHYL O-(4-(METHYLMERCAPTO)PHENYL)-S-N-PROPYLPHOSPHOROTHIONOTHIOLATE see SULPROFOS

ETHYL 4-(METHYLTHIO)-M-TOLYLISOPROPYLPHOSPHORAMIDATE see FENAMIPHOS

o-ETHYL O-(4-(METHYLTHIO)PHENYL)PHOSPHORODITHIOIC ACID S-PROPYL ESTER see SULPROFOS

o-ETHYL O-[4-(METHYLTHIO)PHENYL]PHOSPHORODITHIOIC ACID S-PROPYL ESTER see SULPROFOS

o-ETHYL-O-(4-NITROPHENYL)-BENZENETHIONOPHOSPHONATE see EPN

o-ETHYL-O-(4-NITROPHENYL PHENYL)PHENYLPHOSPHONO-THIOATE see EPN

o-ETHYL-O-p-NITROPHENYL PHENYLPHOSPHONOTHIOATE see EPN

1-ETHYL-1-NITROSOUREA see N-NITROSO-N-ETHYLUREA

N-ETHYL-N-NITROSOUREA see N-NITROSO-N-ETHYLUREA

2-ETHYLOXIRANE see 1,2-BUTYLENE OXIDE

ETHYL(2-(4-PHENOXYPHENOXY)ETHYL)CARBAMATE see FENOXY-CARB

ETHYL (3-(((PHENYLAMINO)CARBONYL)OXY)PHENYL)CARBAMATE see DESMEDIPHAM

4-ETHYL-PHOSPHA-2,6,7-TRIOXABICYCLO(2,2,2)OCTANE see TRI-METHYLOLPROPANE PHOSPHITE

ETHYL SILICON TRICHLORIDE see TRICHLOROETHYLSILANE

S-(2-(ETHYLSULFINYL)ETHYL) O,O-DIMETHYL ESTER PHOSPHORO-THIOIC ACID see OXYDEMETON METHYL

ETHYLTETRAPHOSPHATE see HEXAETHYL TETRAPHOSPHATE

ETHYL TRICHLOROETHYLSILANE see TRICHLOROETHYLSILANE

o-ETHYL-O-2,4,5-TRICHLOROPHENYL ETHYLPHOSPHONOTHIOATE see TRICHLORONATE

ETHYL TRICHLOROPHENYLETHYLPHOSPHONOTHIOATE see TRI-CHLORONATE

ETHYNE see ACETYLENE

ETO see ETHYLENE OXIDE

FENAMINPHOS(DOT) see FENAMIPHOS

FENTIN see TRIPHENYLTIN HYDROXIDE

FENTIN ACETATE see STANNANE, ACETOXYTRIPHENYL-

FENTIN CHLORIDE see TRIPHENYLTIN CHLORIDE

FENTIN HYDROXIDE see TRIPHENYLTIN HYDROXIDE

FERRATE(3-), TRIS(OXALATO)-,TRIAMMONIUM see FERRIC AMMO-NIUM OXALATE

9H-FLUORENE see FLUORENE

N-2-FLUORENYLACETAMIDE see 2-ACETYLAMINOFLUORENE

FLUORHYDRIC ACID see HYDROGEN FLUORIDE

FLUORIC ACID see HYDROGEN FLUORIDE

FLUORINE-19 see FLUORINE

1-FLUORO-1,1,2,2-TETRACHLOROETHANE see HCFC-121

1-FLUORO-1,1,2,2-TETRACHLOROETHANE see 1,1,2,2-TETRACHLORO-1-FLUOROETHANE

1-FLUORO-1,1,1,2-TETRACHLOROETHANE see 1,1,2,2-TETRACHLORO-2-FLUOROETHANE

2-FLUOROACETAMIDE see FLUOROACETAMIDE

2-FLUOROACETIC ACID see FLUOROACETIC ACID

FLUOROACETIC ACID, SODIUM SALT see SODIUM FLUOROACETATE

FLUOROCARBON 1301 see BROMOTRIFLUROMETHANE

FLUORODICHLOROMETHANE see DICHLOROFLUOROMETHANE

FLUOROETHENE see VINYL FLUORIDE

FLUOROTRICHLOROMETHANE see TRICHLOROFLUOROMETHANE

5-FLUOROURACIL see FLUOROURACIL

2-FLUROETHANOL see ETHYLENE FLUOROHYDRIN

FONOFOS (DOT) see FONFOS

FOOD RED 5 see C.I. FOOD RED 5

FOOD RED 15 see C.I. FOOD RED 15

FORMALDEHYDE (solution) see FORMALDEHYDE

FORMALIN see see FORMALDEHYDE

FORMETANATE see FORMETANATE HYDROCHLORIDE

FORMIC ACID, AMIDE, N,N-DIMETHYL- see DIMETHYLFORMAMIDE

FORMIC ACID, METHYL ESTER see METHYL FORMATE

FORMIC ACID, ZINC SALT see ZINC FORMATE

FORMYLIC ACID see FORMIC ACID

FREON 11 see TRICHLOROFLUOROMETHANE

FREON 12 see DICHLORODIFLUOROMETHANE

FREON 20 see CHLOROFORM

FREON 21 see DICHLOROFLUOROMETHANE

FREON 22 see CHLORODIFLUOROMETHANE

FREON 114 see DICHLOROTETRAFLUOROETHANE

FREON 114B2 see DIBROMOTETRAFLUOROETHANE

FREON 115 see MONOCHLOROPENTAFLUOROETHANE

FREON 123 see 2,2-DICHLORO-1,1,1-TRIFLUOROETHANE

FREON 124 see 2-CHLORO-1,1,1,2-TETRAFLUOROETHANE

FREON 124a see 1-CHLORO-1,1,2,2-TETRAFLUOROETHANE

FREON 133a see 2-CHLORO-1,1,1-TRIFLUOROETHANE

FREON 141 see 1,1-DICHLORO-1-FLUOROETHANE

FREON 142b see 1-CHLORO-1,1-DIFLUOROETHANE

FREON 152 see DIFLUOROETHANE

FREON 253fb see 3-CHLORO-1,1,1-TRIFLUOROPROPANE

FULMINATE OF MERCURY see MERCURY FULMINATE

FULMINIC ACID, MERCURY(2+) SALT see MERCURY FULMINATE

FUMIGRAIN see ACRYLONITRILE

FURADAN see CARBOFURAN

2-FURALDEHYDE see FURFURAL

2-FURANACARBOXALDEHYDE see FURFURAL

2,5-FURANDIONE see MALEIC ANHYDRIDE

2-(2-FURANYL)-1H-BENZIMIDAZOLE see FUBERIDAZOLE

FURFURALDEHYDE see FURFURAL

GALLIUM(3+) CHLORIDE see GALLIUM TRICHLORIDE

GALLIUM(III) CHLORIDE see GALLIUM TRICHLORIDE

GAMMA-BHC see LINDANE

GAMMA-HCH see LINDANE

GAMMA-LINDANE see LINDANE

GLACIAL ACETIC ACID see ACETIC ACID

GLACIAL ACRYLIC ACID see ACRYLIC ACID

N-d-GLUCOSYL-(2)-N′-NITROSOMETHYLUREA see D-GLUCOSE, 2-DEOXY-2-[[METHYLNITROSOAMINO)CARBONYL]AMINO]-

GLYCEROL TRINITRATE see NITROGLYCERIN

GLYCINE, N,N′-1,2-ETHANEDIYLBIS(N-(CARBOXYMETHYL)-9CI) see ETHYLENEDIAMINE-TETRAACETIC ACID (EDTA)

GLYCINE, N,N-BIS(CARBOXYMETHYL)- see NITRILOTRIACETIC ACID

GLYCOL METHYL ETHER see 2-METHOXYETHANOL

GLYCOLONITRILE see FORMALDEHYDE CYANOHYDRIN

GRAMOXONE see PARAQUAT DICHLORIDE

GREEN VITRIOL see FERROUS SULFATE

GUINEA GREEN see C.I. ACID GREEN 3

GUINEA GREEN b see C.I. ACID GREEN 3

GUSATHION see AZINPHOS-METHYL

GUTHION see AZINPHOS-METHYL

GYYCOLIC NITRILE see FORMALDEHYDE CYANOHYDRIN

H-ISOINDOLE-1,3(2H)-DIONE,3a,4,7,7a-TETRAHYDRO-2-[(TRICHLOROMETHYL)THIO]-see CAPTAN

HALON 1211 see BROMOCHLORODIFLUOROMETHANE

HALON 1301 see BROMOTRIFLUROMETHANE

HALON 2402 see DIBROMOTETRAFLUOROETHANE

HCFC-21 see DICHLOROFLUOROMETHANE

HCFC-22 see CHLORODIFLUOROMETHANE

HCFC-121 see 1,1,2,2-TETRACHLORO-1-FLUOROETHANE

HCFC-121a see 1,1,2,2-TETRACHLORO-2-FLUOROETHANE

HCFC-124 see 2-CHLORO-1,1,1,2-TETRAFLUOROETHANE

HCFC-124a see 1-CHLORO-1,1,2,2-TETRAFLUOROETHANE

HCFC-123 see 1,1-DICHLORO-1,2,2-TRIFLUOROETHANE

HCFC-123 see 2,2-DICHLORO-1,1,1-TRIFLUOROETHANE

HCFC-123 see DICHLOROTRIFLUROETHANE

HCFC-123a see 1,2-DICHLORO-1,1,2-TRIFLUOROETHANE

HCFC-123b see 1,1-DICHLORO-1,2,2-TRIFLUOROETHANE

HCFC-132b see 1,2-DICHLORO-1,1-DIFLUOROETHANE

HCFC-133a see 2-CHLORO-1,1,1-TRIFLUOROETHANE

HCFC-141b see 1,1-DICHLORO-1-FLUOROETHANE

HCFC-142b see 1-CHLORO-1,1-DIFLUOROETHANE

HCFC-225aa see 2,2-DICHLORO-1,1,1,3,3-PENTAFLUOROPROPANE

HCFC-225ba see 2,3-DICHLORO-1,1,1,2,3-PENTAFLUOROPROPANE

HCFC-225bb see 1,2-DICHLORO-1,1,2,3,3-PENTAFLUOROPROPANE

HCFC 225ca see 3,3-DICHLORO-1,1,1,2,2-PENTAFLUOROPROPANE

HCFC-225cb see 1,3-DICHLORO-1,1,2,2,3-PENTAFLUOROPROPANE

HCFC-225da see 1,2-DICHLORO-1,1,3,3,3-PENTAFLUOROPROPANE

HCFC-225ea see 1,3-DICHLORO-1,1,2,3,3-PENTAFLUOROPROPANE

HCFC-225cc see 1,1-DICHLORO-1,2,2,3,3-PENTAFLUOROPROPANE

HCFC-225eb see 1,1-DICHLORO-1,2,3,3,3-PENTAFLUOROPROPANE

HCFC 253fb see 3-CHLORO-1,1,1-TRIFLUOROPROPANE

α-HCH see α-HEXACHLOROCYCLOHEXANE

β-HCH see β-HEXACHLOROCYCLOHEXANE

δ-HCH see δ-HEXACHLOROCYCLOHEXANE

HCP see HEXACHLOROPHENE

HEMPA see HEXAMETHYLPHOSPHORAMIDE

1,4,5,6,7,8,8-HEPTACHLORO-3A,4,7,7A-TETRAHYDRO-4,7-METHANO-1H-INDENE see HEPTACHLOR

α,β-1,2,3,4,7,7-HEXACHLOROBICLO(2,2,1)HEPTEN-5,6-BIOXYMETHYLENESULFITE see ENDOSULFAN

HEXACHLOROBUTADIENE see HEXACHLORO-1,3-BUTADIENE

δ-1,2,3,4,5,6-HEXACHLOROCYCLOHEXANE-δ see δ-HEXACHLOROCYCLOHEXANE

HEXACHLORCYCLOHEXANE, ALPHA ISOMER see α-HEXACHLOROCYCLOHEXANE

1,2,3,4,5,6-HEXACHLOROCYCLOHEXANE see HEXACHLOROCYCLOHEXANE (ALL ISOMERS)

HEXACHLOROCYCLOHEXANE, (BETA ISOMER) see β-HEXACHLOROCYCLOHEXANE

HEXACHLOROCYCLOHEXANE, (GAMMA ISOMER) see LINDANE

HEXACHLOROCYCLOHEXANE, (DELTA ISOMER) see δ-HEXACHLOROCYCLOHEXANE

HEXACHLOROCYCLOHEXANE, GAMMA see LINDANE

α-HEXACHLOROCYCLOHEXANE see α-HEXACHLOROCYCLOHEXANE

β-HEXACHLOROCYCLOHEXANE see β-HEXACHLOROCYCLOHEXANE

γ-HEXACHLOROCYCLOHEXANE see LINDANE

δ-HEXACHLOROCYCLOHEXANE see δ-HEXACHLOROCYCLOHEXANE

1,2,3,4,10,10-HEXACHLORO-6,7-EPOXY-1,4,4A,5,6,7,8,8A-OCTAHYDRO-1,4-ENDO-EXO-5,8-DI-METHANONAPHTHALENE see DIELDRIN

1,2,3,4,10,10-HEXACHLORO-6,7-EPOXY-1,4,4A,5,6,7,8,8A-OCTAHYDRO-1,4-ENDO-ENDO-1,4,5,8-DIMETHANONAPHTHALENE see ENDRIN

1,1,1,2,2,2-HEXACHLOROETHANE see HEXACHLOROETHANE

HEXACHLOROHEXAHYDRO-ENDO-EXO-DIMETHANONAPHTHALENE see ALDRIN

1,2,3,4,10,10-HEXACHLORO-1,4,4A,5,8,8A-HEXAHYDRO-1,4,5,8-DIMETHANONAPHTHALENE see ALDRIN

1,2,3,4,10,10-HEXACHLORO-1,4,4A,5,8,8A-HEXAHYDRO-1,4:5,8-ENDO-ENDO-DIMETHANONAPHTHALENE see ISODRIN

1,2,3,4,10,10-HEXACHLORO-1,4,4a,5,8,8a-HEXAHYDRO-1,4,5,8-ENDO,EXO-DIMETHANONAPHTHALENE see ALDRIN

1,2,3,4,10,10-HEXACHLORO-1,4,4A,5,8,8A-HEXAHYDRO-EXO-1,4-ENDO-5,8-DIMETHANONAPHTHALENE see ALDRIN

1,4,5,6,7,7-HEXACHLORO-5-NORBORNENE-2,3-DICARBOXYLIC ACID see CHLORENDIC ACID

1,4,5,6,7,7-HEXACHLORO-5-NORBORNENE-2,3-DIMETHANOL, CYCLIC SULFITE see ENDOSULFAN SULFATE

3,4,5,6,9,9-HEXACHLORO-1A,2,2A,3,6,6A,7,7A-OCTAHYDRO-2,7:3,6-DIMETHANO see DIELDRIN

HEXACHLOROPROPYLENE see HEXACHLOROPROPENE

HEXAHYDRO-2H-AZEPINE-2-ONE see CAPROLACTUM

HEXAHYDROBENZENAMINE see CYCLOHEXYLAMINE

HEXAHYDROBENZENE see CYCLOHEXANE

HEXAHYDROPHENOL see CYCLOHEXANOL

HEXAHYDROPYRIDINE see PIPERIDINE

HEXAKIS(2-METHYL-2-PHENYLPROPYL)DISTANNOXANE see FENBUTATIN OXIDE

HEXAMETAPHOSPHATE, SODIUM SALT see SODIUM PHOSPHATE, TRIBASIC

HEXAMETHYLENE DIISOCYANATE, 1,6- see HEXAMETHYLENE-1,6-DIISOCYANATE

HEXAMETHYLENE DIISOCYANATE see HEXAMETHYLENE-1,6-DIISOCYANATE

1,6-HEXAMETHYLENE DIISOCYANATE see HEXAMETHYLENE-1,6-DIISOCYANATE

HEXAMETHYLPHOSPHORIC ACID TRIAMIDE see HEXAMETHYLPHOSPHORAMIDE

HEXANE see n-HEXANE

HEXANE, 1,6-DIISOCYANATO- see HEXAMETHYLENE-1,6-DIISOCYANATE

HEXANE, 1,6-DIISOCYANATO-2,2,4-TRIMETHYL- see 2,2,4-TRIMETHYL HEXAMETHYLENE DIISOCYANATE

HEXANE, 1,6-DIISOCYANATO-2,4,4-TRIMETHYL- see 2,4,4-TRIMETHYLHEXAMETHYLENE DIISOCYANATE

HEXANEDINITRILE see ADIPONITRILE

HEXANEDIOIC ACID, DINITRILE see ADIPONITRILE

HEXANEDIOIC ACID see ADIPIC ACID

1,6-HEXANEDIOIC ACID see ADIPIC ACID

1,6-HEXANOLACTAM see CAPROLACTUM

HEXAZANE see PIPERIDINE

HEXONE see METHYL ISOBUTYL KETONE

HEXYL HYDRIDE see n-HEXANE

HMPT see HEXAMETHYLPHOSPHORAMIDE

HOELON see DICLOFOP METHYL

HYDRAZINE CARBOTHIOAMIDE see THIOSEMICARBAZIDE

HYDRAZINECARBOXAMIDE MONOHYDROCHLORIDE see SEMICARBAZIDE HYDROCHLORIDE

HYDRAZINE, 1,1-DIMETHYL- see 1,1-DIMETHYL HYDRAZINE

HYDRAZINE, 1,2-DIPHENYL- see 1,2-DIPHENYLHYDRAZINE

HYDRAZINE HYDROGEN SULFATE see HYDRAZINE SULFATE

HYDRAZINE, METHYL- see METHYL HYDRAZINE

HYDRAZINE MONOSULFATE see HYDRAZINE SULFATE

HYDRAZOBENZENE see DIPHENYLHYDRAZINE

HYDRAZOBENZENE see 1,2-DIPHENYLHYDRAZINE

HYDRAZOIC ACID, SODIUM SALT see SODIUM AZIDE ($Na(N_3)$)

HYDROCYANIC ACID see HYDROGEN CYANIDE

HYDROCYANIC ACID, POTASSIUM SALT see POTASSIUM CYANIDE

HYDROCYANIC ACID, SODIUM SALT see SODIUM CYANIDE (Na(CN))

HYDRODIMETHYLARSINE OXIDE, SODIUM SALT see SODIUM CACODYLATE

HYDROFLUORIC ACID see HYDROGEN FLUORIDE

HYDROFLUORIC ACID, SODIUM SALT (2:1) see SODIUM BIFLUORIDE

HYDROGEN ARSENIDE see ARSINE

HYDROGEN CARBOXYLIC ACID see FORMIC ACID

HYDROGEN CHLORIDE see HYDROCHLORIC ACID

HYDROGEN, COMPRESSED see HYDROGEN

HYDROGEN DIOXIDE see HYDROGEN PEROXIDE

HYDROGEN FLUORIDE see HYDROFLUORIC ACID

HYDROGEN PHOSPHIDE see PHOSPHINE

HYDROGEN SULFATE see SULFURIC ACID

HYDROPEROXIDE, 1-METHYL-1-PHENYLETHYL- see CUMENE HYDROPEROXIDE

p-HYDROQUINONE see HYDROQUINONE

HYDROSULFURIC ACID see HYDROGEN SULFIDE

HYDROXYACETONITRILE see FORMALDEHYDE CYANOHYDRIN

5-HYDROXY-5-AMINOMETHYLISOXAZOLE see MUSCIMOL

4-HYDROXY-3,5-DIBROMOBENZONITRILE see BROMOXYNIL

3-HYDROXY-N,N-DIMETHYL-, CIS-, DIMETHYL PHOSPHATE see DICROTOPHOS

HYDROXYDIMETHYLARSINE OXIDE see CACODYLIC ACID

1-HYDROXY-2,6-DIMETHYLBENZENE see 2,6-DIMETHYLPHENOL

4-HYDROXY-1,3-DIMETHYLBENZENE see 2,4-DIMETHYLPHENOL

HYDROXYLAMINE,N-NITROSO-N-PHENYL-,AMMONIUM SALT see CUPFERRON

N-(HYDROXYMETHYL)ACRYLAMIDE see N-METHYLOLACRYLAMIDE

1-HYDROXYMETHYLPROPANE see ISOBUTYL ALCOHOL

4-HYDROXYNITROBENZENE see p-NITROPHENOL

4-HYDROXY-3-(3-OXO-1-PHENYLBUTYL)-2H-1-BENZOPYRAN-2-ONE SODIUM SALT (9CI) see WARFARIN SODIUM

3-HYDROXYPHENOL see RESORCINOL

o-HYDROXYPHENOL see CATECHOL

m-HYDROXYPHENOL see RESORCINOL

3-HYDROXY-1-PROPANESULFONIC ACID SULTONE see 1,3-PROPANE SULTONE

3-HYDROXY-1-PROPANESULPHONIC ACID SULTONE see 1,3-PROPANE SULTONE

2-HYDROXYPROPIONITRILE see LACTONITRILE

3-HYDROXY-1-PROPYNE see PROPARGYL ALCOHOL

5(α-HYDROXY-α-2-PYRIDYLBENZYL)-7-(α-2-PYRIDYLBENZYLIDENE)-5-NORBORENE-2,3-DICARBOXIDE see NORBORMIDE

4-HYDROXY-3-(1,2,3,4-TETRAHYDRO-1-NAPTHALENYL)-2H-1-BENZOPYRAN-2-ONE(9CI) see COUMATETRALYL

HYDROXYTOLUENE see CRESOL (MIXED ISOMERS)

3-HYDROXY-4-(2,4-XYLYLAZO)-2,7-NAPHTHALENEDISULFONIC ACID, DISODIUM SALT see C.I. FOOD RED 5

HYPOCHLOROUS ACID, CALCIUM SALT see CALCIUM HYPOCHLORITE

HYPOCHLOROUS ACID, SODIUM SALT see SODIUM HYPOCHLORITE

2,4-IMIDAZOLIDINEDIONE, 5,5-DIPHENYL- see PHENYTOIN

2-IMIDAZOLIDINETHIONE see ETHYLENE THIOUREA

2,2′-IMINOBIS[ETHANOL] see DIETHANOLAMINE

2,2′-IMINODIETHANOL see DIETHANOLAMINE

INORGANIC ARSENIC see ARSENIC

INORGANIC LEAD see LEAD

IODIDEO-(4-CYANOPHENYL)O,O-DIMETHYL PHOSPHOROTHIOATE see CYANOPHOS

IODINE CYANIDE see CYANOGEN

IODOMETHANE see METHYL IODIDE

1-IODOPROPANE see n-PROPYLAMINE

IRON CARBONYL ($Fe(CO)_5$), (TB-5-11)- see IRON PENTACARBONYL

IRON CHLORIDE see FERRIC CHLORIDE

IRON(II) CHLORIDE (1:2) see FERROUS CHLORIDE

IRON(III) CHLORIDE see FERRIC CHLORIDE

IRON, ELEMENTAL see IRON

IRON FLUORIDE see FERRIC FLUORIDE

IRON(III) NITRATE see FERRIC NITRATE

IRON PROTOSULFATE see FERROUS SULFATE

IRON(II) SULFATE(1:1) HEPTAHYDRATE see FERROUS SULFATE

IRON(III) SULFATE see FERRIC SULFATE

IRON(II) SULFATE see FERROUS SULFATE

IRON TRICHLORIDE see FERRIC CHLORIDE

IRON TRINITRATE see FERRIC NITRATE

IRON, TRIS(DIMETHYLCARBAMODITHIOATO-S,S')-, (OC-6-11)- see FERBAM

IRON, TRIS(DIMETHYLCARBAMODITHIOATO-S,S'-) see FERBAM

IRON VITRIOL see FERROUS SULFATE

ISOAMYL HYDRIDE see ISOPENTANE

1,3-ISOBENZOFURANDIONE see PHTHALIC ANHYDRIDE

ISOBUTANAL see ISOBUTYRALDEHYDE

ISOBUTANOIC ACID see ISO-BUTYRIC ACID

ISOBUTANOL see ISOBUTYL ALCOHOL

ISOBUTENE see 2-METHYLPROPENE

ISOBUTYL METHYL KETONE see METHYL ISOBUTYL KETONE

ISOBUTYLENE see 2-METHYLPROPENE

o-((p-ISOCYANATOPHENYL)THIO)PHENYL ISOCYANATE see 2,4'-DIISOCYANATODIPHENYL SULFIDE

ISOCYANIC ACID, 1,5-NAPHTHYLENE ESTER see 1,5-NAPHTHALENE DIISOCYANATE

ISOCYANIC ACID, 2-METHYL-m-PHENYLENE ESTER see TOLUENE-2,6-DIISOCYANATE

ISOCYANIC ACID, METHYLENE(3,5,5-TRIMETHYL-3,1-CYCLOHEXYLENE) ESTER see ISOPHORONE DIISOCYANATE

ISOCYANIC ACID, METHYLPHENYLENE ESTER see TOLUENE 2,4-DIISOCYANATE

ISOCYANIC ACID, p-CHLOROPHENYL ESTER see p-CHLOROPHENYL ISOCYANATE

ISOCYANIC ACID, POLYMETHYLENEPOLYPHENYLENE ESTER see POLYMERIC DIPHENYLMETHANE DIISOCYANATE

β-ISOCYANOTOETHYLMETHACRYLATE see METHACRYLOYLOXYETHYL ISOCYANATE

2-ISOCYANOTOETHYLMETHACRYLATE see METHACRYLOYLOXYETHYL ISOCYANATE

ISOFLUORPHATE see DIISOPROPYLFLUOROPHOSPHATE

1H-ISOINDOLE-1,3(2H)-DIONE, 2-((TRICHLOROMETHYL)THIO)- see FOLPET

ISOLAN see ISOPROPYLMETHYLPYRAZOYL DIMETHYLCARBAMATE

ISOOCTANE see 2,2,4-TRIMETHYLPENTANE

ISOPENTENE see 3-METHYL-1-BUTENE

ISO-PENTYL ACETATE see ISO-AMYL ACETATE

ISOPROPANOL see ISOPROPYL ALCOHOL

m-ISOPROPYLPHENYL METHYL CARBAMATE see PHENOL

2-ISOPROPOXYPHENYL METHYLCARBAMATE see PROPOXUR

2-ISOPROPOXYPHENYL N-METHYLCARBAMATE see PROPOXUR

ISOPROPYHYL METHYLPHOSPHONOFLUORIDATE see SARIN

(1-ISOPROPYL-3-METHYL-1H-PYRAZOL-5-YL)-N,N-DIMETHYL CARBAMATE see ISOPROPYLMETHYLPYRAZOYL DIMETHYLCARBAMATE

ISOPROPYL BENZENE see CUMENE

ISOPROPYL CHLOROCARBONATE see ISOPROPYL CHLOROFORMATE

m-ISOPROPYLPHENOL-N-METHYLCARBAMATE see PHENOL, 3-(1-METHYLETHYL)-, METHYLCARBAMATE

ISOTHIOCYANATOMETHANE see METHYL ISOTHIOCYANATE

ISOTHIOUREA see THIOUREA

ISOCYANIC ACID, 4-METHYL-M-PHENYLENE ESTER see TOLUENE 2,4-DIISOCYANATE

JAPAN RED 5 see C.I. SOLVENT ORANGE 7

KELTHANE see DICOFOL

KETOHEXAMETHYLENE see CYCLOHEXANONE

LANNATE see METHOMYL

LEAD(2+) ACETATE see LEAD ACETATE

LEAD(II) ACETATE see LEAD ACETATE

LEAD ACETATE, BASIC see LEAD SUBACETATE

LEAD(4+) ARSENATE see LEAD ARSENATE (10102-48-4)

LEAD(IV) ARSENATE see LEAD ARSENATE (10102-48-4)

LEAD, BIS(ACETATO-O)TETRAHYDROXYTRI- see LEAD SUBACETATE

LEAD, BIS(OCTADECANOATO)DIOXODI- see LEAD STEARATE

LEAD BORON FLUORIDE see LEAD FLUOBORATE

LEAD(2+) CHLORIDE see LEAD CHLORIDE

LEAD(II) CHLORIDE see LEAD CHLORIDE

LEAD DIACETATE see LEAD ACETATE

LEAD DICHLORIDE see LEAD CHLORIDE

LEAD DIFLUORIDE see LEAD FLUORIDE

LEAD ELEMENTAL see LEAD

LEAD(2+) FLUORIDE see LEAD FLUORIDE

LEAD(II) FLUORIDE see LEAD FLUORIDE

LEAD(II) IODIDE see LEAD IODIDE

LEAD MONOSULFIDE see LEAD SULFIDE

LEAD(2+) NITRATE see LEAD NITRATE

LEAD(II) NITRATE see LEAD NITRATE

LEAD(II) PHOSPHATE see LEAD PHOSPHATE

LEAD STERATE, DIBASIC see LEAD STEARATE

LEAD(II) SULFATE(1:1) see LEAD SULFATE

LEAD SULFOCYANATE see LEAD THIOCYANATE

LEAD TETRAETHYL see TETRAETHYL LEAD

LEAD TETRAMETHYL see TETRAMETHYL LEAD

LEAD(II) THIOCYANATE see LEAD THIOCYANATE

δ-LINDANE see δ-HEXACHLOROCYCLOHEXANE

LITHIUM HYDRIDE (LiH) see LITHIUM HYDRIDE

MALACHITE GREEN see C.I. BASIC GREEN 4

MALACHITE GREEN CHLORIDE see C.I. BASIC GREEN 4

MANGANESE, BIS(DIMETHYLCARBAMODITHIOATO-S,S')- see MANGANESE DIMETHYLDITHIOCARBAMATE

MANGANESE, ELEMENTAL see MANGANESE

MANGANESE, (METHYLCYCLOPENTADIENYL)TRICARBONYL- see MANGANESE TRICARBONYL METHYLCYCLOPENTADIENYL

MBOCA see 4,4'-METHYLENEBIS(2-CHLOROANILINE)

MBT see 2-MERCAPTOBENZOTHIAZOLE

3-MCA see 3-METHYLCHLOANTHRENE

MCPA see METHOXONE

MDI see METHYLBIS(PHENYLISOCYANATE)

MECHLORETHAMINE see NITROGEN MUSTARD

MEK see METHYL ETHYL KETONE

MERCAPTOBENZENE see BENZENETHIOL

MERCAPTOBENZOTHIAZOLE see 2-MERCAPTOBENZOTHIAZOLE

MERCAPTODIMETHUR see METHIOCARB

MERCAPTOETHANE see ETHYL MERCAPTAN

MERCAPTOMETHANE see METHYL MERCAPTAN

MERCAPTOPHOS see FENTHION

MERCURIC SULFOCYANATE see MERCURIC THIOCYANATE

MERCUROUS NITRATE MONOHYDRATE see MERCUROUS NITRATE

MERCURY ACETATE see MERCURIC ACETATE

MERCURY(II) ACETATE see MERCURIC ACETATE

MERCURY, (ACETO-O)PHENYL- see PHENYLMERCURY ACETATE

MERCURY(II) CHLORIDE see MERCURIC CHLORIDE

MERCURY(II) CYANIDE see MERCURIC CYANIDE

MERCURY(I) NITRATE (1:1) see MERCUROUS NITRATE

MERCURY(II) NITRATE (1:2) see MERCURIC NITRATE

MERCURY OXIDE see MERCURIC OXIDE

MERCURY(II) SULFATE (1:1) see MERCURIC SULFATE

MESUROL see METHIOCARB

METADICHLOROBENZENE see 1,3-DICHLOROBENZENE

METALLIC MERCURY see MERCURY

METALLIC NICKEL see NICKEL

METANITROTOLUENE see m-NITROTOLUENE

METAPHENYLENEDIAMINE see 1,3-PHENYLENEDIAMINE

METAPHOSPHORIC ACID, HEXASODIUM SALT see SODIUM PHOSPHATE, TRIBASIC

METAPHOSPHORIC ACID, TRISODIUM SALT see SODIUM PHOSPHATE, TRIBASIC

METHACRYLIC ACID ANHYDRIDE see METHACRYLIC ANHYDRIDE

1-2-METHACRYLIC ACID, ETHYL ESTER see ETHYL METHACRYLATE

METHACRYLIC ACID, METHYL ESTER see METHYL METHACRYLATE

METHACRYLIC CHLORIDE see METHACRYLOYL CHLORIDE

α-METHACRYLONITRILE see METHACRYLONITRILE

((METHACRYLOYL)OXY)TRIBUTYLSTANNANE see TRIBUTYLTIN METHACRYLATE

METHACRYLOYL ANHYDRIDE see METHACRYLIC ANHYDRIDE

METHALLYL CHLORIDE see 3-CHLORO-2-METHYL-1-PROPENE

METHAMINE, N,N-DIMETHYL- see TRIMETHYLAMINE

METHANAMINE, N-METHYL see DIMETHYLAMINE

METHANAMINE, N-METHYL-N-NITROSO- see N-NITROSODIMETHYLAMINE

METHANE, BROMO- see BROMOMETHANE

METHANE, BROMODICHLORO- see DICHLOROBROMOMETHANE

METHANE, BROMOTRIFLUORO- see BROMOTRIFLUROMETHANE

METHANECARBONITRILE see ACETONITRILE

METHANECARBOXAMIDE see ACETAMIDE

METHANE CARBOXYLIC ACID see ACETIC ACID

METHANE, CHLORO- see CHLOROMETHANE

METHANE, CHLOROMETHOXY- see CHLOROMETHYL METHYL ETHER

METHANE, DIAZO- see DIAZOMETHANE

METHANE, DIBROMO- see METHYLENE BROMIDE

METHANE, DICHLORO- see DICHLOROMETHANE

METHANE, DICHLORODIFLUORO- see DICHLORODIFLUOROMETHANE

METHANE, DICHLOROFLUORO- see DICHLOROFLUOROMETHANE

METHANE, IODO see METHYL IODIDE

METHANE, ISOCYANATO- see METHYL ISOCYANATE

METHANE, ISOTHIOCYANATO- see METHYL ISOTHIOCYANATE

METHANE, OXYBIS- see METHYL ETHER

METHANE, OXYBIS(CHLORO)- see BIS(CHLOROMETHYL)ETHER

METHANE, PHENYL- see TOLUENE

METHANESULFENYL CHLORIDE, TRICHLORO- see PERCHLOROMETHYL MERCAPTAN

METHANESULFONIC ACID, METHYL ESTER see METHYLMETHANESULFONATE

METHANESULPHONIC ACID, ETHYL ESTER see ETHYL METHANESULFONATE

METHANESULPHONYL FLUORIDE see METHANESULFONYL FLUORIDE

METHANE, TETRACHLORO- see CARBON TETRACHLORIDE

METHANE, TETRANITRO- see TETRANITROMETHANE

METHANETHIOL see METHYL MERCAPTAN

1-METHANETHIOL see METHYL MERCAPTAN

METHANE, TRIBROMO- see BROMOFORM

METHANE, TRICHLORO- see CHLOROFORM

METHANE, TRICHLOROFLUORO- see TRICHLOROFLUOROMETHANE

METHANE, TRICHLORONITRO- see CHLOROPICRIN

6,9-METHANO-2,4,3-BENZODIOXATHIEPIN, 6,7,8,9,10,10-HEXACHLORO-1,5,5a,6,9,9e-HEXAHYDRO-, 3-OXIDE see ENDOSULFAN

6,9-METHANO-2,4,3-BENZODIOXATHIEPIN, 6,7,8,9,10,10-HEXACHLORO-1,5,5A,6,9,9A-HEXAHYDRO-, 3-OXIDE, (3α, 5Aβ,6α,9α,9Aβ)- see α-ENDOSULFAN

6,9-METHANO-2,4,3-BENZODIOXATHIEPIN, 6,7,8,9,10,10-HEXACHLORO-1,5,5A,6,9,9A-HEXAHYDRO-, 3-OXIDE, 3α, 5Aα,6β,9β,9Aα)- see β-ENDOSULFAN

6,9-METHANO-2,4,3-BENZODIOXATHIEPIN, 6,7,8,9,10,10-HEXACHLORO-1,5,5A,6,9,9A-HEXAHYDRO-, 3-DIOXIDE see ENDOSULFAN SULFATE

6,9-METHANO-2,4,3-BENZODIOXATHIEPIN, 6,7,8,9,10,10-HEXACHLORO-1,5,5A,6,9,9A-HEXAHYDRO-, 3-OXIDE see ENDOSULFAN

4,7-METHANOINDAN, 1,2,3,4,5,6,7,8,8-OCTACHLORO-2,3,3a,4,7,7a-HEXAHYDRO- see CHLORDANE

4,7-METHANOINDENE, 1,4,5,6,7,8,8-HEPTACHLORO-3A,4,7,7A-TETRAHYDRO- see HEPTACHLOR

4,7-METHANO-1H-INDENE, 3A,4,7,7A-TETRAHYDRO- see DICYCLOPENTADIENE

4,7-METHANO-1H-INDENE,1,2,4,5,6,7,8,8-OCTACHLORO-2,3,3A,4,7,7A-HEXAHYDRO- see CHLORDANE

METHANOL, SODIUM SALT see SODIUM METHYLATE

METHANONE, BIS[4-(DIMETHYLAMINO)PHENYL]- see MICHLER'S KETONE

1,3,4-METHENO-2H-CYCLOBUTA (cd)PENTALEN-2-ONE,1,1a,3,3a,4,5,5a,5b,6-DECACHLORO-OCTAHYDRO- see KEPONE

1,2,4-METHENOCYCLOPENTA(CD)PENTALENE-5-CARBOXALDEHYDE, 2,2A,3,3,4,7-HEXACHLORODECAHYDRO-,(1α,2β,2Aβ,4β,4Aβ,5β,6Aβ,6Bβ,7R*)- see ENDRIN ALDEHYDE

METHENYL TRIBROMIDE see BROMOFORM

2-METHOXYANILINE HYDROCHLORIDE see o-ANISIDINE HYDROCHLORIDE

4-METHOXY-1,3-BENZENEDIAMINE SULPHATE see 2,4-DIAMINOSOLE, SULFATE

METHOXY-DDT see METHOXYCHLOR

2-METHOXYETHANOL see 2-METHOXYETHANOL

1-METHOXY-1-METHYL-3-(3,4-DICHLOROPHENYL)UREA see LINURON

2-(4-METHOXY-6-METHYL-1,3,5-TRIAZIN-2-YL)-METHYLAMINO) CARBONYL)AMINO)SULFONYL)-, METHYL ESTER see TRIBENURON METHYL

N-(METHOXY(METHYLTHIO)PHOSPHINOYL)ACETAMIDE see ACEPHATE

p-METHOXY-m-PHENYLENEDIAMINE see 2,4-DIAMINOSOLE

METHYLACETIC ACID see PROPIONIC ACID

METHYLACETIC ANHYDRIDE see PROPIONIC ANHYDRIDE

METHYL ACETONE see METHYL ETHYL KETONE

METHYL ACETYLENE see 1-PROPYNE

3-METHYLACROLEINE see CROTONALDEHYDE, (E)

α-METHYLACRYLONITRILE see METHACRYLONITRILE

2-METHYLACRYLONITRILE see METHACRYLONITRILE

2-METHYLACTONITRILE see 2-METHYLLACTONITRILE

METHYL ALCOHOL see METHANOL

N-METHYLAMINOMETHANETHIONOTHIOLIC ACID SODIUM SALT see METHAM SODIUM

2-METHYLANILINE see o-TOLUIDINE

4-METHYLANILINE see p-TOLUIDINE

o-METHYLANILINE see o-TOLUIDINE

p-METHYLANILINE see p-TOLUIDINE

5-METHYL-o-ANISIDINE see p-CRESIDINE

2-METHYLAZACLYCLOPROPANE see PROPYLENEIMINE

2-METHYLAZIRIDINE see PROPYLENEIMINE

METHYLBENZENE see TOLUENE

2-METHYL-1,2-BENZENEDIAMINE see DIAMINOTOLUENE

α-METHYLBENZENEETHANEAMINE see AMPHETAMINE

N-METHYL-BIS(2-CHLOROETHYL)AMINE see NITROGEN MUSTARD

METHYLBIS(2-CHLOROETHYL)AMINE see NITROGEN MUSTARD

β-METHYLBIVINYL see ISOPRENE

METHYL BROMIDE see BROMOMETHANE

2-METHYL-1,3-BUTADIENE see ISOPRENE

3-METHYL-1,3-BUTADIENE see ISOPRENE

2-METHYLBUTANE see ISOPENTANE

1-METHYLBUTYL ACETATE see sec-AMYL ACETATE

METHYL-t-BUTYL ETHER see METHYL tert-BUTYL ETHER

METHYL-CARBAMIC ACID, ESTER with ESEROLINE see PHYSOSTIGMINE

METHYL CARBONIMIDE see METHYL ISOCYANATE

METHYL CARBONOCHLORIDATE see METHYL CHLOROCARBONATE

METHYL CELLOSOLVE see 2-METHOXYETHANOL

METHYL CHLOROFORM see 1,1,1-TRICHLOROETHANE

METHYL CHLOROFORMATE see METHYL CHLOROCARBONATE

METHYL CHLOROMETHYL ETHER see CHLOROMETHYL METHYL ETHER

2-(2-METHYL-4-CHLOROPHENOXY)PROPANOIC ACID see MECOPROP

METHYLCHLOROSILANE see METHYLTRICHLOROSILANE

METHYLCHOLANTHRENE see 3-METHYLCHLOANTHRENE

20-METHYLCHOLANTHRENE see 3-METHYLCHLOANTHRENE

METHYL CYANIDE EK-488 see ACETONITRILE

METHYLCYCLOPENTADIENYL MANGANESE TRICARBONYL see MANGANESE TRICARBONYL METHYLCYCLOPENTADIENYL

2-METHYLCYCLOPENTADIENYL MANGANESETRICARBONYL see MANGANESE TRICARBONYL METHYLCYCLOPENTADIENYL

METHYL-s-DEMETON see DEMETON-s-METHYL

METHYL 2-(4-(2,4-DICHLOROPHENOXY)PHENOXY)PROPIONATE see DICLOFOP METHYL

1-METHYL-2,4-DINITOBENZENE see 2,4-DINITROTOLUENE

1-METHYL-2,6-DINITOBENZENE see 2,6-DINITROTOLUENE

1-METHYL-3,4-DINITOBENZENE see 3,4-DINITROTOLUENE

METHYLENEDIPHENYL DIISOCYANATE see METHYLBIS(PHENYL-ISOCYANATE)

METHYLENEDIPHENYL DIISOCYANATE (4,4′)- see METHYL-BIS(PHENYLISOCYANATE)

6-METHYL-1,3-DITHIOLO[4,5-B]QUINOXALIN-2-ONE see CHINOMETH-IONAT

2-METHYLE-2-PROPENOIC ACID, ETHYL ESTER see ETHYL METHAC-RYLATE

METHYLENE ACETONE see METHYL VINYL KETONE

METHYLENE BIS(4-CYCLOHEXYLISOCYANATE) see 1,1-METHYLENE BIS(4-ISOCYANATOCYCLOHEXANE)

METHYLENE CHLORIDE see DICHLOROMETHANE

METHYLENE DIBROMIDE see METHYLENE BROMIDE

2,2′-METHYLENEBIPHENYL see FLUORENE

4,4′-METHYLENEBIS(2-CHLOROANILINE) see 4,4′-METHYLENEBIS(2-CHLOROANILINE)

METHYLENEBIS(3-CHLORO-4-AMINOBENZENE) see 4,4′-METHYLENE-BIS(2-CHLOROANILINE)

2,2′-METHYLENEBIS(4-CHLOROPHENOL) see DICHLOROPHENE

4,4′-METHYLENEBIS(ANILINE) see 4,4′-METHYLENEDIANILINE

4,4′-METHYLENEBIS (2-CHLORO-BENZENEAMINE) see 4,4′-METH-YLENEBIS(2-CHLOROANILINE)

4,4′-METHYLENEDI(PHENYLDIISOCYANATE) see METHYLBIS (PHENYLISOCYANATE)

METHYLENEDIANILINE see 4,4′-METHYLENEDIANILINE

(1,2-(METHYLENEDIOXY)-4-PROPYL)BENZENE see DIHYDROSAFROLE

1,2-(METHYLENEDIOXY)-4-PROPENYLBENZENE see ISOSAFROLE

3,4-(METHYLENEDIOXY)-1-PROPENYLBENZENE see ISOSAFROLE

(3,4-METHYLENEDIOXY-6-PROPYLBENZYL) (BUTYL) DIETHYLENE GLYCOL ETHER see PIPERONYL BUTOXIDE

3,4-METHYLENEDIOXY-ALLYLBENZENE see SAFROLE

1-(METHYLETHYL)-ETHYL 3-METHYL-4-(METHYLTHIO)PHENYL-PHOSPHORAMIDATE see FENAMIPHOS

1-METHYLETHYL BENZENE see CUMENE

METHYL ETHYL METHANE see BUTANE

1-METHYLETHYLAMINE see ISOPROPYLAMINE

2-METHYLETHYLEN IMINE see PROPYLENEIMINE

METHYLETHYLENE see PROPYLENE

2-METHYLETHYLENIMINE see PROPYLENEIMINE

METHYLHYDRAZINE see METHYLMERCURIC DICYANAMIDE

1-METHYL HYDRAZINE see METHYL HYDRAZINE

n-METHYL HYDRAZINE see METHYL HYDRAZINE

METHYL HYDRIDE see METHANE

N,N′-((METHYLIMINO)DIMETHYLIDYNE)BIS(2,4-XYLIDINE) see AMITRAZ

3-METHYL-5-ISOPROPYLPHENYL-N-METHYL CARBAMATE see PROMECARB

4-METHYLMERCAPTO-3-METHYLPHENYLDIMETHYLTHIO-PHOSPHATE see FENTHION

METHYLMERCURIC CYANOGUANIDINE see METHYLMERCURIC DICYANAMIDE

METHYLMERCURY DICYANANDIMIDE see METHYLMERCURIC DICYANAMIDE

n-METHYLMETHANAMINE see DIMETHYLAMINE

METHYLMETHANE see ETHANE

METHYL METHANOATE see METHYL FORMATE

METHYL-N-(METHYL(CARBAMOYL)OXY)THIOACETIMIDATE see METHOMYL

METHYL-2-METHYLPROPENOATE see METHYL METHACRYLATE

s-METHYL N-((METHYLCARBAMOYL)OXY)THIOACETIMIDATE see METHOMYL

N-METHYL-N′-NITRO-N-NITROSOGUANIDINE see GUANIDINE, N-METHYL-N′-NITRO-N-NITROSO-

1-METHYL-3-NITRO-1-NITROSOGUANIDINE see GUANIDINE, N-METHYL-N′-NITRO-N-NITROSO-

m-METHYLNITROBENZENE see m-NITROTOLUENE

o-METHYLNITROBENZENE see o-NITROTOLUENE

p-METHYLNITROBENZENE see p-NITROTOLUENE

4-METHYLNITROBENZENE see p-NITROTOLUENE

N-METHYL-N-NITROSOMETHANAMINE see N-NITROSODIMETHYL-AMINE

N-METHYL-N-NITROSOUREA see N-NITROSO-N-METHYLUREA

N-METHYL-N-NITROSOURETHANE see N-NITROSO-N-METHYLURE-THANE

METHYLNITROSOURETHANE see N-NITROSO-N-METHYLURETHANE

METHYLOLACRYLAMIDE see N-METHYLOLACRYLAMIDE

METHYLOXIRANE see PROPYLENE OXIDE

4-METHYLPENTAN-2-ONE see METHYL ISOBUTYL KETONE

2-METHYL-4-PENTANONE see METHYL ISOBUTYL KETONE

4-METHYL-2-PENTANONE see METHYL ISOBUTYL KETONE

METHYLPHENOL see CRESOL (MIXED ISOMERS)

2-METHYLPHENOL see o-CRESOL

METHYLPHENYLDICHLOROSILANE see DICHLOROMETHYLPHENYL-SILANE

4-METHYL-PHENYLENE DIISOCYANATE see TOLUENE 2,4-DIISOCYA-NATE

METHYLPHENYLENE ISOCYANATE see TOLUENEDIISOCYANATE (MIXED ISOMERS)

METHYL-m-PHENYLENE ISOCYANATE see TOLUENEDIISOCYANATE (MIXED ISOMERS)

2-METHYL-m-PHENYLENE ISOCYANATE see TOLUENE-2,6-DIISOCYA-NATE

4-METHYL-PHENYLENE ISOCYANATE see TOLUENE 2,4-DIISOCYA-NATE

METHYLPHOSPHONOFLUORIDIC ACID ISOPROPYL ESTER see SARIN

METHYLPHOSPHONOTHIOIC ACID-O-ETHYL O-(4-(METHYLTHI-O)PHENYL)ESTER (9CI) see PHOSPHONOTHIOIC ACID, METHYL-,O-ETHYL O-(4-(METHYLTHIO)PHENYL)ESTER

METHYLPHOSPHONOTHIOIC ACID-O-(4-NITROPHENYL)-O-PHENYL ESTER see PHOSPHONOTHIOIC ACID, METHYL-,O-(4-NITROPHE-NYL) O-PHENYL) ESTER

2-METHYLPROPAN-1-OL see ISOBUTYL ALCOHOL

2-METHYL-PROPAN-2-OL see tert-BUTYL ALCOHOL

2-METHYLPROPANE see ISOBUTANE

2-METHYL-1-PROPANOL see ISOBUTYL ALCOHOL

2-METHYL-2-PROPANOL see tert-BUTYL ALCOHOL

2-METHYL-2-PROPENE-1,1′-DIOL DIACETATE see METHACROLEIN DI-ACETATE

2-METHYL-2-PROPENENITRILE see METHACRYLONITRILE

METHYL-2-PROPENOATE see METHYL ACRYLATE

2-METHYL-2-PROPENOIC ACID ANHYDRIDE (9CI) see METHACRYLIC ANHYDRIDE

2-METHYL-2-PROPENOYL CHLORIDE see METHACRYLOYL CHLO-RIDE

METHYLPROPIONIC ACID, 2- see ISO-BUTYRIC ACID

2-METHYLPROPIONITRILE see ISOBUTYRONITRILE

1-METHYL PROPYL 2,4-D see 2,4-D sec-BUTYL ESTER

2-METHYLPROPYL ACETATE see ISO-BUTYL ACETATE

1-METHYL PROPYL ACETATE see sec-BUTYL ACETATE

2-METHYLPROPYLAMINE see ISO-BUTYLAMINE

5-METHYL-1H-PYRAZOL-3-YL DIMETHYLCARBAMATE see DIMETILAN

2-METHYLPYRIDINE see 2-METHYLPYRIDINE

N-METHYLPYRROLIDINONE see N-METHYL-2-PYRROLIDONE

N-METHYL-2-PYRROLIDINONE see N-METHYL-2-PYRROLIDONE

(S)-3-(1-METHYL-2-PYRROLIDINYL)PYRIDINE SULFATE (2:1) see NICOTINE SULFATE

1-METHYL-2-PYRROLIDONE see N-METHYL-2-PYRROLIDONE

3-(1-METHYL-2-PYRROLIDYL) PYRIDINE see NICOTINE

6-METHYL-2,3-QUINOXALIN DITHIOCARBONATE see CHINOMETHIONAT

METHYL SULFATE see DIMETHYL SULFATE

METHYL SULFOCYANATE see METHYL THIOCYANATE

METHYL SULFOCYANATE see THIOCYANIC ACID, METHYL ESTER

2-(METHYLTHIO)-ETHANETHIOL-O,O-DIMETHYL PHOSPHOROTHIOATE see PHOSPHOROTHIOIC ACID, O,O-DIMETHYL-5-(2-(METHYLTHIO)ETHYL)ESTER

4-(METHYLTHIO)-3,5-XYLYL-N-METHYLCARBAMATE see METHIOCARB

METHYL THIOCYANATE see THIOCYANIC ACID, METHYL ESTER

METHYLTHIOKYANAT see METHYL THIOCYANATE

6-METHYL-2-THIOURACIL see METHYLTHIOURACIL

METHYL TOLUENE see XYLENE (MIXED ISOMERS)

1,2-METHYLTOLUENE see o-XYLENE

4-METHYLTOLUENE see p-XYLENE

m-METHYLTOLUENE see m-XYLENE

o-METHYLTOLUENE see o-XYLENE

p-METHYLTOLUENE see p-XYLENE

METHYL TRIBROMIDE see BROMOFORM

METHYL VINYL ETHER see VINYL METHYL ETHER

2-METHYL-5-VINYLPYRIDINE see PYRIDINE, 2-METHYL-5-VINYL

MOLECULAR CHLORINE see CHLORINE

MOLYBDENUM(VI) OXIDE see MOLYBDENUM TRIOXIDE

MOLYBDIC ACID ANHYDRIDE see MOLYBDENUM TRIOXIDE

MONOBROMOMETHANE see BROMOMETHANE

MONOBUTYLAMINE see BUTYLAMINE

MONOCHLORBENZENE see CHLOROBENZENEBENZENE

MONOCHLORODIFLUOROMETHANE see CHLORODIFLUOROMETHANE

MONOCHLOROETHANE see CHLOROETHANE

MONOCHLOROMETHANE see CHLOROMETHANE

MONOCHLOROMETHYL METHYL ETHER see CHLOROMETHYL METHYL ETHER

MONOCHLOROTETRAFLUOROETHANE see CHLOROTETRAFLUOROETHANE

MONOCHLOROTRIFLUOROMETHANE see CHLOROTRIFLUOROMETHANE

MONOETHYLAMINE see ETHANAMINE

MONOFLUROETHYLENE see VINYL FLUORIDE

MONOFLUROTRICHLOROMETHANE see TRICHLOROFLUOROMETHANE

MONOMETHYLAMINE see METHANAMINE

MONOMETHYLHYDRAZINE see METHYL HYDRAZINE

MORPHOLINE, 4-NITROSO- see N-NITROSOMORPHOLINE

MURIATIC ACID see HYDROCHLORIC ACID

2-NAPHTHACENECARBOXAMIDE,4-(DIMETHYLAMINO)-1,4,4A,5,5A,6,11,12A-OCTAHYDRO-3,6,10,12,12A-PENTAHYDROXY-6-METHYL-1,11-DIOXO-, MONOHYDROCHLORIDE, (4S (4α,4α,5α,6β,12α))- see TETRACYCLINE HYDROCHLORIDE

5,12-NAPHTHACENEDIONE, 8-ACETYL-10-((3-AMINO-2,3,6-TRIDEOXY-α-L-LYXO-HEXOPYRANOSYL)OXY)-7,8,9,10-TETRAHYDRO-6,8,11-TRIHYDROXY-8-(HYDROXYACETYL)-1-METHOXY HYDROCHLORIDE see DAUNOMYCIN

2-NAPHTHALENAMINE, N,N-BIS(2-CHLOROETHYL)- see CHLORNAPHAZINE

1-NAPHTHALENAMINE see α-NAPHTHYLAMINE

2-NAPHTHALENAMINE see β-NAPHTHYLAMINE

NAPHTHALENE, 2-CHLORO- see 2-CHLORONAPHTHALENE

NAPHTHALENE, 1,5-DIISOCYANATO- see 1,5-NAPHTHALENE DIISOCYANATE

NAPHTHALENE, HEXACHLORO- see HEXACHLORONAPHTHALENE

NAPHTHALENE, OCTACHLORO- see OCTACHLORONAPHTHALENE

para-NAPHTHALENE see ANTHRACENE

1,4-NAPHTHALENEDIONE see 1,4-NAPHTHOQUINONE

2,7-NAPHTHALENEDISULFONIC ACID, 4-AMINO-3-[[4′-[(2,4-DIAMINOPHENYL)AZO][1,1′-BIPHENYL]-4-YL]AZO]-5-HYDROXY-6-(PHENYLAZO)-,DISODIUM SALT see C.I. DIRECT BLACK 38

2,7-NAPHTHALENEDISULFONIC, 3,3′-[[1,1′-BIPHENYL]-4,4′-DIYBIS(AZO)BIS[5-AMINO-4-HYDROXYTETRASODIUM SALT see C.I. DIRECT BLUE 6

2,7-NAPHTHALENEDISULFONIC ACID, 3,3′-((3,3′-DIMETHYL(1,1′-BIPHENYL)-4,4′-DIYL)BIS(AZO))BIS(5-AMINO-4-HYDROXY-, TETRASODIUM SALT see TRYPAN BLUE

2,7-NAPHTHALENEDISULFONIC ACID, 3,3′-((3,3′-DIMETHYL-4,4′-BIPHENYLYLENE)BIS(AZO))BIS(5-AMINO-4′-HYDROXY-, TETRASODIUM SALT see TRYPAN BLUE

NITRIC ACID, NICKEL(2+) SALT see NICKEL NITRATE
NITRIC ACID, NICKEL SALT see NICKEL NITRATE
NITRIC ACID, SILVER(I) SALT see SILVER NITRATE
NITRIC ACID, THALLIUM(1+) SALT see THALLIUM(I) NITRATE
NITRIC ACID, ZINC SALT see ZINC NITRATE
NITRIC ACID, ZIRCONIUM(IV) SALT see ZIRCONIUM NITRATE
2-2′,2″-NITRILOTRIS-DODECYLBENZENESULFONATE (SALT) see TRIETHANOLAMINE DODECYLBENZENE SULFONATE
4-NITRO-2-AMINOTOLUENE see 5-NITRO-o-TOLUENE
ortho-NITROANILINE see o-NITROANILINE
2-NITROANILINE see o-NITROANILINE
para-NITROANILINE see p-NITROANILINE
4-NITROANILINE see p-NITROANILINE
5-NITRO-ortho-ANISIDINE see 5-NITRO-O-ANISIDINE
p-NITROBENZYL CHLORIDE see BENZENE, 1-(CHLOROMETHYL)-4-NITRO-
4′-NITRO-2,4-DICHLORODIPHENYL ETHER see NITROFEN
4-NITRO-2′,4′-DICHLORODIPHENYL ETHER see NITROFEN
NITROGEN MONOXIDE see NITRIC OXIDE
NITROGEN OXIDE (N_2O_4) see NITROGEN DIOXIDE
NITROGEN OXIDE (NO_2) see NITROGEN DIOXIDE
5-NITRO-2-METHOXYANILINE see 5-NITRO-O-ANISIDINE
5-NITRO-2-METHYLANILINE see 5-NITRO-o-TOLUENE
3-NITRO-6-METHYLANILINE see 5-NITRO-o-TOLUENE
N′-NITRO-N-NITROSO-N-METHYLGUANIDINE see GUANIDINE, N-METHYL-N′-NITRO-N-NITROSO-
NITROPENTACHLOROBENZENE see QUINTOZENE
o-NITROPHENOL see 2-NITROPHENOL
3-NITROPHENOL see m-NITROPHENOL
4-NITROPHENOL see p-NITROPHENOL
NITROPHENOLS see NITROPHENOL (MIXED ISOMERS)
N-(4-NITROPHENYL)-N′-(3-PYRIDINYLMETHYL)UREA see PYRIMINIL
p-NITROPHENYL DIEYHYLPHOSPHATE see DIETHYL-p-NITROPHENYL PHOSPHATE
o-NITROPHENYLAMINE see o-NITROANILINE
p-NITROPHENYLAMINE see p-NITROANILINE
β-NITROPROPANE see 2-NITROPROPANE
4-NITROPYRIDINE-1-OXIDE see PYRIDINE, 4-NITRO-, 1-OXIDE
5-NITRO-2-TOLUIDINE see 5-NITRO-o-TOLUENE
5-NITRO-ortho-TOLUIDINE see 5-NITRO-o-TOLUENE
N-NITROSODIBUTYLAMINE see N-NITROSODI-n-BUTYLAMINE
NITROSODIMETHYLAMINE see N-NITROSODIMETHYLAMINE
NITROSODIPROPYLAMINE see N-NITROSODI-N-PROPYLAMINE

4-NITROSODIPHENYLAMINE see p-NITROSODIPHENYLAMINE
N-NITROSODIPROPYLAMINE see N-NITROSODI-N-PROPYLAMINE
N-NITROSO-N-METHYLUREA CARBAMIDE see N-NITROSO-N-METHYLUREA
1-NITROSO-1-METHYLUREA see N-NITROSO-N-METHYLUREA
NITROSOMETHYLURETHANE see N-NITROSO-N-METHYLURETHANE
N-NITROSO-N-METHYLURETHANE see N-NITROSO-N-METHYLURETHANE
4-NITROSOMORPHOLINE see N-NITROSOMORPHOLINE
N'-NITROSONORNICOTINE see N-NITROSONORNICOTINE
para-NITROSOPHENYLANILINE see p-NITROSODIPHENYLAMINE
1-NITROSOPIPERIDINE see n-NITROSOPIPERIDINE
1-NITROSOPYRROLIDINE see N-NITROSOPYRROLIDINE
NITROSYL ETHOXIDE see ETHYL NITRITE
NITROTOLUENE (all isomers) see NITROTOLUENE
NITROTOLUENE (mixed isomers) see NITROTOLUENE
2-NITROTOLUENE see o-NITROTOLUENE
3-NITROTOLUENE see m-NITROTOLUENE
4-NITROTOLUENE see p-NITROTOLUENE
NITROTRICHLOROMETHANE see CHLOROPICRIN
NITROUS ACID ETHYL ESTER see ETHYL NITRITE
NITROUS ACID, SODIUM SALT see SODIUM NITRITE
NITROUS ETHYL ETHER see ETHYL NITRITE
1,2,4,5,6,7,8,8-OCTACHLORO-2,3,3A,4,7,7A-HEXAHYDRO-4,7-METHANO-1H-INDENE see CHLORDANE
1,2,3,4,5,6,7,8-OCTACHLORONAPHTHALENE see OCTACHLORONAPHTHALENE
OCTADECANOIC ACID, LEAD SALT see LEAD STEARATE
OCTADECANOIC ACID, LEAD(II) SALT see LEAD STEARATE
OCTADECANOIC ACID, LEAD SALT, DIBASIC see LEAD STEARATE
OCTAMETHYLPYROPHOSPHORAMIDE see DIPHOSPHORAMIDE, OCTAMETHYL-
OCTANOIC ACID,2,6-DIBROMO-4-CYANOPHENYL ESTER see BROMOXYNIL OCTANOATE
1,3,4,5,6,8,8-OCTOCHLORO-1,3,3a, 4,7,7a-HEXAHYDRO-4,7-METHANOISOBENZOFURAN see ISOBENZAN
OCTYL PHTHALATE, DI-SEC see DI(2-ETHYLHEXYL)PHTHALATE
n-OCTYLPHTHALATE see DI-n-OCTYL PHTHALATE
OLEFIANT GAS see ETHYLENE
OMAL see TRICHLOROPHENOL
ORDRAM see MOLINATE
ORGANORHODIUM COMPLEX PMN-82-147
ORTHODICHLOROBENZENE see o-DICHLOROBENZENE
ORTHONITROTOLUENE see o-NITROTOLUENE
ORTHOPHENYLENEDIAMINE see 1,2-PHENYLENEDIAMINE

ORTHOPHENYLPHENOL see 2-PHENYLPHENOL

ORTHOPHOSPHORIC ACID see PHOSPHORIC ACID ESTER

OSMIUM OXIDE (OsO_4) see OSMIUM TETROXIDE

OSMIUM OXIDE (OsO_4), (T-4)- see OSMIUM TETROXIDE

OSMIUM(IV) OXIDE see OSMIUM TETROXIDE

OUABAINE see OUABAIN

6-OXA-5,7-DISTANNAUNDECANE,5,5,7,7-TETRABUTYL- see BIS(TRIBUTYLTIN) OXIDE

7-OXABICYCLO(2.2.1)HEPTANE-2,3-DICARBOXYLIC ACID, DIPOTASSIUM SALT see DIPOTASSIUM ENDOTHALL

7-OXABICYCLO(2,2,1)HEPTANE-2,3-DICARBOXYLIC ACID see ENDOTHALL

OXACYCLOPENTADIENE see FURAN

OXACYCLOPROPANE see ETHYLENE OXIDE

1,2,4-OXADIAZOLIDINE-3,5-DIONE, 2-(3,4-DICHLOROPHENYL)-4-METHYL- see METHAZOLE

OXADIAZON see OXYDIAZON

OXALIC ACID, AMMONIUM IRON(III) SALT (3:3:1) see FERRIC AMMONIUM OXALATE

OXALIC ACID, DIAMMONIUM SALT see AMMONIUM OXALATE

OXALONITRILE see CYANOGENBROMINE CYANIDE see CYANOGEN BROMIDE

1,4-OXATHIIN-3-CARBOXANILIDE,5,6-DIHYDRO-2-METHYL- see CARBOXIN

1,2-OXATHROLANE 2,2-DIOXIDE see 1,3-PROPANE SULTONE

2H-1,3,2-OXAZAPHOSPHORIN-2-AMINE, N,N-BIS(2-CHLOROETHYL)TETRAHYDRO-, 2-OXIDE see CYCLOPHOSPHAMIDE

1,3,4-OXAZOL-2(3 H)-ONE, 3-(2,4-DICHLORO-5-(1-METHYLETHOXY)PHENYL)-5-(1,1-DIMETHYLETHYL)- see OXYDIAZON

2,4-OXAZOLIDINEDIONE, 3-(3,5-DICHLOROPHENYL)-5-METHYL-5-VINYL- see VINCLOZOLIN

2-OXETANONE see β-PROPIOLACTONE

OXIRANE, (CHLOROMETHYL)- see EPICHLOROHYDRIN

OXIRANE, ETHYL- see 1,2-BUTYLENE OXIDE

OXIRANE see ETHYLENE OXIDE

OXIRANE, METHYL- see PROPYLENE OXIDE

OXIRANE, 2,2′-OXYBIS (METHYLENE) BIS- see DIGLYCIDYL ETHER

OXIRANE, PHENYL- see STYRENE OXIDE

OXIRANE,2,2′-(1,3-PHENYLENEBIS(OXYMETHYLENE))BIS- see DIGLYCIDYL RESORCINOL ETHER

1,1′-OXYBIS(2-CHLORO)ETHANE see BIS(2-CHLOROETHYL)ETHER

1,1′-OXYBIS(4-ISOCYANATOBENZENE) see 4,4′-DIISOCYANATODIPHENYL ETHER

OXYBISMETHANE see METHYL ETHER

2,2′-OXYBIS(METHYLENE)BISOXIRANE see DIGLYCIDYL ETHER

10-10′-OXYBISPHENOXYARSINE see PHENOXARSINE, 10,10′-OXYDI-

OXYCARBON SULFIDE see CARBONYL SULFIDE

4,4′-OXYDIANILINE see 4,4′-DIAMINOPHENYL ETHER

OXYGEN mol (O_3) see OZONE

OXYSULFATOVANADIUM see VANADYL SULFATE

PARACHLOROMETA CRESOL see p-CHLORO-m-CRESOL

PARADICHLOROBENZENE see 1,4-DICHLOROBENZENE

PARANITROTOLUENE see p-NITROTOLUENE

PARAOXON see DIETHYL-p-NITROPHENYL PHOSPHATE

PARAPHENYLENEDIAMINE see p-PHENYLENEDIAMINE

PARAQUAT BIS(METHYL SULFATE) see PARAQUAT METHOSULFATE

PARAQUAT DIMETHYL SULFATE see PARAQUAT METHOSULFATE

PARATHION-METHYL see METHYL PARATHION

PARIS GREEN see CUPRIC ACETOARSENITE

PCB-1016 see AROCLOR 1016

PCB-1221 see AROCLOR 1221

PCB-1232 see AROCLOR 1232

PCB-1254 see AROCLOR 1254

PCB-1242 see AROCLOR 1242

PCB-1248 see AROCLOR 1248

PCB-1260 see AROCLOR 1260

PCBs see POLYCHLORINATED BIPHENYLS

PCNB see QUINTOZENE

PCP see PENTACHLOROPHENOL

PEAR OIL see sec-AMYL ACETATE

PENTABORANE(9) see PENTABORANE

PENTACARBONYLIRON see IRON PENTACARBONYL

PENTACHLORONITROBENZENE see QUINTOZENE

PENTACHLOROPHENATE SODIUM see SODIUM PENTACHLORO-PHENATE

PENTACHLOROPHENOL, SODIUM SALT see SODIUM PENTACHLORO-PHENATE

1-PENTADECANAMINE see PENTADECYLAMINE

n-PENTADECYLAMINE see PENTADECYLAMINE

1-PENTADECYLAMINE see PENTADECYLAMINE

PENTADIENE, 1,3- see 1,3-PENTADIENE

trans-1,3-PENTADIENE see 1,3-PENTADIENE

1,1,1,3,3-PENTAFLUORO-2,2-DICHLOROPROPANE see 2,2-DICHLORO-1,1,1,3,3-PENTAFLUOROPROPANE

PENTAFLUOROMONOCHLOROETHANE see MONOCHLOROPENTA-FLUOROETHANE

PENTAMETHYLENEIMINE see PIPERIDINE

PENTANE, 2,2,4-TRIMETHYL- see 2,2,4-TRIMETHYLPENTANE

n-PENTANE see PENTANE

1-PENTANOL ACETATE see AMYL ACETATE

2-PENTANOL, ACETATE see sec-AMYL ACETATE

PENTYL ACETATE see AMYL ACETATE

tert-PENTYL ACETATE see tert-AMYL ACETATE

PERCHLORETHYLENE see TETRACHLOROETHYLENE

PERCHLOROBENZENE see HEXACHLOROBENZENE

PERCHLOROBUTADIENE see HEXACHLORO-1,3-BUTADIENE

PERCHLOROCYCLOPENTADIENE see HEXACHLOROCYCLOPENTADIENE

PERCHLOROMETHYLMERCAPTAN see PERCHLOROMETHYL MERCAPTAN

PERIETHYLENENAPHTHALENE see ACENAPHTHENE

PERMANGANIC ACID, POTASSIUM SALT see POTASSIUM PERMANGANATE

PEROXIDE, DIBENZOYL see BENZOYL PEROXIDE

PEROXOACETIC ACID see PERACETIC ACID

PHENACHLOR see TRICHLOROPHENOL

PHENANTHRIN see PHENANTHRENE

PHENANTRIN see PHENANTHRENE

PHENARSAZINE OXIDE see PHENOXARSINE, 10,10′-OXYDI-

PHENOL, 2-CHLORO see 2-CHLOROPHENOL

PHENOL, o-CHLORO see 2-CHLOROPHENOL

PHENOL, 2,4-DICHLORO- see 2,4-DICHLOROPHENOL

PHENOL, 4,4′-(1,2-DIETHYL-1,2-ETHENEDIYL)BIS-, (E)- see DIETHYLSTILBESTROL

PHENOL, DIMETHYL- see XYLENOL

PHENOL, 2,4-DIMETHYL- see 2,4-DIMETHYLPHENOL

PHENOL, 2,6-DIMETHYL- see 2,6-DIMETHYLPHENOL

PHENOL, 3,5-DIMETHYL-4-(METHYLTHIO)-, METHYLCARBAMATE see METHIOCARB

PHENOL, 4-(DIMETHYLAMINO)-3,5-DIMETHYL-METHYLCARBAMATE (ESTER) see MEXACARBATE

PHENOL, DINITRO- see DINITROPHENOL

PHENOL, α-DINITRO- see 2,4-DINITROPHENOL

PHENOL, 2-METHYL-4,6-DINITRO- see 4,6-DINITRO-o-CRESOL DINITROPHENOL, MIXED ISOMERS see DINITROPHENOL

PHENOL, 2,4-DINITRO- see 2,4-DINITROPHENOL

PHENOL, 2,5-DINITRO- see 2,5-DINITROPHENOL

PHENOL, 2,6-DINITRO- see 2,6-DINITROPHENOL

PHENOL, 4,4′-ISOPROPYLIDENEDI- see 4,4′-ISOPROPYLIDENEDIPHENOL

PHENOL, METHYL- see CRESOL (MIXED ISOMERS)

PHENOL, 2-METHYL see o-CRESOL

PHENOL, 3-METHYL- see m-CRESOL

PHENOL, 4-METHYL see p-CRESOL

PHENOL, 2,2′-METHYLENEBIS(4-CHLORO- see DICHLOROPHENE

PHENOL, 2-(1-METHYLETHOXY)-, METHYLCARBAMATE see PROPOXUR

PHENOL, 4,4′-(1-METHYLETHYLIDENE)BIS- see 4,4′-ISOPROPYLIDENEDIPHENOL

PHENOL, 2-(1-METHYLHEPTYL)-4,6-DINITRO-, CROTONATE (ESTER) see DINOCAP

PHENOL, 2-(1-METHYLPROPYL)-4,6-DINITRO- see DINITROBUTYL

PHENOL, 2-NITRO- see 2-NITROPHENOL

PHENOL, 3-NITRO- see m-NITROPHENOL

PHENOL, 4-NITRO see p-NITROPHENOL

PHENOL, o-NITRO see 2-NITROPHENOL

PHENOL, PENTACHLORO- see PENTACHLOROPHENOL

PHENOL, 2,3,4,6-TETRACHLORO- see 2,3,4,6-TETRACHLOROPHENOL

PHENOL, 2,2′-THIOBIS[4-CHLORO-6-METHYL- (9CI)- see PHENOL, 2,2′-THIOBIS(4-CHLORO-6-METHYL)-

PHENOL, TRICHLORO- see TRICHLOROPHENOL

PHENOL, 2,3,4-TRICHLORO- see 2,3,4-TRICHLOROPHENOL

PHENOL, 2,3,5-TRICHLORO- see 2,3,5-TRICHLOROPHENOL

PHENOL, 2,3,6-TRICHLORO- see 2,3,6-TRICHLOROPHENOL

PHENOL, 3,4,5-TRICHLORO- see 3,4,5-TRICHLOROPHENOL

PHENOL, 2,4,5-TRICHLORO- see 2,4,5-TRICHLOROPHENOL

PHENOL, 2,4,6-TRICHLORO- see 2,4,6-TRICHLOROPHENOL

PHENOL TRINITRATE see PICRIC ACID

PHENOL, 2,4,6-TRINITRO- see PICRIC ACID

PHENOL,2,4,6-TRINITRO-, AMMONIUM SALT (9CI) see AMMONIUM PICRATE

PHENOLDINITROCRESOL see 4,6-DINITRO-o-CRESOL

PHENOXARSINE, 10,10′-OXYDI- see PHENOXARSINE, 10,10′-OXYDI-

3-PHENOXYBENZYL (1RS)-cis-trans-3-(2,2-DICHLOROVINYL)-2,2-DIMETHYLCYCLOPROPANECARBOXYLATE see PERMETHRIN

m-PHENOXYBENZYL 2,2-DIMETHYL-3-(2-METHYLPROPENYL)CYCLOPROPANECARBOXYLATE see PHENOTHRIN

N-(2-(P-PHENOXY PHENOXY)ETHYL)CARBAMIC ACID see FENOXYCARB

2-(4-PHENOXYPHENOXY)ETHYL-CARBAMIC ACID ETHYL ESTER see FENOXYCARB

1-PHENYL- see ACETOPHENONE

PHENYLACETONITRILE see BENZYL CYANIDE

2-PHENYLACETONITRILE see BENZYL CYANIDE

L-PHENYLALANINE, 4(BIS(2-CHLOROETHYL)AMINO)-see MELPHALAN

PHENYLAMINE see ANILINE

3-PHENYL-5-AMINO-1,2,4-TRIAZOYL-(1)-(N,N′-TETRAMETHYL) DIAMIDOPHOSPHONATE see TRIAMIPHOS

PHENYLARSINEDICHLORIDE see DICHLOROPHENYLARSINE

PHENYLARSONIC ACID see BENZENEARSONIC ACID

PHENYLARSONOUS DICHLORIDE see DICHLOROPHENYLARSINE

4-(PHENYLAZO)-N,N-DIMETHYLANILINE see 4-DIMETHYLAMINO-AZOBENZENE

4-PHENYLAZO ANILINE see 4-AMINOAZOBENZENE

N-PHENYLBENZENAMINE see DIPHENYLAMINE

PHENYL CYANIDE see BENZOIC TRICHLORIDE

PHENYLDICHLOROARSINE see DICHLOROPHENYLARSINE

p-PHENYLENE DIISOCYANATE see 1,4-PHENYLENE DIISOCYANATE

[1,2-PHENYLENEBIS (IMINOCARBONOTHIOYL)] BISCARBAMIC ACID DIETHYL ESTER see THIOPHANATE ETHYL

o-PHENYLENEDIAMINE DIHYDROCHLORIDE see 1,2-PHENYLENEDI-AMINE DIHYDROCHLORIDE

p-PHENYLENEDIAMINE DIHYDROCHLORIDE see 1,4-PHENYLENEDI-AMINE DIHYDROCHLORIDE

PHENYLENEDIAMINE, meta see 1,3-PHENYLENEDIAMINE

PHENYLENEDIAMINE, ortho- see 1,2-PHENYLENEDIAMINE

PHENYLENEDIAMINE, para- see p-PHENYLENEDIAMINE

1,4-PHENYLENEDIAMINE see p-PHENYLENEDIAMINE

3-PHENYLENEDIAMINE see 1,3-PHENYLENEDIAMINE

o-PHENYLENEDIAMINE see 1,2-PHENYLENEDIAMINE

m-PHENYLENEDIAMINE see 1,3-PHENYLENEDIAMINE

o-PHENYLENEPYRENE see INDENO[1,2,3-cd]PYRENE

PHENYLETHANE see ETHYLBENZENE

PHENYLETHYLENE see STYRENE

PHENYLHYDRAZINE MONOHYDROCHLORIDE see PHENYLHYDRA-ZINE HYDROCHLORIDE

PHENYLHYDRAZINIUM CHLORIDE see PHENYLHYDRAZINE HYDRO-CHLORIDE

PHENYLIC ACID see PHENOL

PHENYL MERCAPTAN see BENZENETHIOL

PHENYLMERCURIC ACETATE see PHENYLMERCURY ACETATE

5-(PHENYLMETHYL)-3-FURANYL] METHYL 2,2-DIMETHYL-3-(2-METHYL-1-PROPENYL)CYCLOPROPANECARBOXYLATE see RES-METHRIN

PHENYL METHYL KETONE see ACETOPHENONE

PHENYLMETHYLDICHLOROSILANE see DICHLOROMETHYLPHENYL-SILANE

PHENYLOXIRANE see STYRENE OXIDE

2-PHENYLOXIRANE see STYRENE OXIDE

PHOSPHORIC ACID, 2-DICHLOROETHENYL DIMETHYL ESTER see DICHLORVOS

PHOSPHORIC ACID, 2,2-DICHLOROVINYL DIMETHYL ESTER see DICHLORVOS

PHOSPHORIC ACID, DIMETHYL 1-METHYL-3-(METHYLAMINO)-3-OXO-1-PROPENYL ESTER (E)- see MONOCROPTOPHOS

PHOSPHORIC ACID, DIMETHYL ESTER, ESTER with CIS-3-HYDROXY-N-METHYLCROTONAMIDE see MONOCROPTOPHOS

PHOSPHORIC ACID, DIMETHYL p-(METHYLTHIO)PHENYL ESTER see PHOSPHORIC ACID, DIMETHYL 4-(METHYLTHIO)PHENYL ESTER O,O-DIMETHYL-5-(2-(METHYLTHIO)ETHYL)ESTER

PHOSPHORIC ACID, 3-(DIMETHYLAMINO)-1-METHYL-3-OXO-1-PROPENYLDIMETHYL ESTER, (E)- see DICROTOPHOS

PHOSPHORIC ACID, DISODIUM SALT see SODIUM PHOSPHATE, DIBASIC

PHOSPHORIC ACID, DISODIUM SALT, DODECAHYDRATE see SODIUM PHOSPHATE, DIBASIC

PHOSPHORIC ACID, DISODIUM SALT, HYDRATE see SODIUM PHOSPHATE, DIBASIC

PHOSPHORIC ACID, LEAD(II) SALT (2:3) see LEAD PHOSPHATE

PHOSPHORIC ACID TRIS(2,3-DIBROMOPROPYL)ESTER see TRIS(2,3-DIBROMOPROPYL) PHOSPHATE

PHOSPHORIC ACID, TRISODIUM SALT see SODIUM PHOSPHATE, TRIBASIC

PHOSPHORIC ACID, TRISODIUM SALT, DECAHYDRATE see SODIUM PHOSPHATE, TRIBASIC

PHOSPHORIC ACID, TRISODIUM SALT, DODECAHYDRATE see SODIUM PHOSPHATE, TRIBASIC

PHOSPHORIC ANHYDRIDE see PHOSPHORUS PENTOXIDE

PHOSPHORIC CHLORIDE see PHOSPHORUS OXYCHLORIDE

PHOSPHORIC CHLORIDE see PHOSPHORYL CHLORIDE

PHOSPHOROTHIOIC ACID, O-(4-BROMO-2-CHLOROPHENYL)-O-ETHYL-S-PROPYL ESTER see PROFENOFOS

PHOSPHORDITHIOATE see METHIDATHION

PHOSPHOROFLUORIDIC ACID, BIS(1-METHYLETHYL)ESTER see DIISOPROPYLFLUOROPHOSPHATE

PHOSPHORODITHIOIC ACID S-((tert-BUTYLTHIO)METHYL)-O,O-DIETHYLESTER see TERBUFOS

PHOSPHOROTHIOIC ACID, O-(3-CHLORO-4-METHYL-2-OXO-2H-1-BENZOPYRAN-7-YL) O,O-DIETHYL ESTER see COUMAPHOS

PHOSPHORODITHIOIC ACID-S-(2-CHLORO-1-PHTHALIMIDOETHYL)-O,) DIETHYL ESTER see DIALIFOR1,3-BENZENEDIAMINE,4-METHOXY- see 2,4-DIAMINOSOLE

PHOSPHOROTHIOIC ACID, O-(2-(DIETHYLAMINO)-6-METHYL-4-PYRIMIDINYL)O,O-DIMETHYL ESTER see PIRIMIPHOS METHYL

PHOSPHOROTHIOIC ACID, O,O-DIETHYL-O-(4-NITROPHENYL) ESTER see PARATHION

PHOSPHORODITHIOIC ACID, O,O-DIETHYL ESTER, S,S-DIESTER with METHANEDITHIOL see ETHION

PHOSPHORODITHIOIC ACID, O,O-DIETHYL S-[(ETHYLTHIO)METHYL] ESTER see PHORATE

PHOSPHOROTHIOIC ACID,O,O-DIETHYL O-2-(ETHYLTHIO)ETHYL ESTER, mixed with O,O-DIETHYL S-2-(ETHYLTHIO)ETHYL PHOSPHOROTHIOATE see DEMETON

PHOSPHOROTHIOIC ACID, O,O-DIETHYL O-(2-ISOPROPYL-6-METHYL-4-PYRIMIDINL) ESTER see DIAZINON

O-(6-METHYL-2-(1-METHYLETHYL)-4-PYRIMIDINYL) ESTER see DIAZINON

PHOSPHOROTHIOIC ACID, O,O-DIETHYL O-(P-(METHYLSULFINYL)PHENYL) ESTER see FENSULFOTHION

PHOSPHORODITHIOC ACID, O,O-DIETHYL S-METHYL ESTER see O,O-DIETHYL S-METHYL DITHIOPHOSPHATE

PHOSPHOROTHIOIC ACID,O,O-DIETHYL O-PYRAZINYL ESTER see ZINOPHOS

PHOSPHOROTHIOIC ACID, O-(4-((DIMETHYLAMINO)SULFONYL) PHENYL) O,O-DIMETHYL ESTER see FAMPHUR

PHOSPHORODITHIOIC ACID, O,O-DIMETHYL S-((4-OXO-1,2,3-BENZOTRIAZIN-3(4H)-YL)METHYL)ESTER see AZINPHOS-METHYL

PHOSPHOROTHIOIC ACID, O,O-DIMETHYL O-(3-METHYL-4-(METHYLTHIO)PHENYL) ESTER see FENTHION

PHOSPHOROTHIOIC ACID, O,O-DIMETHYL O-(4-NITROPHENYL) ESTER see METHYL PARATHION

PHOSPHORODITHIOIC ACID-5-5′-1,4-DIOXANE-2,3-DIYL, O,O,O′,O′-TETRAETHYL ESTER see DIOXATHION

PHOSPHORODITHIOIC ACID, O-ETHYL S,S-DIPROPYL ESTER see ETHOPROP

PHOSPHORODITHIOIC ACID, O-ETHYL O-(4-(METHYLTHIO)PHENYL) S-PROPYL ESTER see SULPROFOS

PHOSPHORODITHIONIC ACID,S-2-(ETHYLTHIO)ETHYL-O,O-DIETHYLESTER see DISULFOTON

PHOSPHORODITHIOIC ACID, S,S′-METHYLENE O,O,O′,O′-TETRAETHYL ESTER see ETHION

PHOSPHORODITHIOIC ACID,S-((5-METHOXY-2-OXO-1,3,4-THIADIAZOL-3(2H)-YL)METHYL) O,O-DIMETHYL ESTER S-((5-METHOXY-2-OXO-1,3,4-THIADIAZOL-3(2H)-YL)METHYL)-O,O-DIMETHYL

PHOSPHOROTHIOIC ACID, O,O′-(THIODI-4,1-PHENYLENE)O,O,O′,O′-TETRAMETHYL ESTER see TEMEPHOS

PHOSPHOROTRITHIOUS ACID, TRIBUTYL ESTER see MERPHOS

PHOSPHOROUS CHLORIDE see PHOSPHOROUS TRICHLORIDE

PHOSPHORUS CHLORIDE OXIDE see PHOSPHORYL CHLORIDE

PHOSPHORUS CHLORIDE OXIDE see PHOSPHORUS OXYCHLORIDE

PHOSPHORUS ELEMENTAL, WHITE see PHOSPHORUS

PHOSPHORUS(V) OXIDE see PHOSPHORUS PENTOXIDE

PHOSPHORUS OXYCHLORIDE see PHOSPHORYL CHLORIDE

PHOSPHORUS OXYTRICHLORIDE see PHOSPHORYL CHLORIDE

PHOSPHORUS PENTASULFIDE see SULFUR PHOSPHIDE

PHOSPHORUS PERCHLORIDE see PHOSPHORUS PENTACHLORIDE

PHOSPHORUS SULFIDE see SULFUR PHOSPHIDE

PHOSPHORUS TRICHLORIDE see PHOSPHOROUS TRICHLORIDE

PHOSPHORUS YELLOW see PHOSPHORUS

PHOSPHORUS (YELLOW OR WHITE) see PHOSPHORUS

PHOSPHORYL CHLORIDE see PHOSPHORUS OXYCHLORIDE

PHOSPHORYL TRICHLORIDE see PHOSPHORUS OXYCHLORIDE

PHOSTOXIN see ALUMINUM PHOSPHIDE

PHSOSTOL SALICYLATE SALICYLIC ACID with PHYSOSTIGMINE (1:1) see PHYSOSTIGMINE, SALICYLATE (1:1)

PHTHALIC ACID, DIOCTYL ESTER see DI-n-OCTYL PHTHALATE

PHTHALIMIDOMETHYL O,O-DIMETHYL PHOSPHORODITHIOATE see PHOSMET

PHYGON see DICHLONE

PHYSOSTOL see PHYSOSTIGMINE

PICHLORAM see PICLORAM

2-PICOLINE see 2-METHYLPYRIDINE

o-PICOLINE see 2-METHYLPYRIDINE

PICRIC ACID, AMMONIUM SALT see AMMONIUM PICRATEAMMONIUM FLUOROSILICATE see AMMONIUM SILICOFLUORIDE

N,N′-[1,4-PIPERAZINEDIYLBIS(2,2,2-TRICHLOROETHYLIDENE)] BISFORMAMIDE see TRIFORINE

1,1′-PIPERAZINE-1,4-DIYLDI-(N-(2,2,2-TRICHLOROETHYL) FORMAMIDE) see TRIFORINE

PIPERIDINE, 1-NITROSO see n-NITROSOPIPERIDINE

2,6-PIPERIDINEDIONE, 4-[2-3,5-DIMETHYL-2-OXOCYCLOHEXYL)-2-HYDROXYETHYL]-,[IS-[1α(S*),3α,5β]- see CYCLOHEXIMIDECYCLOHEXANAMINE see CYCLOHEXYLAMINE

PLUMBANE, TETRAETHYL- see TETRAETHYL LEAD

PLUMBANE, TETRAMETHYL- see TETRAMETHYL LEAD

POLY(OXY-1,2-ETHANEDIYL),α-[2-[BIS(2-AMINOETHYL)METHYLAMMONIO]ETHYL]-ω-HYDROXY-, N,N′-DICOCOACYL DERIVATIVES, METHYL SULFATES see CYFLUTHRIN

POLYCHLORINATED BIPHENYL (AROCLOR 1016) see AROCLOR 1016

POLYCHLORINATED BIPHENYL (AROCLOR 1221) see AROCLOR 1221

POLYCHLORINATED BIPHENYL (AROCLOR 1232) see AROCLOR 1232

POLYCHLORINATED BIPHENYL (AROCLOR 1242) see AROCLOR 1242

POLYCHLORINATED BIPHENYL (AROCLOR 1248) see AROCLOR 1248

POLYCHLORINATED BIPHENYL (AROCLOR 1254) see AROCLOR 1254

POLYCHLORINATED BIPHENYL (AROCLOR 1260) see AROCLOR 1260

POLYCHLORINATED CAMPHENE see TOXAPHENE

POLYMERIC MDI see POLYMERIC DIPHENYLMETHANE DIISOCYANATE

POLYMETHYLENE POLYPHENYL POLYISOCYANATE see POLYMERIC DIPHENYLMETHANE DIISOCYANATE

POLYMETHYLENE POLYPHENYLENE ISOCYANATE see POLYMERIC DIPHENYLMETHANE DIISOCYANATE

POLYMETHYL POLYPHENYL POLYISOCYANATE see POLYMERIC DIPHENYLMETHANE DIISOCYANATE

POLY[OXY(METHYL)-1,2-ETHANEDIYL],α-(2,4-DICHLOROPHENOXY)ACETYL-ω-BUTOXY- see 2,4-D ESTERS

PONCEAU MX see C.I. FOOD RED 5

POTASSIUM DIHYDROGEN ARSENATE see POTASSIUM ARSENATE

POTASSIUM FLUOZIRCONATE see ZIRCONIUM POTASSIUM FLUORIDE

POTASSIUM HYDRATE see POTASSIUM HYDROXIDE

POTASSIUM HYDROGEN ARSENATE see POTASSIUM ARSENATE

POTASSIUM METHYLDITHIOCARBAMATE see POTASSIUM N-METHYLDITHIOCARBAMATE

PROP-2-YN-1-OL see PROPARGYL ALCOHOL

PROPADIENE see 1,2-PROPADIENE

n-PROPANAL see PROPIONALDEHYDE

PROPANAL see PROPIONALDEHYDE

1-PROPANAL see PROPIONALDEHYDE

PROPANAL, 2-METHYL- see ISOBUTYRALDEHYDE

PROPANAL, 2-METHYL-2-(METHYTHIO)-,O-((METHYLAMINO)CARBONYL)OXIME see ALDICARB

PROPANAL,2-METHYL-2-(METHYLSULFONYL)-O-[(METHYLAMINO)CARBONYL]OXIME see ALDICARB SULFONE

PROPANAMINE see n-PROPYLAMINE

1-PROPANAMINE, 2-METHYL- see ISO-BUTYLAMINE

1-PROPANAMINE, n-PROPYL- see DIPROPYLAMINE

1-PROPANAMINE, N-NITROSO-N-PROPYL- see N-NITROSODI-N-PROPYLAMINE

2-PROPANAMINE see ISOPROPYLAMINE

2-PROPANEAMINE, 2-METHYL- see tert-BUTYLAMINE

1-PROPANECARBOXYIC ACID see BUTYRIC ACID

PROPANE-2-CARBOXYLIC ACID see ISO-BUTYRIC ACID

PROPANE, 1-CHLORO-2,3-EPOXY- see EPICHLOROHYDRIN

PROPANE, 2-CHLORO see ISOPROPYL CHLORIDE

PROPANE, 1,2-DIBROMO-3-CHLORO- see 1,2-DIBROMO-3-CHLOROPROPANE

PROPANE, DICHLORO- see DICHLOROPROPANE

PROPANE, 1,1-DICHLORO- see 1,1-DICHLOROPROPANE

PROPANE, 1,1-DICHLORO-1,2,2,3,3-PENTAFLUORO- see 1,1-DICHLORO-1,2,2,3,3-PENTAFLUOROPROPANE

PROPANE, 1,1-DICHLORO-1,2,3,3,3-PENTAFLUORO- see 1,1-DICHLORO-1,2,3,3,3-PENTAFLUOROPROPANE

PROPANE, 1,2-DICHLORO see 1,2-DICHLOROPROPANE

PROPANE, 1,2-DICHLORO-1,1,2,3,3-PENTAFLUORO- see 1,2-DICHLORO-1,1,2,3,3-PENTAFLUOROPROPANE

PROPANE, 1,3-DICHLORO-1,1,2,2,3-PENTAFLUORO- see 1,3-DICHLORO-1,1,2,2,3-PENTAFLUOROPROPANE

PROPANE, 1,3-DICHLORO-1,1,2,3,3-PENTAFLUORO- see 1,3-DICHLORO-1,1,2,3,3-PENTAFLUOROPROPANE

PROPANE, 2,3-DICHLORO-1,1,1,2,3-PENTAFLUORO- see 2,3-DICHLORO-1,1,1,2,3-PENTAFLUOROPROPANE

PROPANE, DICHLOROPENTAFLUORO- see DICHLOROPENTAFLUOROPROPANE

PROPANE, 2,2-DIMETHYL- see 2,2-DIMETHYLPROPANE

PROPANEDINITRILE see MALONONITRILE

PROPANEDIOIC ACID, DITHALLIUM SALT see THALLOUS MALONATE

PROPANE, 1,2-EPOXY- see PROPYLENE OXIDE

PROPANE, 2-METHOXY-2-METHYL- see METHYL tert-BUTYL ETHER

PROPANE, 2-METHYL see ISOBUTANE

PROPANENITRILE see PROPIONITRILE

PROPANENITRILE, 2-METHYL- see ISOBUTYRONITRILE

PROPANENITRILE, 2-HYDROXY-2-METHYL- see 2-METHYLLACTONITRILE

PROPANE, 2-NITRO see 2-NITROPROPANE

PROPANE, 2,2′-OXYBIS(1-CHLORO- see BIS(2-CHLORO-1-METHYLETHYL) ETHER 1,3-DICHLOROACETONE see BIS(CHLOROMETHYL)KETONE

PROPANE SULTONE see 1,3-PROPANE SULTONE

1,2,3-PROPANETRICARBOXYLIC ACID, 2-HYDROXY-, AMMONIUM SALT see AMMONIUM CITRATE, DIBASIC

PROPANE, 1,2,3-TRICHLORO- see 1,2,3-TRICHLOROPROPANE

1,2,3-PROPANETROL, TRINITRATE see NITROGLYCERIN

PROPANOIC ACID, 2-(2,4-DICHLOROPHENOXY)- see 2,4-DP

PROPANOIC ACID, 2,2-DICHLORO- see 2,2-DICHLOROPROPIONIC ACID

PROPANOIC ACID, 2-METHYL- see ISO-BUTYRIC ACID

PROPANOIC ACID, 2-(2,4,5-TRICHLOROPHENOXY)- see SILVEX (2,4,5-TP)

PROPANOIC ACID, 2-(2,4,5-TRICHLOROPHENOXY)-, ISOOCTYL ESTER see 2,4,5-TP ESTERS

PROPANOIC ACID, 2-(4-((5-(TRIFLUOROMETHYL)-2-PYRIDINYL) OXY)PHENOXY)-,BUTYL ESTER see FLUAZIFOP-BUTYL

1-PROPANOL, 2,3-DIBROMO-, PHOSPHATE (3:1) see TRIS(2,3-DIBROMOPROPYL) PHOSPHATE

2-PROPANOL see ISOPROPYL ALCOHOL

1-PROPANOL, 2-METHYL- see ISOBUTYL ALCOHOL

2-PROPANOL, 2-METHYL- see tert-BUTYL ALCOHOL

2-PROPANONE see ACETONE

2-PROPANONE, 1-BROMO see BROMOACETONE

2-PROPANONE, 1,3-DICHLORO- see BIS(CHLOROMETHYL)KETONE

PROPARGIL see PROPARGITE

PROPELLANT 124 see 2-CHLORO-1,1,1,2-TETRAFLUOROETHANE

2-PROPEN-1-AMINE see ALLYLAMINE

2-PROPEN-1-OL see ALLYL ALCOHOL

2-PROPEN-1-ONE see ACROLEIN

2-PROPENAL see ACROLEIN

2-PROPENAMIDE see ACRYLAMIDE

2-PROPENAMIDE, N-(HYDROXYMETHYL)- see N-METHYLOLACRYL-AMIDE

2-PROPENAMINE see ALLYLAMINE

PROPENE see PROPYLENE

1-PROPENE see PROPYLENE

1-PROPENE, 1-CHLORO- see 1-CHLOROPROPYLENE

1-PROPENE, 2-CHLORO- see 2-CHLOROPROPYLENE

1-PROPENE, 3-CHLORO-2-METHYL- see 3-CHLORO-2-METHYL-1-PRO-PENE

1-PROPENE, 3-CHLORO see ALLYL CHLORIDE

3-PROPENE, 3-CHLORO see ALLYL CHLORIDE

1-PROPENE, DICHLORO- see DICHLOROPROPENE

PROPENE, 1,3-DICHLORO- see 1,3-DICHLOROPROPYLENE

1-PROPENE, 1,3-DICHLORO- mixed with 1,2-DICHLOROPROPANE see DI-CHLOROPROPANE - DICHLOROPROPENE (MIXTURE)

1-PROPENE, 1,3-DICHLORO-, (E)- see trans-1,3-DICHLOROPROPENE

1-PROPENE, 2,3-DICHLORO- see 2,3-DICHLOROPRENE

PROPENE, 2,3-DICHLORO see 2,3-DICHLOROPRENE

1-PROPENE, 1,1,2,3,3-HEXACHLORO- see HEXACHLOROPROPENE

1-PROPENE, 2-METHYL see 2-METHYLPROPENE

2-PROPENENITRILE see ACRYLONITRILE

2-PROPENENITRILE, 2-METHYL- see METHACRYLONITRILE

PROPENE OXIDE see PROPYLENE OXIDE

2-PROPENITRILE see ACRYLONITRILE

2-PROPENOIC ACID see ACRYLIC ACID

2-PROPENOIC ACID, BUTYL ESTER see BUTYL ACRYLATE

2-PROPENOIC ACID, ETHYL ESTER see ETHYL ACRYLATE

2-PROPENOIC ACID, 1-METHYL-, ETHYL ESTER see ETHYL METHAC-RYLATE

2-PROPENOIC ACID, 2-METHYL-, METHYL ESTER see METHYL METH-ACRYLATE

2-PROPENOIC ACID, METHYL ESTER see METHYL ACRYLATE

2-PROPENOYL CHLORIDE see ACRYLYL CHLORIDE

5-(1-PROPENYL)-1,3-BENZODIOXOLE see ISOSAFROLE

PROPENYL ALCOHOL see ALLYL ALCOHOL

PROPENYL CHLORIDE see 1-CHLOROPROPYLENE

2-PROPENYL CHLORIDE see 2-CHLOROPROPYLENE

3-PROPENYL CHLORIDE see ALLYL CHLORIDE

2-PROPENYLAMINE see ALLYLAMINE

PROPINE see 1-PROPYNE

PROPIOLACTONE, β- see β-PROPIOLACTONE

3-PROPIOLACTONE see β-PROPIOLACTONE

PROPIONALDEHYDE,2-METHYL-2-(METHYLTHIO)-,O-(METHYLCAR-BAMOYL)OXIME see ALDICARB

PROPIONANILIDE, 3′,4′-DICHLORO- see PROPANIL

PROPIONIC ACID ANHYDRIDE see PROPIONIC ANHYDRIDE

PROPIONIC ACID, 2-(4-CHLORO-2-METHYLPHENOXY) see MECOPROP

PROPIONIC ACID, 2-(4-((6-CHLORO-2-BENZOXAZOLYL)OXY) PHENOXY)-,ETHYLESTER, (+-)- see FENOXAPROP ETHYL

PROPIONIC ACID 3,4-DICHLOROANILIDE see PROPANIL

PROPIONIC NITRILE see PROPIONITRILE

PROPIONITRILE, 3-CHLORO- see 3-CHLOROPROPIONITRILE

1-PROPYENE, 3-BROMO- see PROPARGYL BROMIDE

2-PROPYLAMINE see ISOPROPYLAMINE

PROPYLAMINE, 1-METHYL see sec-BUTYLAMINE

PROPYL CHLOROCARBONATE see PROPYL CHLOROFORMATE

n-PROPYL CHLOROFORMATE see PROPYL CHLOROFORMATE

PROPYL 2,4-D ESTER see 2,4-D PROPYL ESTER

PROPYL 2,4-D see 2,4-D PROPYL ESTER

PROPYLENE ALDEHYDE see CROTONALDEHYDE

PROPYLENE CHLORIDE see 1,2-DICHLOROPROPANE

PROPYLENE DICHLORIDE see DICHLOROPROPANE

PROPYLENE DICHLORIDE see 1,2-DICHLOROPROPANE

PROPYLENE EPOXIDE see PROPYLENE OXIDE

1-PROPYLENE see PROPYLENE

1,2-PROPYLENIMINE see PROPYLENEIMINE

PROPYL-ETHYLBUTYLTHIOCARBAMATE see PEBULATE

PROPYLETHYLENE see 1-PENTENE

PROPYLIDENE CHLORIDE see 1,1-DICHLOROPROPANE

4-PROPYL-1,2-(METHYLE NEDIOXY)BENZENE see DIHYDROSAFROLE

DI-n-PROPYLNITROSAMINE see N-NITROSODI-N-PROPYLAMINE

PROPYL OXIRANE see 1,2-BUTYLENE OXIDE

n-PROPYL-1-PROPANAMINE see DIPROPYLAMINE

2-PROPYN-1-OL see PROPARGYL ALCOHOL

PROPYNE see 1-PROPYNE

PSEUDOACETIC ACID see PROPIONIC ACID

PSEUDOBUTYLENE see 2-BUTENE

PSEUDOCUMENE see 1,2,4-TRIMETHYLBENZENE

PURATRONIC CHROMIUM TRIOXIDE see CHROMIC ACID

PUROSTROPHAN see OUABAIN

β-PYRENE see PYRENE

PYRENE, 1-NITRO- see 1-NITROPYRENE

PYRETHRIN I see PYRETHRINS

PYRETHRIN II see PYRETHRINS

PYRETHROIDS see PYRETHRINS

PYRETHRUM see PYRETHRINS

3,6-PYRIDAZINEDIONE, 1,2-DIHYDRO- see MALEIC HYDRAZIDE

4-PYRIDINAMINE see 4-AMINOPYRIDINE

PYRIDINE, 4-AMINO- see 4-AMINOPYRIDINE

2-PYRIDINE CARBOXYLIC ACID, 4-AMINO-3,5,6-TRICHLORO- see PICLORAM

PYRIDINE, 2-CHLORO-6(TRICHLOROMETHYL)- see NITRAPYRIN

2,5-PYRIDINEDICARBOXYLIC ACID, DIPROPYL ESTER see DIPROPYL ISOCINCHOMERONATE

PYRIDINE, 2-METHYL- see 2-METHYLPYRIDINE

PYRIDINE, 3-(1-METHYL-2-PYRROLIDINYL)-, (S)- see NICOTINE

PYRIDINE, 3-(1-NITROSO-2-PYRROLIDINYL)-, (S)- see N-NITROSONORNICOTINE

4-PYRIDYLAMINE see 4-AMINOPYRIDINE

N-3-PYRIDYLMETHYL-N′-p-NITROPHENYLUREA see PYRIMINIL

2,4(1H,3H)-PYRIMIDINEDIONE,5-[BIS(2-CHLOROETHYL)AMINO]- see URACIL MUSTARD

2,4-(1H,3H)-PYRIMIDINEDIONE,5-BROMO-6-METHYL-3(1-METHYLPROPYL), LITHIUM SALT see BROMACIL, LITHIUM SALT

2,4(1H,3H)-PYRIMIDINEDIONE,5-CHLORO-3-(1,1-DIMETHYL)-6-METHYL- see TERBACIL

2(1H)-PYRIMIDINONE, TETRAHYDRO-5,5-DIMETHYL-,(3-(4-(TRIFLUOROMETHYL)PHENYL)-1-(2-(4-(TRIFLUOROMETHYL) PHENYL)ETHENYL)-2-PROPENYLIDENE)HYDRAZONE see HYDRAMETHYLNON

4(1H)-PYRIMIDIONE, 2,3-DIHYDRO-6-METHYL-2-THIOXO- see METHYLTHIOURACIL

PYROCATECHOL see CATECHOL

PYROPHOSPHORODITHIOIC ACID, TETRAETHYL ESTER see SULFOTEP

PYROSULFURIC ACID see OLEUM (FUMING SULFURIC ACID)

PYRROLIDINE, 1-NITROSO- see N-NITROSOPYRROLIDINE

2-PYRROLIDINONE, 1-METHYL- see N-METHYL-2-PYRROLIDONE

QUINTOCENE see QUINTOZENE

R-123 see 2,2-DICHLORO-1,1,1-TRIFLUOROETHANE

R-123a see 1,2-DICHLORO-1,1,2-TRIFLUOROETHANE

R 124 see 2-CHLORO-1,1,1,2-TETRAFLUOROETHANE

11411 RED see C.I. FOOD RED 15

REFRIGERANT 124 see 2-CHLORO-1,1,1,2-TETRAFLUOROETHANE

REGLONE see DIQUAT

RESORCINOL DIGLYCIDYL ETHER see DIGLYCIDYL RESORCINOL ETHER

RHODAMINE 6G EXTRA BASE see C.I. BASIC RED 1

RHODAMINE 6G see C.I. BASIC RED 1

RHODAMINE B 500 see C.I. FOOD RED 15

RHODAMINE B see C.I. FOOD RED 15

ROZOL see CHLOROPHACIONONE

SAL AMMONIAC see AMMONIUM CHLORIDE

SCHRADAN see DIPHOSPHORAMIDE, OCTAMETHYL-

9,10-SECOERGOSTA-5,7,10(19),22-TETRAEN-3-β-OL see ERGOCALCIFEROL

SELENIOUS ACID, DISODIUM SALT see SODIUM SELINITE

SELENIUM CHLORIDE OXIDE see SELENIUM OXYCHLORIDE

SELENIUM DIOXIDE see SELENIOUS ACID

SELENIUM(IV) DIOXIDE (1:2) see SELENIUM DIOXIDE

SELENIUM DISULFIDE see SELENIUM SULFIDE

SELENIUM(IV) DISULFIDE see SELENIUM SULFIDE

SELENIUM ELEMENTAL see SELENIUM

SELENIUM HYDRIDE see HYDROGEN SELENIDE

SELENIUM OXIDE see SELENIUM DIOXIDE

SENCOR see METRIBUZIN

1-SERINE DIAZOACETATE see AZASERINE

1-SERINE DIAZOACETATE (ESTER) see AZASERINE

SEVIN see CARBARYL

SILANE, CHLOROTRIMETHYL- see TRIMETHYLCHLOROSILANE

SILANE, DICHLORO- see DICHLOROSILANE

SILANE, DICHLORODIMETHYL- see DIMETHYLDICHLOROSILANE

SILANE, METHYLTRICHLORO- see METHYLTRICHLOROSILANE

SILANE, TETRAMETHYL- see TETRAMETHYLSILANE

SILANE, TRICHLORO- see TRICHLOROSILANE

SILANE, TRICHLOROETHYL- see TRICHLOROETHYLSILANE

SILANE, TRICHLOROMETHYL- see METHYLTRICHLOROSILANE

SILANE, TRICHLOROPHENYL- see TRICHLOROPHENYLSILANE

SILANE, TRIETHOXY- see TRIETHOXYSILANE

SILICANE see SILANE

SILICATE(2-), HEXAFLUORO-, ZINC (1:1) see ZINC SILICOFLUORIDE

SILICOCHLOROFORM see TRICHLOROSILANE

SILICON TETRAHYDRIDE see SILANE

SILVER ELEMENTAL see SILVER

SILVER(I) NITRATE see SILVER NITRATE

SILVER METAL see SILVER

SILVER POTASSIUM CYANIDE see POTASSIUM SILVER CYANIDE

SILVEX see SILVEX (2,4,5-TP)

SODA LYE see SODIUM HYDROXIDE

SODIUM 2-METHOXY-3,6-DICHLOROBENZOATE see SODIUM DICAMBA

SODIUM 2-PHENYLPHENATE see SODIUM O-PHENYLPHENOXIDE

SODIUM 2,4-DICHLOROPHENOXYACETATE see 2,4-D SODIUM SALT

SODIUM 3,6-DICHLORO-2-METHOXYBENZOATE see SODIUM DICAMBA

SODIUM BISULFIDE see SODIUM HYDROSULFIDE

SODIUM CHROMATE(VI) see SODIUM CHROMATE

SODIUM DICHROMATE see SODIUM BICHROMATE

SODIUM DICHROMATE(VI) see SODIUM BICHROMATE

SODIUM DIMETHYLAMINOCARBODITHIOATE see SODIUM DIMETHYLDITHIOCARBAMATE

SODIUM DISULFITE see SODIUM BISULFITE

SODIUM ELEMENTAL see SODIUM

SODIUM FLUOACETIC ACID see SODIUM FLUOROACETATE

SODIUM FLUORIDE ($Na(HF_2)$) see SODIUM BIFLUORIDE

SODIUM HEXAMETAPHOSPHATE see SODIUM PHOSPHATE, TRIBASIC

SODIUM HYDRATE see SODIUM HYDROXIDE

SODIUM HYDROFLUORIDE see SODIUM FLUORIDE

SODIUM HYDROGEN DIFLUORIDE see SODIUM BIFLUORIDE

SODIUM HYDROGEN FLUORIDE see SODIUM BIFLUORIDE

SODIUM HYDROGEN SULFITE see SODIUM BISULFITE

SODIUM METAARSENITE see SODIUM ARSENITE

SODIUM METAL see SODIUM

SODIUM METHOXIDE see SODIUM METHYLATE

SODIUM METHYLDITHIOCARBAMATE see METHAM SODIUM

SODIUM PENTACHLOROPHENOL see SODIUM PENTACHLOROPHENATE

SODIUM PENTOBARBITURATE see PENTOBARBITOL SODIUM

SODIUM O-PHENYLPHENATE see SODIUM O-PHENYLPHENOXIDE

SODIUM PHOSPHATE, DIBASIC MONOHYDRATE see SODIUM PHOSPHATE, DIBASIC

SODIUM PHOSPHATE, TRISODIUM SALT, DECAHYDRATE see SODIUM PHOSPHATE, TRIBASIC

SODIUM SULFHYDRATE see SODIUM HYDROSULFIDE

SODIUM SULFIDE see SODIUM HYDROSULFIDE

SODIUM TELLURATE(IV) see SODIUM TELLURITE

SODIUM TRIMETAPHOSPHATE see SODIUM PHOSPHATE, TRIBASIC

SOLVENT ORANGE 7 see C.I. SOLVENT ORANGE 7

SOLVENT YELLOW 3 see C.I. SOLVENT YELLOW 3

SOLVENT YELLOW 14 see C.I. SOLVENT YELLOW 14

SOLVENT YELLOW 34 see C.I. SOLVENT YELLOW 34

SPIRIT OF HARTSHORN see AMMONIA

S,S,S-TRIBUTYL PHOSPHOROTRITHIOITE see MERPHOS

S,S′,S-TRIBUTYL PHOSPHOROTRITHIOITE see MERPHOS

STANNANE, CHLOROTRIPHENYL- see TRIPHENYLTIN CHLORIDE

STANNANE, FLUOROTRIBUTYL- see TRIBUTYLTIN FLUORIDE

STANNANE, TETRAETHYL- see TETRAETHYLTIN

STANNANE, TRIBUTYL(METHACRYLOYLOXY)- see TRIBUTYLTIN METHACRYLATE

STANNANE, TRIBUTYLFLUORO- see TRIBUTYLTIN FLUORIDE

STEARIC ACID, CADMIUM SALT see CADMIUM STEARATE

STEARIC ACID, LEAD(II) SALT see LEAD STEARATE

STEARIC ACID, LEAD SALT, DIBASIC see LEAD STEARATE

STEARIC ACID, LEAD SALT see LEAD STEARATE

STIBINE, TRIBROMO- see ANTIMONY TRIBROMIDE

STIBINE, TRICHLORO- see ANTIMONY TRICHLORIDE

STIBINE, TRIFLUORO- see ANTIMONY TRIFLUORIDE

STREPTOZOCIN see D-GLUCOSE, 2-DEOXY-2-[[METHYLNITROSOAMINO)CARBONYL]AMINO]-

STRONTIUM CHROMATE(VI) see STRONTIUM CHROMATE

STRYCHININE SULFATE see STRYCHNINE, SULFATE

STRYCHNIDIN-10-ONE see STRYCHNINE

STRYCHNIDIN-10-ONE, 2,3-DIMETHOXY- see BRUCINE

STRYCHNIDIN-10-ONE SULFATE (2:1) see STRYCHNINE, SULFATE

STRYCHNINE SULFATE (2:1) see STRYCHNINE, SULFATE

STYRENE MONOMER see STYRENE

SUDAN I see C.I. SOLVENT YELLOW 14

SUDAN II see C.I. SOLVENT ORANGE 7

SUDAN ORANGE R see C.I. SOLVENT YELLOW 14

SUDAN ORANGE SUDAN ORANGE RPA see C.I. SOLVENT ORANGE 7

SULFAMIC ACID, MONOAMMONIUM SALT see AMMONIUM SULFAMATE

SULFONYL FLUORIDE see SULFURYL FLUORIDE

SULFUR CHLORIDE (DI) see SULFUR MONOCHLORIDE

SULFUR CHLORIDE see SULFUR MONOCHLORIDE

SULFUR HYDRIDE see HYDROGEN SULFIDE

SULFURIC ACID (FUMING) see OLEUM (FUMING SULFURIC ACID)

SULFURIC ACID, AMMONIUM NICKEL(II) SALT (2:2:1) see NICKEL AMMONIUM SULFATE

SULFURIC ACID, CHROMIUM(3+) SALT (3:2) see CHROMIC SULFATE

SULFURIC ACID, COPPER(2+) SALT (1:1) see CUPRIC SULFATE

SULFURIC ACID, COPPER(II) SALT (1:1) see CUPRIC SULFATE

SULFURIC ACID, DIETHYL ESTER see DIETHYL SULFATE

SULFURIC ACID, DIMETHYL ESTER see DIMETHYL SULFATE

SULFURIC ACID, DITHALLIUM(I) SALT(8CI,9CI) see THALLIUM(I) SULFATE

SULFURIC ACID, IRON(II) SALT (1:1) see FERROUS SULFATE

SULFURIC ACID, LEAD(II) SALT(1:1) see LEAD SULFATE

SULFURIC ACID, LEAD(II) SALT(1:1) see LEAD SULFATE

SULFURIC ACID, MERCURY(II) SALT (1:1) see MERCURIC SULFATE

SULFURIC ACID, MIXTURE WITH SULFUR TRIOXIDE see OLEUM (FUMING SULFURIC ACID)

SULFURIC ACID, NICKEL(II) SALT see NICKEL SULFATE

SULFURIC ACID, THALLIUM SALT see THALLIUM SULFATE

SULFURIC ACID, ZINC SALT (1:1) see ZINC SULFATE

SULFURIC ACID, ZIRCONIUM(IV) SALT(2:1) see ZIRCONIUM SULFATE

SULFURIC OXIDE see SULFUR TRIOXIDE

SULFURIC OXYFLUORIDE see VIKANE

SULFUR SELENIDE see SELENIUM SULFIDE

SULFUR TETRAFLUORIDE see SULFUR FLUORIDE (SF4), (T-4)-

SULFUROUS ACID, 2-CHLOROETHYL 2-[4-(1,1-DIMETHYLETHYL)PHENOXY]-1-METHYLETHYL ESTER see ARAMITE

SULFUROUS ACID, 2-(4-(1,1-DIMETHYLETHYL)PHENOXY) CYCLOHEXYL 2-PROPYNYL ESTER see PROPARGITE

SULFUROUS ACID, MONOSODIUM SALT see SODIUM BISULFITE

SULFUROUS OXIDE see SULFUR DIOXIDE

SULPHURYL DIFLUORIDE see SULFURYL FLUORIDE

SUNAPTIC ACID B see NAPHTHENIC ACID

SUNAPTIC ACID C see NAPHTHENIC ACID

SYSTOX see DEMETON

2,4,5-T see 2,4,5-T ACID (93-76-5)

2,4,5-T ACID AMINE see 2,4,5-T AMINES (2008-46-0)

2,4,5-T ACID AMINE see 2,4,5-T AMINES (3813-14-7)

2,4,5-T ACID AMINE see 2,4,5-T AMINES (6369-97-7)

2,4,5-T ACID AMINE see 2,4,5-T AMINES (1319-72-8)

2,4,5-T ACID AMINE see 2,4,5-T AMINES (6369-96-6)

TAR CAMPHOR see NAPHTHALENE

2,4,5-T-n-BUTYL ESTER see 2,4,5-T ESTERS (93-79-8)

2,4,5-T, n-BUTYL ESTER see 2,4,5-T ESTERS (61792-07-2)

2,4,5-T, ETHYLHEXYL ESTER see 2,4,5-T ESTERS (1928-47-8)

2,3,7,8-TCDD see 2,3,7,8-TETRACHLORODIBENZO-p-DIOXIN (TCDD)

P,P-TDE see DDD

TDE see DDD

2,6-TDI see TOLUENE-2,6-DIISOCYANATE

TECHNICAL CHLORINATED CAMPHENE, 67-69% CHLORINE see TOXAPHENE

TEDP see SULFOTEP

TELLURIUM, ELEMENTAL see TELLURIUM

TELLURIUM FLUORIDE see TELLURIUM HEXAFLUORIDE

TELLUROUS ACID, DISODIUM see SODIUM TELLURITE

TETAN see TETRANITROMETHANE

2,4,5-T ETHYLHEXYL ESTER see 2,4,5-T ESTERS

TETRAAMINE COPPER SULFATE see CUPRIC SULFATE, AMMONIA-TEDBUTANEDIOIC ACID, 2,3-DIHYDROXY-[R-(R*,R*)]-, COPPER(2+)SALT (1:1) see CUPRIC TARTRATE

TETRACARBONYLNICKEL see NICKEL CARBONYL

TETRACHLORETHANE see 1,1,2,2-TETRACHLOROETHANE

1,1,2,2-TETRACHLORO-1-FLUOROETHANE see HCFC-121

2,4,5,6-TETRACHLORO-1,3-BENZENEDICARBONITRILE see CHLOROTHALONIL

TETRACHLOROBENZENE see 1,2,4,5-TETRACHLOROBENZENE

2,3,6,7-TETRACHLORODIBENZO-para-DIOXIN see 2,3,7,8-TETRACHLORODIBENZO-p-DIOXIN (TCDD)

TETRACHLORODIPHENYLETHANE see DDD

1,1,2,2-TETRACHLOROETHENE see TETRACHLOROETHYLENE

1,1,2,2-TETRACHLOROETHYLENE see TETRACHLOROETHYLENE

TETRACHLOROMETHANE see CARBON TETRACHLORIDE

O,O,O,O-TETRAETHYL-DIPHOSPHATE see TEPP

TETRAETHYL DITHIOPYROPHOSPHATE see SULFOTEP

TETRAETHYL PYROPHOSPHATE see TEPP

TETRAETHYLDITHIONOPYROPHOSPHATE see SULFOTEP

1,1,1,2-TETRAFLUORO-2-CHLOROETHANE see 2-CHLORO-1,1,1,2-TETRAFLUOROETHANE

1,1,2,2-TETRAFLUORO-2-CHLOROETHANE see 1-CHLORO-1,1,2,2-TETRAFLUOROETHANE

1,1,2,2-TETRAFLUORO-1,2-DICHLOROETHANE see DICHLOROTETRAFLUOROETHANE

TETRAFLUORO BORATE(1-), LEAD (2+) see LEAD FLUOBORATE

1,1,2,2-TETRAFLUOROETHYLENE see TETRAFLUOROETHYLENE

N-(5,6,7,9)-TETRAHYDRO-1,2,3,10-TETRAMETHOXY-9-OXOBENZO(a)HEPTALEN-7-YL)-ACETAMIDE see COLCHICINE

TETRAHYDRO-3,5-DIMETHYL-2H-1,3,5-THIADIAZINE-2-THIONE, ION(1-), SODIUM see DAZOMET, SODIUM SALT

TETRAHYDRO-3,5-DIMETHYL-2H-1,3,5-THIADIAZINE-2-THIONE see DAZOMET

TETRAHYDRO-2H-3,5-DIMETHYL-1,3,5-THIADIAZINE-2-THIONE see DAZOMET

TETRAHYDRO-5,5-DIMETHYL-2(1H)-PYRIMIDINONE[3-[4-(TRIFLUOROM ETHYL)PHENYL]-1-[2-[4-(TRIFLUOROMETHYL)PHENYL] ETHENYL]-2-PRO PENYLIDENE]HYDRAZONE see HYDRAMETHYLNON

TETRAHYDROFURAN see FURAN, TETRAHYDRO

TETRAMETHYL O,O′-THIO-DI-p-PHENYLENEPHOSPHOROTHIOATE see TEMEPHOS

O,O,O′,O′-TETRAMETHYL O,O′-THIODI-p-PHENYLENE PHOSPHOROTHIOATE see TEMEPHOS

2,2,3,3-TETRAMETHYLCYCLOPROPANE CARBOXYLIC ACID CYANO(3-PHENOXYPHENYL)METHYL ESTER see FENPROPATHRIN

4,4′-TETRAMETHYLDIAMINODIPHENYLMETHANE see 4,4′-METHYLENEBIS (N,N-DIMETHYL)BENZENAMINE

TETRAMETHYLPLUMBANE see TETRAMETHYL LEAD

TETRAPHOSPHORIC ACID, HEXAETHYL ESTER see HEXAETHYL TETRAPHOSPHATE

THALLIUM(I) CHLORIDE see THALLIUM CHLORIDE (TlCl)

THALLIUM, ELEMENTAL see THALLIUM

THALLIUM MALONITE see THALLOUS MALONATE

THALLIUM MONOSELENIDE see SELENIOUS ACID, DITHALLIUM(1+) SALT

THALLIUM(III) OXIDE see THALLIC OXIDE

THALLIUM SELENIDE see SELENIOUS ACID, DITHALLIUM(1+) SALT

THALLIUM(1+) SULFATE (2:1) see THALLIUM(I) SULFATE

THALLOUS ACETATE see THALLIUM(I) ACETATE

THALLOUS CARBONATE see THALLIUM(I) CARBONATE

THALLOUS CHLORIDE see THALLIUM CHLORIDE (TlCl)

THALLOUS NITRATE see THALLIUM(I) NITRATE

THALLOUS SULFATE see THALLIUM SULFATE

THALLOUS SULFATE see THALLIUM(I) SULFATE

THEMET(R) see PHORATE

2H-1,3,5-THIADIAZINE-2-THIONE, TETRAHYDRO-3,5-DIMETHYL-, ION(1-), SODIUM see DAZOMET, SODIUM SALT

2-(THIAZOL-4-YL)BENZIMIDAZOLE see THIABENDAZOLE

2-THIAZOLE-4-YL BENZIMIDAZOLE see THIABENDAZOLE

2-(4-THIAZOLYL)-1H-BENZIMIDAZOLE see THIABENDAZOLE

THIMET see PHORATE

2-THIO-3,5-DIMETHYLTETRAHYDRO-1,3,5-THIADIAZINE see DAZOMET

2-THIO-6-METHYL-1,3-PYRIMIDIN-4-ONE see METHYLTHIOURACIL

1,1′-THIOBIS(2-CHLOROETHANE) see MUSTARD GAS

THIOCARBONOHYDRAZIDE see THIOCARBAZIDE

THIOCYANATOETHANE see ETHYLTHIOCYANATE

THIOCYANIC ACID, AMMONIUM SALT see AMMONIUM THIOCYANATE

THIOCYANIC ACID, METHYL ESTER see METHYL THIOCYANATE

THIODAN see α-ENDOSULFAN

THIODAN see ENDOSULFAN

α-THIODAN see α-ENDOSULFAN

β-THIODAN see β-ENDOSULFAN

THIODIPHOSPHORIC ACID, TETRAETHYL ESTER see SULFOTEP

THIOMETHANOL see METHYL MERCAPTAN

THIONAZIN see ZINOPHOS

THIOPEROXYDICARBO NICDIAMIDE $(((H_2N)C(S))_2S_2)$, TETRAMETHYL- see THIRAM

THIOPEROXYDICARBONIC DIAMIDE, TETRAMETHYL- see THIRAM

THIOPHENOL see BENZENETHIOL

3-THIOSEMICARBAZIDE see THIOSEMICARBAZIDE

THIOSEMICARBAZONE ACETONE see ACETONE THIOSEMICARBAZIDE

THIOUREA, 1-NAPHTHALENYL- see ANTU

2-THIOUREA see THIOUREA

THIURAM see THIRAM

THORIUM(IV) OXIDE see THORIUM DIOXIDE

THORIUM OXIDE see THORIUM DIOXIDE

TIN, BIS(TRIBUTYL)-,OXIDE see BIS(TRIBUTYLTIN) OXIDE

TIN, TETRAETHYL- see TETRAETHYLTIN

2,4,5-T, ISOOCTYL ESTER see 2,4,5-T ESTERS (25168-15-4)

TITANIUM CHLORIDE see TITANIUM TETRACHLORIDE

TITANIUM CHLORIDE $(TiCl_4)$ (T-4)- see TITANIUM TETRACHLORIDE

TITANIUM(IV) CHLORIDE see TITANIUM TETRACHLORIDE

O,O′-TOLIDINE see 3,3′-DIMETHYLBENZIDINE

o-TOLIDINE see 3,3′-DIMETHYLBENZIDINE

3,3′-TOLIDINE see 3,3′-DIMETHYLBENZIDINE

o-TOLIDINE DIHYDROCHLORIDE see 3,3′-DIMETHYLBENZIDINE DIHYDROCHLORIDE

o-TOLIDINE DIHYDROFLUORIDE see 3,3-DIMETHYLBENZIDINE DIHYDROFLUORIDE

TOLUENE-2,4-DIAMINE see 2,4-DIAMINOTOLUENE

TOLUENE-3,4-DIAMINE see DIAMINOTOLUENE

TOLUENE-AR,AR′-DIAMINE see DIAMINOTOLUENE (MIXED ISOMERS)

TOLUENE, 2,4-DINITRO- see 2,4-DINITROTOLUENE

TOLUENE, 2,6-DINITRO- see 2,6-DINITROTOLUENE

TOLUENE, 3,4-DINITRO- see 3,4-DINITROTOLUENE

TOLUENE DIISOCYANATE see TOLUENE 2,4-DIISOCYANATE

2,4-TOLUENE DIISOCYANATE see TOLUENE 2,4-DIISOCYANATE

TOLUENE, DINITRO- see DINITROTOLUENE (MIXED ISOMERS)

TOLUENEDIISOCYANATE (UNSPECIFIED ISOMERS) see TOLUENEDIISOCYANATE (MIXED ISOMERS)

p-TOLUIDINE,α,α,α-TRIFLUORO-2,6-DINITRO-N,N-DIPROPYL- see TRIFLURALIN

4-TOLUIDINE see p-TOLUIDINE

TOLUIDINE, ortho- see o-TOLUIDINE

ortho-TOLUIDINE see o-TOLUIDINE

para-TOLUIDINE see p-TOLUIDINE

TOLUIDINE HYDROCHLORIDE, ortho- see o-TOLUIDINE HYDROCHLORIDE

ortho-TOLUIDINE HYDROCHLORIDE see o-TOLUIDINE HYDROCHLORIDE

TOLUIDINE, para- see p-TOLUIDINE

o-TOLUOL see o-CRESOL4-CRESOL see p-CRESOL

3,4-TOLUYLENEDIAMINE see DIAMINOTOLUENE

TOLYENE DIISOCYANATE see TOLUENEDIISOCYANATE (MIXED ISOMERS)

m-TOLYENE DIISOCYANATE see TOLUENE-2,6-DIISOCYANATE

4-(o-TOLYLAZO)-o-TOLUIDINE see C.I. SOLVENT YELLOW 3

TOLYLENEDIAMINE see DIAMINOTOLUENE (MIXED ISOMERS)

TOLYLENE 2,6-DIISOCYANATE see TOLUENE-2,6-DIISOCYANATE

TOLYLENE ISOCYANATE see TOLUENEDIISOCYANATE (MIXED ISOMERS)

o-TOLYL THIOUREA see THIOUREA, (2-METHYLPHENYL)-

TORDON see PICLORAM

2,4,5-TP see SILVEX (2,4,5-TP)

2,4,5-TP ACID see SILVEX (2,4,5-TP)

2,4,5-TP ACID ESTERS see 2,4,5-TP ESTERS

1,2-trans-DICHLOROETHYLENE see 1,2-DICHLOROETHYLENE

TRI-(2-CHLOROETHYL)AMINE see TRIS(2-CHLOROETHYL)AMINE

TRI-N-BUTYLSTANNYLMETHACRYLATE see TRIBUTYLTIN METHACRYLATE

TRIATOMIC OXYGEN see OZONE

3,5,7-TRIAZA-1-AZONIATRICYCLODECANE-1-(3-CHLORO-2-PROPENYL)-,CHLORIDE see 1-(3-CHLORALLYL)-3,5,7-TRIAZA-1-AZONIAADAMANTANE CHLORIDE

as-TRIAZIN-5(4H)-ONE,4-AMINO-6-tert-BUTYL-3-(METHYLTHIO)- see METRIBUZIN

1,2,4-TRIAZIN-5(4H)-ONE,4-AMINO-6-(1,1-DIMETHYLETHYL)-3-(METHYLTHIO)- see METRIBUZIN

1,3,5-TRIAZIN-2-AMINE,4,6-DICHLORO-N-(2-CHLOROPHENYL)- see ANILAZINE

1,3,5-TRIAZINE-2,4-DIAMINE,N-ETHYL-N′-(1-METHYLETHYL)-6-(METHYLTHIO)- see AMETRYN

TRICHLOROMETHYLSULFENYL CHLORIDE see PERCHLOROMETHYL MERCAPTAN

TRICHLOROMONOFLUOROMETHANE see TRICHLOROFLUORO-METHANE

TRICHLOROMONOSILANE see SILANE, TRICHLORO-

TRICHLORONAT see TRICHLORONATE

TRICHLORONITROMETHANE see CHLOROPICRIN

(2,4,5-TRICHLOROPHENOXY)-, ACETIC ACID SODIUM SALT see 2,4,5-T SALTS

2-(2,4,5-TRICHLOROPHENOXY)PROPANOIC ACID see SILVEX (2,4,5-TP)

α-(2,4,5-TRICHLOROPHENOXY)PROPIONIC ACID see 2,4,5-TP ESTERS

2,4,5-TRICHLOROPHENOXYACETIC ACID see 2,4,5-T ACID (93-76-5)

((3,5,6-TRICHLORO-2-PYRIDYL)OXY)ACETIC ACID, comp. with TRIETHYLAMINE (1:1) see TRICLOPYR TRIETHYLAMMONIUM SALT

TRICHLOROSILANE see SILANE, TRICHLORO-

1,1,2-TRICHLOROTRIFLUOROETHANE FREON 113

1,1,2-TRICHLORO-1,2,2-TRIFLUOROETHANE FREON 113

TRIETHANOLAMINE DODECYLBENZENE SULFONATE see TRIETHANOLAMINE DODECYLBENZENE SULFONATE

TRIFLUOROBORANE see BORON TRIFLUORIDE

TRIFLUOROBROMOETHYLENE see BROMOTRIFLUORETHYLENE

TRIFLUOROBROMOMETHANE see BROMOTRIFLUROMETHANE

1,1,2-TRIFLUORO-2-CHLOROETHYLENE see TRIFLUOROCHLORO-ETHYLENE

TRIFLUOROCHLOROMETHANE see CHLOROTRIFLUOROMETHANE

1,1,1-TRIFLUORO-2,2-DICHLOROETHANE see 2,2-DICHLORO-1,1,1-TRI-FLUOROETHANE

1,1,2-TRIFLUORO-1,2-DICHLOROETHANE see 1,2-DICHLORO-1,1,2-TRI-FLUOROETHANE

α,α,α-TRIFLUORO-2,6-DINITRO-N,N-DIPROPYL-p-TOLUIDINE see TRI-FLURALIN

2-[4-[[5-(TRIFLUOROMETHYL)-2-PYRIDINYL]OXY]-PHENOXY]PROPANOIC ACID, BUTYL ESTER see FLUAZIFOP-BUTYL

3-(TRIFLUOROMETHYL)ANILINE see BENZENEAMINE, 3-(TRIFLUOROMETHYL)-

m-(TRIFLUOROMETHYL)BENZENAMINE see BENZENEAMINE, 3-(TRI-FLUOROMETHYL)-

TRIFLUOROMONOCHLOROETHYLENE see TRIFLUOROCHLORO-ETHYLENE

TRIFLUOROMONOCHLOROMETHANE see CHLOROTRIFLUORO-METHANE

2,4,6-TRIFLUORO-s-TRIAZINE see CYANURIC FLUORIDE

TRIFLUOROVINYLBROMIDE see BROMOTRIFLUORETHYLENE

3,4,5,-TRIMETHOXYBENZOIC ACID see RESPERINE

α,α′,α″-TRIMETHYLAMINETRICARBOXYLIC ACID see NITRILOTRIACETIC ACID

2,4,6-TRIMETHYLANILINE see ANILINE, 2,4,6-TRIMETHYL-

2,4,6-TRIMETHYLBENZENAMINE see ANILINE, 2,4,6-TRIMETHYL-

1,3,4-TRIMETHYLBENZENE see 1,2,4-TRIMETHYLBENZENE

3,5,5-TRIMETHYL-2-CYCLOHEXENE-1-ONE see ISOPHRONE

TRIMETHYLENE DICHLORIDE see 1,3-DICHLOROPROPANE

TRIMETHYLENE see CYCLOPROPANE

TRIMETHYLSILYL CHLORIDE see TRIMETHYLCHLOROSILANE

TRIMETHYLSTANNYL CHLORIDE see TRIMETHYLTIN CHLORIDE

2,4,6-TRIMETHYL-1,3,5-TRIOXANE see PARALDEHYDE

TRIMUSTINE HYDROCHLORIDE see TRIS(2-CHLOROETHYL)AMINE

TRINITROBENZENE see 1,3,5-TRINITROBENZENE

sym-TRINITROBENZENE see 1,3,5-TRINITROBENZENE

TRINITROPHENOL see PICRIC ACID

2,4,6-TRINITOPHENOL see PICRIC ACID

1,3,5-TRIOXANE,2,4,6-TRIMETHYL see PARALDEHYDE

TRIOXYMETHYLENE see PARAFORMALDEHYDE

TRIPHENYLTIN ACETATE see STANNANE, ACETOXYTRIPHENYL-

TRIPHOSPHORIC ACID, PENTASODIUM SALT see SODIUM PHOSPHATE, TRIBASIC

TRIS(AZIRIDINO)-1,4-BENZOQUINONE see TRIAZIQUONE

2,3,5-TRIS(AZIRIDINYL)-p-BENZOQUINONE see TRIAZIQUONE

TRIS(DIMETHYLCARBAMODITHIOATO-S,S′)IRON see FERBAM

TRISODIUM PHOSPHATE see SODIUM PHOSPHATE, TRIBASIC

TRISODIUM TRIFLUORIDE see SODIUM FLUORIDE

2,4,5-T SODIUM SALT see 2,4,5-T AMINES (6369-96-6)

UNSYMMETRICAL DIMETHYLHYDRAZINE see 1,1-DIMETHYL HYDRAZINE

URACIL, 5-FLUORO- see FLUOROURACIL

URANIUM ACETATE see URANYL ACETATE

URANIUM BIS(ACETO-O)DIOXO- see URANYL ACETATE

URANIUM BIS(NITRATO-O)DIOXO-, (T-4) see URANYL NITRATE

URANIUM, BIS(NITRATO-O,O′)DIOXO-,(OC-6-11)- see URANYL NITRATE

URANIUM OXYACETATE see URANYL ACETATE

UREA, 2-(5-tert-BUTYL-1,3,4-THIADIAZOL-2-YL)-1,3-DIMETHYL- see TEBUTHIURON

UREA, N′-(4-CHLOROPHENYL)-N,N-DIMETHYL- see MONURON

UREA,1-(p-CHLOROPHENYL)-3-(2,6-DIFLUOROBENZOYL)- see DIFLUBENZURON

UREA, 3-(3,4-DICHLOROPHENYL)-1,1-DIMETHYL- see DIURON

UREA, N′-(3,4-DICHLOROPHENYL)-N,N-DIMETHYL- see DIURON

UREA, N,N-DIMETHYL-N′-[3-(TRIFLUOROMETHYL)PHENYL]- see FLUOMETURON

UREA, N-ETHYL-N-NITROSO- see N-NITROSO-N-ETHYLUREA

UREA, 1-METHYL-1-NITROSO- see N-NITROSO-N-METHYLUREA

UREA, N-METHYL-N-NITROSO- see N-NITROSO-N-METHYLUREA

UREA, SELENO- see SELENOUREA

UREA, 2-THIO-1-(THIOCARBAMOYL)- see DITHIOBIURET

VALINOMICIN see VALINOMYCIN

VANADATE (V031-), AMMONIUM see AMMONIUM VANADATE

VANADIC ACID, AMMONIUM SALT see AMMONIUM VANADATE

VANADIUM, ELEMENTAL see VANADIUM

VANADIUM OXIDE see VANADIUM PENTOXIDE

VANADIUM(V) OXIDE FUME see VANADIUM PENTOXIDE

VANADIUM, OXO[SULFATO(2-)-O]- see VANADYL SULFATE

VAT YELLOW 4 see C.I. VAT YELLOW 4

VENTOX see ACRYLONITRILE

VICTORIA GREEN B see C.I. BASIC GREEN 4

VIENNA GREEN see CUPRIC ACETOARSENITE

VINYL ACETATE MONOMER see VINYL ACETATE

VINYLBENZENE see STYRENE

VINYL CARBINOL see ALLYL ALCOHOL

VINYL-2-CHLOROETHYL ETHER see 2-CHLOROETHYL VINYL ETHER

VINYL CYANIDE see ACRYLONITRILE

VINYL CYANIDE, PROPENENITRILE see ACRYLONITRILE

VINYLETHYLENE see 1,3-BUTADIENE

VINYL METHYL KETONE see METHYL VINYL KETONE

VINYL TRICHLORIDE see 1,1,2-TRICHLOROETHANE

VIROSIN see ANTIMYCIN

VYDATE see OXAMYL

VYDATE OXAMYL INSECTICIDE/NEMATOCIDE see OXAMYL

WHITE ARSENIC see ARSENIC TRIOXIDE

WHITE ASBESTOS see ASBESTOS (Friable)

WHITE PHOSPHORUS see PHOSPHORUS

XYLENE see XYLENE (MIXED ISOMERS)

1,2-XYLENE see o-XYLENE

1,3-XYLENE see m-XYLENE

1,4-XYLENE see p-XYLENE

XYLENES (TOTAL) see XYLENE (MIXED ISOMERS)

XYLENES, N.O.S. see XYLENE (MIXED ISOMERS)

XYLENOLS, MIXED see XYLENOL

m-XYLOL see m-XYLENE

YELLOW PHOSPHORUS see PHOSPHORUS

YOHIMBAN-16-CARBOXYLIC ACID, 11,17-DIMETHOXY-18-(3,4,5-TRIMETHOXYBENXOYL)OXY-, METHYL ESTER,(3β, 16β, 17α, 18β, 20α)-see RESPERINE

ZACTRAN see MEXACARBATE

ZECTANE see MEXACARBATE

ZINC AMMONIATE ETHYLENEBIS(DITHIOCARBAMATE)-POLY (ETHYLENETHIURAM DISULFIDE) see METIRAM

ZINCATE(2-), TETRACHLORO-, DIAMMONIUM, (T-4)- see ZINC AMMONIUM CHLORIDE

ZINCATE(3-), TETRACHLORO-, TRIAMMONIUM, (T-4)- see ZINC AMMONIUM CHLORIDE

ZINCATE (2-), TETRACHLORO-, DIAMMONIUM SALT see ZINC AMMONIUM CHLORIDE

ZINC CHLORIDE FUME see ZINC CHLORIDE

ZINC DIACETATE see ZINC ACETATE

ZINC DIBROMIDE see ZINC BROMIDE

ZINC DICYANIDE see ZINC CYANIDE

ZINC DIFORMATE see ZINC FORMATE

ZINC DITHIONITE see ZINC HYDROSULFITE

ZINC FLUORIDE (ZnF_2) see ZINC FLUORIDE

ZINC FLUOSILICATE see ZINC SILICOFLUORIDE

ZINC HEXAFLUOROSILICATE see ZINC SILICOFLUORIDE

ZINC, METALLIC see ZINC

ZINC METIRAM see METIRAM

ZINC PHOSPHIDE (conc. > 10%) see ZINC PHOSPHIDE

ZINC PHOSPHIDE (Zn_3P_2), when present at concentrations greater than 10% see ZINC PHOSPHIDE

ZINC PHOSPHIDE (Zn_3P_2) when present at concentrations of 10% or less see ZINC PHOSPHIDE (conc. >= 10%)

ZIRCONATE(2-), HEXAFLUORO-, DIPOTASSIUM, (OC-6-11)- see ZIRCONIUM POTASSIUM FLUORIDE

ZIRCONIUM CHLORIDE see ZIRCONIUM TETRACHLORIDE

ZIRCONIUM CHLORIDE ($ZRCl_4$), (T-4)- see ZIRCONIUM TETRACHLORIDE

ZIRCONIUM(IV) CHLORIDE (1:4) see ZIRCONIUM TETRACHLORIDE

ZIRCONIUM(IV) SULFATE(1:2) see ZIRCONIUM SULFATE

CAS INDEX

76-14-2 see DICHLOROTETRAFLUOROETHANE
76-15-3 see MONOCHLOROPENTAFLUOROETHANE
76-44-8 see HEPTACHLOR
76-87-9 see TRIPHENYLTIN HYDROXIDE
77-47-4 see HEXACHLOROCYCLOPENTADIENE
77-73-6 see DICYCLOPENTADIENE
77-78-1 see DIMETHYL SULFATE
77-81-6 see TABUN
78-00-2 see TETRAETHYL LEAD
78-34-2 see DIOXATHION
78-48-8 see S,S,S-TRIBUTYLTRITHIOPHOSPHATE
78-53-5 see AMITON
78-59-1 see ISOPHRONE
78-71-7 see OXETANE, 3,3-BIS(CHLOROMETHYL)-
78-78-4 see ISOPENTANE
78-79-5 see ISOPRENE
78-81-9 see iso-BUTYLAMINE
78-82-0 see ISOBUTYRONITRILE
78-83-1 see ISOBUTYL ALCOHOL
78-84-2 see ISOBUTYRALDEHYDE
78-87-5 see 1,2-DICHLOROPROPANE
78-88-6 see 2,3-DICHLOROPRENE
78-92-2 see sec-BUTYL ALCOHOL
78-93-3 see METHYL ETHYL KETONE
78-94-4 see METHYL VINYL KETONE
78-97-7 see LACTONITRILE
78-99-9 see 1,1-DICHLOROPROPANE
79-00-5 see 1,1,2-TRICHLOROETHANE
79-01-6 see TRICHLOROETHYLENE
79-06-1 see ACRYLAMIDE
79-09-4 see PROPIONIC ACID
79-10-7 see ACRYLIC ACID
79-11-8 see CHLOROACETIC ACID
79-19-6 see THIOSEMICARBAZIDE
79-21-0 see PERACETIC ACID
79-22-1 see METHYL CHLOROCARBONATE
79-31-2 see iso-BUTYRIC ACID
79-34-5 see 1,1,2,2-TETRACHLOROETHANE
79-38-9 see TRIFLUOROCHLOROETHYLENE
79-44-7 see DIMETHYLCARBAMOYL CHLORIDE
79-46-9 see 2-NITROPROPANE
80-05-7 see 4,4′-ISOPROPYLIDENEDIPHENOL
80-15-9 see CUMENE HYDROPEROXIDE
80-62-6 see METHYL METHACRYLATE
80-63-7 see METHYL 2-CHLOROACRYLATE
81-07-2 see SACCHARIN
81-81-2 see WARFARIN
81-88-9 see C.I. FOOD RED 15
82-28-0 see 1-AMINO-2-METHYLANTHRAQUINONE
82-66-6 see DIPHACIONE
82-68-8 see QUINTOZENE
83-32-3 see ACENAPHTHENE
84-66-2 see DIETHYL PHTHALATE
84-74-2 see DIBUTYL PHTHALATE
85-00-7 see DIQUAT
85-01-8 see PHENANTHRENE

109-61-5 see PROPYL CHLOROFORMATE
109-66-0 see PENTANE
109-67-1 see 1-PENTENE
109-73-9 see BUTYLAMINE
109-77-3 see MALONONITRILE
109-86-4 see 2-METHOXYETHANOL
109-89-7 see DIETHYLAMINE
109-92-2 see VINYL ETHYL ETHER
109-95-5 see ETHYL NITRITE
109-99-9 see FURAN, TETRAHYDRO
110-00-9 see FURAN
110-16-7 see MALEIC ACID
110-17-8 see FURMARIC ACID
110-19-0 see iso-BUTYL ACETATE
110-54-3 see n-HEXANE
110-57-4 see trans-1,4-DICHLORO-2-BUTENE
110-75-8 see 2-CHLOROETHYL VINYL ETHER
110-80-5 see 2-ETHOXYETHANOL
110-82-7 see CYCLOHEXANE
110-86-1 see PYRIDINE
110-89-4 see PIPERIDINE
111-42-2 see DIETHANOLAMINE
111-44-4 see BIS(2-CHLOROETHYL)ETHER
111-54-6 see ETHYLENEBISDITHIOCARBAMIC ACID
111-69-3 see ADIPONITRILE
111-91-1 see BIS(2-CHLOROETHOXY)METHANE
114-26-1 see PROPOXUR
115-02-6 see AZASERINE
115-07-1 see PROPYLENE
115-10-6 see METHYL ETHER
115-11-7 see 2-METHYLPROPENE
115-21-9 see TRICHLOROETHYLSILANE
115-26-4 see DIMEFOX
115-28-6 see CHLORENDIC ACID
115-29-7 see ENDOSULFAN
115-32-2 see DICOFOL
115-90-2 see FENSULFOTHION
116-06-3 see ALDICARB
116-14-3 see TETRAFLUOROETHYLENE
117-79-3 see 2-AMINOANTHRAQUINONE
117-80-6 see DICHLONE
117-81-7 see DI(2-ETHYLHEXYL)PHTHALATE
117-84-0 see DI-n-OCTYL PHTHALATE
118-74-1 see HEXACHLOROBENZENE
119-38-0 see ISOPROPYLMETHYLPYRAZOYL DIMETHYLCARBAMATE
119-90-4 see 3,3′-DIMETHOXYBENZIDINE
119-93-7 see o-TOLIDINE
119-93-7 see 3,3′-DIMETHYLBENZIDINE
120-12-7 see ANTHRACENE
120-36-5 see 2,4-DP
120-58-1 see ISOSAFROLE
120-71-8 see p-CRESIDINE
120-80-9 see CATECHOL
120-82-1 see 1,2,4-TRICHLOROBENZENE
120-83-2 see 2,4-DICHLOROPHENOL
121-14-2 see 2,4-DINITROTOLUENE

121-21-1 see PYRETHRINS
121-29-9 see PYRETHRINS
121-44-8 see TRIETHYLAMINE
121-69-7 see N,N-DIMETHYLANILINE
121-75-5 see MALATHION
122-09-8 see BENZENEETHANAMINE, α,α-DIMETHYL-
122-14-5 see FENITROTHION
122-34-9 see SIMAZINE
122-39-4 see DIPHENYLAMINE
122-66-7 see 1,2-DIPHENYLHYDRAZINE
123-31-9 see HYDROQUINONE
123-33-1 see MALEIC HYDRAZIDE
123-38-6 see PROPIONALDEHYDE
123-61-5 see 1,3-PHENYLENE DIISOCYANATE
123-62-6 see PROPIONIC ANHYDRIDE
123-63-7 see PARALDEHYDE
123-72-8 see BUTYRALDEHYDE
123-73-9 see CROTONALDEHYDE, (E)
123-86-4 see BUTYL ACETATE
123-91-1 see 1,4-DIOXANE
123-92-2 see iso-AMYL ACETATE
124-04-9 see ADIPIC ACID
124-40-3 see DIMETHYLAMINE
124-41-4 see SODIUM METHYLATE
124-48-1 see CHLORODIBROMOMETHANE
124-65-2 see SODIUM CACODYLATE
124-73-2 see DIBROMOTETRAFLUOROETHANE
124-87-8 see PICROTOXIN
126-72-7 see TRIS(2,3-DIBROMOPROPYL) PHOSPHATE
126-98-7 see METHACRYLONITRILE
126-99-8 see CHLOROPRENE
127-18-4 see TETRACHLOROETHYLENE
127-82-2 see ZINC PHENOLSULFONATE
128-03-0 see POTASSIUM DIMETHYLDITHIOCARBAMATE
128-04-1 see SODIUM DIMETHYLDITHIOCARBAMATE
128-66-5 see C.I. VAT YELLOW 4
129-00-0 see PYRENE
129-06-0 see WARFARIN SODIUM
130-15-4 see 1,4-NAPHTHOQUINONE
131-11-3 see DIMETHYL PHTHALATE
131-52-2 see SODIUM PENTACHLOROPHENATE
131-74-8 see AMMONIUM PICRATE
131-89-5 see 2-CYCLOHEXYL-4,6-DINITROPHENOL
132-27-4 see SODIUM O-PHENYLPHENOXIDE
132-64-9 see DIBENZOFURAN
133-06-2 see CAPTAN
133-07-3 see FOLPET
133-90-4 see CHLORAMBEN
134-29-2 see o-ANISIDINE HYDROCHLORIDE
134-32-7 see α-NAPHTHYLAMINE
135-20-6 see CUPFERRON
136-45-8 see DIPROPYL ISOCINCHOMERONATE
137-26-8 see THIRAM
137-41-7 see POTASSIUM N-METHYLDITHIOCARBAMATE
137-42-8 see METHAM SODIUM
138-93-2 see DISODIUM CYANODITHIOIMIDOCARBONATE

139-13-9 see NITRILOTRIACETIC ACID
139-25-3 see 3,3′-DIMETHYLDIPHENYLMETHANE-4,4′-DIISOCYANATE
139-65-1 see 4,4′-THIODIANILINE
140-29-0 see BENZYL CYANIDE
140-57-8 see ARAMITE
140-76-1 see PYRIDINE, 2-METHYL-5-VINYL
140-88-5 see ETHYL ACRYLATE
141-32-2 see BUTYL ACRYLATE
141-66-2 see DICROTOPHOS
141-78-6 see ETHYL ACETATE
142-28-9 see 1,3-DICHLOROPROPANE
142-59-6 see NABAM
142-71-2 see CUPRIC ACETATE
142-84-7 see DIPROPYLAMINE
143-33-9 see SODIUM CYANIDE (Na(CN))
143-50-0 see KEPONE
144-49-0 see FLUOROACETIC ACID
145-73-3 see ENDOTHALL
148-79-8 see THIABENDAZOLE
148-82-3 see MELPHALAN
149-30-4 see 2-MERCAPTOBENZOTHIAZOLE
149-74-6 see DICHLOROMETHYLPHENYLSILANE
150-50-5 see MERPHOS
150-68-5 see MONURON
151-38-2 see METHOXYETHYLMERCURIC ACETATE
151-50-8 see POTASSIUM CYANIDE
151-56-4 see ETHYLENEIMINE
152-16-9 see DIPHOSPHORAMIDE, OCTAMETHYL-
156-10-5 see p-NITROSODIPHENYLAMINE
156-60-5 see 1,2-DICHLOROETHYLENE
156-62-7 see CALCIUM CYANAMIDE
189-55-9 see DIBENZ[a,i]PYRENE
189-64-0 see DIBENZO(a,h)PYRENE
191-24-2 see BENZO[ghi]PERYLENE
191-30-0 see DIBENZO(a,l)PYRENE
192-65-4 see DIBENZO(a,e)PYRENE
193-39-5 see INDENO[1,2,3-cd]PYRENE
194-59-2 see 7H-DIBENZO(c,g)CARBAZOLE
205-82-3 see BENZO[j]FLUORANTHENE
205-99-2 see BENZO[b]FLUORANTHENE
206-44-0 see FLUORANTHENE
207-08-9 see BENZO[k]FLUORANTHENE
208-96-8 see ACENAPHTHYLENE
218-01-9 see CHRYSENE
224-42-0 see DIBENZ(a,j)ACRIDINE
225-51-4 see BEN[c]ACRIDINE
226-36-8 see DIBENZ(a,h)ACRIDINE
297-78-9 see ISOBENZAN
297-97-2 see ZINOPHOS
298-00-0 see METHYL PARATHION
298-02-2 see PHORATE
298-04-4 see DISULFOTON
300-62-9 see AMPHETAMINE
300-76-5 see NALED
301-04-2 see LEAD ACETATE
301-12-2 see OXYDEMETON METHYL-

302-01-2 see HYDRAZINE
303-34-4 see LASIOCARPINE
305-03-3 see CHLORAMBUCIL
306-83-2 see 2,2-DICHLORO-1,1,1-TRIFLUOROETHANE
309-00-2 see ALDRIN
311-45-5 see DIETHYL-p-NITROPHENYL PHOSPHATE
314-40-9 see BROMACIL
315-18-4 see MEXACARBATE
316-42-7 see EMETINE, DIHYDROCHLORIDE
319-84-6 see α-HEXACHLOROCYCLOHEXANE
319-85-7 see β-HEXACHLOROCYCLOHEXANE
319-86-8 see δ-HEXACHLOROCYCLOHEXANE
327-98-0 see TRICHLORONATE
329-71-5 see 2,5-DINITROPHENOL
330-54-1 see DIURON
330-55-2 see LINURON
333-41-5 see DIAZINON
334-88-3 see DIAZOMETHANE
353-42-4 see BORON TRIFLUORIDE COMPOUND with METHYL ETHER
353-50-4 see CARBONIC DIFLUORIDE
353-59-3 see BROMOCHLORODIFLUOROMETHANE
354-11-0 see 1,1,2,2-TETRACHLORO-2-FLUOROETHANE
354-14-3 see 1,1,2,2-TETRACHLORO-1-FLUOROETHANE
354-23-4 see 1,2-DICHLORO-1,1,2-TRIFLUOROETHANE
354-25-6 see 1-CHLORO-1,1,2,2-TETRAFLUOROETHANE
357-57-3 see BRUCINE
359-06-8 see FLUOROACETYL CHLORIDE
371-62-0 see ETHYLENE FLUOROHYDRIN
379-79-3 see ERGOTAMINE TARTRATE
422-44-6 see 1,2-DICHLORO-1,1,2,3,3-PENTAFLUOROPROPANE
422-48-0 see 2,3-DICHLORO-1,1,1,2,3-PENTAFLUOROPROPANE
422-56-0 see 3,3-DICHLORO-1,1,1,2,2-PENTAFLUOROPROPANE
431-86-7 see 1,2-DICHLORO-1,1,3,3,3-PENTAFLUOROPROPANE
460-19-5 see CYANOGEN
460-35-5 see 3-CHLORO-1,1,1-TRIFLUOROPROPANE
463-49-0 see 1,2-PROPADIENE
463-58-1 see CARBONYL SULFIDE
463-82-1 see 2,2-DIMETHYLPROPANE
465-73-6 see ISODRIN
470-90-6 see CHLORFENVINFOS
492-80-8 see C.I. SOLVENT YELLOW 34
494-03-1 see CHLORNAPHAZINE
496-72-0 see DIAMINOTOLUENE
502-39-6 see METHYLMERCURIC DICYANAMIDE
504-24-5 see 4-AMINOPYRIDINE
504-60-9 see 1,3-PENTADIENE
505-60-2 see MUSTARD GAS
506-61-6 see POTASSIUM SILVER CYANIDE
506-64-9 see SILVER CYANIDE
506-68-3 see CYANOGEN BROMIDE
506-77-4 see CYANOGEN CHLORIDE
506-78-5 see CYANOGEN IODIDE
506-87-6 see AMMONIUM CARBONATE
506-96-7 see ACETYL BROMIDE
507-55-1 see 1,3-DICHLORO-1,1,2,2,3-PENTAFLUOROPROPANE
509-14-8 see TETRANITROMETHANE

594-42-3 see PERCHLOROMETHYL MERCAPTAN
597-64-8 see TETRAETHYLTIN
598-31-2 see BROMOACETONE
598-73-2 see BROMOTRIFLUORETHYLENE
606-20-2 see 2,6-DINITROTOLUENE
608-73-1 see HEXACHLOROCYCLOHEXANE (ALL ISOMERS)
608-93-5 see PENTACHLOROBENZENE
609-19-8 see 3,4,5-TRICHLOROPHENOL
610-39-9 see 3,4-DINITROTOLUENE
612-82-8 see 3,3′-DIMETHYLBENZIDINE DIHYDROCHLORIDE
612-83-9 see 3,3′-DICHLOROBENZIDINE DIHYDROCHLORIDE
614-78-8 see THIOUREA, (2-METHYLPHENYL)-
615-05-4 see 2,4-DIAMINOSOLE
615-28-1 see 1,2-PHENYLENEDIAMINE DIHYDROCHLORIDE
615-53-2 see N-NITROSO-N-METHYLURETHANE
621-64-7 see N-NITROSODI-N-PROPYLAMINE
624-18-0 see 1,4-PHENYLENEDIAMINE DIHYDROCHLORIDE
624-64-6 see 2-BUTENE, (E)-
624-83-9 see METHYL ISOCYANATE
625-16-1 see tert-AMYL ACETATE
626-38-0 see sec-AMYL ACETATE
627-11-2 see CHLOROETHYL CHLOROFORMATE
627-20-3 see 2-PENTENE, (Z)-
628-63-7 see AMYL ACETATE
628-86-4 see MERCURY FULMINATE
630-10-4 see SELENOUREA
630-20-6 see 1,1,1,2-TETRACHLOROETHANE
630-60-4 see OUABAIN
631-61-8 see AMMONIUM ACETATE
636-21-5 see o-TOLUIDINE HYDROCHLORIDE
639-58-7 see TRIPHENYLTIN CHLORIDE
640-19-7 see FLUOROACETAMIDE
644-64-4 see DIMETILAN
646-04-8 see 2-PENTENE, (E)-
675-14-9 see CYANURIC FLUORIDE
676-97-1 see METHYL PHOSPHONIC DICHLORIDE
680-31-9 see HEXAMETHYLPHOSPHORAMIDE
684-93-5 see N-NITROSO-N-METHYLUREA
689-97-4 see VINYL ACETYLENE
692-42-2 see DIETHYLARSINE
696-28-6 see DICHLOROPHENYLARSINE
709-98-8 see PROPANIL
732-11-6 see PHOSMET
757-58-4 see HEXAETHYL TETRAPHOSPHATE
759-73-9 see N-NITROSO-N-ETHYLUREA
759-94-4 see ETHYL DIPROPYLTHIOCARBAMATE
760-93-0 see METHACRYLIC ANHYDRIDE
764-41-0 see 1,4-DICHLORO-2-BUTENE
765-34-4 see GLYCIDYLALDEHYDE
786-19-6 see CARBOPHENOTHION
812-04-4 see 1,1-DICHLORO-1,2,2-TRIFLUOROETHANE
814-49-3 see DIETHYL CHLOROPHOSPHATE
814-68-6 see ACRYLYL CHLORIDE
815-82-7 see CUPRIC TARTRATE
822-06-0 see HEXAMETHYLENE-1,6-DIISOCYANATE
823-40-5 see DIAMINOTOLUENE

824-11-3 see TRIMETHYLOLPROPANE PHOSPHITE
834-12-8 see AMETRYN
842-07-9 see C.I. SOLVENT YELLOW 14
872-50-4 see N-METHYL-2-PYRROLIDONE
900-95-8 see STANNANE, ACETOXYTRIPHENYL-
919-86-8 see DEMETON-s-METHYL
920-46-7 see METHACRYLOYL CHLORIDE
924-16-3 see N-NITROSODI-n-BUTYLAMINE
924-42-5 see N-METHYLOLACRYLAMIDE
930-55-2 see N-NITROSOPYRROLIDINE
933-75-5 see 2,3,6-TRICHLOROPHENOL
933-78-8 see 2,3,5-TRICHLOROPHENOL
944-22-9 see FONFOS
947-02-4 see PHOSFOLAN
950-10-7 see MEPHOSFOLAN
950-37-8 see METHIDATHION
957-51-7 see DIPHENAMID
959-98-8 see α-ENDOSULFAN
961-11-5 see TETRACHLORVINPHOS
989-38-8 see C.I. BASIC RED 1
991-42-4 see NORBORMIDE
998-30-1 see TRIETHOXYSILANE
999-81-5 see CHLORMEQUAT CHLORIDE
1024-57-3 see HEPTACHLOR EPOXIDE
1031-07-8 see ENDOSULFAN SULFATE
1031-47-6 see TRIAMIPHOS
1066-30-4 see CHROMIC ACETATE
1066-33-7 see AMMONIUM BICARBONATE
1066-45-1 see TRIMETHYLTIN CHLORIDE
1072-35-1 see LEAD STEARATE
1111-78-0 see AMMONIUM CARBAMATE
1114-71-2 see PEBULATE
1116-54-7 see N-NITROSODIETHANOLAMINE
1120-71-4 see 1,3-PROPANE SULTONE
1122-60-7 see NITROCYCLOHEXANE
1124-33-0 see PYRIDINE, 4-NITRO-, 1-OXIDE
1129-41-5 see METOLCARB
1134-23-2 see CYCLOATE
1163-19-5 see DECABROMODIPHENYL OXIDE
1185-57-5 see FERRIC AMMONIUM CITRATE
1194-65-6 see DICHLOBENIL
1300-71-6 see XYLENOL
1303-28-2 see ARSENIC PENTOXIDE
1303-32-8 see ARSENIC DISULFIDE
1303-33-9 see ARSENIC TRISULFIDE
1306-19-0 see CADMIUM OXIDE
1309-64-4 see ANTIMONY TRIOXIDE
1310-58-3 see POTASSIUM HYDROXIDE
1310-73-2 see SODIUM HYDROXIDE
1313-27-5 see MOLYBDENUM TRIOXIDE
1314-20-1 see THORIUM DIOXIDE
1314-32-5 see THALLIC OXIDE
1314-56-3 see PHOSPHORUS PENTOXIDE
1314-62-1 see VANADIUM PENTOXIDE
1314-80-3 see SULFUR PHOSPHIDE
1314-84-7 see ZINC PHOSPHIDE (conc.<= 10%)

1314-84-7 see ZINC PHOSPHIDE
1314-87-0 see LEAD SULFIDE
1319-72-8 see 2,4,5-T AMINES
1319-77-3 see CRESOL (MIXED ISOMERS)
1320-18-9 see 2,4-D PROPYLENE GLYCOL BUTYL ETHER ESTER
1321-12-6 see NITROTOLUENE
1327-52-2 see ARSENIC ACID
1327-53-3 see ARSENIC TRIOXIDE
1330-20-7 see XYLENE (MIXED ISOMERS)
1332-07-6 see ZINC BORATE
1332-21-4 see ASBESTOS (Friable)
1333-74-0 see HYDROGEN
1333-83-1 see SODIUM BIFLUORIDE
1335-32-6 see LEAD SUBACETATE
1335-87-1 see HEXACHLORONAPHTHALENE
1336-21-6 see AMMONIUM HYDROXIDE
1336-36-3 see POLYCHLORINATED BIPHENYLS
1338-23-4 see METHYL ETHYL KETONE PEROXIDE
1338-24-5 see NAPHTHENIC ACID
1341-49-7 see AMMONIUM BIFLUORIDE
1344-28-1 see ALUMINUM OXIDE (fibrous forms)
1397-94-0 see ANTIMYCIN
1420-07-1 see DINOTERB
1464-53-5 see DIEPOXYBUTANE
1558-25-4 see TRICHLORO(CHLOROMETHYL)SILANE
1563-38-8 see CARBOFURAN PHENOL
1563-66-2 see CARBOFURAN
1582-09-8 see TRIFLURALIN
1600-27-7 see MERCURIC ACETATE
1615-80-1 see HYDRAZINE, 1,2-DIETHYL-
1622-32-8 see ETHANESULFONYL CHLORIDE, 2-CHLORO-
1634-04-4 see METHYL tert-BUTYL ETHER
1642-54-2 see DIETHYLCARBAMAZINE CITRATE
1646-88-4 see ALDICARB SULFONE
1649-08-7 see 1,2-DICHLORO-1,1-DIFLUOROETHANE
1689-84-5 see BROMOXYNIL
1689-99-2 see BROMOXYNIL OCTANOATE
1717-00-6 see 1,1-DICHLORO-1-FLUOROETHANE
1746-01-6 see 2,3,7,8-TETRACHLORODIBENZO-p-DIOXIN (TCDD)
1752-30-3 see ACETONE THIOSEMICARBAZIDE
1762-95-4 see AMMONIUM THIOCYANATE
1836-75-5 see NITROFEN
1861-40-1 see BENFLURALIN
1863-63-4 see AMMONIUM BENZOATE
1888-71-7 see HEXACHLOROPROPENE
1897-45-6 see CHLOROTHALONIL
1910-42-5 see PARAQUAT DICHLORIDE
1912-24-9 see ATRAZINE
1918-00-9 see DICAMBA
1918-02-1 see PICLORAM
1918-16-7 see PROPACHLOR 2-CHLORO-N-(1-METHYLETHYL)-N-PHENYLACETAMIDE
1928-38-7 see 2,4-D METHYL ESTER
1928-43-4 see 2,4-D 2-ETHYLHEXYL ESTER
1928-47-8 see 2,4,5-T ESTERS
1928-61-6 see 2,4-D PROPYL ESTER

1929-73-3 see 2,4-D BUTOXYETHYL ESTER
1929-82-4 see NITRAPYRIN
1937-37-7 see C.I. DIRECT BLACK 38
1982-47-4 see CHLOROXURON
1982-69-0 see SODIUM DICAMBA
1983-10-4 see TRIBUTYLTIN FLUORIDE
2001-95-8 see VALINOMYCIN
2008-41-5 see BUTYLATE
2008-46-0 see 2,4,5-T AMINES
2032-65-7 see METHIOCARB
2074-50-2 see PARAQUAT METHOSULFATE
2097-19-0 see PHENYLSILATRANE
2104-64-5 see EPN
2155-70-6 see TRIBUTYLTIN METHACRYLATE
2164-07-0 see DIPOTASSIUM ENDOTHALL
2164-17-2 see FLUOMETURON
2212-67-1 see MOLINATE
2223-93-0 see CADMIUM STEARATE
2231-57-4 see THIOCARBAZIDE
2234-13-1 see OCTACHLORONAPHTHALENE
2238-07-5 see DIGLYCIDYL ETHER
2275-18-5 see PROTHOATE
2300-66-5 see DIMETHYLAMINE DICAMBA
2303-16-4 see DIALLATE
2303-17-5 see TRIALLATE
2312-35-8 see PROPARGITE
2439-01-2 see CHINOMETHIONAT
2439-10-3 see DODINE
2497-07-6 see OXYDISULFOTON
2524-03-0 see DIMETHYL CHLOROTHIOPHOSPHATE
2540-82-1 see FORMOTHION
2545-59-7 see 2,4,5-T ESTERS
2556-36-7 see 1,4-CYCLOHEXANE DIISOCYANATE
2570-26-5 see PENTADECYLAMINE
2587-90-8 see PHOSPHOROTHIOIC ACID
2602-46-2 see C.I. DIRECT BLUE 6
2631-37-0 see PROMECARB
2636-26-2 see CYANOPHOS
2642-71-9 see AZINPHOS-ETHYL
2655-15-4 see 2,3,5-TRIMETHYLPHENYL METHYLCARBAMATE
2665-30-7 see PHOSPHONOTHIOIC ACID, METHYL-, O-(4-NITROPHENYL) O-PHENYL) ESTER
2699-79-8 see SULFURYL FLUORIDE
2702-72-9 see 2,4-D SODIUM SALT
2703-13-1 see PHOSPHONOTHIOIC ACID, METHYL-, O-ETHYL O-(4-(METHYLTHIO)PHENYL)ESTER
2757-18-8 see THALLOUS MALONATE
2763-96-4 see MUSCIMOL
2764-72-9 see DIQUAT
2778-04-3 see ENDOTHION
2832-40-8 see C.I. DISPERSE YELLOW 3
2837-89-0 see 2-CHLORO-1,1,1,2-TETRAFLUOROETHANE
2921-88-2 see CHLORPYRIFOS
2944-67-4 see FERRIC AMMONIUM OXALATE
2971-38-2 see 2,4-D CHLOROCROTYL ESTER
3012-65-5 see AMMONIUM CITRATE, DIBASIC

7287-19-6 see PROMETHRYN
7421-93-4 see ENDRIN ALDEHYDE
7428-48-0 see LEAD STEARATE
7429-90-5 see ALUMINUM (fume or dust)
7439-89-6 see IRON
7439-92-1 see LEAD
7439-96-5 see MANGANESE
7439-97-6 see MERCURY
7440-02-0 see NICKEL
7440-22-4 see SILVER
7440-23-5 see SODIUM
7440-28-0 see THALLIUM
7440-31-5 see TIN
7440-36-0 see ANTIMONY
7440-38-2 see ARSENIC
7440-39-3 see BARIUM
7440-41-7 see BERYLLIUM
7440-43-9 see CADMIUM
7440-47-3 see CHROMIUM
7440-48-4 see COBALT
7440-50-8 see COPPER
7440-62-2 see VANADIUM
7440-66-6 see ZINC (FUME OR DUST)
7440-66-6 see ZINC
7446-08-4 see SELENIUM DIOXIDE
7446-09-5 see SULFUR DIOXIDE
7446-09-5 see SULFUR DIOXIDE (ANHYDROUS)
7446-11-9 see SULFUR TRIOXIDE
7446-14-2 see LEAD SULFATE
7446-18-6 see THALLIUM(I) SULFATE
7446-27-7 see LEAD PHOSPHATE
7447-39-4 see CUPRIC CHLORIDE
7487-94-7 see MERCURIC CHLORIDE
7488-56-4 see SELENIUM SULFIDE
7550-45-0 see TITANIUM TETRACHLORIDE
7558-79-4 see SODIUM PHOSPHATE, DIBASIC
7580-67-8 see LITHIUM HYDRIDE
7601-54-9 see SODIUM PHOSPHATE, TRIBASIC
7631-89-2 see SODIUM ARSENATE
7631-90-5 see SODIUM BISULFITE
7632-00-0 see SODIUM NITRITE
7637-07-2 see BORON TRIFLUORIDE
7645-25-2 see LEAD ARSENATE
7646-85-7 see ZINC CHLORIDE
7647-01-0 see HYDROCHLORIC ACID
7647-18-9 see ANTIMONY PENTACHLORIDE
7664-38-2 see PHOSPHORIC ACID
7664-39-3 see HYDROFLUORIC ACID (conc. 50% OR GREATER, OR
7664-39-3 see HYDROGEN FLUORIDE
7664-41-7 see AMMONIA
7664-93-9 see SULFURIC ACID
7681-49-4 see SODIUM FLUORIDE
7681-52-9 see SODIUM HYPOCHLORITE
7696-12-0 see TETRAMETHRIN
7697-37-2 see NITRIC ACID
7699-45-8 see ZINC BROMIDE

7789-43-7 see COBALTOUS BROMIDE
7789-61-9 see ANTIMONY TRIBROMIDE
7790-94-5 see CHLOROSULFONIC ACID
7791-12-0 see THALLIUM CHLORIDE (TlCl)
7791-21-1 see CHLORINE OXIDE
7791-23-3 see SELENIUM OXYCHLORIDE
7803-51-2 see PHOSPHINE
7803-55-6 see AMMONIUM VANADATE
7803-62-5 see SILANE
8001-35-2 see TOXAPHENE
8001-58-9 see CREOSOTE
8003-19-8 see DICHLOROPROPANE - DICHLOROPROPENE (MIXTURE)
8003-34-7 see PYRETHRINS
8014-95-7 see OLEUM (FUMING SULFURIC ACID)
8065-48-3 see DEMETON
9006-42-2 see METIRAM
9016-87-9 see POLYMERIC DIPHENYLMETHANE DIISOCYANATE
10025-73-7 see CHROMIC CHLORIDE
10025-78-2 see TRICHLOROSILANE
10025-78-2 see SILANE, TRICHLORO-
10025-87-3 see PHOSPHORYL CHLORIDE
10025-87-3 see PHOSPHORUS OXYCHLORIDE
10025-91-9 see ANTIMONY TRICHLORIDE
10026-11-6 see ZIRCONIUM TETRACHLORIDE
10026-13-8 see PHOSPHORUS PENTACHLORIDE
10028-15-6 see OZONE
10028-22-5 see FERRIC SULFATE
10031-59-1 see THALLIUM SULFATE
10034-93-2 see HYDRAZINE SULFATE
10039-32-4 see SODIUM PHOSPHATE, DIBASIC
10043-01-3 see ALUMINUM SULFATE
10045-89-3 see FERROUS AMMONIUM SULFATE
10045-94-0 see MERCURIC NITRATE
10049-04-4 see CHLORINE DIOXIDE
10049-05-5 see CHROMOUS CHLORIDE
10061-02-6 see trans-1,3-DICHLOROPROPENE
10099-74-8 see LEAD NITRATE
10101-53-8 see CHROMIC SULFATE
10101-63-0 see LEAD IODIDE
10101-89-0 see SODIUM PHOSPHATE, TRIBASIC
10102-06-4 see URANYL NITRATE
10102-18-8 see SODIUM SELINITE
10102-20-2 see SODIUM TELLURITE
10102-43-9 see NITRIC OXIDE
10102-44-0 see NITROGEN DIOXIDE
10102-45-1 see THALLIUM(I) NITRATE
10102-48-4 see LEAD ARSENATE
10108-64-2 see CADMIUM CHLORIDE
10124-50-2 see POTASSIUM ARSENITE
10124-56-8 see SODIUM PHOSPHATE, TRIBASIC
10140-65-5 see SODIUM PHOSPHATE, DIBASIC
10140-87-1 see ETHANOL, 1,2-DICHLORO-, ACETATE
10192-30-0 see AMMONIUM BISULFITE
10196-04-0 see AMMONIUM SULFIDE
10210-68-1 see COBALT CARBONYL
10222-01-2 see 2,2-DIBROMO-3-NITRILOPROPIONAMIDE

14484-64-1 see FERBAM
14639-97-5 see ZINC AMMONIUM CHLORIDE
14639-98-6 see ZINC AMMONIUM CHLORIDE
14644-61-2 see ZIRCONIUM SULFATE
14763-77-0 see COPPER CYANIDE
15271-41-7 see BICYCLO[2.2.1]HEPTANE-2-CARBONITRILE, 5-CHLORO-6-((((METHYAMINO)CARBONYL)OXY)IMINO)-,(1ST-(1-α,2-β,4-α,5-α,6e))-
15339-36-3 see MANGANESE DIMETHYLDITHIOCARBAMATE
15646-96-5 see 2,4,4-TRIMETHYLHEXAMETHYLENE DIISOCYANATE
15699-18-0 see NICKEL AMMONIUM SULFATE
15739-80-7 see LEAD SULFATE
15950-66-0 see 2,3,4-TRICHLOROPHENOL
15972-60-8 see ALACHLOR
16071-86-6 see C.I. DIRECT BROWN 95
16543-55-8 see N-NITROSONORNICOTINE
16721-80-5 see SODIUM HYDROSULFIDE
16752-77-5 see METHOMYL
16871-71-9 see ZINC SILICOFLUORIDE
16919-19-0 see AMMONIUM SILICOFLUORIDE
16923-95-8 see ZIRCONIUM POTASSIUM FLUORIDE
16938-22-0 see 2,2,4-TRIMETHYL HEXAMETHYLENE DIISOCYANATE
16984-48-8 see FLUORIDE
17702-41-9 see DECABORANE(14)
17702-57-7 see FORMPARANATE
17804-35-2 see BENOMYL
18883-66-4 see D-GLUCOSE, 2-DEOXY-2-[[(METHYLNITROSOAMI-NO)CARBONYL]AMINO]-
19044-88-3 see ORYZALIN
19287-45-7 see DIBORANE
19624-22-7 see PENTABORANE
19666-30-9 see OXYDIAZON
20325-40-0 see 3,3′-DIMETHOXYBENZIDINE DIHYDROCHLORIDE
20354-26-1 see METHAZOLE
20816-12-0 see OSMIUM TETROXIDE
20830-75-5 see DIGOXIN
20830-81-3 see DAUNOMYCIN
20859-73-8 see ALUMINUM PHOSPHIDE
21087-64-9 see METRIBUZIN
21548-32-3 see FOSTHIETAN
21609-90-5 see LEPTOPHOS
21725-46-2 see CYANAZINE
21908-53-2 see MERCURIC OXIDE
21923-23-9 see CHLORTHIOPHOS
22224-92-6 see FENAMIPHOS
22781-23-3 see BENDIOCARB
22961-82-6 see BENDIOCARB PHENOL
23135-22-0 see OXAMYL
23422-53-9 see FORMETANATE HYDROCHLORIDE
23505-41-1 see PIRIMFOS-ETHYL
23564-05-8 see THIOPHANATE-METHYL
23564-06-9 see THIOPHANATE ETHYL
23950-58-5 see PRONAMIDE
24017-47-8 see TRIAZOFOS
24934-91-6 see CHLORMEPHOS
25154-54-5 see DINITROBENZENE (MIXED ISOMERS)

25154-55-6 see NITROPHENOL (MIXED ISOMERS)
25155-30-0 see SODIUM DODECYLBENZENESULFONATE
25167-67-3 see BUTENE
25167-80-0 see CHLOROPHENOLS
25167-82-2 see TRICHLOROPHENOL
25168-15-4 see 2,4,5-T ESTERS
25168-26-7 see 2,4-D ISOOCTYL ESTER
25311-71-1 see ISOFENPHOS
25321-14-6 see DINITROTOLUENE (MIXED ISOMERS)
25321-22-6 see DICHLOROBENZENE (MIXED ISOMERS)
25550-58-7 see DINITROPHENOL
26002-80-2 see PHENOTHRIN
26264-06-2 see CALCIUM DODECYLBENZENESULFONATE
26419-73-8 see TRIPATE
26471-62-5 see TOLUENEDIISOCYANATE (MIXED ISOMERS)
26628-22-8 see SODIUM AZIDE (Na(N_3))
26638-19-7 see DICHLOROPROPANE
26644-46-2 see TRIFORINE
26952-23-8 see DICHLOROPROPENE
27137-85-5 see TRICHLORO(DICHLOROPHENYL)SILANE
27176-87-0 see DODECYLBENZENESULFONIC ACID
27314-13-2 see NORFLURAZON
27323-41-7 see TRIETHANOLAMINE DODECYLBENZENE SULFONATE
27774-13-6 see VANADYL SULFATE
28057-48-9 see d-trans-ALLETHRIN
28249-77-6 see THIOBENCARB
28300-74-5 see ANTIMONY POTASSIUM TARTRATE
28347-13-9 see XYLYLENE DICHLORIDE
28407-37-6 see C.I. DIRECT BLUE 218
28772-56-7 see BROMADIOLONE
29232-93-7 see PIRIMIPHOS METHYL
30525-89-4 see PARAFORMALDEHYDE
30558-43-1 see A2213
30560-19-1 see ACEPHATE
30674-80-7 see METHACRYLOYLOXYETHYL ISOCYANATE
31218-83-4 see PROPETAMPHOS
32534-95-5 see 2,4,5-TP ESTERS
33089-61-1 see AMITRAZ
33213-65-9 see β-ENDOSULFAN
34014-18-1 see TEBUTHIURON
34077-87-7 see DICHLOROTRIFLUROETHANE
35367-38-5 see DIFLUBENZURON
35400-43-2 see SULPROFOS
35554-44-0 see IMAZALIL
35691-65-7 see 1-BROMO-1-(BROMOMETHYL)-1,3-PROPANEDICARBON-ITRILE
36355-01-8 see POLYBROMINATED BIPHENYLS
36478-76-9 see URANYL NITRATE
37211-05-5 see NICKEL CHLORIDE
38661-72-2 see 1,3-BIS(METHYLISOCYANATE)CYCLOHEXANE
38727-55-8 see DIETHATYL ETHYL
39156-41-7 see 2,4-DIAMINOSOLE, SULFATE
39196-18-4 see THIOFANOX
39289-28-6 see DICHLOROFLUOROMETHANE
39300-45-3 see DINOCAP

39515-41-8 see FENPROPATHRIN
40487-42-1 see PENDIMETHALIN N-(1-ETHYLPROPYL)-3,4-DIMETHYL-2,6-DINITROBENZENAMINE
41198-08-7 see PROFENOFOS
41766-75-0 see 3,3-DIMETHYLBENZIDINE DIHYDROFLUORIDE
42504-46-1 see ISOPROPANOLAMINE DODECYLBENZENE SULFONATE
42874-03-3 see OXYFLUORFEN
43121-43-3 see TRIADIMEFON
50471-44-8 see VINCLOZOLIN
50782-69-9 see PHOSPHONOTHIOIC ACID, METHYL-, S[2-[BIS(1-METHYLETHYL)AMINOETHYL) O-ETHYL ESTER (1:1)
51235-04-2 see HEXAZINONE
51338-27-3 see DICLOFOP METHYL
51630-58-1 see FENVALERATE
52628-25-8 see ZINC AMMONIUM CHLORIDE
52645-53-1 see PERMETHRIN
52652-59-2 see LEAD STEARATE
52740-16-6 see CALCIUM ARSENITE
53404-19-6 see BROMACIL, LITHIUM SALT
53404-37-8 see 2,4-D 2-ETHYL-4-METHYLPENTYL ESTER
53404-60-7 see DAZOMET, SODIUM SALT
53467-11-1 see 2,4-D ESTERS
53469-21-9 see AROCLOR 1242
53558-25-1 see PYRIMINIL
55290-64-7 see DIMETHIPIN
55299-18-8 see DIPHENYLHYDRAZINE
55406-53-6 see 3-IODO-2-PROPYNYL BUTYLCARBAMATE
55488-87-4 see FERRIC AMMONIUM OXALATE
56189-09-4 see LEAD STEARATE
57213-69-1 see TRICLOPYR TRIETHYLAMMONIUM SALT
58270-08-9 see ZINC, DICHLORO(4,4-DIMETHYL-5((((METHYLAMINO)CARBONYL)OXY)IMINO)PENTANENITRILE)-,(T-4)-
59669-26-0 see THIODICARB
60168-88-9 see FENARIMOL
60207-90-1 see PROPICONAZOLE
61792-07-2 see 2,4,5-T ESTERS
62207-76-5 see COBALT, ((2,2′-(1,2-ETHANEDIYLBIS(NITRILOMETHYLIDYNE))BIS(6-FLUOROPHENOLATO))(2)-
62476-59-9 see ACIFLUORFEN, SODIUM SALT
63938-10-3 see CHLOROTETRAFLUOROETHANE
64902-72-3 see CHLORSULFURON
64969-34-2 see 3,3′-DICHLOROBENZIDINE SULFATE
65762-25-6 see 1-CHLORO-1,1-DIFLUOROETHANE
66441-23-4 see FENOXAPROP ETHYL
67485-29-4 see HYDRAMETHYLNON
68085-85-8 see CYHALOTHRIN
68359-37-5 see CYFLUTHRIN
69409-94-5 see FLUVALINATE
69806-50-4 see FLUAZIFOP-BUTYL
71751-41-2 see ABAMECTIN
72178-02-0 see FOMESAFEN
72490-01-8 see FENOXYCARB
73666-77-0 see CHLORODIFLUOROMETHANE
74051-80-5 see SETHOXYDIM
75790-84-0 see 4-METHYLDIPHENYLMETHANE-3,4-DIISOCYANATE
75790-87-3 see 2,4′-DIISOCYANATODIPHENYL SULFIDE